Chemical Spectroscopy and Photochemistry in the Vacuum-Ultraviolet

NATO ADVANCED STUDY INSTITUTES SERIES

Proceedings of the Advanced Study Institute Programme, which aims
at the dissemination of advanced knowledge and
the formation of contacts among scientists from different countries

The series is published by an international board of publishers in conjunction
with NATO Scientific Affairs Division

A	Life Sciences	Plenum Publishing Corporation
B	Physics	London and New York
C	Mathematical and Physical Sciences	D. Reidel Publishing Company Dordrecht and Boston
D	Behavioral and Social Sciences	Sijthoff International Publishing Company Leiden
E	Applied Sciences	Noordhoff International Publishing Leiden

Series C – Mathematical and Physical Sciences

Volume 8 – Chemical Spectroscopy and Photochemistry in the Vacuum-Ultraviolet

Chemical Spectroscopy and Photochemistry in the Vacuum-Ultraviolet

Proceedings of the Advanced Study Institute, held under the Auspices of NATO and the Royal Society of Canada, August 5–17, 1973, Valmorin, Quebec, Canada

edited by

C. SANDORFY, *Université de Montréal*

P. J. AUSLOOS, *U.S. National Bureau of Standards*

M. B. ROBIN, *Bell Laboratories*

D. Reidel Publishing Company

Dordrecht-Holland / Boston-U.S.A.

Published in cooperation with NATO Scientific Affairs Division

Library of Congress Catalog Card Number 73–91209

ISBN-13: 978-94-010-2155-5   e-ISBN-13: 978-94-010-2153-1
DOI: 10.1007/978-94-010-2153-1

Published by D. Reidel Publishing Company
P.O. Box 17, Dordrecht, Holland

Sold and distributed in the U.S.A., Canada, and Mexico
by D. Reidel Publishing Company, Inc.
306 Dartmouth Street, Boston, Mass. 02116, U.S.A.

CONTENTS

PREFACE   XV

LIST OF PARTICIPANTS   XVII

VACUUM ULTRAVIOLET AND PHOTOELECTRON SPECTROSCOPY – HISTORICAL
BACKGROUND AND SURVEY
W.C. PRICE   1
– V.U.V. Spectra of Polyatomic Molecules   1
– Photoionization   7
– Photoelectron Spectroscopy   8
– References   11

ASPECTS OF MOLECULAR RYDBERG STATES
M.B. ROBIN   13
– References   23

PHOTOELECTRON SPECTROSCOPY
D.W. TURNER   25
– Introduction.– Electrons and Photons   25
– Some Experimental Aspects of Electron Energy Analysis   26
– Light Sources   27
– Photoelectron Bands Designated According to the Ionic
  States   28
– Vibrational Fine Structure   30
  Vibrational Fine Structure in the Third Band of
      Formaldehyde   32
  Vibrational Fine Structure in the Ketene Photoelectron
      Spectrum   32
  Vertical Ionization   34
– Interpretative Methods Involving Molecular Orbital Cal-
  culations   35
– Correlations between P.E. Spectral Data and Other
  Physical Parameters   38
– Photoelectron Spectra of Transient Species   39
– Photoelectron Spectra from Solids   40
– References   42

ABSOLUTE INTENSITIES AND CLASSIFICATION OF TRANSITIONS IN
ELECTRON IMPACT SPECTROSCOPY
E.N. LASSETTRE   43
– Introduction   43
– Generalized Oscillator Strengths   43
– Determination of Oscillator Strengths   46
  Oscillator Strengths Obtained by Extrapolation   46
  Optical Oscillator Strengths from Forward Scattering   48

The names of those who actually delivered the lectures are underlined.

- Classification of Transitions. Dipole-Quadrupole Behavior   51
  Franck-Condon Principle. Invariance in Relative Intensities 58
  Singlet-Triplet Transitions   60
  Strongly Forbidden Transitions at Low Kinetic Energies   60
  Identification of Triplets by Positive Ion Scattering   61
- Recent Results   62
  Deviations from Born Approximation at Small Scattering
        Angles. Selection Rules   62
  Singlet-Triplet Excitation at High Kinetic Energy.
        $1^1S \rightarrow 2^3S$ Transition in Helium   63
  Theoretical Study of Singlet-Triplet Selection Rules   65
  Other Singlet-Triplet Transitions   65
  Calculation of Singlet-Triplet Energy Differences   66
  Excited Triplet of H_2O 66
  Other Recent Researches   68
- Addendum on Benzene   68
- References   70

PHOTOIONIZATION OF FREE RADICALS
J. BERKOWITZ   75
- Methods of Production   77
- Available Data   78
  Hydrocarbon and Fluorocarbon Radicals   79
  Atmospheric Radicals   82
  $O(^3P)$ 82
  $N(^4S)$ 85
  OH   86
  O_3 and HNO_3 88
- References   91

PHOTOELECTRON SPECTROSCOPY OF MOLECULAR BEAMS
J. BERKOWITZ   93
- Introduction and Chemical Studies   93
- Physical Studies   99
  Comparison of Photoelectron Intensities and Franck-
        Condon Factors in the Photoionization of H_2,
        HD and D_2 101
  Rotational Band Shapes in the Photoelectron Spectra
        of HF and DF   103
  Spin-Orbit Coupling and the Ratio of Photoionization
        Cross Sections in Some Atoms   107
- Summary and Conclusions   108
- References   109

HIGH RESOLUTION VACUUM ULTRAVIOLET SPECTROSCOPY OF
DIATOMIC MOLECULES
A.E. DOUGLAS   113
  - Instrumentation   113
    Spectrographs   114
    Sources   116
  - Spectra   118
    Spectra of Valence States   119
    Spectra of Rydberg States   120
    BH   123
    Nitric Oxide   123
    Nitrogen   124
    Fluorine   125
    Hydrogen Fluoride   125
  - Conclusions   127
  - References   127

OPTICAL PROPERTIES OF LIQUIDS IN THE VACUUM UV
R.D. BIRKHOFF, R.N. HAMM, M.W. WILLIAMS, E.T. ARAKAWA
and L.R. PAINTER   129
  - Introduction   129
  - Experimental Background   130
  - Theory   131
  - Carbon Ring Liquids   132
  - Water   132
  - Other Liquids   139
  - Discussion   142
  - References   146

THE FAR ULTRAVIOLET ABSORPTION SPECTRA OF ORGANIC
MOLECULES: LONE PAIRS AND DOUBLE BONDS
C. SANDORFY   149
  - Introduction   149
  - Halogen Derivatives of Methane   150
  - Halogen Derivatives of Ethane   159
  - Ethylene Derivatives   162
  - Benzene Derivatives   165
  - Molecules Containing Divalent Sulfur   169
  - References   173

THE ETHANE PROBLEM
C. SANDORFY   177
  - The Ground State   177
  - The Ion   179
  - The Ultraviolet Spectra   181
  - Possible Weak Transitions   186
  - The Primary Steps   188
  - References   189

A POT-POURRI OF ULTRAVIOLET AND PHOTOELECTRON SPECTRA OF
IODIDES
D.R. SALAHUB and R.A. BOSCHI   191
- PE Spectrum of Vinyl Iodide   191
- UV Spectra of Alkyl Iodides   192
- PE Spectra of Fluoroiodoethanes   193
- References   194

BEYOND THE ORBITAL APPROXIMATION
R. DAUDEL   197
- The Need for a Language Adapted to Very Elaborate
        Wave-Functions   197
- The Loge Theory   198
- Change in the Localizability of Electrons during
        Ionization   203
- The Localizability of the Electron as a Starting
        Point to Calculate an Elaborate Wave-Function   205
- Loge Theory and Exciton Localizability   209
- References   210

NATURAL AND MAGNETIC CIRCULAR DICHROISM SPECTROSCOPY IN
THE VACUUM ULTRAVIOLET
O. SCHNEPP   211
- Introduction   211
  General   211
  Natural Circular Dichroism   212
  Magnetic Circular Dichroism   213
- Instrumentation   214
- The Natural CD Spectra of Benzene Chromophores   216
- The Natural CD Spectra of Ethylene Chromophores   220
- MCD Spectra of Benzene and Toluene   221
-- References   222

$(H_3C)_2C=C(CH_3)_2$, $(H_3C)_2BN(CH_3)_2$, $(H_3C)BF$ AND $(H_3C)_3B$.
FURTHER EVIDENCE FOR A $\pi \rightarrow \sigma^*$ ASSIGNMENT OF THE OLEFIN
UV "MYSTERY" BAND
W. FUSS and H. BOCK   223
- Helium (I) PE Spectra   225
- Far UV Spectra   226
- Concluding Remarks   231
- References   234

PHOTOELECTRON SPECTRA OF MODERATE SIZED MOLECULES
D.C. FROST, F.G. HERRING and C.A. McDOWELL   237
- Introduction   237
- The Interpretation of the PE Spectra of Structurally
        Similar Molecules   239

- A Comparison of the Bonding of First and Second Row
    p-Block Elements   240
- Photoelectron Studies of the Bonding of Particular
    Elements   242
- The Investigation of "Through Space" and "Through
    Bond" Interactions   242
- The Correlation of PES with UV Spectra and Other
    Techniques   243
- Theoretical Studies   244
- References   245

HAM - A SEMI-EMPIRICAL MO THEORY
L. ÅSBRINK, C. FRIDH and E. LINDHOLM   247
- Introduction   247
- The HAM Method   247
- Shielding (preliminary form)   250
- Electrostatic Interaction (preliminary form)   250
- Results   251
- References   251

AB INITIO CALCULATIONS FOR EXCITED STATES OF MOLECULES
S.D. PEYERIMHOFF and R.J. BUENKER   257
- Introduction   257
- Geometrical Aspects.   258
   Use of Ab Initio Walsh-Type Diagrams   261
   Some Examples   264
- Calculation of Vertical Transition Energies   267
- Consideration of the Vibrational Structure of
       Electronic Transitions   278
- References and Footnotes   285

THRESHOLD ELECTRON-IMPACT SPECTROSCOPY
H.H. BRONGERSMA   287
- Introduction   287
- Interaction of Electrons and Molecules   288
- Experimental Techniques   290
- Applications of Threshold Methods   294
   Saturated Hydrocarbons   295
   Olefins   296
   Other Unsaturated Hydrocarbons   298
   Molecules Containing a Carbonyl Group   299
   Molecules Containing a Nitrogen Atom   300
   Other Molecules   301
- Conclusions   302
- References   303

SPECTRA OF FREE RADICALS AND MOLECULAR IONS PRODUCED BY
VACUUM ULTRAVIOLET PHOTOLYSIS IN LOW-TEMPERATURE MATRICES
D.E. MILLIGAN and M.E. JACOX   305
- General Principles   305
- Stabilization and Spectra of Free Radicals   307
- Stabilization and Spectra of Molecular Ions   309
- References   314

RYDBERG STATES OF DIATOMIC AND POLYATOMIC MOLECULES
V. McKOY and T. BETTS   317
- Introduction   317
- Atomic Calibration   320
- Results   321
- Conclusions   334
- References   335

THEORY OF INTRAVALENCY AND RYDBERG TRANSITIONS IN MOLECULES
O. SINANOGLU   337
- Contents   337
- Questions   338
- Theory of Excited States, Non-Closed Shell States   340
  Semi-Internal Orbitals   341
  The "Charge Wave Function, Ψ_c" of a Non-Closed Shell State 342
  Energy of a Non-Closed Shell State   348
- Relation of the Intravalency and Rydberg States of
        Spectra   350
- Classification of Transitions into Intravalency,
        Pre-Rydberg and Rydberg Types   353
- Spectral Form of the Charge Wave Function of an
        Intravalency State   356
- The Charge Wave Function of a Pre-Rydberg State   357
- Electric Dipole Oscillator Strengths of Intravalency($\bar{V}$)
        and Pre-Rydberg(pR) Transitions   361
- Inner Ionization Potentials; Excited Core Rydberg
        Series Limits   365
- Theory of Excited States of Molecules.- Many Center
        Aspects   366
  Intravalency and Pre-Rydberg States of Molecules
        with RHF-MO(SCF)/NCMET   366
  The Intravalency AO Pool   369
  Molecular Semi-Internal Orbitals from the $\bar{V}$-AO-Pool;
        Problems of the United Atom   373
  The Spectral Charge Wave Function of a Molecular
        Intravalency State ($\bar{V}$) and of a Pre-Rydberg
        State(pR)   379
- What is a Rydberg Orbital ?   380
- References and Footnotes   381
- Appendix   384

ON THE ASSIGNMENT OF MOLECULAR RYDBERG SERIES
P. HOCHMANN, H.-T. WANG, W.S. FELPS, S. FOSTER and
S.P. McGLYNN   385
- Introduction   385
- The Rydberg Equation   386
- Series Identification   388
- Conclusion   390
- References   392

ON MOLECULAR RYDBERG TERM VALUES
S. FOSTER, P. HOCHMANN and S.P. McGLYNN   395
- Introduction   395
- Results   398
- Conclusions   400
- References   401

CORRELATION OF MOLECULAR AND RARE GAS TERM VALUES
J.D. SCOTT, G.C. CAUSLEY and B.R. RUSSELL   403
- References   406

ON THE POSSIBILITIES OF STUDYING THE ELECTRONIC STRUCTURE
OF ORGANIC MOLECULES THROUGH THE ANALYSIS OF THE WAVE
FUNCTION
O. CHALVET and R. CONSTANCIEL   407
- References   412

THE EARLY YEARS OF PHOTOCHEMISTRY IN THE VACUUM ULTRAVIOLET
W. GROTH   415
- Oxygen   422
- Nitrogen   422
- Hydrogen   424
- Carbon Dioxide   424
- Carbon Monoxide   425
- Water Vapour   426
- Organic Compounds   426

PHOTOIONIZATION AND FRAGMENTATION OF POLYATOMIC MOLECULES
W.A. CHUPKA   433
- Experimental   433
- General Considerations   435
- Photoionization and Energy Deposition   436
- Dissociation Processes and the Determination of Thermo-
        chemical Quantities   439
- Diatomic Molecules   442
- Small Polyatomic Molecules   442

- Large Polyatomic Molecules   446
  Statistical Theory of Mass Spectra   448
  Radiationless Transitions   450
  Recent Tests of QET   452
  Benzene and Some Derivatives   455
  Kinetic Energy Release in Fragmentation   459
- Conclusion   460
- References   461

FAR ULTRAVIOLET PHOTOCHEMISTRY OF ORGANIC COMPOUNDS
P. AUSLOOS and S.G. LIAS   465
- Direct Methods of Determining Photofragments   467
- Indirect Methods of Determining Photofragments   468
- Alkanes and Cycloalkanes   470
- Alkenes   478
- Carbonyl Compounds   480
- References   481

ENERGY PARTIONING IN THE PHOTOCHEMISTRY OF ALKANES
R.D. KOOB   483
- Photophysical and Photochemical Description   483
- Physical Methods for Determining Energy Partioning   485
- Chemical Methods   486
- Experimental Results and Discussion   488
- Conclusion   491
- References   492

RECENT STUDIES OF THE FLUORESCENCE FROM SOME HYDROCARBON
MOLECULES
SANFORD LIPSKY   495
- Introduction   495
- Experimental   495
- Saturated Hydrocarbons   498
- Aromatic Molecules   505
- Simple Olefins   507
- References and Footnotes   510

PRODUCTION OF ELECTRONICALLY EXCITED SPECIES IN PHOTO-
DISSOCIATION OF SIMPLE MOLECULES IN THE VACUUM ULTRAVIOLET
H. OKABE   513
- Introduction   513
- Apparatus   514
- Application of Threshold Measurements   514
- Nature of the Primary Process   518
  The Spin Conservation Rule   518
  Predissociation   519
  Configuration of an Excited State   520
- References   522

ENERGY PARTITIONING IN THE PHOTOFRAGMENTS
C. VERMEIL   525
- Introduction   525
- Internal Energy of Photofragments   526
- Translational Energy   527
- Hot Photofragments   530
- Other Methods of Excitation   531
- References   533

GENERATION OF COHERENT LIGHT IN THE VACUUM ULTRAVIOLET
S.C. WALLACE   535
- References   541

SYNCHROTRON RADIATION AS A LIGHT SOURCE
J.W. TAYLOR   543
- Introduction   543
- Machines for the Production of Light   544
- Basic Equations Relating to Photon Energy and
       Intensity   545
- The Polarization of Synchrotron Radiation   549
- Possible Machine Improvements   549
- Some Experiments Utilizing the Pulse and Polarization
       Characteristics of Synchrotron Radiation   553
- References   555

OPTICAL STUDIES OF MOLECULAR CRYSTALS IN THE VACUUM
ULTRAVIOLET USING SYNCHROTRON RADIATION
E.E. KOCH   559
- Introduction   559
- Experimental Aspects   561
- Excitonic Excitations in Solid Rare Gases   563
- Atmospheric Molecules   565
- Organic Molecular Crystals   565
- References   568

PHOTOCHEMISTRY OF PLANETARY ATMOSPHERES AND INTERSTELLAR
MOLECULES
L.J. STIEF   571
- Planetary Atmospheres   571
   The HO_2 Mechanisms   573
   Ozone Mechanism   576
   Ionic Mechanism   577
- Interstellar Molecules   577
   The Interstellar Medium   578
   The Lifetime of Interstellar Molecules   580
- References   582

SUBJECT INDEX   585

PREFACE

It is probably safe to predict that the future of chemistry is linked to the excited states of molecules and to other short lived species, ions and free radicals. Molecules have only one ground state but many excited states. However large the scope of normal, ground state chemistry might be, above and beyond it lies the world of excited states, each one having its own chemistry. The electronic transitions leading to the excited states, either discrete of continuous, are examined in molecular electronic spectroscopy. Electronic spectroscopy is the queen of all spectroscopies: for if we have the resolution we have everything. Unfortunately, the chemist who is interested in the structure and reactions of larger molecules must often renounce all that information. The spectra are complex and often diffuse; resolution does not always help. To understand such spectra he must look at whole families of molecules; to some extent structural analogies help. Let us call this chemical spectroscopy and handle it with care.

In order to understand the properties of molecules we also need theory. We know that molecular problems are, in principle, soluble by the methods of quantum mechanics. Present time quantum chemistry is able to provide a nearly accurate description of not too large molecules in their ground states. It is probably again safe to predict that the future of quantum chemistry is connected with molecular excited states or, generally spoken, the accurate handling of the open-shell problem.

The purpose of the Valmorin Advanced Study Institute was to bring together spectroscopists, photochemists and theoretical chemists and make them work together on excited states, on ions, on radicals. Thus, in a sense, our meeting was interdisciplinary. (Although the writer of these lines has a natural aversion for slogans). There are many meetings of spectroscopy, of photochemistry, of quantum chemistry. There would have been no justification for having another. But only photochemists go to the photochemistry meetings, only spectroscopists attend spectroscopy meetings, only theoreticians go to quantum chemistry conferences. We often ignore each other's problems; in many cases we have difficulties in understanding each other's approach and language. Hence, the coming together of researchers of different branches can be very fruitful. A realistic common denominator is needed, however.

In our case this common denominator was the vacuum ultraviolet; that high frequency part of the ultraviolet spectrum in which most electronic transitions become Rydberg transitions and escalate towards one of the ionization potentials. The study of

the phenomena which occur when matter interacts with high frequency ultraviolet radiation requires many methods and techniques:
optical absorption and fluorescence measurements, electron impact
methods, photoelectron spectroscopy, circular dichroism, photoionization, matrix isolation techniques, vacuum ultraviolet photochemistry, advanced calculations.

Our Advanced Study Institute was a trivalent – spectroscopic,
photochemical, theoretical – meeting of chemists and physicists.

I should like to take this opportunity to express my deep
appreciation to my two colleagues on the Organizing Committee:
Dr. P.J. AUSLOOS from the U.S. National Bureau of Standards and
Dr. M.B. ROBIN from Bell Laboratories for their invaluable help.
Without them it would have been impossible to muster the impressive slate of lecturers and guestspeakers of which we were truly
proud.

Our very sincere thanks are due to the Royal Society of
Canada and to the North-Atlantic Treaty Organization for spon-
soring our Advanced Study Institute. We wish to express our
thanks to Dr. A.E. DOUGLAS from the National Research Council of
Canada and to Dr. J.R. WHITEHEAD from the Ministry of Science and
Technology of Canada, Canadian member of the Science Committee of
NATO for delivering inaugural addresses in the name of the Royal
Society of Canada and the North-Atlantic Treaty Organization
respectively and to Professor K.J. LAIDLER, from the University
of Ottawa, chairman of the Symposium Committee of the Royal
Society of Canada.

Our very sincere thanks are due to all our lecturers,
guestspeakers and other participants who accepted our invitations
and made our meeting successful.

Our Advanced Study Institute took place at the Far Hills Inn,
Valmorin, Québec in the beautiful setting of the Laurentian
Mountains. During our deliberations we took constant inspiration
from nature – and nature is great in Canada. Our sincere thanks
are due to the general manager of the Far Hills Inn, Mr. E.A.
LUCAS and his collaborators who did everything to make us feel at
home.

Last, but not least, we should like to express our thanks to
Mme Lucie LECOMTE for her excellent secretarial work.

Montreal, November 30, 1973.

C. SANDORFY.

L. ÅSBRINK, Royal Institute of Technology, Physics Department, S-100 44 Stockholm 70, Sweden.

G.H. ATKINSON, National Bureau of Standards, Room A253, Bldg 222 Washington, D.C. 20234, USA.

P.J. AUSLOOS, National Bureau of Standards, A-265 Chemistry Building, Washington, D.C. 20234, USA.

P.J. BALDWIN, Queen Mary College, Physics Department, Mile End Road, London El 4NS, Great-Britain.

J. BERKOWITZ, Argonne National Laboratory, 9700 Cass Avenue, Argonne, Illinois 60439, USA.

M. BERRY, University of Wisconsin, Department of Chemistry, Madison, Wisconsin 53706, USA.

R.D. BIRKHOFF, Oak Ridge National Lab., Health Physics Division, Oak Ridge, Tennessee 37830, USA.

H. BOCK, Universität Frankfurt, 6 Frankfurt/Main, Theodor Stern-Kai 7, Germany.

R.A.A. BOSCHI, TWP-SANDOZ AG, Dornacherstrasse 210, 4053 Basel, Switzerland.

M. BRITH, Bar-Ilan University, Chemistry Department, Ramat-Gan, Israel.

H.H. BRONGERSMA, Philips Research Laboratories, Eindhoven, The Netherlands.

G.C. CAUSLEY, Department of Chemistry, North Texas State University, Denton, Texas 76203, USA.

O. CHALVET, Centre de Mécanique Ondulatoire Appliquée, 23, rue du Maroc, Paris 19e, France.

O. CHESHNOVSKY, Tel-Aviv University, Institute of Chemistry, Tel-Aviv, Israel.

W. CHUPKA, Argonne National Laboratories, 9700 S. Cass Avenue, Argonne, Illinois 60439, USA.

G.J. COLLIN, Secteur Sciences Pures, Université du Québec à Chicoutimi, Chicoutimi, Québec, Canada.

P. DAUDEL, Institut du Radium, 11, rue Pierre et Marie Curie, Paris 5e, France.

R. DAUDEL, Centre de Mécanique Ondulatoire Appliquée, 23, rue du Maroc, 75019 Paris, France.

J. DOUCET, Université de Montréal, Département de Chimie, C.P. 6128, Montréal H3C 3J7, Québec, Canada.

A.E. DOUGLAS, National Research Council, Division of Physics, Ottawa K1A 0S1, Ontario, Canada.

G. DUROCHER, Université de Montréal, Département de Chimie, C.P. 6128, Montréal H3C 3J7, Québec, Canada.

S. FLISZAR, Université de Montréal, Département de Chimie, C.P. 6128, Montréal H3C 3J7, Québec, Canada.

Th. FÖRSTER, Institut für Physikalische Chemie der Universität, Pfaffenwaldring 55, 7000 Stuttgart 70, Germany.

C. FRIDH, Royal Institute of Technology, Physics Department,
    S-100 44 Stockholm, Sweden.
A. GEDANKEN, Tel-Aviv University, Chemistry Department, Tel-Aviv,
    Israel.
R. GILBERT, Université de Montréal, Département de Chimie,
    C.P. 6128, Montréal H3C 3J7, Québec, Canada.
M.S. GORDON, North Dakota State University, Chemistry Department,
    Fargo, North Dakota 58102, USA.
D. GOUTIER, Université de Montréal, Département de Chimie,
    C.P. 6128, Montréal H3C 3J7, Québec, Canada.
W.E. GROTH, Institut für Physikalische Chemie der Universität
    Bonn, 53 Bonn, Wegelerstrasse 12, Germany.
J.A. HERMAN, Université Laval, Département de Chimie, Québec,
    P.Q., Canada.
F.G. HERRING, University of British Columbia, Department of
    Chemistry, Vancouver 8, B.C., Canada.
F. HIRAYAMA, University of Minnesota, Department of Chemistry,
    Minneapolis, Minn. 55455, USA.
C.S. HOFFMAN, E.I. Du Pont de Nemours and Company, Organic
    Chemicals Department, Wilmington, Delaware 19898, USA.
R.H. HUEBNER, Argonne National Laboratory, Bldg 203, Rm C-129,
    Argonne, Illinois 60439, USA.
M.E. JACOX, National Bureau of Standards, Washington, D.C. 20234,
    USA.
W.C. JOHNSON, Oregon State University, Biochemistry Department,
    Corvallis, Oregon 97331, USA.
B. KLEMM, National Aeronautics and Space Administration, Goddard
    Space Flight Center, Greenbelt, Maryland 20771, USA.
R. KLEIN, Laboratoire Curie, 11, rue Pierre et Marie Curie,
    Paris 5e, France.
C.E. KLOTS, Oak Ridge National Laboratory, Health Physics Divi-
    sion, P.O. Box X, Oak Ridge, Tennessee 37830, USA.
E.E. KOCH, Deutsches Elektronen-Synchrotron DESY-F41, 2000
    Hamburg 52, Notkestieg 1, Germany.
J. KONSTANTATOS, N.R.C. Democritos, Greek Atomic Energy Commis-
    sion, Athens, Greece.
R.D. KOOB, North Dakota State University, Department of Chemistry,
    Fargo, North Dakota 58102, USA.
E.N. LASSETTRE, Carnegie-Mellon University, Center for Special
    Studies, 4400 Fifth Avenue, Pittsburgh, Pennsylvania 15213,
    USA.
L. LECOMTE, Université de Montréal, Département de Physique,
    C.P. 6128, Montréal H3C 3J7, Québec, Canada.
E. LINDHOLM, Royal Institute of Technology, Physics Department,
    S-100 44 Stockholm 70, Sweden.
S. LIPSKY, University of Minnesota, Department of Chemistry,
    Minneapolis, Minnesota 55455, USA.
B.A. LOMBOS, Sir George Williams University, Department of Elec.
    Eng'g, 1435 Drummond, Montréal, Québec, Canada.

M. LUCKEN, Sterling Chemical Laboratory, Yale University,
New Haven, Connecticut, USA.

D. LYKE, Department of Chemistry, University of Western Ontario,
London, Ontario, Canada.

R. MACAULAY, Université de Montréal, Département de Chimie,
C.P. 6128, Montréal H3C 3J7, Québec, Canada.

S.P. McGLYNN, Louisiana State University, Chemistry Department,
Baton Rouge, Louisiana 70803, USA.

V. McKOY, California Institute of Technology, Department of
Chemistry, Bldg 148-72, Pasadena, California 91109, USA.

R.A.N. McLEAN, Université de Montréal, Département de Chimie,
C.P. 6128, Montréal H3C 3J7, Québec, Canada.

I. MESSING, Tel-Aviv University, Institute of Chemistry,
Ramat-Aviv, Israel.

D.E. MILLIGAN, National Bureau of Standards, Washington, D.C.
20234, USA.

D. MINTZ, California Institute of Technology, Department of
Chemistry, Pasadena, California 91109, USA.

H. OKABE, National Bureau of Standards, Washington, D.C. 20234,
USA.

Z. OPHIR, Tel-Aviv University, Chemistry Department, Tel-Aviv,
Israel.

J.C. PERSON, Argonne National Laboratory, Bldg 203, Argonne,
Illinois 60439, USA.

S.D. PEYERIMHOFF, Universität Bonn, Lehrstuhl für Theoretische
Chemie, 53 Bonn, Wegelerstrasse 12, Germany.

A.A. PLANCKAERT, Université de Montréal, Département de Chimie,
C.P. 6128, Montréal H3C 3J7, Québec, Canada.

W.C. PRICE, University of London, King's College, Department of
Physics, Strand, London WC2R 2LS, Great-Britain.

A. PULLMAN, Institut de Biologie Physico-Chimique, 13 rue Pierre
et Marie Curie, Paris 5e, France.

B. PULLMAN, Institut de Biologie Physico-Chimique, 13 rue Pierre
et Marie Curie, Paris 5e, France.

J.F. RENDINA, McPherson Instrument Corp., 530 Main Street,
Acton, Massachusetts 01720, USA.

M.B. ROBIN, Bell Laboratories, 600 Mountain Ave, Murray-Hill,
New Jersey 07974, USA.

W. ROTHMAN, University of Houston, Department of Chemistry,
Houston, Texas 77004, USA.

Y. ROUSSEAU, Université de Montréal, Département de Chimie,
C.P. 6128, Montréal H3C 3J7, Québec, Canada.

R. ROUTHIER, Université de Montréal, Département de Chimie,
C.P. 6128, Montréal H3C 3J7, Québec, Canada.

B.R. RUSSELL, North Texas State University, Chemistry Department,
Denton, Texas 76203, USA.

D.R. SALAHUB, University of Waterloo, Department of Applied
Mathematics, Waterloo, Ontario, Canada.

C. SANDORFY, Université de Montréal, Département de Chimie,
    C.P. 6128, Montréal H3C 3J7, Québec, Canada.
P. SAUVAGEAU, Université de Montréal, Département de Chimie,
    C.P. 6128, Montréal H3C 3J7, Québec, Canada.
A.A. SCALA, Worcester Polytechnic Institute, Worcester, Mass.
    01609, USA.
O. SCHNEPP, University of Southern California, Department of
    Chemistry, University Park, Los Angeles, California 90007,
    USA.
J.D. SCOTT, North Texas State University, Chemistry Department,
    Denton, Texas 76203, USA.
O. SINANOĞLU, Yale University, Sterling Chemistry Laboratory,
    225 Prospect St., New Haven, Connecticut 06520, USA.
P.A. SNYDER, Oregon State University, Department of Biochemistry,
    Corvallis, Oregon 97331, USA.
L.J. STIEF, National Aeronautics and Space Administration,
    Goddard Space Flight Center (691), Greenbelt, Maryland
    20771, USA.
J.W. TAYLOR, University of Wisconsin, Department of Chemistry,
    1101 University Avenue, Madison, Wisconsin 53706, USA.
D.W. TURNER, Oxford University, Physical Chemistry Laboratory,
    South Parks Road, Oxford OX1 3OZ, Great-Britain.
C. VERMEIL, E.S.P.C.I., 10 rue Vauquelin, Paris 5e, France.
S.C. WALLACE, Thomas J. Watson Research Center (IBM), P.O. Box
    218, Yorktown Heights, New York 10598, USA.
K.H. WELGE, Universität Bielefeld, Institut für Physik,
    Bielefeld, Viktoriastrasse 44, Germany.
C.NGUYEN XUAN, Università di Roma, Laboratorio di Metodologie
    Avanzate, Inorganiche del C.N.R., c/o Ist di Chimica
    Generale ed Inorganica, P. le delle Scienze, 5 Rome, Italy.
C. ZAULI, Università degli Studi, Viale Risorgimento 4,
    40136 Bologna, Italy.

VACUUM ULTRAVIOLET AND PHOTOELECTRON SPECTROSCOPY - HISTORICAL
BACKGROUND AND SURVEY

W.C. Price

King's College
London WC2

The early work of Schumann and Lyman at the beginning of the
century which enabled spectroscopic studies to be extended into
the energy range above 6 electron volts is too well known to require any description here. Once the field had been opened up
exploration developed along several different paths according to
the particular interests of various investigators. Emission
studies giving data on ionized atoms were carried out using hot
spark sources and the absorption spectra of gases and solids were
investigated using both line and continuous background radiation.
While initially interest centred on the discovery of the absorption
bands predicted by classical dispersion theory to lie in this
spectral region it soon became clear that absorption spectroscopy
in the vacuum ultraviolet was destined to become a major technique
for the investigation of the electronic structure of matter. In
order to confine the present survey within reasonable bounds and
to deal mainly with the interests of this summer school it is proposed to limit our considerations to the absorption and photoionization of polyatomic molecules in the short wavelength region
below 2000A.

V.U.V. SPECTRA OF POLYATOMIC MOLECULES

In the twenties detailed studies of the spectra of diatomic
molecules had established a sound understanding of the spectra and
electronic structures of diatomic molecules. This was a necessary
preliminary to an extension of the field to molecules containing
more than two atoms. For these polyatomic molecules a considerable
amount of empirical spectral data had accumulated but no basic
theory of the nature of the electronic transitions associated with

the spectra was available although the infra-red and Raman spectra
were fairly well understood. The first break-through came when
Dieke and Kistiakowsky[1] succeeded in analysing the rotational fine
structure of the near ultraviolet absorption spectrum of formaldehyde and showed that the transition moment associated with the
system is perpendicular to the CO axis and in the plane of the
molecule. However the information concerning electronic structure
derivable from near ultraviolet spectra is very limited as it also
was in the case of the near u.v. spectrum of benzene which was
interpreted soon afterwards by Sklar[2]. It has turned out that the
studies of molecular spectra in the vacuum ultraviolet region have
been the most fruitful means of revealing their electron archi-
tecture. In 1935 Price[3] reported the analysis of the absorption
spectrum of acetylene in the region 1520-1000A. He showed that
the bands could be arranged into two Rydberg series which con-
verged to a common limit at 11.40 eV (1087A) - see Fig. 1. The
identification of the 'purely' electronic bands was made by the

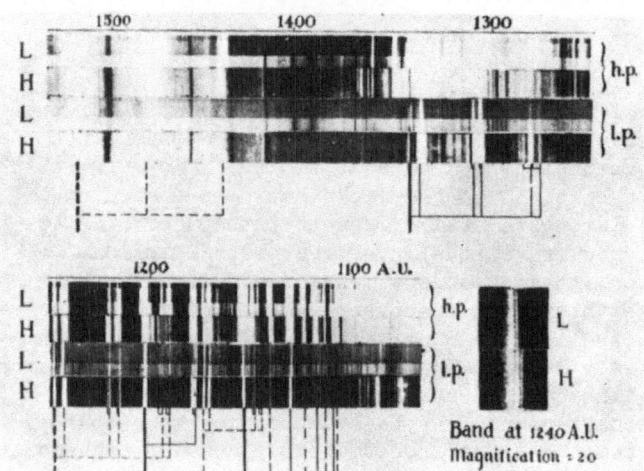

Fig. 1. The absorption spectrum of acetylene in range 1520-1000A

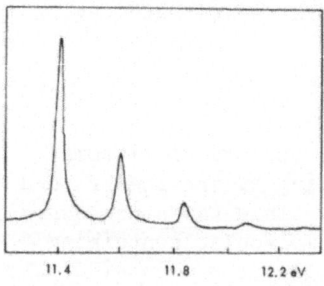

Fig. 2. The photoelectron spectrum of acetylene (first band).

use of partially deuterated acetylene, the heavy water needed for its preparation having just become available at the time. Bands of this type corresponding to the removal of a bonding electron show a small blue isotopic shift on deuteration which results from the change in the difference of zero point energy between the ground and the excited state. This enables them to be distinguished from bands which are due to vibrationally excited states accompanying the main electronic transitions. These latter bands show larger shifts in the opposite direction the magnitudes of which depend upon whether they are H or CH mass dependent. From these shifts it can be deduced that the main vibration accompanying the electronic transitions is that of the symmetrical stretching vibration of the triple bond. The fact that one of the Rydberg series was found to consist of double and the other of single headed bands fitted in with the conclusion that an electron from the (π_u^4) shell was being excited. The photoelectron spectrum of this molecule[4], Fig. 2, which was obtained later confirms the interpretation of the general features of the spectrum.

A somewhat similar spectrum was obtained for ethylene though only one strong series was found in this case (Refs. 3 and 5; Figs. 3 and 4). The spectrum was clearly identifiable with the excitation

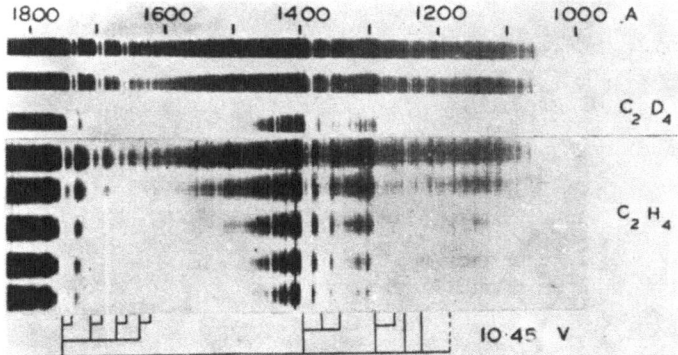

Fig. 3. The absorption spectrum of ethylene (C_2H_4 and C_2D_4).

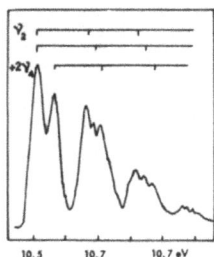

Fig. 4. The photoelectron spectrum of C_2H_4 (first band).

of a π electron from the double CC bond leading to ionization at
10.51 eV. The natural molecule to be studied after acetylene and
ethylene was ethane but the very diffuse bands found for this
molecule were not amenable to analysis in Rydberg series. A similarly disappointing situation was found for methane where the bands
were even more diffuse than those of ethane. However many other
small molecules were found to give Rydberg bands and in particular
benzene exhibited a series of sharp bands which were shown to
correspond to π excitation leading to ionization at 9.25 eV [6] (Fig.
5). Fairly sharp Rydberg series were generally found for small
molecules with non-bonding lone pair p type electrons such as in
H_2O, H_2S, HCl, HBr, HI, CO_2, CS_2, H_2CO, CH_3I. The series of the
molecules containing heavy atoms were frequently found to go to
two different limits with the separation expected for the spin-
orbit splitting of the ion (Fig. 6). These spectra were very

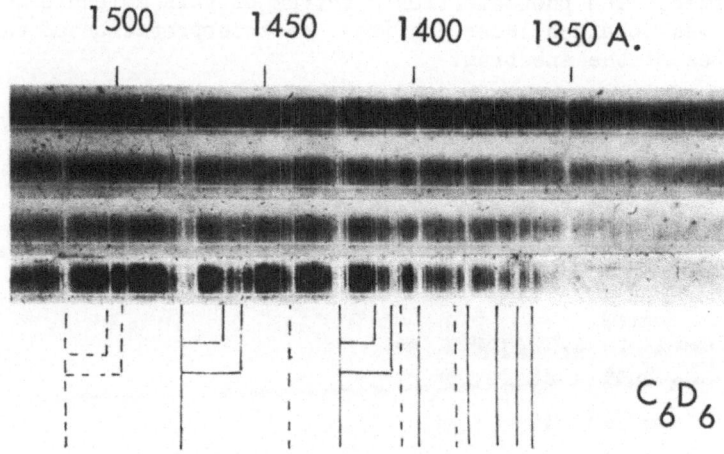

Fig. 5.   The absorption spectrum of benzene.

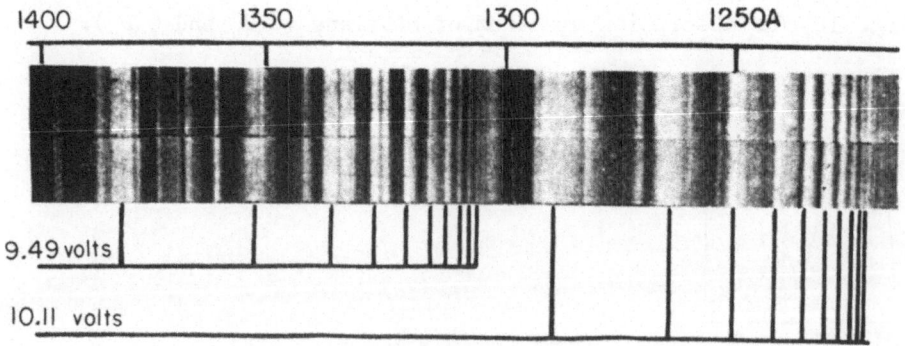

Fig. 6.   The absorption spectrum of methyl iodide.

important at the time since they provided experimental confirmation of the molecular orbital theories of Mulliken and others who were extending the orbital-shell concepts for atoms (single positive centre) to systems containing several positive centres i.e. molecules. At a later period, about 1960, Herzberg combining flash photolysis techniques with vacuum ultraviolet spectroscopy observed Rydberg series for a large number of free radicals as well as small molecules in an extensive range of investigations. Space limitation prevents their discussion here and the reader is referred to his standard work on the electronic spectra of polyatomic molecules (Ref. 7) for further information on these and other spectra.

Although many fairly large molecules such as butadiene, diacetylene, furan, thiophene etc. gave spectra which were sharp enough to enable accurate values of their first ionization potentials to be determined it was found that bands generally became diffuse for larger molecules e.g. in going to the higher members of an homologous series such as the alkyl iodides. The diffuseness of bands which are at longer wavelengths than the first ionization potential mainly arises from the dissociation of the excited parent molecule into neutral fragments. Additional possibilities which can result in band broadening are operative above the onset of ionization. Only occasionally were sharp band systems found at wavelengths shorter than that corresponding to the first ionization potential i.e. for electrons which come from the inner orbitals. The systems found for CS_2 by Tanaka and his associates (Refs. 8, 9 and 10; Figs. 7, 8 and 9) are examples of such exceptional cases. In general bands occurring in the energy range above the first ionization potential were found to be very broad and diffuse. Causes for this are the conversion of the energy of

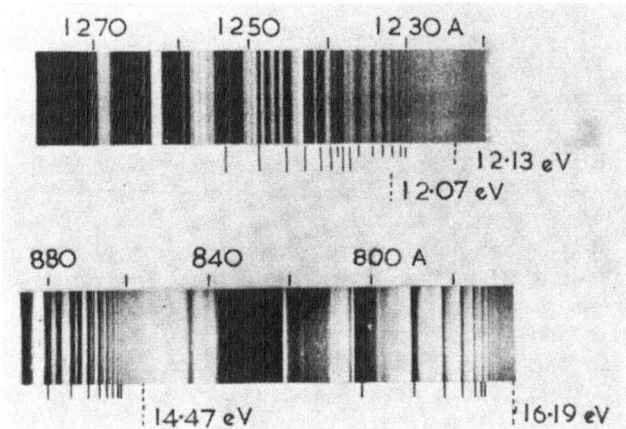

Fig. 7. Absorption systems in carbon disulfide.

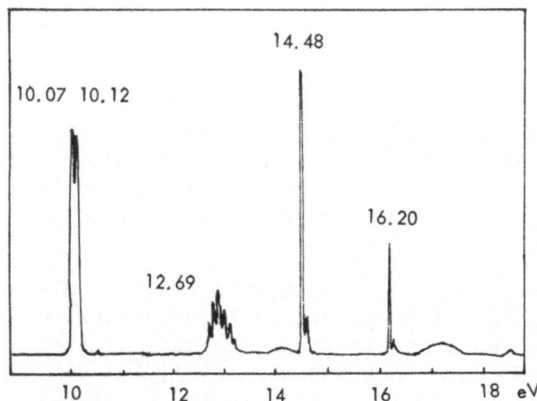

Fig. 8. Photoelectron bands of CS$_2$ related to absorption systems of Fig. 7.

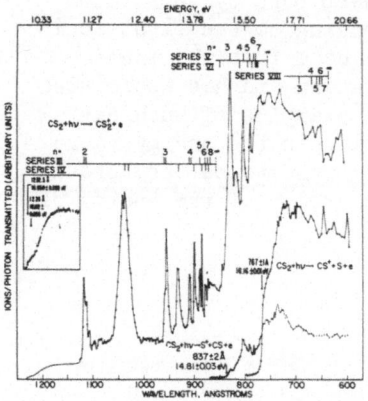

Fig. 9. Photoionization efficiency curve of CS$_2$ after Dibeler and Walker (Ref. 10).

these 'superexcited' states into ionization of the outer electron by autoionization or by the dissociation of the molecule into neutral or ionized fragments. The problem of how to obtain the ionization potentials of molecules the bands of which are so diffuse as to make them intractable to spectroscopic analysis was solved in two stages both of which were developed by looking at the phenomena associated with the interaction of a photon with a molecule other than the probability of its absorption. Such additional aspects of the interaction are of course the production of photoelectrons and positive ions. Onset of ionization in the gas can be looked for as it is illuminated with photons of increasing energy. This was the method of photoionization developed by Watanabe and his associates[11]. It was further developed into a mass spectrometric photoionization efficiency technique by Hurzeler, Inghram, Morrison and others[12]. A more specific study of the energies of the photoelectron ejected in the process was initiated by the work of Vilesov, Kurbatov and Terenin[13]. This was greatly improved upon by the development of He I photoelectron spectroscopy by Turner and his associates[14] and He II spectroscopy by Price[15]. These developments came after periods of frustration arising from the difficulties which were then preventing further progress of investigations into molecular electronic structures. The success of photon impact over electron impact studies arises from the greater precision with which the energy of the photon can be defined. However the increasing precision of electron monochromators and their use in electron energy loss studies of collisions with molecules by Simpson[16], Lassettre[17] and others has provided valuable supplementary information in this field. A brief outline of the photoionization developments will now be presented.

PHOTOIONIZATION

By using a vacuum ultraviolet monochromator in conjunction with a hydrogen source and scanning a range of wavelengths over an exit slit behind which was mounted an ionization chamber containing the gas (Fig. 10) it was found possible to determine very precisely the wavelength at which the ionization set in. The

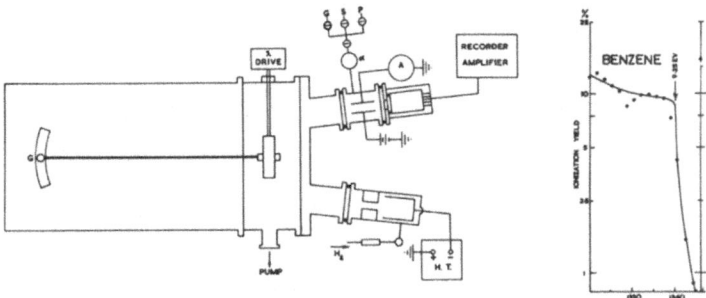

Fig. 10. V.U.V. monochromator and photoionization cell for Watanabe type i.p. studies.

first ionization potentials of many hundreds of molecules whose
v.u.v. spectra were completely diffuse were determined this way[11].
In particular methane, ethane and a large number of homologous
series of all sorts of organic compounds which had vapour pressures
of a few microns at room temperatures were assigned first ionization potentials with accuracies of 0.1 to 0.01 eV. From this body
of data it was possible to show how various substituents in a molecule affected the ionization potentials of the electrons in the
chromophoric group associated with the first i.p. Although in
several cases some vibrational structure could be detected near
the onset by virtue of sudden increases in ion current associated
with the autoionization of these vibration levels of the ion this
was very exceptional. Also it was not possible to determine higher ionization potentials since as the photon energy was increased
sudden further increases in ion current occurred at the energy of
the first strong Rydberg band going to a second i.p. as a result
of the autoionization of this band. These increases usually started about 2 eV lower than the second i.p. and autoionization of the
higher Rydberg bands further contributes to the ionization curve
at higher photon energies. This hides any feature of the curve
which might be used to indicate the onset of a second ionization
potential and is responsible for the failure of all resonance
methods including mass spectrophotometric and electron impact studies to give accurate data on inner ionization potentials or the
vibrational patterns associated with them.

PHOTOELECTRON SPECTROSCOPY

In 1961 a paper appeared by Vilesov, Kurbatov and Terenin[13]
in which the parallel plate ion chamber of Watanabe was replaced
by a retarding potential analyser so arranged as to measure the
energies of photoelectrons arising from photons entering from a
monochromator and ionizing the vapour in the ion chamber. As a
lithium fluoride window was used to isolate the monochromator it
was not possible to use photons of energies greater than 11 eV.
Because of the weakness of their source which necessitated the use
of monochromator band widths of 15A and also because their electron
analyser was of low efficiency they were only able to achieve a
resolution of about 0.20 eV on the energies of the photoelectrons
ejected from the molecules they studied. In spite of these limitations they observed some higher ionization potentials for a number
of aromatic hydrocarbons and methyl anilines at values which were
subsequently confirmed by more accurate techniques. In a discussion with Dr. Turner in 1962 on the problem of how to observe ionization from deeper orbitals the author suggested modifying the
Vilesov technique by the use of the strong helium resonance line
as a source of high energy (21.2 eV) photons thereby dispensing
with the monochromator and employing differential pumping on this
source to avoid the limitations imposed by the lithium fluoride

window. The suggestion was taken up by Dr. Turner and he developed it into the highly successful technique of photoelectron spectroscopy[14]. This technique revealed for the first time and in a
very direct way the inner orbitals of molecules about which all
other methods had failed to give any information. For example
prior to the advent of this technique there was no knowledge of
the inner orbitals of the water molecule and it was an exciting
moment when the photoelectron spectrum revealed their existence
for the first time and showed that they had the properties which
theory predicted. Another exciting moment was when by using the
resonance line of He II (40.1 eV) Price[15] was able to observe ionization from the inner a_1, 2sC orbital of methane. The vast amount
of fundamental information given by this technique in the last decade has given a great boost to the theoretical attempts for calculating molecular properties and has added greatly to our understanding of the valence orbital structure of molecules. Almost
simultaneously with this development and quite independently
Siegbahn[18] and his group at Uppsala developed the technique of
X ray photoelectron spectroscopy which gave complementary information about the core electrons as well as valence electrons in both
gaseous molecules and solids. A recent paper has extended this
technique even to liquids.

Recent developments in photoelectron spectroscopy include
studies of angular dependence, variation of cross section with
photoelectron energy, the photoelectron spectra of radicals and
atoms, the study at high temperatures of large aromatic molecules
and ionic molecules such as the alkali halides using molecular beam
techniques as well as the use of synchrotron radiation to increase
the energy range of all studies. We shall conclude this chapter
with a brief account of how the method reveals the difference between covalent binding such as in the halogen acids and the ionic
binding characteristic of alkali halide molecules.

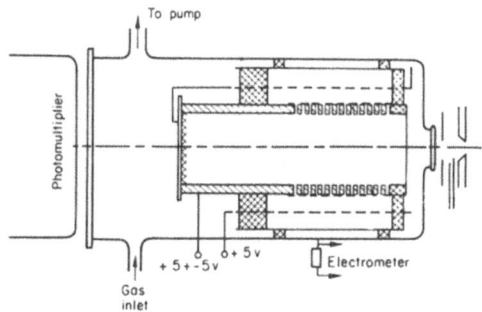

Fig. 11. Attachment for measuring photoelectron energies –
Vilesov, Kurbatov and Terenin (Ref. 13).

If it were possible to obtain the photoelectron spectrum of
I⁻ this would be found to consist of two lines corresponding to
the ^2P and $^2P_{\frac{1}{2}}$ states of I separated by 0.94 eV i.e. $^3/2$ times
the spin-orbit coupling constant. The $^2P_{3/2}$ line would appear at
3.3 eV (the electron affinity of I) and would be twice as strong
as the $^2P_{\frac{1}{2}}$ line which would be found at 3.3 + 0.94 eV. The spectrum would be exactly analogous to that of an inert gas. Let us
now consider the effect of bringing up a positive ion to the I⁻.
The $^2P_{3/2}$ state would split into pπ $^2\Pi_{3/2}$ and pσ $^2\Sigma$. In first approximation they should each have equal cross sections equal also to
the cross section of the $^2\Pi_{\frac{1}{2}}$ state. While all three will grad-
ually increase in ionization potential with the approach of the
positive charge the pσ electrons will begin to resonate one at a
time across to the positive ion i.e. resonance between I⁻ + M⁺
($^1\Sigma$ ionic) and I + M ($^1\Sigma$ covalent). An incident photon would find
an electron either on the I⁻ or between it and the positive ion if
the exchange were slow enough compared with the photon interaction
time. This would mean a splitting of the $^2\Sigma$ band into two bands
one of which was nearly coincident with $^2\Pi_{3/2}$ and the other shifted
to higher ionization energies because of the bonding character it
derived from the exchange. The former could be called the ionic
orbital band and the latter the covalent orbital band and the
relative intensities of these afford a measure of their occupancies
and reflect the fractional ionic character of the bond. The

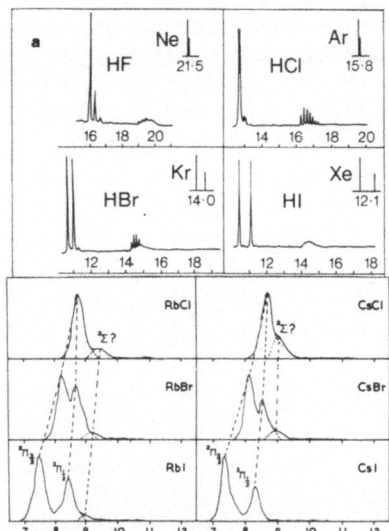

Fig. 12(a). Photoelectron spectra of the halogen acids and the
isoelectronic inert gases. (b) Photoelectron spectra of the
halides of rubidium and caesium.

features predicted above have been found in the spectra of the rubidium and caesium halides by Potts, Price and Williams[19]. They are illustrated in Fig. 10 and will be discussed in detail elsewhere. Similar results have also been obtained by Berkowitz[20] on the more easily vapourised thallium halides.

REFERENCES

1. G.H. Dieke and G.B. Kistiakowsky, Phys. Rev. 45, 4 (1934.
2. A.L. Sklar, J. Chem. Phys. 5, 669 (1937).
3. W.C. Price, Phys. Rev. 47, 444 (1935).
4. C. Baker and D.W. Turner, Proc. Roy. Soc. A308, 19 (1968).
5. W.C. Price and W.T. Tutte, Proc. Roy. Soc. A174, 207 (1940).
6. W.C. Price and R.W. Wood, J. Chem. Phys. 3, 439 (1935).
7. G. Herzberg, Electronic Spectra of Polyatomic Molecules (Van Nostrand, New York, 1966).
8. Y. Tanaka, A.S. Jursa and F.J. LeBlanc, J. Chem. Phys. 32, 1205 (1960).
9. D.W. Turner, Molecular Photoelectron Spectroscopy (Wiley – Interscience, New York, 1970).
10. V.H. Dibeler and J.A. Walker, International Mass Spectroscopy Conference, Berlin (1967).
11. K. Watanabe and T. Nakayama, J. Chem. Phys. 29, 48 (1958).
12. H. Hurzeler, M.G. Inghram and J.D. Morrison, J. Chem. Phys. 27, 313 (1957).
13. F.I. Vilesov, B.C. Kurbatov and A.N. Terenin, Dokl. Akad. Nauk SSSR, 138, 1329 (1961), and Soviet Physics – Doklady, 8, 883 (1962).
14. D.W. Turner and M.I. Al-Joboury, J. Chem. Phys. 37, 3007 (1962) and J. Chem. Soc. 5141 (1963).
15. W.C. Price, Molecular Spectroscopy IV (Ed. P.W. Hepple, Inst. Petroleum, London, 1969) p.221, and Phil. Trans. Roy. Soc. A268, 59 (1970).
16. J.A. Simpson, Rev. Sci. Inst. 35, 68 (1964).
17. E.N. Lassettre, A. Skeberle and M.A. Dillon, J. Chem. Phys. 46, 4536 (1967).
18. K. Siegbahn et al, Electron Spectroscopy for Chemical Analysis (Nova Acta Reg. Soc. Sci. Uppsala, Ser. IV, 20 1967).
19. A.W. Potts, W.C. Price and T.A. Williams – to be published.
20. J. Berkowitz, J. Chem. Phys. 56, 2766 (1972) and 57, 3194 (1972).

ASPECTS OF MOLECULAR RYDBERG STATES

M. B. Robin

Bell Laboratories
Murray Hill, New Jersey, 07974, U.S.A.

Spectroscopy in the region beyond 50,000 cm^{-1} is difficult for both technical and fundamental reasons. With regard to the latter, consider a typical molecule with a set of occupied valence shell MO's, ψ, and a corresponding set of virtual valence shell MO's, ψ^*, Fig. 1. [1] In such a molecule, it is

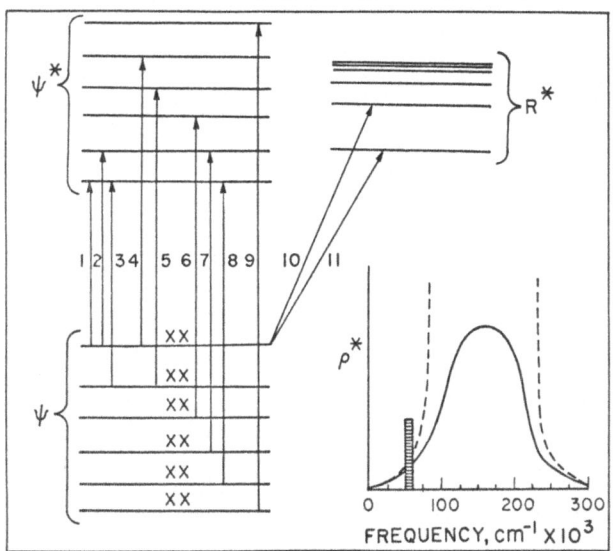

Fig. 1: Possible transitions in a hypothetical polyatomic molecule; shown schematically at lower right is the density of excited valence shell (—) and valence shell plus Rydberg (---) excited states as a function of excitation frequency.

almost certain that transition 1 will be the lowest, with 2 and/or 3 following it. There is some slight ambiguity here in the ordering of states, but that is what makes spectroscopy interesting. Similarly, transitions 8 and 9 must be the highest, almost without doubt. But it is in the intermediate region, where transitions 4, 5, 6 and 7 all have about the same energy, that one encounters more transitions than can be coped with. This is shown as a plot of excitation energy <u>versus</u> density of excited states in the Figure. Roughly speaking, it is at about 50,000 cm^{-1} that the density of states becomes large in polyatomic molecules, and our ability to interprete correspondingly decreases.

But the situation worsens. Beyond the valence shell, there are sets of MO's composed of AO's of higher principal quantum number, the Rydberg AO's, R*. Excitations to these generally commence near 50,000 cm^{-1}, and change the density curve to that shown by the dashed line, Fig. 1. Thus there are several good reasons for not going beyond 50,000 cm^{-1}, and most spectroscopists do not. For the adventuresome few who do jump the 50,000 cm^{-1} barrier, the most important pressing problem with respect to understanding the photochemistry and spectroscopy in this region is that of distinguishing Rydberg from valence shell excited states; if we cannot distinguish between $\pi \rightarrow \pi^*$ and $\pi \rightarrow 7s$, then we will make some very bad mistakes in interpretation.

Until quite recently, it was understood that the frequencies of sharp Rydberg excitations followed the well-known atom-like formula $h\nu = I.P. - R/(n-\delta)^2$. Many are just coming to realize however, that Rydberg excitations can also be very broad and featureless, with only the first member of the series visible, and with a frequency which is usually far off the formula even if a series were identifiable. Therefore other means of identifying Rydbergs are needed.

Physically, a molecular Rydberg orbital is distinguished from a valence shell orbital by its considerably larger size, and so is much more susceptible to perturbation from the outside. As an illustration of how advantage can be taken of this to differentiate the two types of states, consider the 2600 Å band of benzene, Fig. 2. The transition in question is a $\pi \rightarrow \pi^*$ valence shell excitation, and upon application of 136 atm of He, there is no detectable change in the band under moderate resolution. In fact, all valence shell transitions when pressurized to <u>ca</u>. 100 atm with a second gas yield spectra indistinguishable from those taken at near-zero pressure. This perturbation which is negligibly weak for valence shell excited states is severe for Rydberg states. For example, as the 5p → 6s Rydberg excitation of methyl iodide is pressurized, Fig. 3, the band is progressively broadened to the high-frequency side, finally achieving a width at the highest pressure about 50 times that at zero pressure. This

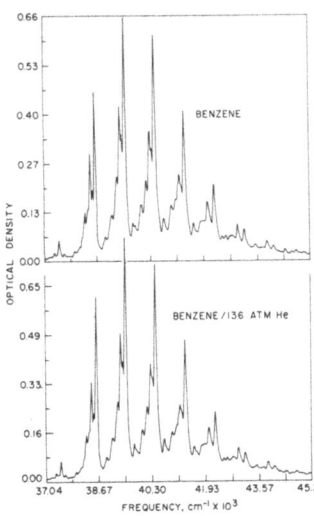

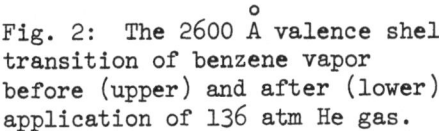

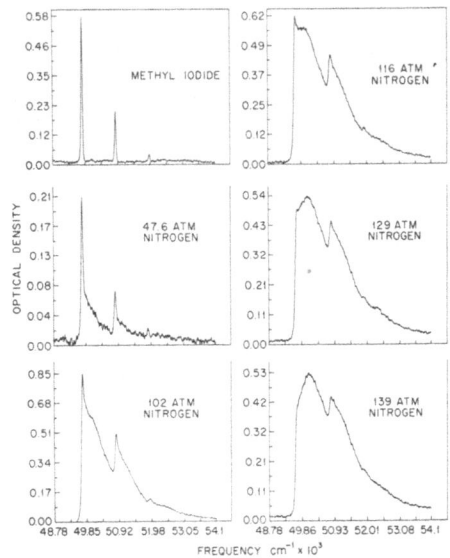

Fig. 2: The 2600 Å valence shell
transition of benzene vapor
before (upper) and after (lower)
application of 136 atm He gas.

Fig. 3: The asymmetrical
broadening of the 5p → 6s
Rydberg transition of methyl
iodide under nitrogen gas
perturbation.

totally symmetric broadening to the high-frequency side upon
modest pressurization with a second gas is characteristic of
transitions to big-orbit upper states.

The pressure perturbation works best with sharp lines; if
the band in question is already very broad and unstructured, then
the perturbation must be made more drastic to effect the test,
i.e. the absorber is placed in a solution, crystal, or matrix.
Under these conditions, it is not yet clear what happens, but
Rice and Jortner present this picture. If the absorber is in a
medium of high electron mobility, as in the solid or liquid rare
gases, then the lowest molecular Rydberg state is still defined,
but with an absorption which is about 8000 cm^{-1} higher in frequency and perhaps 2-10 times wider. In organic solutions and
organic crystals in which the mobility is low, there is a very
strong scattering of the optical electron so that the lifetime
of the state is very short and the band therefore so broadened
as to be undetectable. Both of these situations contrast with
the case of valence shell excitations, where the gas-to-condensed
phase transition gives a polarization red shift. By way of
illustration, consider the amide spectra given in Fig. 4, where
the full curves are the vapor spectra. In formamide and N,N-
dimethyl formamide, the most intense features at 50,000-60,000 cm^{-1}
are the valence shell N → V_1 transitions, which are seen to persist

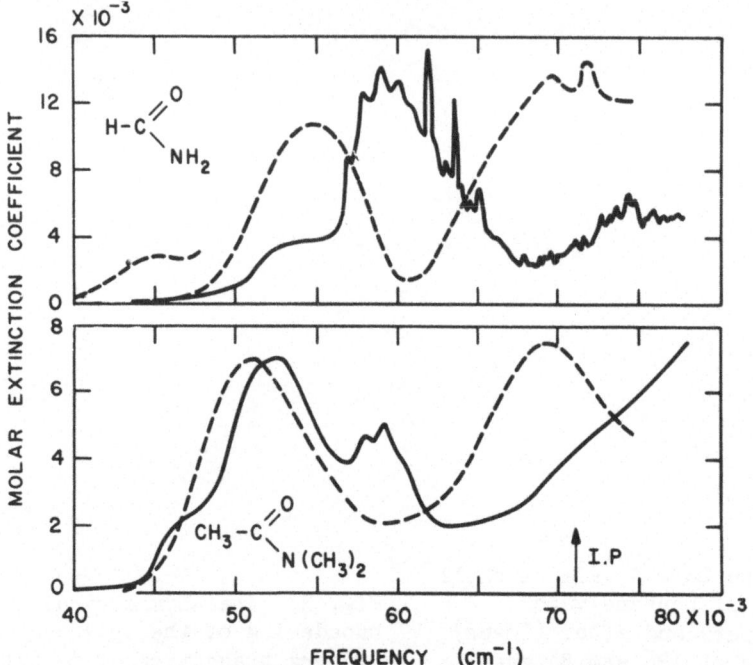

Fig. 4: Absorption spectra of two amides taken in the gas
phase (—) and in the condensed phase at 24°K (---).

relatively unperturbed in the solid films (dashed lines). Just
to the high-frequency side of the N → V_1 maxima, sharper Rydberg
excitations are seen in the gas phase, which as expected, do not
survive the transition into the condensed phase. The weaker,
much broader bands on the low-frequency sides of N → V_1 do not
have the usual shapes of Rydberg excitations, but in the condensed
phase, they too are obliterated, thus demonstrating their Rydberg
character. The very weak band at 45,000 cm^{-1} in formamide is the
n → π* valence shell excitation.

Another feature which can occasionally be put to good use
involves the photoelectron spectrum. Since the photoelectron
spectrum yields a Franck-Condon envelope appropriate to the
optical electron at infinity, while the optical spectrum
involves the same electron being placed in a finite but still very
large nonbonding orbital, it is no surprise then that the Franck-
Condon envelope of a Rydberg excitation can closely resemble the
photoelectron band towards which it would converge. As an
example, consider the case of HNCO, Fig. 5. The first two ionization potentials of this molecule involve the lone pair orbital
on nitrogen and the pi orbital of the CO group, the Franck-Condon
factors being noticeably different for the two ionization processes.
In the optical spectrum, exactly these shapes are seen again

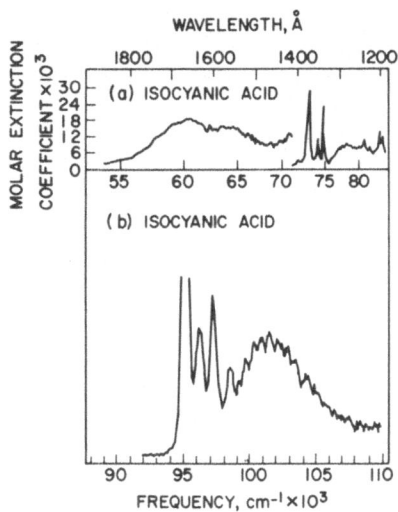

Fig. 5: Optical absorption spectrum of isocyanic acid (a) and the first two bands of its photoelectron spectrum (b).

at 70,000–80,000 cm^{-1}, showing that these bands are Rydberg excitations going to the first two ionization potentials. On the other hand, there is no reason to expect the photoelectron band envelope to resemble that of an optical valence shell excitation, and no such resemblance is ever observed in polyatomic molecules.

Having discussed briefly some physical tests for distinguishing Rydberg and valence shell excitations, let us now consider the energetic aspects of Rydberg excitations. For this, it is most useful to consider the term value defined in the following way: measure the ionization potential of a particular electron in a molecule using photoelectron spectroscopy, and then optically measure the excitation energy needed to lift that same electron from its original orbital into a Rydberg orbital. The term value is then the difference between the ionization and bound-excitation energies, and is in effect, the Rydberg orbital ionization potential. For a given principal quantum number, the discussion of the term value is equivalent to a discussion of δ in the Rydberg formula, for the term value is $R/(n-\delta)^2$.

First, consider the n=3 term values of the first row atoms, Fig. 6. As one goes from H to F, the 3d term value stays very nearly constant at the hydrogenic value, implying $\delta \sim 0.0-0.1$. In the same series, the situation is rather different for 3p orbitals, for here the term values increase, and for the atoms between C and F, a δ of 0.6-0.8 fits the data. Finally, 3s orbitals show the strongest deviation from the H atom value, corresponding to $\delta = 1.0-1.3$. The deviations from the H atom term values are due to exchange and penetration effects. For the Li atom, calculations show that the exchange binding is larger than that due to penetration, but that in heavier atoms it is the

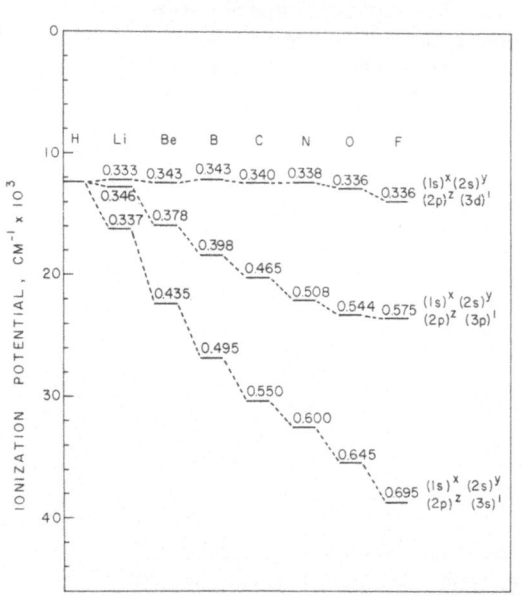

Fig. 6: Rydberg term
values of the first-row
atoms, averaged over the
multiplet structure. The
numbers refer to the effective nuclear charges (X1/3)
of the Rydberg orbitals.

penetration energy which is the major factor in determining δ.
Physically, if the Rydberg orbital is one with appreciable
density close to the nucleus (3s > 3p > 3d), then the effective
nuclear charge that it feels will be larger than 1.0, and so the
Rydberg ionization potential will be larger than that of hydrogen.

A very useful idea is presented by Edlen and Mulliken, who
point out that the Rydberg orbital must remain orthogonal to
the core orbitals. If the Rydberg orbital in question is ns,
then it is orthogonal to (n-1)p and (n-1)d by virtue of the
angular variations of the wave functions. However, for orbitals
in the core having the same symmetry as the Rydberg orbital
(called precursors) this won't work, and so it is necessary to
Schmit orthogonalize so that the Rydberg orbital partakes of
core wave function character, which of course is more tightly
binding. Thus, the presence of precursor core orbitals leads to
penetration of the core and $\delta \neq 0$.

Certain of the concepts mentioned above for atoms can be
carried over to molecules, in which penetration still is the
major cause of nonzero δ's, with the magnitude of the effect
depending upon the effective nuclear charges of the constituent
atoms. Consider first the 3s, 3p and 3d term values in a series
of alkylated oxygen compounds in which the originating MO is the
lone pair on the oxygen atom, Fig. 7. As water is alkylated to
form the various alcohols and ethers, the 3d term values remain
very close to the hydrogenic value of 12,190 cm^{-1}. The 3p orbitals,
having more penetration than 3d, have term values of ca. 20,000 cm^{-1}.

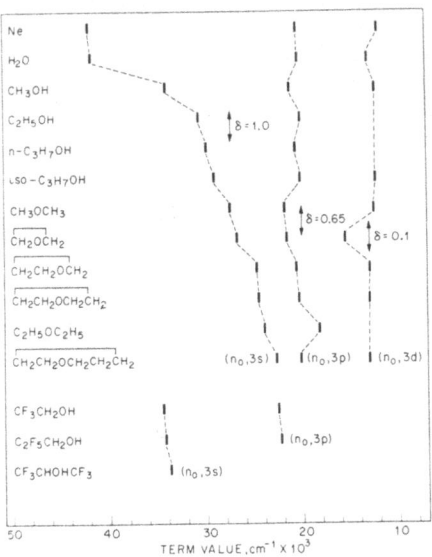

Fig. 7: Rydberg term values of the alkylated and fluorinated oxo compounds.

It is the 3s orbitals which show the largest variations, starting at very high values but decreasing by 50% upon alkylation, to converge to $\sim$21,000 cm^{-1} in the highly alkylated ethers. If these compounds instead are fluorinated, the corresponding 3p term values increase slightly to 22,000 cm^{-1}, while the 3s term values are much higher than in their alkyl counterparts. The arrows in Fig. 7 represent the term values corresponding to the δ values quoted above as being characteristic of the heavier atoms in the first row, and one sees that for 3s, some term values are considerably higher and others considerably lower than the "expected" values, and one could easily get the impression that there was no rhyme or reason to the deviation of the 3s term value from $\delta=1.0$. Actually, the deviation is regular when considered as a function of the size of the alkyl group. It is also clear that the idea that if the spectra of related molecules are adjusted so that their ionization potentials coincide, then the Rydberg excitations will also coincide, is acceptable for 3p and 3d Rydberg orbitals, but can be quite erroneous for 3s orbitals.

Term value studies similar to that in Fig. 7 have been made for the alkylated and fluorinated amines, sulfides, olefins, halides, ketones, amides, etc., etc. with very nearly the same results: the term values of the 3p and 3d Rydberg orbitals are independent of alkyl substitution at 20,000 and 13,000 cm^{-1}, respectively, whereas 3s may start at any high value but converges to 21,000 cm^{-1} for highly alkylated central chromophores. This is summarized in part in Fig. 8 where the ground state ionization potentials as a function of the number of CH_x groups in the molecule are shown in the upper part of the figure and the 3s excited state ionization potentials (term values) are shown below.

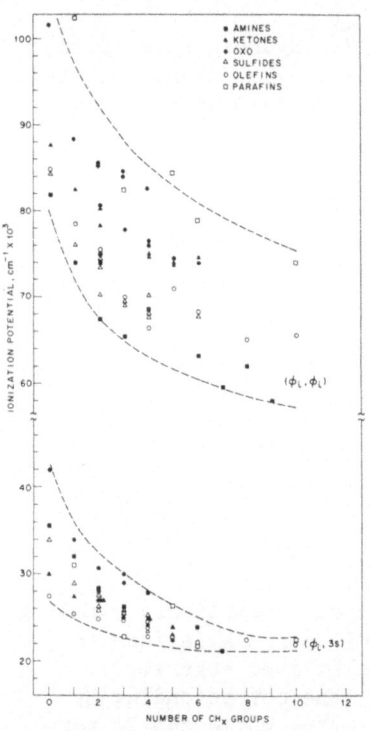

Fig. 8: Comparison of the ground
state (upper) and 3s excited state
(lower) ionization potentials as a
function of the number of CH_x groups
attached to the central chromophore.

One sees that as the alkylation proceeds, the ground state ionization potential decreases as expected, but there is no other correlation, since the limiting ionization potential of an amine will
have no relationship to that of a ketone, for example. For the
3s excited state ionization potentials there is the expected spread
in values at low alkylation, but for highly alkylated systems,
there is a very definite convergence to the value 21,000 cm^{-1}.
Though the ground state ionization potential is strongly dependent
upon the central atom but shows a slight drop with alkylation, in
the excited 3s state, the ionization potential reaches a steady
value with large alkyl groups which is independent of the central
atom!

By way of explanation, in water, 3s is an oxygen orbital with
the high penetration characteristic of atoms with high effective
nuclear charge. In methanol, the Rydberg MO now has C atom
character as well, an atom with far lower penetration (Fig. 6),
and so the term value drops. As the alkyl group(s) become even
larger, the Rydberg orbital contains more C atom contribution and
thus less central atom character. In the limit of very large
alkyl groups, the central atom contribution is swamped by that of
the alkyl group, and so the term value is that of the alkyl group,
rather than that of the central atom. In such a case the positive
charge is pinned to a specific center, but the optical electron is

delocalized over the entire alkyl group so that the excited state
somewhat resembles a Wannier exciton in a solid. Since transitions
to such Rydberg states will show a very strong charge-transfer
character, it is seen that the alkyl groups are highly chromophoric in these transitions, being the terminating MO's. In the
case of transitions to 3p or 3d, the atomic term values do not
change very much from C to F (Fig. 6) and so in molecules made
up of these atoms, the term values will not shift since the
penetration at the constituent atoms does not vary.

If the molecule is fluorinated rather than alkylated, one
expects an increased 3s term value since the penetration at F is
larger than at C, N or O, but again converging to a common value
for highly fluorinated species. This too is observed, the fluorinated term value limit being 36,000 cm^{-1} for transitions to 3s.

Another interesting aspect of the 3s Rydberg term values is
that they are very nearly independent of the originating MO in
a molecule. The water spectrum presents a very good example of
this. The first ionization potential of water is 102,000 cm^{-1} and
the corresponding Rydberg excitation is centered at 60,000 cm^{-1},
yielding a term value of 42,000 cm^{-1}. The optical transition is a
broad, dissociating one because the 3s orbital is strongly mixed
with the σ^*(O-H) antibonding valence shell orbital of the same
symmetry. Higher ionization potentials are found at 119,000 and
149,000 cm^{-1}, and corresponding to them, broad Rydberg excitations
to 3s are observed at 77,000 and 102,000 cm^{-1}. The fourth ionization potential of water is 260,000 cm^{-1}, leading to the expectation
of a broad band to 3s centered at 218,000 cm^{-1}, as yet unobserved.
Since the 3s-3p separation is also constant (21,000 cm^{-1}) sharp
transitions to 3p should appear 21,000 cm^{-1} beyond each of the
continua to 3s.

The constancy of the 3s term values in a molecule can be
pushed into the X-ray region as well. Thus the 1s ionization
potential of methane is 2,345,500 cm^{-1} according to the ESCA
spectrum, while in the X-ray absorption spectrum a weak line at
2,315,900 cm^{-1} and a strong line at 2,324,800 cm^{-1} are observed.
The term values for these lines, 29,600 and 20,700 cm^{-1} identify
them as transitions to 3s and 3p, respectively, with term values
very close to those observed at over 2,000,000 cm^{-1} lower frequency, 32,100 and ca. 20,000 cm^{-1}.

One can quickly derive a simple formula for calculating the
3s term value in a polyatomic molecule of N atoms. Assume first
that the molecular term value is a sum of atomic term values
weighted by the fraction of each type of atom

$$TV_{mol} = \sum_i \rho \, TV_{Atom\ i}$$

Moreover, let

$$TV_{atom\ i} = \frac{R\delta Z_i}{9} + Q_i$$

where $R/9$ is the hydrogenic 3s term value, δZ_i is the partial charge at center i, and Q_i expresses the binding due to penetration at center i. Next, let $\delta Z_i = n_i/N$ and $\rho_i = n_i/N$, so that

$$TV_{mol} = \sum_i \frac{n_i}{N} \left(\frac{R}{9N} + Q_i\right)$$

which assumes that both positive hole and the optical electron are equally spread over all centers of the molecule. Once the penetration terms are picked empirically (Q_C = 21,000; Q_N = 25,000; Q_O = 30,000; Q_F = 32,000 cm^{-1}), the 3s term value is easily calculated, independent of molecular geometry or chemical structure. The success of this simple formula is illustrated in Table I.

TABLE I

Molecule	3s Term Value	
	calc.	obs.
CH_4	33,200	31,500
C_4H_{10}	24,000	25,200
$C_{10}H_{15}$	22,200	22,500
H_2O	42,200	41,800
CH_3OCH_3	28,000	25,700
$C_4H_9OC_4H_9$	22,400	22,000
$N(CH_3)_3$	25,100	24,400
HF	44,200	42,700
CF_4	32,300	30,000
CF_3CH_2OH	30,000	34,500

REFERENCES

[1] Literature references to all of the concepts discussed in this article may be found in M. B. Robin, "Higher Excited States of Polyatomic Molecules," Vol. I and II, Academic Press, New York, 1974.

PHOTOELECTRON SPECTROSCOPY

D.W. Turner

Physical Chemistry Laboratory,
South Parks Road,
Oxford.

INTRODUCTION - ELECTRONS AND PHOTONS

Whilst the free electron possesses energy ($\frac{1}{2}$mv) and momentum (mv) which may take any value, electrons bound in atomic or molecular orbits possess discrete energies and momenta. The subject of electron spectroscopy as a new branch of molecular spectroscopy derives from the possibility of exchange between bound and free electrons. The discrete changes in the free electron's energy and momentum which may be detected by electronic means, reflects their energy levels in the bound condition. Thus bombardment of molecules by free electrons causes these to be slowed by discrete amounts and we have the subject of electron energy loss spectroscopy. This gives essentially the same information as optical absorption spectroscopy but with different selection rules. We distinguish this from transition of one electron from the bound state to the free condition which has no direct analogy in optical spectroscopy and which requires a source of exactly defined energy greater than the ionization energy. Photons have proved the most satisfactory source, though metastable atoms and more recently high energy electrons have also been employed. As will be seen in the attempt to reach higher and higher energy resolution, the ejection of electrons by photons has proved to be of almost universal application and to possess some special advantages and constitutes a new branch of molecular spectroscopy, photoelectron spectroscopy[1]. It is applicable in principle to the study of all states of matter, though in practice the most subtle detail is only obtained in the examination of vapours, that is to say, of isolated molecules.

The defining relationship is as follows:

$$E = h\nu - I_i - \Delta E_{vib} - \Delta E_{rot} \pm \Delta E_{trans} \quad .$$

Here E is the energy with which the electron is expelled by photons of energy $h\nu$, and I represents the ionization energy of electrons in their different shells and subshells. Note that here we define ionization energy by reference to the energy needed to generate an electron-ion pair where the ion may be in its ground or some excited state, and this is to be contrasted with higher ionization energies which envisage multiple ionization, that is the ejection of more than one electron and the formation of a doubly or higher positively charged ion.

SOME EXPERIMENTAL ASPECTS OF ELECTRON ENERGY ANALYSIS

Photoelectron spectroscopy using uv sources has been described as a high resolution spectroscopy[2] on the grounds that the natural limit set by fundamental processes such as rotation and translation of molecules can be reached. This limit (kT = 26 meV at room temperature) of something less than 10 meV can be reached experimentally using any of a wide variety of electron energy analysers, using both magnetic or electrostatic analysis. The most important experimental limitation is set by the spatial uniformity of the potentials of the analyser internal surfaces. Notably the potential of the chamber in which ionization occurs and the slits between which electrons are constrained to pass. As an example of the ease with which such potentials are altered, we show the contact potentials of a graphite surface as a function of distance in which the effect of polishing, scratching and touching with a finger produced very large changes (of the order of 50 meV or greater) are seen.

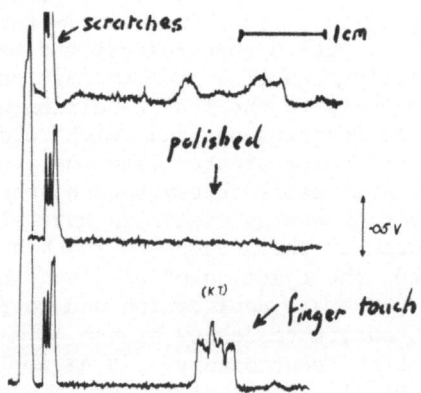

Contact Potentials of Graphite Surfaces.

(recorded using radioactive tipped probe).

This is a source of instrumental broadening which may vary from time to time with the same instrument and from sample to sample since it is very sensitive to surface adsorption. As in other branches of spectroscopy, however, it may be at least in part removed by deconvolution methods using suitable reference materials to provide an instrument function. An example of this shown here (from earlier work by D.W. Turner and A.R. Muir, unpublished) is the deconvolution of a part of the hydrogen photoelectron spectrum using argon to provide an instrument function.

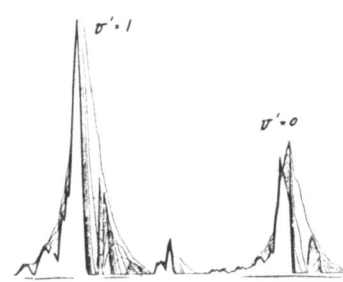

This was an iterative deconvolution programme and could not be carried too far being sensitive to a noise content of both. In applying such deconvolution procedures it is important to remember that there is an irreducible linewidth arising from translation motion which varies with mass, temperature and ejected electron kinetic energy[3].

$$\Delta E_{trans} = 0.75(ET/M)^{\frac{1}{2}} \text{ meV}$$

Advantage has been taken of the reduction of the translational width with electron kinetic energy in a very elegant experiment by Lindholm and Asbrink in which the rotational fine structure in the hydrogen spectrum has been directly resolved by making use of the neon resonance line instead of (as here) the helium resonance line much lower kinetic energies (approximately 1eV) were involved.

LIGHT SOURCE

Vacuum ultraviolet emissions from the rare gases have provided the most commonly used forms of excitation, and of these the helium resonance line (HeI-584A) has been most widely employed. The following table gives the energy in electron volts of the principal lines of the rare gases and hydrogen, which has been found to be useful in decreasing order of importance.

HeI	21.2
HeII	40.8
NeI	16.65, 16.83
ArI	11.62, 11.83
H Lyman α	10.2

The helium discharge emits substantially pure HeI radiation at higher pressures and can be made to emit substantially pure HeII radiation at low pressures, by using controlled energy electron excitation approach. The two emissions can also be separated by the use of filters. In the case of the other rare gases and hydrogen, however, minority vacuum ultraviolet emissions cannot

readily be filtered out, and failure to recognise the presence of
these minority emissions may cause misinterpretation in spectra
unless care is taken. An example of this from unpublished work by
C.R. Brundle in the author's laboratory is given below. Here the
highest energy band of the NO spectrum is compared when helium
and argon are used in the light source. It will be seen that in
the latter case the spectrum is unusually complex in that there
are photoelectrons associated not only with the two argon resonance
lines, but also with hydrogen and other unidentified photons.

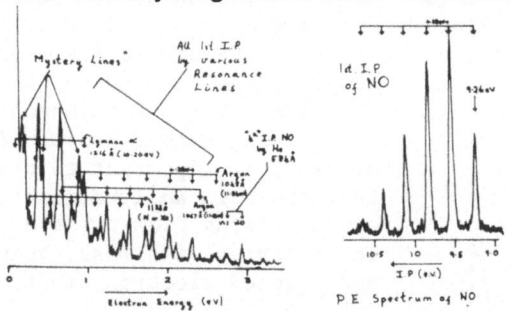

P.E. Spectrum of NO using Argon and Helium Radiation

PHOTOELECTRON BANDS DESIGNATED ACCORDING TO THE IONIC STATES

     We shall see later that it is especially convenient from the
point of view of the chemist to describe the different bands in
the photoelectron spectrum by referring to the orbital which is
losing an electron. This allows direct comparison with the orbital
concept of molecular structure. It is, however, strictly more
accurate to refer to the state of the ion which is left behind,
the bands in the photoelectron spectrum then bearing ionic state
terms. The singlet molecules of low symmetry ionized to a series
of such terms which correspond to configurational excitation of a
ground state molecular ion. It will be obvious that certain
transitions between these states will be allowed, and the spacings
between the corresponding bands in the photoelectron spectrum
gives the energy difference and hence the wavelengths at which
fluorescence may be expected to be observable. Herzberg[4] has shown
how the electron impact induced fluorescence of carbon dioxide can
be identified with the CO_2^+ ion by comparison with the photoelectron
spectrum of CO_2. This same fluorescence emission can indeed be
seen using an image intensifier in the ionization region of a
photoelectron spectrometer, and we are attempting to establish
coincidences between these fluorescent photons and the appropriate
photoelectrons.

Image Intensifier Picture of Ionization Regions in HeI
Photoelectron Spectrometer with (left) and without (right) CO_2.

The optical spectrum of this emission in the ultra-violet is
shown below

$^2\Sigma_u^+(0,0,0) \rightarrow {}^2\Pi_{g\frac{3}{2}/\frac{1}{2}}$ $(0,0,0), (B \rightarrow X)$ $\underline{CO_2^+ \text{ Fluorescence}}$

(P. Cutmore & D.W. Turner)

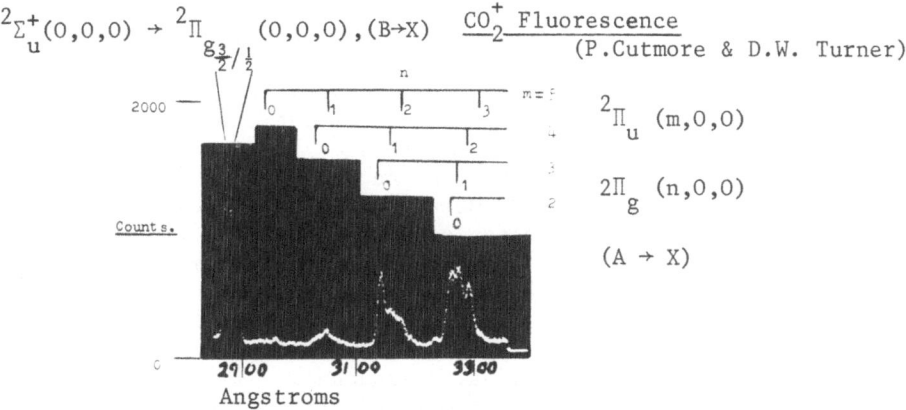

$^2\Pi_u$ (m,0,0)

$2\Pi_g$ (n,0,0)

$(A \rightarrow X)$

Angstroms

This consists of two series of bands, one of which corresponds to
the $\Pi \rightarrow \Pi$ transition and one to the $\Sigma \rightarrow \Pi$ transition. In the latter
case the spin-orbit splitting in the ground state of the ion is
particularly obvious. The two sets of transitions are indicated
by $\Longrightarrow$ on the photoelectron spectrum shown below.

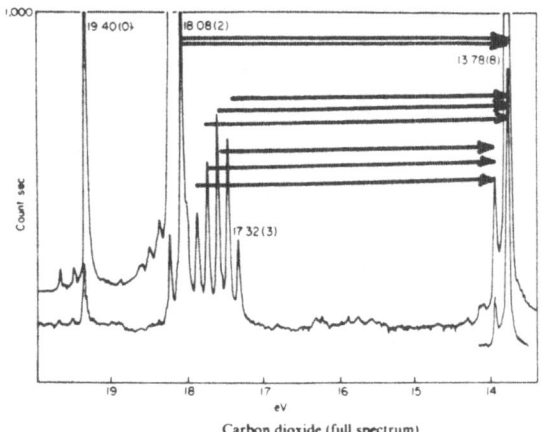

Carbon dioxide (full spectrum)

VIBRATIONAL FINE STRUCTURE

One of the most intriguing features of the photoelectron
spectra of many simple organic molecules is the wealth of and
variations in the vibrational fine structure associated with many
of the bands. In this lies a potentially powerful interpretative
method since the character of the vibration which is excited
reveals the change in equilibrium geometry between the molecule and
the ion in the electronic states in question. There are two
important parameters to be derived. One is the symmetry classification of the vibration, often this may be related to a vibration
of the parent molecule unless major symmetry changes occur, the
other is the strength with which the higher vibrational quantum
levels are excited, in other words, the length of the Franck-
Condon progression. The latter parameter is conveniently measured
by the difference in energy between the adiabatic and vertical
ionization energies, and its importance lies in a possible
connection between change in geometry and change in force constant.
In optical spectroscopy we have Mecke's Rule and it was noticed
quite early[1] that in photoelectron spectra there is a roughly
linear relationship between the adiabatic-vertical ionization
energy difference (ΔE) and the change in vibrational frequency
($\omega''/\omega' - 1$). Though no great weight should be placed upon this
correlation it may occasionally help us to choose between otherwise equally plausible vibrational assignments. In the case of
the photoelectron spectrum of diazobicyclooctane (DABCO) for
example, the first two bands relate to the in-phase and out-of-
phase combinations of nitrogen lone pair orbitals ($n_+ n_-$).
Heilbronner and co-workers[5] have shown how using extended Hückel
type calculations, the largest changes in nuclear coordinates
corresponding to each ionization can be evaluated. From this the
probable vibrational mode which would satisfy these could be
predicted. It was concluded that the n_+ ionization leads to the
excitation of ν_4 and n_- to the excitation of ν_6 as tabulated
below.

	ion	molecule	ion	molecule
n_+ leads to ν_4	780 cm^{-1} ←	965 cm^{-1}	780 cm^{-1}	965 cm^{-1}
n_- leads to ν_6	530 cm^{-1} ←	600 cm^{-1}	530 cm^{-1}	

We note, however, that the two bands though exhibiting long
vibrational progressions are not of the same width, the second is
a much longer progression, when according to the correlation noted
above it should in fact be shorter. The predicted ΔE values from
the ratios in vibrational frequency being 0.2 and 0.15 eV
respectively, suggesting that ν_4 is excited in each case ω''/ω'
for the second band being 1.82 leading to an expected ΔE of 0.9 eV
as observed.

The photoelectron spectra of (a) Ketene (with sections to expanded scale, (b) Formaldehyde (part of) and (c) Formaldehyde d$_2$ (part of).

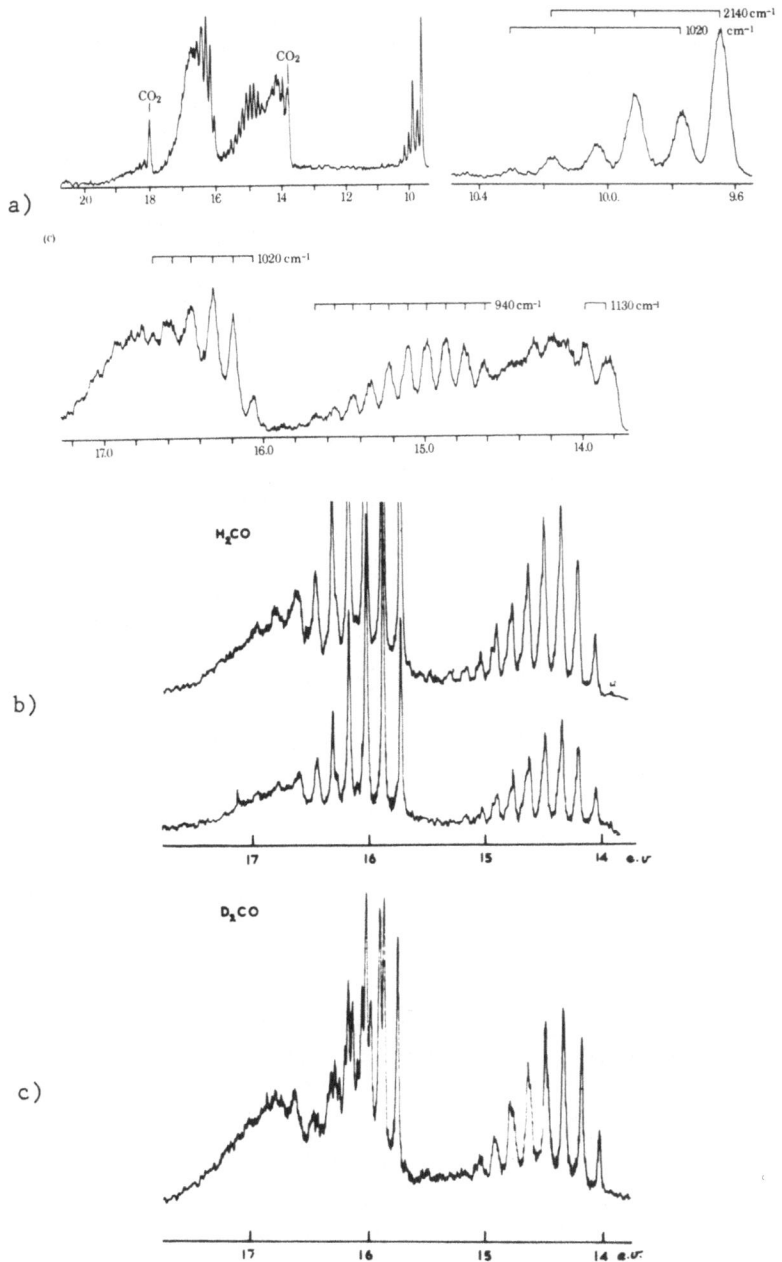

Vibrational Fine Structure in the Third Band of Formaldehyde.

One of the most powerful tools for identification of the mode
of vibration in any spectroscopy, of course, is the effect of
deuteration. CH modes may be expected to be easily identified by
the reduction in frequency which deuteration brings about. This
is not always straightforward, however, particularly in cases
where the frequencies are close in value to another molecular
vibration, not involving hydrogen. The possibility that deuteration then may create or destroy a near resonance condition can
cause changes more drastic than simply an alteration of a
progression interval. An example of this is provided by the third
band of the formaldehyde photoelectron spectrum[1]. Upon deuteration
this changes from a regular progression to a complicated band
system which is seen to be the superposition of two progressions.
This arises from the strong interaction between ν_1 and ν_2 which
though in formaldehyde itself remain essentially CH stretching and
CO stretching modes, in deuteroformaldehyde are very strongly
mixed.

Vibrational Fine Structure in the Ketene Photoelectron Spectrum.

Ketene is perhaps the most extensive and beautifully
structured photoelectron spectrum yet recorded, and although this
has not yet been fully analysed[1], some conclusions can be drawn
from the absence of deuteration effects in the first and second
bands, and an effect in the fourth band reminiscent of that
described above for formaldehyde. We can in any case make a
reasonable guess as to the nature of the vibrations seen from an
application of the ΔE versus $(\omega''/\omega' - 1)$ relationship cited above.
This leads to the following conclusions

Band	4	3	2	1
Vibration	ν_3	ν_3	$\nu_4 + 2\nu_2$	ν_2, ν_4
	(CH$_2$ def.)	(CH$_2$ def.)	(C-C-O str)	(C-C-O str)

Since the hydrogen atoms lie on the nodal surfaces of the π
molecular orbitals, we may conclude that the first two bands in
which no hydrogen vibration is excited are most probably the π
molecular orbitals. Furthermore it is possible that since in the
next two bands there is evidently much C-H bonding, these may be
identified with orbitals related to the π molecular orbitals of
carbon dioxide in which the binding energy has been increased by
movement of the proton positive charges into regions of high
electron density. In carbon dioxide of course such π molecular
orbitals are doubly degenerate and only one pair (π_u and π_g) are

affected. It is instructive to compare therefore ketene with
carbon dioxide (and with allene) noting that the effect of taking
a proton from one of the terminal atoms of the chain out to one
side is to slightly loosen that π orbital in whose nodal surface
it moves, and to very much tighten the corresponding orthogonal
π orbital. We may estimate the magnitude of the loosening and
tightening to be expected by comparisons made with the photoelectron spectra of suitably related pairs of molecules, oxygen and
formaldehyde for example and formaldehyde and ethylene. The essential steps of this comparison are tabulated below.

Energy levels from P.E. Spectra:	Compare with :
$O = C = O$ $CH_2 = C = O$ $CH_2 = C = CH_2$	$O_2(^1\Delta_g) \rightarrow H_2CO \rightarrow CH_2 = CH_2$

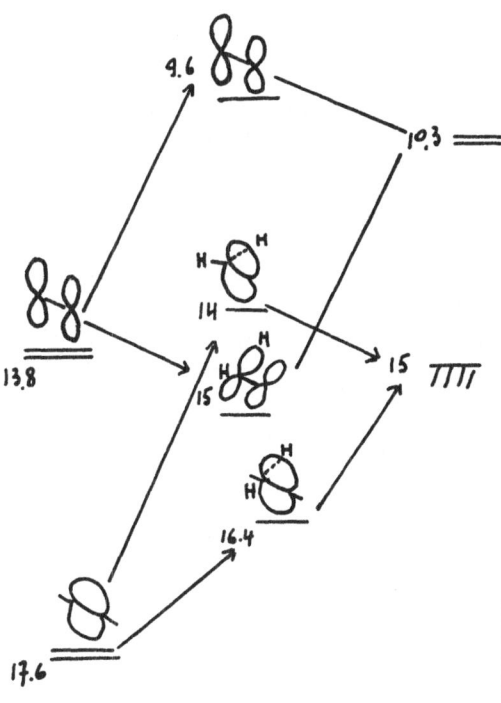

π_g 11.5 → ~ 7(π^*) → 3.5

π_u ~17 → 14.7(π) → 10.6

π_g 11.5 → 10.9(n_o) → 12.5

π_u ~17 → 16.4(σ) → 14-16

A more satisfying approach has been developed by Hollas which is effectively the reverse of that outlined for DABCO. Here dimensional changes are calculated from the intensities in the vibrational fine structure of a photoelectron spectrum, the proportion of the flux associated with the ith vibrational mode in terms of its ratio to the 0-0 transition is shown to relate directly to the normal coordinate change Δq_i which is in turn related to changes in bond length. In sufficiently simple cases the changes in bond length can thus be deduced quantitatively from the photoelectron spectral intensities. This procedure was applied to the spectra of the cyanogen halides, the long progression in the B state of the ion (the third band) leading as one would expect to a large change of the carbon halogen bond distance, consistent with the strongly π bonding character of this molecular orbital $(1\Pi,\ Cl\ 3_p + C2_p + N2_p)$.

Vertical Ionization

It is worth noting that there is a pitfall in taking the most intense peak in a photoelectron band as representing the energy for vertical ionization and that instrumental band width is of significance. Ionization from the ground vibrational state of the molecule to that of the molecular ion $(0,0,0,....)$ occurs at a unique energy subject only to a rotational and translation broadening, whereas vertical ionization may be accomplished by transition into a number of different combinations of vibrational modes. Whether or not this structure is integrated to give a correctly weighted estimate of the probability within a particular energy range depends upon the instrumental resolution in relation to the fineness of the vibrational structure. This is illustrated schematically below in a hypothetical example with two vibrational modes, each being excited up to 4 quanta.

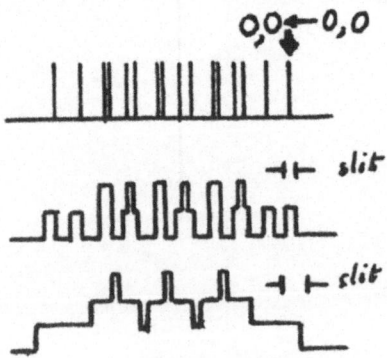

Such considerations tell us that the vertical ionization energy
of benzene is nearer 9.4 than 9.24, the peak of the very sharp
0-0 transition which is strong in the high resolution results,
though of relatively low integrated intensity as is clear when
this is compared with the earlier low resolution data[1,6].

INTERPRETATIVE METHODS INVOLVING MOLECULAR ORBITAL CALCULATIONS

Since for closed shell molecules the negatives of molecular
orbital energies may be equated with ionization energies by
Koopmans theorem, many attempts have been made to find the best
computational technique to fit photoelectron spectral data.
However the results for any one case taken in isolation have
rarely been convincing and more satisfactory procedures have
involved the comparison within series of related structures. It
is by no means always necessary to use the full armoury of
theoretical chemistry and sometimes chemically significant results
can be deduced by careful choice of model compounds and employment
of the technique of group orbital interactions. In this approach
a complex structure is broken down into smaller constituent parts
and the interactions between these parts produced by a comparison
between the photoelectron spectra of the complete molecular
structure with the photoelectron spectra of simple molecules
containing the part structures. This is essentially a
perturbation approach. We have shown how in the case of some
benzenoid compounds basis functions involving the initially
localised π molecular orbitals of the benzene ring explain some
of the main features of photoelectron spectra of, for example,
biphenyls[7], styrene[8] and related structures[9]. The photoelectron
structure of biphenyls show many and often overlapping bands, but
there are certain common features. In some biphenyls the dihedral angle is already known from electron diffraction, and when
this data is made use of it becomes clear that the first four
bands : examples of which are shown below

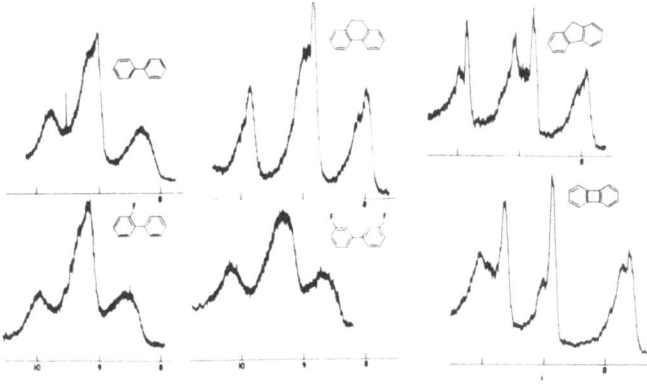

can be understood as arising from the interaction between the
e_{1g} orbitals of the benzene rings in which varying degrees of
through space interaction dependent upon the dihedral angle, cause
a splitting as indicated in the following diagram:

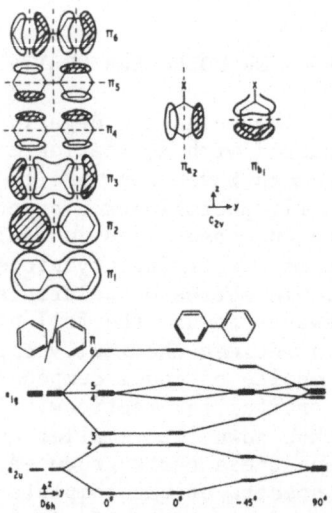

These first four bands are to be understood then as the
interaction of ψ_2 and ψ_3 of the isolated benzene rings, and we can
evaluate an interaction parameter B

$$2B = \Delta E = \varepsilon(\pi_6 - \pi_3)$$

in the planar case and for the non planar compound, $B_\theta = B\cos\theta$.
B is very similar in magnitude to that appropriate to other
directly linked interacting π moieties.

Splitting of π levels (ΔE) found from photoelectron spectra

compound	$\Delta E/eV = 2B$	$r/\overset{\circ}{A}$
HC≡C-C≡CH	2.45	1.38
N≡C-C≡N	2.46	1.38
HC≡C-C≡N	2.43	1.38
H_2C=CH-CH=CH_2(t)	2.41	1.48
H_2C=CH-C N	2.40	1.44
Ph·C≡CH	2.21	1.42
Ph·C≡CN	2.20	1.42
Ph·CH=CH_2	2.07	1.48
fluorene	1.98	1.48

Together with biphenyls, stilbenes and styrenes provide classical examples of steric inhibition of resonance. In styrenes, however, few determinations of the dihedral angle has been carried out, though it has been shown that in polymerization steric considerations are predominant factors. Here again the interaction pattern of the π energy levels on rotation of one of the phenyl groups about the bond linking to the ethylene group can be considered on a group orbital basis. The basis set are the semi-localized orthogonal π m.o's of the phenyl groups $\pi_{a_{2u}}$, $\pi_{e_{1g}}(b_1)$ and $\pi_{e_{1g}}(a_2)$ and $\pi_{C=C}$ of the ethylene moiety. Direct through-space overlap is predominantly between this latter orbital and $\pi_{e_{1g}}(b_1)$ of the phenyl group and gives rise to a splitting (ΔE) between the resultant molecular orbitals π_2 and π_4 (the first and third bands in the photoelectron spectrum).

$$\Delta E = \varepsilon(\pi_4 - \pi_2) = [(A_{C=C} - A_2)^2 + 4B_\theta^2]^{\frac{1}{2}}$$

The difference between this and the earlier simpler expression for the biphenyls, arising of course from the asymmetry of the basis set energies in the present example. The dihedral angles that are inferred are listed below:

Dihedral Angles (θ) of Substituted Phenylethylenes

phenylethylene derivative	this work (a)	(b)	other workers (c)	(cf.ref.8) (d)
phenylethylene	0	0	26	26
trans-β-methyl	0	0	26	26
o-methyl	22	31,33[e]		
α-methyl	38	33	32	36
o,o'-methyl	55	54,50		
trans-β-phenyl	0	0	17	17
cis-β-phenyl	35	28	43	34
α-phenyl	40	43	33	49
o,o'-dichloro	~46			
pentafluoro	~28			
n = 2*	0	0		
n = 3	0	0	20[e]	
n = 4	60		40[e]	

(a) absorption spectroscopy;   (b) semi-empirical calculation;   (c) reinterpretation of data in (c) ;   (d) scale models,

CORRELATIONS BETWEEN P.E. SPECTRAL DATA AND OTHER PHYSICAL PARAMETERS

Since the physical and chemical behaviour of molecules must arise from the dispositions, energies and polarisabilities of the molecular electrons, it can confidently be predicted that photoelectron spectroscopy will provide a considerable insight into chemical behaviour. A number of examples of such correlations have already been provided. One of the simplest and most easy to understand and noted quite early on[1], was a relation between σ bond ionization energy in halides and bond strength. More recently linear correlations have been established between halogen p lone pair ionization energies and Pauling halogen electronegativity for the halogen acids[10], halobenzenes[10] and halothiaphenes [11]. In the diatomic halogen molecules analogous correlations have been reported with atomic ionization energies and halogen p orbital Mulliken electronegativity[12]. This concept has been extended to make correlations between phosphorous lone-pair ionization energy in trivalently bonded phosphorous compounds and INDO phosphorous charge, the basis of an interpretative critique[13]. These of course represent examples where short-range direct interactions are important. In more complex molecules quite remote charge transfer effects can be detected in suitable cases for example in parasubstituted bromobenzenes where correlations with the extensive literature of Hammett constants can be attempted. An example of such a correlation is shown below.

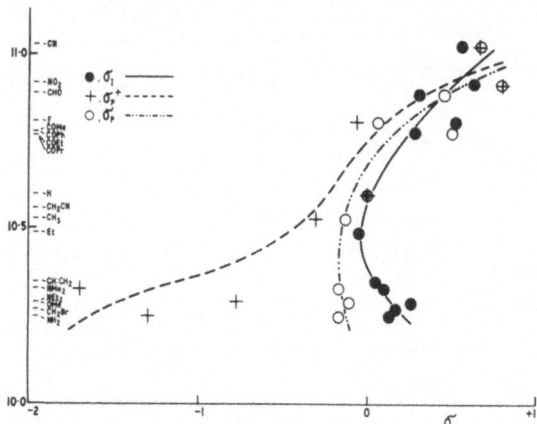

Ionization Energies of Br $4p_y$ electrons in p-bromo benzenes. Br R and Hammett σ values for different substituents −R.

PHOTOELECTRON SPECTRA OF TRANSIENT SPECIES

There have been a number of attempts to record the P.E. spectra of simple free radicals by, for example, admitting the products of electrical discharges in low pressure gases to the chamber of a photoelectron spectrometer[14] or by studying the products of pyrolytic reactions[15]. In most cases difficulties arise from the overlap of bands where the spectrum of the wanted species tends to be swamped under the strong photoelectron spectra from the main stable entities present. In certain cases, therefore, only one or two of the expected photoelectron bands have been detected. Detailed information is available on CS, SO, $O_2(^1\Delta_g)$ and NF_2 and somewhat more fragmentary information about the CH_3 radical and a number of atomic species including oxygen, nitrogen, hydrogen and the halogens. An example is given here of the unstable molecule HBS which has been produced by T.P. Fehlner in the author's laboratory from the reaction between H_2S and hot boron. Three bands can be identified, one of which exhibits a short vibrational sequence of peaks whose widths suggests a splitting into doublet components separated by about 300 cm^{-1} which we attribute to spin orbit coupling. This band thus corresponds to ionization from the π orbital of HBS which is partially localised on sulphur. It is of interest to compare the partial spectrum of HBS with those of two isoelectronic species. The first two bands in the spectrum of CS are remarkably similar to those of HBS in both energy and Franck-Condon envelope, except that the order of σ and π orbitals are reversed. In CS the first band at 11.33 eV results from ionization from a non-bonding σ orbital, while the second at 12.79 eV results from ionization from a bonding π orbital. On the other hand the highest π and σ orbital of HCN appear to be of nearly the same energy. Both HBS and CBS can be related to HCP by the movement of one nuclear charge unit, CS by its transfer from the terminal position into the P nucleus and HBS by its transfer from C to P. In the first row compound (CO, HCN, HBO) the CO-HCN difference is mainly one of reduced π orbital ionization energy. The HBO-HCN difference is unknown, however by reference to the methyl derivatives of HCN where the π ionization is hardly affected, the difference may be expected to be mainly in reduced σ ionization energy. Thus we might expect HCP to behave in similar fashion having nearly the same σ and π ionization energies near 11.2 eV.

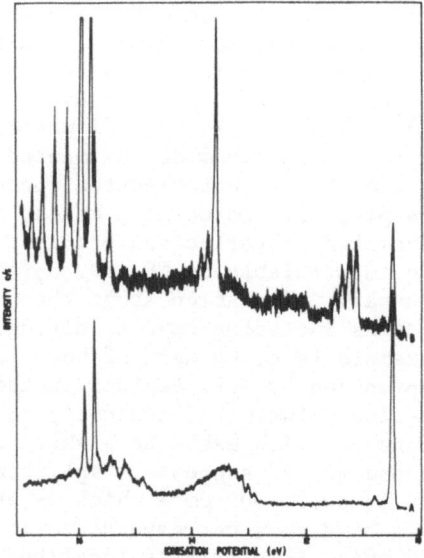

Photoelectron spectra of (above) the products of reaction between
H_2S and hot boron and below H_2S over cold boron. Note: both include the admixture of argon as calibrant.

PHOTOELECTRON SPECTRA FROM SOLIDS

In contrast to the wealth of fine detail seen in the photoelectron spectrum from vapours, solids might seem to provide
somewhat unpromising materials for the high resolution spectro-
scopist. X-ray photoelectron spectra of solids have indicated
band structures whose widths are of the order of 1 eV or greater
in the valence shell as well as in the inner shell, which of
course only the X-ray photons can reach. In addition insulating
materials bring problems of their own, since failure to disperse
uniformally the generated positive charge can give rise to an
additional line broadening. It is furthermore often said that
it requires the high energy electron characteristic of X-ray
photoelectron spectra to avoid overwhelming difficulties with the
escape of very slow electrons through surface contaminant layers
and that ultra clean vacuum conditions are crucial to the employment of lower photon energies. Such low energy studies, however,
are of considerable importance since they give direct information
about the valence band structure and hence about the density of
states. The photoelectron spectra which are obtained when very
low energy photons are used, that is energies close to the

threshold for photoelectron ejection, shows quite large changes
in photoelectron band structure with photon energy. This is
strongly reminiscent of what is also observed in the photoelectron spectra of gases, where autoionization processes are usually
held to be responsible. As in gases where the use of higher photon energies eventually renders such processes insignificant, so
in solids the use of the helium resonance radiation seems to approach a condition where the photoelectron spectrum quite closely
reflects the density of states. In some cases it is possible to
give an atomic designation to particular bands indicating that
the electrons are associated with rather well localised orbitals,
the lattice broadening then only being a rather secondary effect.
Such seems to be the case in the alkali halides, notably the
iodides. The highest valence level is rather well resolved by
the HeI photoelectron spectrometer into a near 2:1 intensity
ratio doublet confirming that electrons are coming from the 5p
shell of the iodide ion. The spin doublet splitting in the
neutral iodine atom left by this ionization is   0.94 eV (spectroscopic). It is interesting to compare this spectrum with that
of gaseous xenon which is isoelectronic.

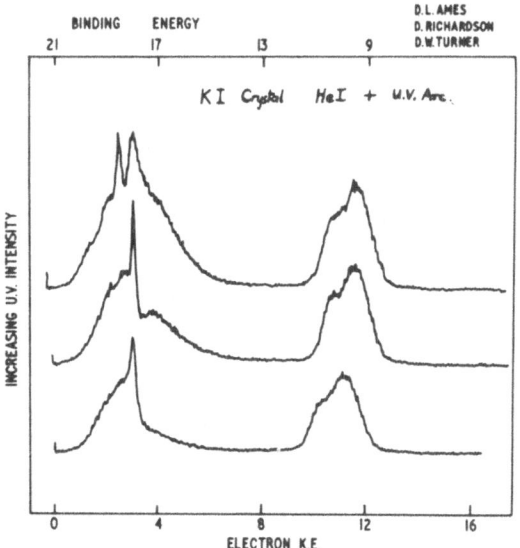

The HeI photoelectron spectrum from potassium iodide crystals
showing the effect of simultaneous irradiation with a high pressure mercury-Xenon arc.

We have found that it is quite possible to obtain satisfactory photoelectron spectra from solids provided sufficiently high sample temperatures can be attained even in comparatively inferior vacuum systems. This may be due to a combination of reduced adsorption of contaminants, vapours and steady evaporation of the surface. We have also noted that the effect of u.v. irradiation is quite marked in assisting the clean-up process on a freshly inserted sample and in improving resolution on insulating specimens presumably by promoting charge redistribution within the crystal. A most interesting secondary effect has been noticed also in sodium, potassium and cesium iodide, the clearest example being provided by potassium iodide, where sharp spikes have been found whose intensity reflects that of the near u.v. radiation. This is shown in the figure above where the three spectra correspond to increasing levels of u.v. light intensity. The effect is reversible and from the sharpness of the peaks may relate to rather loosely bound surface species. The possibility that excitons and the near ultraviolet emission lines of the helium lamp are involved cannot be excluded.

REFERENCES

1. D.W. Turner, A.D. Baker, C. Baker and C.R. Brundle, "Molecular Photoelectron Spectroscopy", Wiley 1970.
2. D.W. Turner, Nature, 213, 795 (1967).
3. D.W. Turner, Phil. Trans. Roy. Soc. Lond. A268, 7 (1970).
4. G. Herzberg, Chem. Soc. Centenary Lecture, 1971.
5. E. Heilbronner and K.A. Muszkat, J. Amer. Chem. Soc. 92, 3818 (1970).
6. A.D. Baker, D.P. May and D.W. Turner, J. Chem. Soc. B22, (1968).
7. J.P. Maier and D.W. Turner, Disc. Faraday Soc. 54, 149 (1972).
8. J.P. Maier and D.W. Turner, J. Chem. Soc., Faraday Trans. II, 69, 196 (1973).
9. J.P. Maier and D.W. Turner, J. Chem. Soc., Faraday Trans. II.
10. A.D. Baker, D. Betteridge, N.R. Kemp and R.E. Kirby, Int. J. Mass. Spectrom. Ion Phys. 4, 90 (1970).
11. A.D. Baker, D. Betteridge, N.R. Kemp and R.E. Kirby, Analyt. Chem. 42, 1064 (1970).
12. S. Evans and A.F. Orchard, Inorg. Chim. Acta 5, 81 (1971).
13. P. Baybutt, M.F. Guest and I.M. Hillier, Mol. Phys. 25, 1025 (1973).
14. N. Jonathan, A. Morris, M. Okucha, K.J. Ross and D.J. Smith, Disc. Faraday Soc. 54, 48 (1972).
15. A.B. Cornford, D.C. Frost, F.G. Herring and C.A. McDowell, Disc. Faraday Soc. 54, 56 (1972).

ABSOLUTE INTENSITIES AND CLASSIFICATION OF TRANSITIONS IN ELECTRON IMPACT SPECTROSCOPY[*]

Edwin N. Lassettre

Center for Special Studies and Department of Chemistry
Carnegie-Mellon University, 4400 Fifth Avenue,
Pittsburgh, Pennsylvania 15213

INTRODUCTION

If we define Electron Impact Spectroscopy as that phase of inelastic electron scattering which deals with resolved (electron) spectra, then the subject is still far too large to be surveyed in a one-hour speech. If, however, attention is confined to the study of absolute (in contrast to relative) intensities, the scope is reduced to the point where a survey of recent research is possible. A brief discussion of selection rules, which identify transitions of low intensity, can also be included. The similarity between electron impact and ultraviolet spectroscopy is most pronounced for electrons of high initial kinetic energy and this survey will be primarily, but not exclusively, devoted to investigations of this type. It will not be possible to survey the subject of resonances, first discovered by G. J. Schulz.[1] Resonances have, however, been the subject of many excellent reviews the latest of which are due to Schulz.[2,3]

GENERALIZED OSCILLATOR STRENGTHS

The inelastic scattering of electrons can be measured by, among other quantities, the collision amplitude F_n. First order perturbation theory (the Born approximation) gives for this quantity, in atomic units

$$F_n = -\frac{1}{2\pi} \int e^{i\vec{K}\cdot\vec{r}} V_{on}(d\vec{r}) \tag{1}$$

where $\vec{K}$ is the change in wave vector $\vec{k}_o - \vec{k}_n$, V_{no} is the matrix element of the interaction potential, V, of the incoming electron with the scatterer

$$V_{no} = \langle \psi_n | V | \psi_o \rangle \tag{2}$$

and ψ_o, ψ_n are the eigenfunctions of the scatterer in the initial and final states. We note in passing that when the incident electron is far from the scatterer, then

$$V_{no} \cong \frac{\mu_{no} \cdot \vec{r}}{r^3} + \ldots \tag{3}$$

where $\vec{\mu}_{no}$ is the matrix element of electric dipole moment.

Bethe[4] showed, by integrating over the coordinates of the incident electron, that (see also Ref. 108, Appendix B)

$$F_n = \frac{2\varepsilon}{K^2} \tag{4}$$

where for a molecule,

$$\varepsilon = \langle \psi_n | \sum_j N_j e^{i\vec{K} \cdot \vec{\rho}_j} - \sum_i e^{i\vec{K} \cdot \vec{r}_i} | \psi_o \rangle . \tag{5}$$

Here $\vec{\rho}_j$ is a position vector to the jth nucleus of atomic number N_j and $\vec{r}_i$ is a position vector to the ith electron. If eq. (5) is expanded in a power series in K, the result is

$$\varepsilon = \sum_{\ell=o}^{\infty} i^\ell K^\ell \varepsilon_\ell \tag{6}$$

where

$$\varepsilon_\ell = \frac{1}{\ell!} \langle \psi_n | \sum_j N_j z_j^\ell - \sum_i z_i^\ell | \psi_o \rangle \tag{7}$$

and Z_j, Z_i are the Z coordinates of the nuclei and electrons, respectively, in a coordinate system with the Z axis parallel to $\vec{K}$. Using eq. (6), we find, since $\varepsilon_o = 0$ due to orthogonality,

$$F_n = \frac{2i\varepsilon_1}{K} - 2\varepsilon_2 - 2i\varepsilon_3 K + \ldots \tag{8}$$

Obviously, ε_1 is the component of the matrix element of electric dipole moment, $\vec{\mu}_{no}$ of eq. (3), parallel to $\vec{K}$. When K is small, the first term on the right of (8) is dominant and the collision amplitude is proportional to the matrix element of electric dipole moment. This shows that the Fourier integral on the right of (1) is determined, when K is small, primarily by the asymptotic expression for the integrand at large values of r. We can see from the following argument that this result is physically sensible. Viewed classically, a collision with small K means one in which the momentum changes but little and hence the passing electron is subject to only a small force. This is the case for a distant collision. It is physically obvious for such a collision that the force acting on the distant electron is that for an electric dipole. When the matrix element of electric dipole moment vanishes, then the amplitude, at small K, is expected to be greatly reduced. This conclusion, from a purely classical argument, is in complete accord with eq. (8). The fact that an amplitude given by a Fourier transform, like eq. (1), has a magnitude which is largely determined (at small K) by the asymptotic behavior of the integrand is of considerable interest and will be used later in other contexts.

In order to exhibit the analogy between electron impact and optical spectra, it is useful to deal with a quantity, f, which we refer to as the generalized oscillator strength. This quantity is defined by the following relation (in atomic units)

$$f = \frac{W}{2} \frac{k_o}{k_n} K^2 \sigma \tag{9}$$

where k_o, k_n are the wave vectors for incident and scattered electrons, W is the excitation energy and σ is the differential collision cross section. In terms of the collision amplitude, F_n, the cross section is

$$\sigma = (k_n/k_o)|F_n|^2 \quad . \tag{10}$$

When the Born approximation holds, then eq. (4) is applicable and this gives another formula for f

$$f = 2W\epsilon\epsilon^*/K^2 \quad . \tag{11}$$

Bethe[4] was the first to introduce the concept of generalized oscillator strength. He defined the quantity by eq. (11). For purposes of experimental determination, however, eq. (9) is more convenient since all quantities on the right of (9) can be determined by experiment. We have adopted this definition.[5,6] If the Born approximation (Eq. 4) does not hold, then (11) and (9) are not equivalent.

When the Born approximation holds, it follows from (11) that as $K \to 0$

$$\lim. \; f \;\; = \;\; 2W\epsilon_1\epsilon_1^* \;\; . \tag{12}$$

Since ϵ_1 is the matrix element of electric dipole moment, the right side of eq. (12) is the optical oscillator strength, designated hereafter by f_0. This property of f was first discovered by Bethe.[4] When the Born approximation does *not* hold, it was shown by Lassettre Skerbele and Dillon[7] that the limit (12) is still obtained as $K \rightarrow 0$. In a later research, Bonham[8] obtained the same result using a different analysis. In a later publication, Bonham[9] concluded that his proof applied only to bound states. In a recent review, Inokuti[10] tested the limit theorem on recent collision cross section calculations. He concluded that it was compatible with all of the theoretical results.

DETERMINATION OF OSCILLATOR STRENGTHS

A method has been developed for the determination of absolute collision cross sections which depends on the construction of one apparatus for the measurement of absolute elastic cross sections[11] and a second apparatus[12] for the determination of relative intensities at high resolution. The latter instrument has been modified to permit accurate measurement of the beam current at each scattering angle. The modified version has been described in detail in a recent review.[13] An entirely different method, which depends on the use of a single instrument, has been developed by Simpson and Kuyatt at the National Bureau of Standards. The first absolute measurements on helium at $\theta = 5°$ were reported by Chamberlain, Kuyatt and Mielczarek.[14] An entirely different instrument has been developed by Boersch, Geiger and their collaborators.[15-17] A method for the determination of f_0, at very high kinetic energies, has been developed by Geiger.[18-22]

Optical oscillator strengths have been determined by measuring generalized oscillator strengths as a function of momentum change and extrapolating to K = 0. Methods have also been developed by which oscillator strengths can be calculated from electron impact spectra obtained at zero scattering angle. Results from both methods are described below.

Oscillator Strengths Obtained by Extrapolation

Table I contains limiting oscillator strengths obtained by extrapolating, to K = 0, data in which f is determined as a function of K^2. The table contains oscillator strengths for helium[23,24] mercury,[25-27] carbon monoxide,[28,29] nitrogen,[30] water,[31] and ammonia.[32] The oscillator strength for the fourth positive bands of carbon monoxide is of particular interest. A large discrepancy was found between the extrapolated oscillator strengths obtained

TABLE I: Limiting Oscillator Strengths

Substance	Transition	f_o(electron impact)	Reference
He	$1^1S \rightarrow 2^1P$	0.269 ± 0.010	23
	$1^1S \rightarrow 2^1S$	0	24
Hg	$6^1S_o \rightarrow 6^1P_1$	1.11 ± 0.10	25,27
	$6^1S_o \rightarrow 6^3P_1$	0.0285 ± 0.0040	26
	$6^1S_o \rightarrow 6p'^3P_1$	0.704 ± 0.070	26
	$6^1S_o \rightarrow 7p'^1P_1$	0.067 ± 0.009	26
CO	$A^1\Pi \leftarrow X^1\Sigma^+$; $v' = 2$	0.0429 ± 0.0010	28,29
	$A^1\Pi \leftarrow X^1\Sigma^+$; $\Sigma_{v'} f_{v'}$	0.195	28,29
	$B^1\Sigma^+ \leftarrow X^1\Sigma^+$; $v' = 0$	0.015 ± 0.004	
	$C^1\Sigma^+ \leftarrow X^1\Sigma^+$; $v' = 0$	0.163 ± 0.015	
N_2	$a'^1\Pi_g \leftarrow X^1\Sigma_g^+$	0	30
	$a''^1\Sigma_g^+ \leftarrow X^1\Sigma_g^+$	0	30
H_2O	$\tilde{A}^1B_1 \leftarrow \tilde{X}^1A_1$	0.060 ± 0.006	31
NH_3	$\tilde{A}^1A_2'' \leftarrow \tilde{X}^1A_2''$	0.070 ± 0.007	32

from electron impact measurements[33] and those calculated from life-time measurements.[34] A redetermination of the electron impact value using a new experimental technique[28] did not resolve the disagreement. The value of f_0 obtained in the new determinations, 0.195, is that shown in Table I while the value obtained by Hesser from lifetime measurements[34] was 0.094, a discrepancy of a factor of two. No satisfactory explanation was found for this discrepancy until the experiments of Mumma, Stone and Zipf[35] on emission intensities demonstrated that the electronic transition moment changes linearly with internuclear distance. They re-analyzed the data of Hesser and obtained an f value of 0.15 which agrees much better with the electron impact value. G. M. Lawrence[36] also re-analyzed Hesser's data and found 0.17 for the oscillator strength, in even better agreement with the electron impact value. The remaining discrepancy is no doubt accounted for by perturbations first suggested by A. E. Douglas[37] and elaborated by Rhodes[38] and others. For molecules, the relation between os-cillator strength and lifetime is not a simple and straightforward one if partially resolved spectra are employed. This was emphasized

by Douglas[37] who also described the main mechanisms by which different relationships can be understood. New determinations by
Chervenak and Anderson[39] and Imhof and Read[40] introduce new dis-
crepancies, the former because their lifetime measurements disagree with other investigations and the latter because a quadratic
(rather than linear) dependence of transition moment on internuclear
distance is indicated. The optical determinations of Mumma, Stone
and Zipf[35] and the electron impact determinations of Lassettre and
Skerbele[28] agree quantitatively in providing a linear dependence
of transition moment on internuclear distance. This supports the
interpretation by Mumma, et. al.[35]

Early oscillator strength determinations, at low resolution, on
nitrogen[41] and oxygen[42] have been shown to be in error, probably
due to streaming errors in the McLeod gauge used in the determinations. The error can be corrected (see Ref. 30) by multiplying
each oscillator strength by 0.754. With this renormalization,
the values of n - 1 (n the refractive index) can be calculated
and compared with experiment. The largest discrepancy for N_2, O_2
and H_2O is 3%. These data are reviewed in Reference 13 which
contains corrected values of the oscillator strengths.

Oscillator strength determinations at very high kinetic energies have been carried out by Geiger and co-workers.[43,44] Oscillator strengths for the rare gases are shown in Table II. Geiger
and Schmoranzer[45] have published a high resolution study of H_2,
HD and D_2 at 34 keV. The transition probabilities are expressed
as dipole strengths. Very high resolution spectra of N_2 and O_2
have also been published.[46,47]

Optical Oscillator Strengths from Forward Scattering

When electrons are scattered inelastically without change in
direction, the momentum change is a minimum. When the excitation
energy W is less than 0.2 times the kinetic energy, E, of the
incident electron the following approximate equations hold[48]

$$K^2 = 8\bar{E}[\sin^2 \frac{1}{2} \theta + (\frac{W}{4E})^2] \quad . \qquad (13)$$

Here $\bar{E}$ is the average of the electron energy before and after
collision, θ is the scattering angle. All quantities are in
atomic units. When $\theta = 0$

$$K^2 = \frac{W^2}{2\bar{E}} \qquad (14)$$

when W = 10 eV and E = 500 eV, eq. (14) gives $K^2 = 0.0037$. This is
so small that f differs negligibly from f_o for typical transitions.

TABLE II: Oscillator Strengths at High Energies

Substance	Transition	f_{exp}	E
He[a)	$1^1S \rightarrow 2^1P$	0.312 ± 0.04	25 KeV
	$1^1S \rightarrow 3^1P$	0.0898 ± 0.006	--
Ne[b)	$2p^6\,{}^1S_o \rightarrow 2p^5 3s\,{}^3P_1$ $\rightarrow 2p^5 3s\,{}^1P_1$	0.140 ± 0.01	25 KeV
Ar[b)	$3p^6\,{}^1S_o \rightarrow 3p^5 4s\,{}^3P_1$ $\rightarrow 3p^5 4s\,{}^1P_1$	0.233 ± 0.02	25 KeV
Kr[b)	$4p^6\,{}^1S_o \rightarrow 4p^5 5s\,{}^1P_1$ $\rightarrow 4p^5 5s\,{}^1P_1$	0.346 ± 0.06	25 KeV

a)   J. Geiger, Zeit. Physik 175, 530 (1963).
b)   J. Geiger, Zeit. Physik 177, 138 (1963).

In the case of the $1^1S \rightarrow 2^1P$ transition in helium, for example, f differs from f_o by less than 0.5%. Hence, using eq. (9), we find, neglecting small quantities,

$$\sigma = \frac{4\overline{E}f_o}{W^3} \left[\frac{1}{1 + (\theta/\alpha)^2} \right] \tag{14}$$

$$\alpha = W/2\overline{E} \quad . \tag{15}$$

In the determination of zero angle spectra, the scattering angle is obviously not exactly zero but is contained in a small range around $\theta = 0$. The magnitude of this range (the angular resolution) depends on the sizes and placement of slits or pinholes. The magnitude of this range is very important because α of eq. (15) is small. For example, when W = 10 eV and E = 500 eV, $\alpha = 0.0101$ radians or 0.58°. Let σ_o be the value of σ at $\theta = 0$ and σ_α be the value at $\theta = \alpha$. Then $\sigma_o/\sigma_\alpha = 2$. Obviously, account must be taken of the angular spread around $\theta = 0$.

The scattered current at $\theta = 0$ is proportional to the average, $\overline{\sigma}$, of the collision cross section, the average being taken over the angular spread around $\theta = 0$.

$$\overline{\sigma} = 4\overline{E}f_o \beta/W^3 \tag{16}$$

where

$$\beta = \langle \frac{1}{1 + (\theta/\alpha)^2} \rangle_{av.} \quad . \tag{17}$$

For a system in which scattering angle is limited by pinholes, let θ_0 be the maximum angle through which scattering can occur and still reach the pinhole at the entrance to the analyzer. Then

$$\beta = (\alpha/\theta_0)^2 \ln[1 + (\theta_0/\alpha)^2] \quad . \tag{18}$$

If the scattering angle is limited by slits whose lengths are much greater than their widths, like the spectrometer of Ref. 12, then let η be the maximum angle subtended at the center of the collision chamber by the slit length. (Account must be taken of the electrostatic lens of Ref. 12 between collision chamber and slit.) Then

$$\beta = (\alpha/\eta) \mathrm{Tan}^{-1}(\eta/\alpha) \quad . \tag{19}$$

From eq. (16), we find

$$f_o = w^3 \bar{\sigma}/4\beta\bar{E} \quad . \tag{20}$$

Harshbarger and Lassettre[49] determined η of eq. (19) from studies on mixtures of helium and carbon monoxide at $\theta = 0°$. Since oscillator strengths for the $1^1S \rightarrow 2^1P$ transition in helium and that for the fourth positive bands of carbon monoxide are known, the value of η was adjusted to make the ratio of apparent oscillator strengths equal to the experimentally observed ratio. Similar experiments were done using mixtures of mercury and carbon monoxide. In this way, $\eta = 0.020 \pm 0.002$ was obtained. Drastic changes in lens voltages would change this value but since essentially the same focussing conditions are used in determining all spectra at $\theta = 0°$, the above value can be applied with minimum error to any of our spectra. In a recent study of acetone, Huebner, Celotta, Mielczarek and Kuyatt[50] gave $\theta_0 = 0.02$ radians for their apparatus.

Lassettre and collaborators have determined oscillator strengths by mixing two gases, one a substance of known oscillator strength (such as helium) and the other a gas whose oscillator strengths are to be determined. From the mixture spectrum, the oscillator strengths can be calculated using eq. (20) with β given by (19). In order to avoid changes in composition each gas must be expanded to low pressures through a separate leak. The mixing occurs at low pressures. In our apparatus, the pressure changes in the collision chamber are measured with a capacitance manometer. A recent example is the study of methane and carbon tetrafluoride.[49] Another example, which requires a function different from either (18) or (19) for β, is given in Ref. 29.

Kuyatt and collaborators have developed the technique of oscillator strength determination at $\theta = 0°$ using an apparatus similar to that of Ref. 14. For their apparatus β is given by eq. (18) with $\theta_0 = 0.02$. Instead of using a test gas (like helium) they normalize at one point in the continuum to a known oscillator strength determined by optical methods. Acetone provides a recent example.[50] Formaldehyde[51] and nitrous oxide[52] have also been studied.

CLASSIFICATION OF TRANSITIONS

Dipole-Quadrupole Behavior

Using eqs. (6), (7) and (11), we obtain for f the power series

$$f = f_o + f_2 K^2 + f_4 K^4 + \ldots \tag{21}$$

where

$$f_{2n} = 2W \sum_{\ell=1}^{2n+1} (-1)^{\ell+n+1} \varepsilon_\ell \varepsilon_{2n+2-\ell} \tag{22}$$

It follows from (9) that

$$\sigma = \frac{4k_n}{k_o} \left[\frac{\varepsilon_1^2}{K^2} + (\varepsilon_2^2 - 2\varepsilon_1\varepsilon_3) + (\varepsilon_3^2 - 2\varepsilon_2\varepsilon_4 + \varepsilon_1\varepsilon_5)K^2 \right.$$

$$\left. + \ldots \right] \tag{23}$$

Selection rules apply to the quantities ε_ℓ of eq. (22). These selection rules manifest themselves in the behavior of the generalized oscillator strength. If, for example, $\varepsilon_1 = 0$ but $\varepsilon_2 \neq 0$ then the first nonvanishing term in eq. (21) is proportional to K^2. If f be extrapolated to $K = 0$, the result should be $f_o = 0$. This point has been tested, and confirmed, in a study of the $1^1S \rightarrow 2^1S$ transition in helium.[24] It has been applied to the 12.26 eV transition in nitrogen by Lassettre, Skerbele and Meyer.[53] It was shown that the transition was forbidden and $^1\Sigma_g^+$ was selected as the probable term symbol for the upper state. Lutz and Dressler[54] observed the 12.26 eV excitation in the ultraviolet absorption spectrum by working at high enough pressures to obtain a collision induced dipole moment. From the observed rotation spectrum, they proved that the term symbol for the excited state is $^1\Sigma_g^+$. This is the first instance in which electron impact spectroscopy was responsible for both the discovery of a new state (made independently by Heideman, Kuyatt and Chamberlain[55] and by Meyer and Lassettre[56]) and its tentative (and subsequently confirmed) identification.[53]

Kevin J. Ross and collaborators are conducting a systematic study of electron impact spectra of metal vapors.[57-59] They have studied a transition in cesium[60] for which $\varepsilon_1 = 0$, $\varepsilon_2 = 0$ but $\varepsilon_3 \neq 0$. This is the only such transition which has been observed.

Read and Whiterod[61] have given a systematic classification, based on symmetry groups, of the dependence on K of the leading term in the expansion (23). Classification of vibronic transitions has also been considered.[62]

Application of methods for the identification of transitions based on the behavior of f at small values of K has not been highly successful for polyatomic molecules because the vibration spectrum must be resolved to be sure of finding forbidden transitions. This is frequently difficult because the levels are closely spaced and have substantial thermal widths. For example, kT is about 0.028 eV at laboratory temperatures and breadths of this magnitude are generally expected due to unresolved rotational structure. Much more serious is broadening due to predissociation, a pervasive effect in the ultraviolet spectra of polyatomic molecules. Under these conditions, eqs. (21) and (23) have only limited utility. In certain instances, however, these relations are a useful qualitative guide. We consider in the following paragraph an example in the spectrum of CO_2.

Electron impact spectra of CO_2 (see Ref. 12) are shown in Figure 1. The region from 7 eV to 10 eV is of special interest. The spectrum labeled UV is an ultraviolet spectrum. It is apparent from Figure 1 that the electron impact spectrum at $\theta = 0°$ closely resembles the ultraviolet spectrum in the range 7–10 eV even to the vibrational levels where these are resolvable in electron impact. In both electron impact and the ultraviolet there are two general maxima underlying the vibrational fine structure. The maxima occur at 8.52 and 9.30 eV, respectively.[63] Winter, Bender and Goddard[64] have recently reported the results of an elaborate quantum mechanical calculation of energy states of CO_2. Two states, $^1\Delta_u$ at 8.38 eV and $^1\Sigma_g$ at 9.23 eV, had excitation energies nearly matching the two observed peaks at 8.52 eV and 9.30 eV. Hence, the term symbols $^1\Delta_u$ and $^1\Sigma_g$, respectively, were assigned to the observed states. This is contrary to the assignment made by Lassettre and Shiloff.[65] Despite the fact that the calculated and observed excitation energies are in good agreement there are strong arguments against the above assignments. These are outlined below.

In attaching the term symbols $^1\Delta_u$ and $^1\Pi_g$ to the excitations at 8.52 eV and 9.30 eV an appeal is made (implicitly at least) to the Franck-Condon principle which states that *a vertical excitation from the ground state equilibrium configuration gives rise to the transition of maximum intensity.* This principle works very well

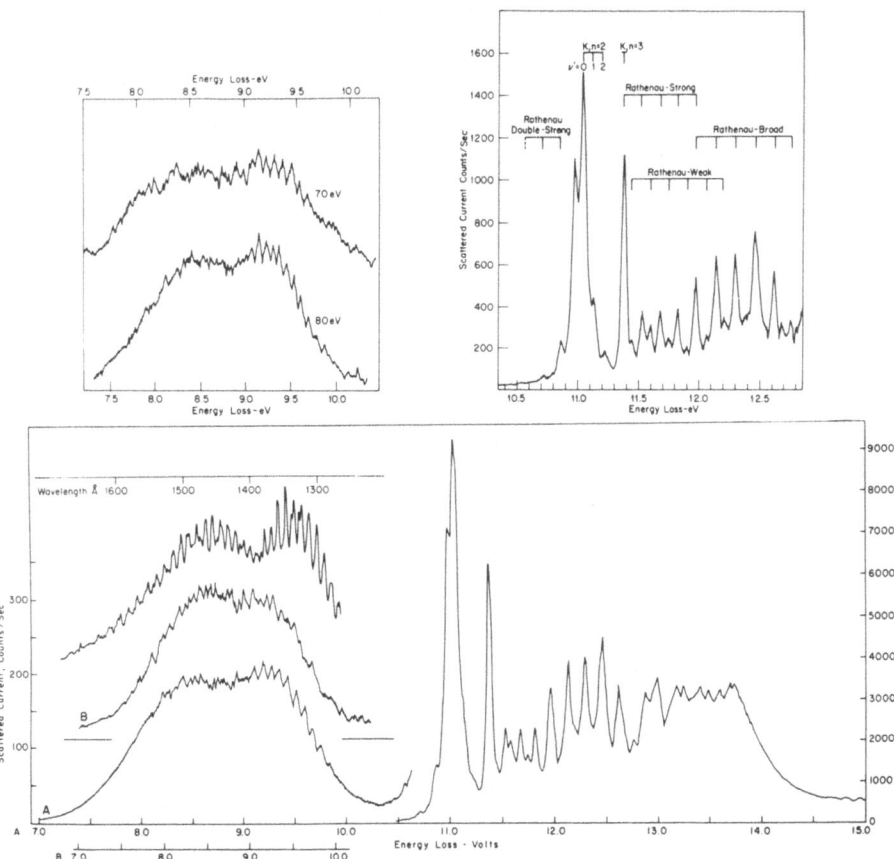

Figure 1.   Electron impact spectra of CO_2 at $\theta = 0°$ and kinetic
            energies 48 (lower) and 58 (upper right) by Lassettre
            et al. (12).   Curve A is continuation of 48 eV spec-
            trum while B is a new spectrum at 28 eV kinetic energy.
            Spectrum labeled UV is ultraviolet spectrum.   Note
            spectra at 70 and 80 eV at upper left.

for electric dipole allowed vertical excitations but both of the transitions $^1\Delta_u \leftarrow X^1\Sigma_g^+$ and $^1\Pi_g \leftarrow X^1\Sigma_g^+$ are forbidden. Hence, we need to consider how the Franck-Condon principle needs to be modified for forbidden transitions. For this purpose, we consider diatomic and polyatomic molecules separately.

Since vibration cannot change the symmetry of a diatomic molecule, a transition which is forbidden in vertical excitation at one internuclear distance is forbidden at all other internuclear distances. It follows that the Franck-Condon principle needs no modification for forbidden transitions in diatomic molecules. The Franck-Condon factor has the same functional form regardless of whether the moment involved in radiation absorption or emission is electric dipole, magnetic dipole, or electric quadrupole. In the case of electron impact excitation, the same holds regardless of the order, ℓ, of the first nonvanishing moment ε_ℓ. The dependence of the appropriate moment on internuclear distance will be significant, of course, but if the dependence is linear then the Franck-Condon factor can be supplemented by the r-centroid along lines developed in detail by R. W. Nicholls and his collaborators.[66]

We can think of eq. (23) in two different contexts. First, we can regard σ as giving the collision cross section for a hypothetical excitation of a molecule with nuclei fixed in a given configuration. Or second, we can regard σ as the cross section for an actual excitation from one quantum state to another taking account of nuclear motion (e.g., vibration) as well as electronic motion. The moments ε_ℓ have different significances in the two cases. We can distinguish the first case by showing explicitly the dependence on the nuclear coordinates q as e.g., $\varepsilon_\ell(q) = \varepsilon_\ell'(q_o) + \varepsilon_\ell''(\Delta q)$ where $\Delta q = q - q_o$ and q_o designates the ground state equilibrium configuration. For an actual excitation, we can use the symbol ε_ℓ without a nuclear variable. Obviously, $\varepsilon'(q_o)$ and $\varepsilon''(\Delta q)$ refer respectively to the value of ε_ℓ for a vertical excitation from the ground state equilibrium configuration and the change, $\varepsilon_\ell''(\Delta q)$, in $\varepsilon_\ell(q)$ which accompanies a change Δq in configuration. The two quantities ε_ℓ and $\varepsilon_\ell(q)$ are related by the equation

$$\varepsilon_\ell = \langle \phi_n(q) | \varepsilon_\ell(q) | \phi_o(q) \rangle \tag{24}$$

where the matrix element integral extends over nuclear coordinates q and ϕ_n, ϕ_o are the nuclear (vibration) eigenfunctions in excited and ground states, respectively. This assumes separability of electronic and nuclear motion but this is inevitable in discussing the Franck-Condon principle. Rotation, which is unresolved, is ignored.

Note that substituting $\varepsilon_\ell(q) = \varepsilon_\ell'(q_o) + \varepsilon_\ell''(\Delta q)$ and assuming $\varepsilon_\ell''(\Delta q) = \sum_m \eta_m \Delta q_m$ (Δq_m is the mth normal coordinate) gives

$$\varepsilon_\ell = \langle \phi_n | \phi_o \rangle \varepsilon_\ell'(q_o) + \sum_m \langle \phi_n | \Delta q_m | \phi_o \rangle \eta_m . \qquad (25)$$

Obviously, the first term on the right of (25) is ε_ℓ' and the second is ε_ℓ''. The quantity $|\langle \phi_n | \phi_o \rangle|^2$ is the usual Franck-Condon factor. We note in passing that the function, ϕ_o, on the right of a bracket enclosure is a vibrational function for the ground electronic state and that on the left, ϕ_n, is a vibrational function for the excited electronic state. Since the vibrational potential functions are different in ground and excited states ϕ_o and ϕ_n are not orthogonal. By a well known argument, the maximum value of $|\langle \phi_n | \phi_o \rangle|$ occurs at that state n for which the highest maxima of ϕ_o and ϕ_n coincide. At laboratory temperatures, nearly all molecules of CO_2 will be in the ground vibrational state and hence ϕ_o has its highest (and only) maximum at the ground state equilibrium configuration $\Delta q_m = 0$. The desired state n (corresponding to the transition of maximum intensity) is that closest to the intersection of a vertical line with the upper state potential surface. When $\varepsilon_\ell'(q_o)$ vanishes, it follows from (25) that intensities are determined by terms like $\langle \phi_n | \Delta q_m | \phi_o \rangle$. The maximum value of this matrix element can be located (approximately) by extending a line vertically upward from the configuration at which $\Delta q_m \phi_o$ is a maximum. This will occur at a value of Δq_m different from zero, i.e., from a strained ground state configuration. Having explained these conventions, we return to the main argument.

Even when ε_1' vanishes, the leading term on the right of (23) will still be dominant at small enough K, if ε_1'' does not vanish, because of the factor $1/K^2$. Hence, at $\theta = 0°$ and high kinetic energy, the electron impact spectrum will be dominated by allowed transitions even though these become allowed only because of vibrational distortion from the ground state equilibrium configuration. This differs from the diatomic case where ε_1' and ε_1'' must both vanish if one vanishes. Lassettre and Shiloff[65] showed, by an application of the methods of Herzberg and Teller (extended to electron impact excitation by Read[61]), that an expansion of the electronic eigenfunction of the $^1\Delta_u$ state as a power series in the nuclear displacement coordinates, around the equilibrium configuration q_o, contained no linear terms which are dipole connected to the ground state eigenfunction but the expansion of the $^1\Pi_g$ eigenfunction contained two linear terms which are dipole connected to the ground state eigenfunction. Hence, for the transition $^1\Delta_u \leftarrow X^1\Sigma_g^+$ both ε_1' and ε_1'' vanish but for the transition $^1\Pi_g \leftarrow X^1\Sigma_g^+$ ε_1' vanishes but ε_1'' does not. Hence, referring back to (23), it is obvious that for the transition

$$^1\Delta_u \leftarrow X^1\Sigma_g^+$$

$$\sigma = \frac{4k_n}{k_o} [\epsilon_2^2 - 2\epsilon_1\epsilon_3 + \dots] \tag{26}$$

but for the transition

$$^1\Pi_g \leftarrow X^1\Sigma_g^+$$

$$\sigma = \frac{4k_n}{k_o} [\frac{(\epsilon_1'')^2}{K^2} + (\epsilon_2^2 - 2\epsilon_1\epsilon_3) + \dots] \quad . \tag{27}$$

As the scattering angle increases, however, K^2 also increases rapidly (like $\sin^2 \theta/2$, in fact) and the term $(\epsilon_1'')^2/K^2$, which is small to begin with because ϵ_1'' is small, becomes unimportant in comparison to $\epsilon_2^2 - 2\epsilon_1\epsilon_3$ and eq. (26) and (27) both begin with this same term. For small Δq it is obvious that $\epsilon_\ell''(\Delta q)$ is small compared to $\epsilon_\ell'(q_o)$ if the latter does not vanish. As a first approximation, therefore, we may consider the quantity $\epsilon_2^2 - 2\epsilon_1\epsilon_3$ for a vertical excitation. Since $\epsilon_1'(q_o)$ vanishes, we need only consider $\epsilon_2'(q_o)$. We find that this quantity vanishes for the transition $^1\Delta_u \leftarrow X^1\Sigma_g^+$ but does not vanish for $^1\Pi_g \leftarrow X^1\Sigma_g^+$. Thus, we expect that at some angle large enough to eliminate the term ϵ_1^2/K^2 the spectrum should contain only one peak. In Figure 2 are

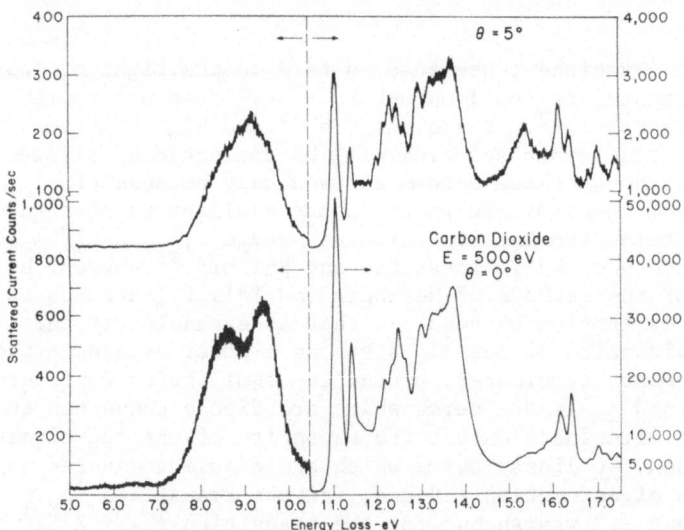

Figure 2.   Electron impact spectra of CO_2 at two scattering angles and 500 eV kinetic energy.

shown two, previously unpublished, electron spectra of CO_2 taken
with the instrument of Ref. 12. Resolution has been deliberately
reduced to gain intensity. The top spectrum at $\theta = 5°$ shows only
one peak at 9.19 eV in contrast to the bottom spectrum at $\theta = 0°$
which contains two peaks at 8.52 eV and 9.30 eV, respectively.
Due to the breadths of the peaks, these values are accurate to
only about 0.05 eV. According to the above argument, this peak
should arise by excitation to $^1\Pi_g$. The excitation energy, 9.23 eV
calculated by Winter, et. al., is in excellent agreement with the
observed value 9.19 ± 0.05 eV. Note that as θ goes from $0°$ to
$5°$, at an incident energy of 500 eV, K^2 increases by a factor of
94. Hence, since ε_1'' is small to begin with, it is plausible
that $(\varepsilon_1'')^2/K^2$ becomes negligible in (27).

We must still account for two peaks in the spectrum at $\theta = 0°$
(see Fig. 2). Because both ε_1' and ε_1'' vanish (to the first order
in Δq) for excitation to $^1\Delta_u$, it is not satisfactory to assign one
peak to this excited state. For excitation to $^1\Pi_g$, on the other
hand, ε_1' vanishes but ε_1'' does not (to terms linear in Δq). For
such a transition, it has already been pointed out that intensities
are determined by terms like $\langle \phi_n | \Delta q_m | \phi_o \rangle$ (see eq. 25) and the
transition of maximum intensity is obtained by vertical excitation
from a *strained* ground state configuration. To be specific, we
consider the bending vibration of frequency ν_2. Designate the
displacement coordinate by X instead of Δq_m. Since the vibration
is doubly degenerate[67] we get one factor X from the matrix element
$\langle \phi_n | X | \phi_o \rangle$ and another from the volume element. Hence, the following factor appears in the matrix element

$$X^2 e^{-\frac{\beta}{2} X^2}$$

where

$$\beta = 2\pi\nu_2 \mu/h$$

and μ is the reduced mass. The function (28) has a maximum at a
value X_m given by

$$X_m = \sqrt{\frac{h}{\pi\mu\nu_2}}$$

The frequency in wavenumbers is 672.2 cm^{-1} and $\mu = 8.727$ mass units.
Hence,

$$X_m = 0.269 \text{ Å}$$

Using the C–O distance, 1.159 Å, we find that the apex angle of
distorted CO_2 is 153.9°, a displacement of 26.1° from the linear
configuration. In such a distorted configuration, a $^1\Pi_g$ excited
state will be split into two components and so will $^1\Delta_u$.
Lassettre and Shiloff showed (using the Herzberg–Teller theory of
selection rules) that (to terms linear in the Δq's) vertical
excitation to both components of the split $^1\Pi_g$ state are allowed
but to neither component of the split $^1\Delta_u$. Moreover, they determined generalized oscillator strengths at these two points in
the spectrum and showed that the trends with K^2 were parallel and
did not extrapolate to zero. This alone eliminates $^1\Delta_u$ as one
component. The only satisfactory alternative is to assign the
two peaks to transitions to the split components of $^1\Pi_g$. It is
apparent that the calculation of excitation energies is not, by
itself, enough to provide reliable assignments for forbidden
transitions in ultraviolet absorption spectra or in electron
impact spectra obtained at high kinetic energy. Intensities must
also be considered, qualitatively at least.

The above argument for CO_2 illustrates a general principle
which is applicable to forbidden transitions in polyatomic molecules. The principle may be stated as follows. *In comparing
calculated excitation energies for forbidden transitions to
excitation energies at peak intensities in electron impact spectra
for purposes of assigning term symbols, it is best to compare at
scattering angles of a few degrees and high kinetic energies
(hundreds of volts).* This is a general principle which is well
worth following in future comparisons.

Franck–Condon Principle.   Invariance in Relative Intensities.

The Franck–Condon principle, that the most intense transitions correspond to vertical excitations from the ground state
equilibrium configuration, is dependent for its validity on the
great disparity in mass (and velocity) between nuclei and electrons.
Hence, the principle applies to electron impact spectra at any
scattering angle. The more sophisticated version of the Franck–
Condon principle, (developed especially by R. W. Nicholls[68] and
co-workers) which relates the distribution of intensities among

transitions to overlap integrals over products of vibrational functions (Franck–Condon factors) and r-centroids is also applicable to electron scattering. Within the framework of the Born approximation, an approximate demonstration was given by Craggs and Massey.[69] This demonstration was extended by Nicholls.[68] It has been demonstrated by experiment, however, that Franck–Condon factors still describe the intensity distribution under conditions where the Born approximation does not hold.[12] The theories of Craggs and Massey and the more detailed theory of Nicholls indicate that relative intensities for vibrational progressions within the same electronic transition, when compared at fixed momentum change, are the same at all momentum changes. At high kinetic energy and scattering angles above 1°, this is equivalent to the statement that relative intensity is independent of angle. It is found by experiment[12] that relative intensities in such progressions are independent of both scattering angle and kinetic energy of incident electrons over very wide ranges of both angle and energy. Exceptions to this generalization are expected when kinetic energy of the incident electron is within a resonance. Such behavior has been encountered by Trajmar, Williams and Cartwright[70] for the transition $a^3\Pi \leftarrow X^1\Sigma^+$ in CO. At 10 eV incident energy, they find that relative intensities do not agree with Franck–Condon factors and change with scattering angle. Sanche and Schulz[71] have shown that there is a resonance in CO centered at 10.04 eV. For diatomic molecules, the Franck–Condon principle has been extensively studied. Relative intensities have been determined experimentally[12] and discussed in a recent review.[72]

Polyatomic systems have also been studied (see Refs. 12 and 72) but the theory is less developed. Harshbarger[73] has calculated Franck–Condon factors for a transition in NH_3 and another[74] in NH_3^+.

An interesting application of the Franck–Condon principle is the classification of transitions in polyatomic molecules. If the vibrational structures of two electronic transitions overlap in a spectrum, the problem arises of associating the vibrational peaks with the proper electronic transition. This can be frequently accomplished by determining spectra at several scattering angles. Those vibrational peaks associated with the same electronic transition retain the same relative intensities while two peaks associated with different electronic transitions usually change in relative intensity. Experiments of this type have revealed some unexpected results for polyatomic molecules. Benzene is a notable example. (See Ref. 12.)

The invariance of relative intensities of vibrational components of an electronic transition seems dependent, in the treatments of Craggs and Massey[69] and in that of Nicholls,[68] on the

separability of nuclear and electronic motions. This raises the
question of whether such invariance will prevail when the excited
state is degenerate since in that case the Jahn–Teller effect leads
to extensive mixing of electronic and nuclear motions. Harshbarger
and Lassettre[49] have discussed this point in connection with the
angular dependence of the spectra of CH_4 and CF_4. They concluded,
with widely used approximations, that relative intensities were
still invariant over small ranges of scattering angle. Additional
experimental and theoretical study would be worthwhile.

Singlet–Triplet Transitions

Singlet–triplet transitions have been observed by Lassettre
and collaborators[12],[75-77] at kinetic energies in the range 33–100
eV and scattering angles less than 20° for the most part in diatomic molecules. Trajmar, Kupperman and their collaborators have
systematically studied singlet–triplet transitions at low kinetic
energies and large scattering angles.[78-83] The trend in relative
intensity with scattering angle has been used by them to identify
singlet–triplet transitions. Doering and collaborators[84-87] have
also used electron scattering through large angles at low kinetic
energies in the study of singlet–triplet transitions. The investigations of Trajmar, et. al., have been successful in detecting new excited triplet states and the trend with angle of relative
intensities has been suggestive, but not definitive, in identifying the states. The field is developing rapidly.

Strongly Forbidden Transitions at Low Kinetic Energies

Cartwright, Trajmar, Williams and Huestis[88] have recently
studied the scattering by O_2 with excitation of the $^1\Delta_g$ and $b^1\Sigma_g^+$
states. It was found that intensity for the transition
$b^1\Sigma_g^+ \leftarrow X^3\Sigma_g^-$ was very low at small angles and increased as the
scattering angle increased over the range 4° to 32°. In this and
a subsequent paper by Goddard, Huestis, Cartwright and Trajmar[89]
it was shown, by an application of group theory, that only odd
spherical harmonics occur in the scattered wave and hence the
scattered amplitude vanishes at $\theta = 0°$. In essence, the method
is as follows: the system of colliding electron plus molecule
constitutes a system with the same symmetry group as the neutral
molecule and hence every irreducible representation of that
symmetry group which occurs in the system incident electron plus
molecule must also occur in the system scattered electron plus
molecule. Since both incident electron and scattered electron are
far from the scatterer, a direct product representation is applicable in both cases. Expanding the incident plane wave into partial
waves by expansion in a Legendre series, in the usual way, makes
it easy to determine which irreducible representations are present

in the incident sytem. Since the same representations are present in the scattered system, and the state of the molecule is known, the nonvanishing Legendre terms in the scattered wave can be found immediately and the missing components identified. Hence, the angles at which the amplitude vanishes are identified.

Selection rules like these are especially useful under scattering conditions where the Born approximation does not hold. In the case of the $b^1\Sigma_g^+ \leftarrow X^3\Sigma_g^-$ transition in O_2, the Born cross section vanishes identically.[90] Nevertheless, the transition is observed[88] at very low kinetic energies, 15 eV. The above selection rules, which apply exactly (neglecting spin-orbit coupling) still show that the transition is forbidden at some angles.

Identification of Triplets by Positive Ion Scattering

An interesting method for the identification of triplets has been used by Moore and Doering.[91] The method involves excitation by H^+ and H_2^+ projectiles. Obviously, H^+ cannot bring about a singlet-triplet transition by electron exchange since H^+ contains no electron. However, electron exchange can take place (and frequently does) when H_2^+ is the projectile since this ion contains an electron. An example (taken from Ref. 91) is shown in Figure 3 where positive ion impact spectra of ethylene are shown. The top spectrum, obtained by impact with H^+, shows no excitation below 7 eV while the bottom spectrum, taken with H_2^+, contains a broad excitation with peak at 4.5 eV. The peak at 4.5 eV is obviously due to excitation of a triplet. When applicable, *definitively* identifies a singlet-triplet transition in contrast to the study of relative intensities in electron impact (Section "Singlet-Triplet Transitions") which depends on a plausibility argument. Doering has recently

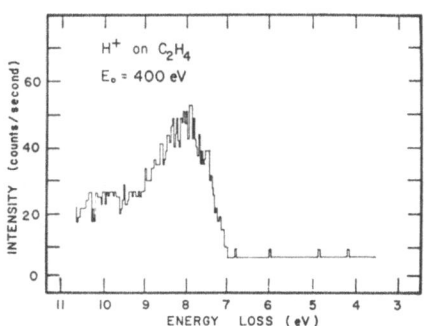

J. H. Moore and J. P. Doering, J. Chem. Phys. 52, 1692 (1970).

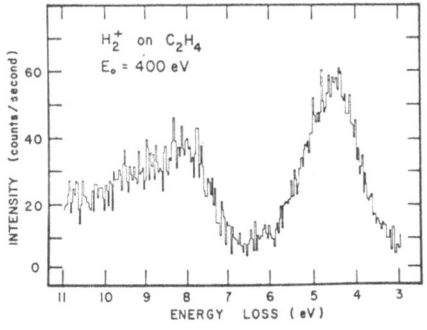

Figure 3. C_2H_4. Two spectra H^+ and H_2^+ at $\theta = 0°$.

published[92] a review of his research on positive ion scattering.

RECENT RESULTS

Deviations from Born Approximation at Small Scattering Angles.
Selection Rules.

Recent investigations of generalized oscillator strengths have
shown that when the term symbols in ground and excited state are
the same, then curves of f vs $(\Delta P)^2$ at different kinetic energies
lie on different curves but when term symbols are different data
at different energies lie on a common curve. This is illustrated
in Figure 4 where data are shown for two forbidden transitions[30]
in N_2 and a transition in carbon monoxide.[23] Another transition in
CO has also been studied for which the f curve changes abruptly at
small momentum changes.[93] These results imply that the terms in the
collision amplitude representing deviations from the Born approximation are proportional to the matrix element of an operator which
is invariant under all operations of the symmetry group of the
scatterer. No exact expression for such an operator has been obtained. However, Lassettre[94] developed an operator series the first
term of which has the desired property and a qualitative argument
was given which indicated that the first term was the most important. The same selection rule was obtained by Winifred M. Huo by an
entirely different method.[95] She has studied an effective potential
function V_{no} which gives for the T matrix the following expression

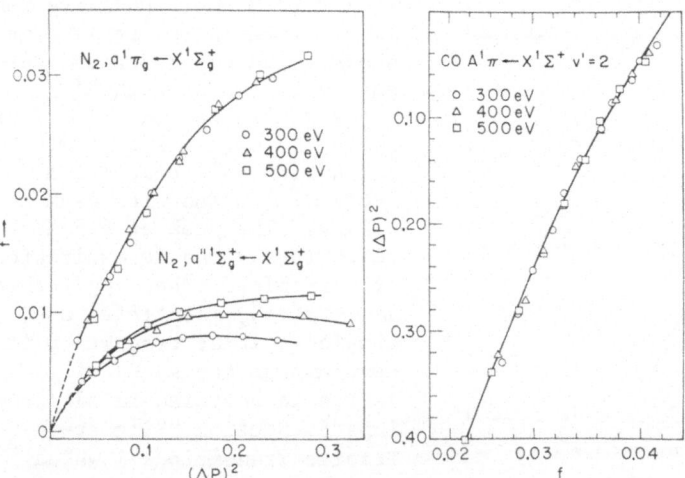

Figure 4.   f vs $(\Delta P)^2$ for two transitions in N_2 and
             one transition in CO.

$$T_{no} = \langle e^{i\vec{k}_n \cdot \vec{r}} | \tilde{V}_{no} | e^{i\vec{k}_o \cdot \vec{r}} \rangle .$$

This expression has the same form as the Born approximation but V_{no} is not the matrix element of interaction potential, as in the Born approximation, but is a function which gives (neglecting exchange) the exact T matrix. Although V_{no} is a complicated function, its asymptotic expansion in inverse powers of the distance from the scatterer can be obtained. The first nonvanishing term of the non-Born part of V_{no} varies like the inverse fourth power of the distance and is proportional to a quantity referred to as the "transition polarizability". Small scattering angles implies a distant collision and this in turn means that the potential function at large distances is the important quantity in the small angle scattering. Huo showed that the transition polarizability vanishes unless the term symbol is the same in ground and excited state. This means that the Born approximation should be satisfied at small angles except when the term symbol is unchanged in excitation.

At very large scattering angles, it is likely that the Born approximation is never satisfied except at very high kinetic energies. In mercury, large deviations have been found for the transition $6^1S_0 \rightarrow 6^1P_1$ by Hanne and Kessler[96] and these have been confirmed by Skerbele and Lassettre.[97] These deviations are due to interaction of the incoming electron with the nucleus and were first explicitly pointed out by Geltman and Hidalgo[98] in a treatment of inelastic electron scattering by atomic hydrogen. The higher the charge on the nucleus the smaller the angle at which the deviations are noticeable. The high energy limit has been discussed by Huo.[99]

Singlet-Triplet Excitation at High Kinetic Energy. $1^1S \rightarrow 2^3S$ Transition in Helium.

A new electron spectrometer is being developed (by Michael A. Dillon and E. N. Lassettre), at Carnegie-Mellon University, for the study of ionization continua using coincidence counting techniques to identify electrons scattered and ejected (ionized) in the same collision. This application requires (in addition to short resolving times) that very high scattered currents be obtained. This has been achieved with sufficient success (by means of properly designed lens chains and a unique collision cell) so that very small collision cross sections can be studied in conventional scattering experiments on the excitation (without ionization) of higher quantum states. In particular, the transition $1^1S \rightarrow 2^3S$ can be resolved from $1^1S \rightarrow 2^1S$ and still have high enough scattered currents to determine collision cross sections. Results at 200 eV and 400 eV are given in Figure 5 as a function of $(\Delta P)^2$. (ΔP is the momentum change.) A notable feature, not observed at high kinetic energy in any previous research, is the abrupt change in the cross section curve at $(\Delta P)^2 \cong 2$. Note also the strong similarity between the curves at 200 eV and 400 eV in Figure 5. This similarity might be expected from the approximation of Ochkur and Brattsev[100] but their functional form and energy dependence differ greatly from Figure 5.

At 200 eV, at angles beyond the minimum in Figure 5 (20°), the experimental curve closely resembles, in shape, that calculated by Joachain and Van den Eynde[101] from the first Born approximation with exchange. At smaller angles, the theoretical and experimental curves are entirely different. At angles greater than 20°, the calculated cross sections exceed the observed in magnitude even though the general trends with angle are similar. From calculations of direct scattering in helium[102] it is known that the shape of a cross section curve is less sensitive to error in the wavefunctions than its absolute magnitude. Assuming that the same is true for exchange scattering, we conclude that the collision cross section beyond 20° is mainly that of the first order Born approximation with exchange while that with $\theta < 20°$ is largely due to higher order terms in the Born series. It seems probable that at $\theta = 0°$, the contribution of the first Born term is negligible.

Failure of the first Born approximation with exchange at small scattering angles was also indicated by the investigations of Vriens, Simpson and Mielczarek[103] on the $1^1S \rightarrow 2^3S$ transition in helium. They studied the cross section at kinetic energies as high as 225 eV and angles to 15°, not high enough to reveal the abrupt change in Figure 5. Steelhammer and Lipsky[104] showed

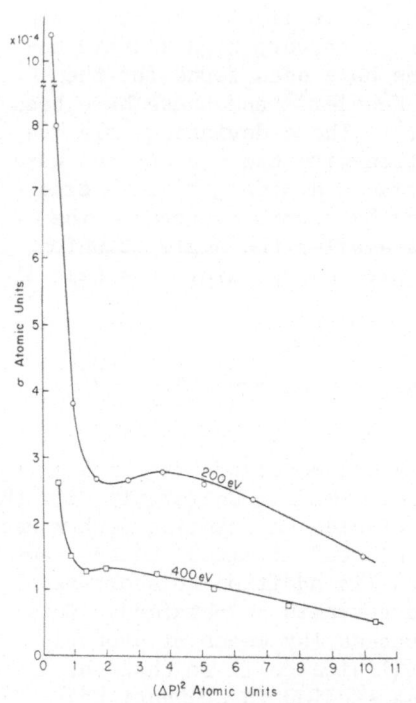

Figure 5. σ vs $(\Delta P)^2$ for $1^1S \rightarrow 2^3S$ transition in helium.
Michael A. Dillon.

that inclusion of nuclear terms in the exchange amplitude accounted for a maximum in the cross section at $\theta = 0°$ thus improving the agreement between theory and experiment at low kinetic energy. However, Skerbele, Harshbarger and Lassettre[105] determined the ratio of cross section σ_T for the $1^1S \to 2^3S$ transition to that σ_S for the $1^1S \to 2^1S$ transition at $\theta = 0°$ and kinetic energies of 300, 400 and 500 eV. The best available theory gives for σ_T/σ_S a value at least two orders of magnitude less than the experimental ratio at 500 eV. More accurate theories are essential and are being developed as described in the next section.

Theoretical Study of Singlet-Triplet Selection Rules.

The collision cross section studies described in the preceding section show that the second and higher order terms in the Born series with exchange (assuming that it converges) are necessary to account for the observed collision cross sections for the $1^1S \to 2^3S$ transition in helium at small scattering angles. Hence, an investigation has been undertaken (by Winifred M. Huo) of the second order terms in the Born approximation with exchange. One result of immediate interest is obtained by expanding the second order terms in inverse powers of P_0, the momentum of the incident electron, using an integration by parts similar to that employed in obtaining Ochkur's equation. The leading term in this series, which varies like $1/P_0^2$, is the only important term at high kinetic energies. It has the property that (at small momentum changes) it vanishes unless orbital term symbols in the ground and excited states are the same. Orbital term symbols represent irreducible representations of the symmetry group acting on positional coordinates (but not spins). Hence, the geometrical symmetry of the system is involved. The selection rule applies to the first term of an asymptotic series and not to the entire amplitude. Nevertheless, it is obvious that vanishing of this term will have a major influence on the entire amplitude at high kinetic energy. The asymptotic expansion and the selection rule apply not merely to the $1^1S \to 2^3S$ transition in helium but to any transition in any substance no matter how complex. Hence, a principle of very general applicability is involved. Numerical calculations are in progress.

Other Singlet-Triplet Transitions.

The suggestion, in the preceding two sections, that singlet-triplet transitions with the same term symbol occur with abnormally high cross sections in forward scattering explains an apparent anomaly in the electron impact spectrum of nitrogen. In 1966, Meyer and Lassettre[106] reported the observation of a weak transition at 11.86 eV with incident electrons of kinetic energy 400 eV. The excitation energy (11.86 eV) agrees well with the assumption that the E state of nitrogen (a triplet) is involved. At the time, it seemed so unusual to observe a singlet-triplet transition at such high energies in forward scattering that we suggested that the

excited state was, in fact, a singlet. Trajmar, Rice and Kupperman[81] studied the transition at low kinetic energy (35 eV) and
large scattering angles (to 80°). They concluded that the trend
with angle was consistent with the behavior of a singlet-triplet
transition. However, they gave no explanation for observation of
the transition at 400 eV. Since the term symbol for the ground
state is $X^1\Sigma_g^+$ and for the excited state is $E^3\Sigma_g^+$, the theory of
the preceding section supplies the answer. The orbital term sym-
bols in ground and excited states are the same and hence the transition is a member of a class whose cross sections in forward scattering are expected to be abnormally high at high kinetic energies.
A systematic study of such transitions would be of considerable
interest.

Calculation of Singlet-Triplet Energy Differences.

    A method for the calculation of triplet state energies has
been developed by Lassettre and Dillon.[107] Consider an atom or
molecule in which one electron has been excited from a closed shell
ground state. The excited configuration gives rise to a singlet
of energy E_S and a triplet of energy E_T. The energy difference is
given by the formula

$$E_S - E_T = \frac{1}{\pi g W} \int_0^\infty K^2 f_B dK \quad . \tag{29}$$

Here f_B is the generalized oscillator strength, in Born approximation, for excitation of the singlet, W is the excitation energy
of the singlet and g is the degeneracy of the excited state singlet.
In the above equation, for simplicity, the momentum change is designated by K instead of ΔP. The generalized oscillator strength,
f_B, is that for vertical excitation of the singlet and the energy
E_T of (29) is the vertical excitation energy of the triplet state,
i.e., the excitation energy at the peak intensity. If f_B is determined by the experiment, the kinetic energy must be high in order
that the Born approximation be valid. Equation (29) is based on
a result obtained recently by Lassettre.[108]

    Since eq. (29) is deduced using an extreme one-electron
model, it is only an approximation. Its accuracy has been tested
on transitions for which both $E_S - E_T$ and f are known. The results
for several such cases are shown in Table III taken from Ref. 107.
Agreement is especially good when the oscillator strengths are
obtained from experimental data. (Table III at end of Addendum.)

Excited Triplet of H_2O.

    In electron impact spectra, obtained at low kinetic energy, an
energy loss in H_2O is observed beginning at about 4 eV and extending upward to the strong transition centered at 7.4 eV. This energy

loss was first observed by Schulz[110] using the trapped electron method and confirmed by Compton, et. al.[111] with an entirely different threshold method. Knoop, et. al.,[112] using a variant of the trapped electron method with a well depth of about a volt, observed the same transition and concluded that it could not be due to an impurity, contrary to the conclusion of Azria and Fiquet-Fayard.[113] Skerbele, Dillon and Lassettre[114] also observed, with incident electrons of 55 eV kinetic energy, weak scattering beginning at ~ 4 eV. The same transition was also studied by Trajmar, Williams and Kupperman[82,83] who determined relative intensities as a function of angle and definitely concluded that the transition at 4.5 eV was due to excitation of a $^3B_1$ state, a suggestion which was first made by Schulz[110] and assumed by subsequent investigators except Knoop, et. al. who favored $^3A_1$ as the term symbol.

In the interim, theoretical calculations of the energy of the $^3B_1$ state led to the conclusion that the state was not bound relative to either $OH(^2\Pi)$ + $H(^2S)$ or $O(^3P)$ + $H_2(^1\Sigma_g^+)$. In particular, this was the outcome of a recent calculation by Hosteny, Hinds, Wahl and Krauss[115] using multiconfiguration SCF methods. If this is confirmed by subsequent research, then no excitation energy below 5.07 eV can be due to excitation of the $^3B_1$ state. This would pose a problem since several investigators have observed scattering below this limit.

In view of this situation, equation (29) has been applied by Lassettre and Skerbele[116] to calculate $E_S - E_T$ from measured oscillator strengths for the 7.4 eV transition of H_2O. Since all of the measurements were made before equation (29) was discovered, they are not ideally suited to the evaluation of the integral on the right of (29). We have used the data, and some assumptions, to place plausible upper and lower bounds on the integral. We report the result as the midpoint of the upper and lower bounds and an uncertainty which represents the spread. That is

$$E_S - E_T = 0.58 \pm 0.42 \text{ eV} . \qquad (30)$$

This represents a band within which $E_S - E_T$, calculated from (29), is presumed to lie. The value 0.58 eV does not represent our estimate of the most probable value. It may lie anywhere in the range but is probably on the low side. However, the range (30) is not broad enough to include the transition at 4.5 eV. It is believed that the data exclude the possibility that 4.5 eV can be the excitation energy for a <u>vertical</u> transition to the $^3B_1$ state.

Knoop, et. al.[112] observed a peak at 7.2 eV and stated that the energy scale was accurate enough so that the peak was not the same as the 7.4 eV peak observed by many investigators at higher kinetic energies. They attributed the 7.2 eV peak to the transition $^3B_1 \leftarrow X^1A_1$. It would follow that $E_S - E_T \cong 0.2$ eV a value consistent with (30) by a narrow margin. The observation of Trajmar,

Williams and Kupperman[83] seem consistent with the conclusion of
Knoop, et. al. since the 7.4 eV peak shifts to 7.2 eV as the scattering angle changes from 7.5° to 77.5° with incident electrons of
20.6 eV kinetic energy. This is the expected behavior if the
transition $^1B_1 \leftarrow X^1A_1$ decreases in intensity more rapidly than
$^3B_1 \leftarrow X^1A_1$ as the angle increases. Conceivably, the triplet envelope could have a long tail extending to 4 eV and this would be consistent with the observations by Trajmar, et. al.[83] since they
showed that the intensity ratio of the 4.5 eV to the 7.4 eV transition is leveling off at increasing scattering angle as though
approaching a constant value. If, at large angle, the peak has
shifted to 7.2 eV and is in actuality the peak for the triplet, then
the ratio would be that for two points on the envelope of the same
transition and this should certainly be a constant. However, this
is all unsupported conjecture since equation (29) provides us only
with the vertical excitation energy. It tells us nothing about the
envelope shape. For the present, we can do no more until further
data are collected.

Other Recent Researches.

There have been many recent contributions to electron impact
spectroscopy which are omitted from this review. I must note in
passing, however, without discussion the interesting research of
Van der Wiel and Brion on ionization continua using fast coincidence techniques.[117-119]

ADDENDUM ON BENZENE

After the above had been written, the author discovered, at
the NATO Advanced Study Institute, that a substantial interest in
the benzene spectrum still persists, especially that region in the
vicinity of 6.2 - 6.5 eV. Because of this, it seems worthwhile to
briefly discuss the electron impact spectrum of benzene in this
region and to review, in the light of our present knowledge, the
interpretation of the spectrum.

It was mentioned in the Section entitled "Franck-Condon
Principle. Invariance in Relative Intensities" that the spectrum
of benzene had been studied as a function of angle. The trend with
angle in the region 6.0 - 6.6 eV is shown in Figure 6 which is
taken from Ref. 12. Note especially Figures 6a and 6b which show
that the transitions at 6.10 (which appears as a shoulder in most
spectra) and 6.20 eV change with angle in such fashion that the
relative intensities are unchanged with angle. Similarly, the
transitions at 6.31, 6.41 and 6.53 retain the same relative intensities. The two sets, however, change drastically in intensity
relative to each other. Hence, we infer that the 6.10 and 6.20 eV
transitions belong to one electronic transition and those at 6.31,
6.41 and 6.53 eV belong to a different electronic transition. This

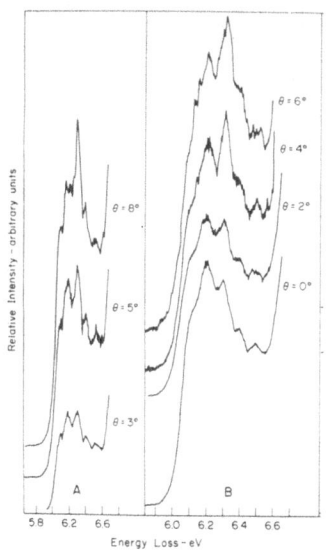

Figure 6. Portions of electron-impact spectra of benzene at (A) 50 eV and (B) 90 eV.

was stated in Ref. 12 (see page 5092). These experiments do not determine the term symbols but $^1B_{1u}$ and $^1E_{2g}$ are frequently mentioned in this connection. Vertical excitation from the equilibrium ground state configuration to each of these is (electric dipole) forbidden. Despite this, the excitations could appear weakly because of electric dipole moments induced by vibrational distortion. The author has been reminded by M. B. Robin that the transitions also become allowed as the result of electronic configuration mixing due to deviations from the Born–Oppenheimer approximation (separability of electronic and nuclear motions). In fact, the "borrowing" of intensity from the strong transition $^1E_{1u} \leftarrow X^1A_{1g}$ by the second of the above mechanisms is the accepted explanation for the appearance of the transition (or transitions) in the ultraviolet spectrum.[120,121] The most accurate calculations of Buenker and Peyerimhoff[122] (Fig. 1b of Ref. 122) show that the term symbols $^1B_{1u}$ and $^1E_{2g}$ are reasonable assignments.

TABLE III: Singlet–Triplet Energy Differences

Substance	Transitions	$E_S - E_T$ (eV) Experiment	Theory	Source of Generalized Oscillator Strengths
He	$1^1S \rightarrow 2^{1,3}P$	0.253	0.310	Ref. 109.
			0.312	Polynomial fitted to f's of Ref. 109 with $K^2 \leq 2$.
			0.317	Ditto but $K^2 \leq .5$.
			0.262	Experimental f's from Ref. 103.
He	$1^1S \rightarrow 2^{1,3}S$	0.796	0.664	Ref. 109.
	$1^1S \rightarrow 3^{1,3}P$	0.080	0.093	Ref. 109.
	$1^1S \rightarrow 3^{1,3}S$	0.202	0.180	Ref. 109.
Co	$X^{1+} \rightarrow A^{1,3}$	2.1	2.24	Ref. 108.
		2.1	2.09	Ref. 108. Renormalized.

REFERENCES

*Supported by the United States Air Force Office of Scientific Research, Contract No. AFOSR-73-2244B.
1. G. J. Schulz, Phys. Rev. 125, 229 (1962).
2. G. J. Schulz, Rev. Mod. Phys., in press (1973).
3. G. J. Schulz, Rev. Mod. Phys., in press (1973).
4. H. A. Bethe, Ann. Physik. 5, 325 (1930).
5. E. N. Lassettre, Radiation Research, Suppl. 1, 530-546 (1959).
6. E. N. Lassettre and S. A. Francis, J. Chem. Phys. 40, 1208 (1964).
7. E. N. Lassettre, A. Skerbele and M. A. Dillon, J. Chem. Phys. 50, 4486 (1969).
8. R. A. Bonham, J. Chem. Phys. 53, 3726 (1970).
9. R. A. Bonham, J. Chem. Phys. 54, 5445 (1971).
10. M. Inokuti, Rev. Mod. Phys. 43, 297 (1971).
11. J. P. Bromberg, J. Chem. Phys. 50, 3906 (1969).
12. E. N. Lassettre, A. Skerbele, M. A. Dillon and K. J. Ross, J. Chem. Phys. 48, 5066 (1968).
13. E. N. Lassettre and A. Skerbele, Section 7.2 of "Experimental Methods in Molecular Physics", edited by D. E. Williams, Academic Press, New York, 1973.
14. G. E. Chamberlain, C. E. Kuyatt and S. R. Mielczarek, Phys. Rev. A2, 1905 (1970).
15. H. Boersch, J. Geiger and H. Hellwig, Phys. Letters 3, 64 (1962).
16. H. Boersch, J. Geiger and W. Stickel, Z. Physik. 180, 415 (1964).
17. H. Boersch, J. Geiger and M. Topchowsky, Phys. Letters 17, 266 (1965).
18. J. Geiger, Z. Physki. 175, 530 (1963).
19. J. Geiger, Z. Physik. 177, 138 (1964).
20. J. Geiger, Z. Physik. 181, 413 (1964).
21. J. Geiger and M. Topchowsky, Z. Naturforsch. 21a, 626 (1966).
22. J. Geiger and H. Schmoranzer, J. Mol. Spectry. 32, 39 (1969).
23. E. N. Lassettre, A. Skerbele and M. A. Dillon, J. Chem. Phys. 52, 2797 (1970).
24. A. Skerbele and E. N. Lassettre, J. Chem. Phys. 45, 1977 (1968).
25. A. Skerbele and E. N. Lassettre, J. Chem. Phys. 52, 2708 (1970).
26. A. Skerbele and E. N. Lassettre, J. Chem. Phys. 56, 845 (1972).
27. A. Skerbele and E. N. Lassettre, J. Chem. Phys. 58, 2887 (1973).
28. E. N. Lassettre and A. Skerbele, J. Chem. Phys. 54, 1597 (1971).
29. V. D. Meyer and E. N. Lassettre, J. Chem. Phys. 54, 1698 (1971).
30. A. Skerbele and E. N. Lassettre, J. Chem. Phys. 53, 3806 (1970).
31. E. N. Lassettre and A. Skerbele, J. Chem. Phys., submitted.
32. W. R. Harshbarger, A. Skerbele and E. N. Lassettre, J. Chem. Phys. 54, 3784 (1971).
33. V. D. Meyer, A. Skerbele and E. N. Lassettre, J. Chem. Phys. 43, 805 (1965).
34. J. E. Hesser, J. Chem. Phys. 48, 2518 (1968).
35. M. J. Mumma, E. J. Stone and E. C. Zipf, J. Chem. Phys. 54, 2627 (1971).

36. G. M. Lawrence, Bull. Am. Phys. Soc. 16, 204 (1971).
37. A. E. Douglas, J. Chem. Phys. 45, 1007 (1966).
38. W. H. Rhodes, J. Chem. Phys. 50, 2885 (1969).
39. J. G. Chervenak, J. Opt. Soc. Am. 61, 952 (1971).
40. R. E. Imhof and F. H. Read, Chem. Phys. Letters 11, 362 (1971).
41. S. M. Silverman and E. N. Lassettre, J. Chem. Phys. 42, 3420 (1965).
42. S. M. Silverman and E. N. Lassettre, J. Chem. Phys. 50, 2922 (1964).
43. J. Geiger, Zeit. Physik. 175, 530 (1963).
44. J. Geiger, Zeit. Physik. 177, 138 (1963).
45. J. Geiger and H. Schmoranzer, J. Mol. Spectry. 32, 39 (1969).
46. J. Geiger and B. Schröder, J. Chem. Phys. 50, 7 (1969).
47. J. Geiger and B. Schröder, J. Chem. Phys. 49, 740 (1968).
48. E. N. Lassettre and S. A. Francis, J. Chem. Phys. 40, 1208 (1964).
49. W. R. Harshbarger and E. N. Lassettre, J. Chem. Phys. 58, 1505 (1973).
50. R. H. Huebner, Celotta, S. R. Mielczarek and C. E. Kuyatt, J. Chem. Phys., in press.
51. M. J. Weiss, C. E. Kuyatt and S. R. Mielczarek, J. Chem. Phys. 54, 4147 (1971).
52. M. J. Weiss, S. R. Mielczarek and C. E. Kuyatt, J. Chem. Phys. 54, 1412 (1971).
53. E. N. Lassettre, A. Skerbele and V. D. Meyer, J. Chem. Phys. 45, 3214 (1966).
54. B. Lutz and K. Dressler, Phys. Rev. Letters 19, 1219 (1967).
55. H. G. M. Heideman, C. E. Kuyatt and G. E. Chamberlain, J. Chem. Phys. 44, 355 (1966).
56. V. D. Meyer and E. N. Lassettre, J. Chem. Phys. 44, 2535 (1966).
57. I. V. Hertel and K. J. Ross, J. Phys. B1, 697 (1968).
58. W. R. Newell, K. J. Ross and J. B. P. Wickes, VIIth International Conference on Physics of Electronic and Atomic Collisions, Abstracts of Papers, North Holland Publishing Co., Amsterdam (1971), p. 99.
59. N. Johnson, A. Morris, K. J. Ross and D. J. Smith, J. Chem. Phys. 54, 4954 (1971).
60. I. V. Hertel and K. J. Ross, J. Chem. Phys. 50, 536 (1969).
61. F. H. Read and Whiterod, Proc. Phys. Soc. (London) 82, 434 (1963).
62. F. H. Read, Proc. Phys. Soc. (London) 83, 619 (1964).
63. V. D. Meyer and E. N. Lassettre, J. Chem. Phys. 42, 3436 (1965).
64. N. W. Winter, C. F. Bender and W. A. Goddard, III, "Theoretical Assignments of Low-Lying States of Carbon Dioxide," Chem. Phys. Letters, 20, 489 (1973).
65. E. N. Lassettre and J. C. Shiloff, J. Chem. Phys. 43, 560 (1965).
66. R. W. Nicholls and A. L. Stewart, in Atomic and Molecular Processes, edited by D. R. Bates, Academic Press, New York, 1962.

67. G. Herzberg, "Infrared and Raman Spectra", Van Nostrand, New York, 1945. See Chapter 2.
68. R. W. Nicholls, J. Quant. Spectry. Radiation Transfer 2, 433 (1962).
69. J. D. Craggs and H. S. W. Massey, in Handbuch der Physik, edited by S. Flügge (Springer, Berlin, 1959) Vol. 37, p. 314.
70. S. Trajmar, W. Williams and D. C. Cartwright, VIIth International Conference on the Physics of Electronic and Atomic Collisions, Abstracts of Papers, North Holland Publishing Co., Amsterdam, 1971.
71. L. Sanché and G. J. Schulz, Phys. Rev. Letters 26, 943 (1971).
72. E. N. Lassettre, Can. J. Chem. 47, 1733 (1969).
73. W. R. Harshbarger, J. Chem. Phys. 53, 903 (1970).
74. W. R. Harshbarger, J. Chem. Phys. 56, 177 (1972).
75. A. Skerbele, M. A. Dillon and E. N. Lassettre, J. Chem. Phys. 46, 4162 (1967).
76. A. Skerbele, M. A. Dillon and E. N. Lassettre, J. Chem. Phys. 46, 4161 (1967).
77. A. Skerbele, M. A. Dillon and E. N. Lassettre, J. Chem. Phys. 49, 3543 (1968).
78. J. K. Rice, A. Kupperman and S. Trajmar, J. Chem. Phys. 48, 945 (1968).
79. S. Trajmar, J. K. Rice, P. S. P. Wei and A. Kupperman, Chem. Phys. Letters 1, 703 (1968).
80. A. Kupperman, J. K. Rice and S. Trajmar, J. Phys. Chem. 72, 3894 (1968).
81. S. Trajmar, J. K. Rice and A. Kupperman, Advan. Chem. Phys. 18, 15 (1970).
82. S. Trajmar, W. Williams and A. Kupperman, J. Chem. Phys. 54, 2274 (1971).
83. S. Trajmar, W. Williams and A. Kupperman, J. Chem. Phys. 58, 252 (1973).
84. J. P. Doering, J. Chem. Phys. 45, 1065 (1966).
85. J. P. Doering, J. Chem. Phys. 46, 1194 (1967).
86. A. J. Williams and J. P. Doering, J. Chem. Phys. 51, 2859 (1969).
87. J. P. Doering, J. Chem. Phys. 51, 2866 (1969).
88. D. C. Cartwright, S. Trajmar, W. Williams and D. L. Huestis, Phys. Rev. Letters 27, 704 (1971).
89. W. A. Goddard III, D. L. Huestis, D. G. Cartwright and S. Trajmar, Chem. Phys. Letters 11, 329 (1971).
90. E. N. Lassettre and M. E. Krasnow, J. Chem. Phys. 40, 1248 (1964).
91. J. H. Moore and J. P. Doering, J. Chem. Phys. 52, 1692 (1970).
92. J. P. Doering, "Inelastic Scattering of Positive Ions", to be published in Berichte der Bunsengesellschaft.
93. A. Skerbele and E. N. Lassettre, J. Chem. Phys. 53, 424 (1971).
94. E. N. Lassettre, J. Chem. Phys. 53, 3801 (1970).
95. W. M. Huo, J. Chem. Phys. 56, 3468 (1972).
96. F. Hanne and J. A. Kessler, Phys. Rev. A5, 2457 (1972).

97. A. Skerbele and E. N. Lassettre, J. Chem. Phys. 58, 2887 (1973).
98. S. Geltman and M. B. Hidalgo, J. Phys. B4, 1299 (1971).
99. W. M. Huo, J. Chem. Phys. 57, 4800 (1972). (see page 4806).
100. V. I. Ochkur and V. F. Brattsev, Optics and Spectros. 19, 274 (1965).
101. C. J. Jochain and R. K. Van den Eynde, Physica 46, 8 (1970).
102. E. N. Lassettre and E. A. Jones, J. Chem. Phys. 40, 1218 (1964).
103. L. Vriens, J. A. Simpson and S. R. Mielczarek, Phys. Rev. 165, 7 (1968).
104. J. C. Steelhammer and S. Lipsky, J. Chem. Phys. 53, 1445 (1970).
105. A. Skerbele, W. R. Harshbarger and E. N. Lassettre, J. Chem. Phys. 58, 4285 (1973).
106. V. D. Meyer and E. N. Lassettre, J. Chem. Phys. 44, 2535 (1966).
107. E. N. Lassettre and M. A. Dillon, J. Chem. Phys., in press.
108. E. N. Lassettre, J. Chem. Phys. 57, 4357 (1972).
109. Y.-K. Kim and M. Inokuti, Phys. Rev. 175, 176 (1968).
110. G. J. Schulz, J. Chem. Phys. 33, 1661 (1960).
111. R. N. Compton, R. H. Huebner, P. W. Reinhardt and L. G. Christophorou, J. Chem. Phys. 48, 901 (1968).
112. (a) F. W. E. Knoop, H. H. Brongersma and L. J. Osterhoff, Chem. Phys. Letters 13, 20 (1972).
    (b) F. W. E. Knoop, Ph.D. Thesis, Rijks University, Leiden, The Netherlands, 1972.
113. R. Azria and F. Fiquet-Fayard, C. R. Acad. Sci. Paris B273, 944 (1971).
114. A. Skerbele, M. A. Dillon and E. N. Lassettre, J. Chem. Phys. 49, 5042 (1968).
115. R. P. Hosteny, A. R. Hinds, A. C. Wahl and M. Krauss, "MCSCF Calculations on the Lowest Triplet State of H_2O", in press.
116. E. N. Lassettre and A. Skerbele, J. Chem. Phys., to be submitted.
117. M. J. van der Wiel and C. E. Brion, J. Electron Spectros. 1, 309 (1972/73).
118. M. J. van der Wiel and C. E. Brion, J. Electron Spectros. 443 (1973).
119. M. J. van der Wiel and C. E. Brion, J. Electron Spectros. 1, 439 (1973).
120. G. Herzberg, "Electronic Spectra of Polyatomic Molecules", Van Nostrand, New York, 1966, pp. 559-60.
121. J. R. Platt and H. B. Klevens, Chem. Rev. 41, 301 (1947).
122. S. D. Peyerimhoff and R. J. Buenker, Theoret. Chim. Acta 19, 1 (1970).

PHOTOIONIZATION OF FREE RADICALS

J. Berkowitz

Argonne National Laboratory
Argonne, Illinois 60439

Since this is the first lecture at this Institute on photo-ionization and free radicals, permit me to take a few minutes to outline my concept of these terms.

Under photoionization I include the use of a continuum or quasi-continuum light source, from which successive nominal wavelengths λ_i, having band widths $\Delta\lambda_i$, are selected by a mono-chromator. In general, these wavelengths will be in the vacuum ultraviolet, although in some cases they can extend above 1850 Å and to shorter wavelengths in the x-ray region. The selected band of photons is permitted to interact with gaseous atoms or mole-cules, and the resulting <u>ions</u> (if any) are detected. (In photo-electron spectroscopy, it is the electrons which are analyzed and detected.) I wish to further restrict the term photoionization here to imply mass analysis of the resulting ions, at pressures sufficiently low that bi-molecular phenomena such as chemi-ioniza-tion, collision induced ionization and ion-molecule reactions do not occur.

The study of the wavelength (or photon energy) dependence of normalized photoion intensity resulting from the photoionization of stable, volatile molecules has provided a wealth of data in the past 16 years on
a) the onset of formation of the parent molecular ion, usually correlated with the adiabatic first ionization potential of the molecule.
b) The onset of formation of one or more fragment ions. This threshold, together with that in a), is closely connected with the energy required to decompose the molecular ion into the particular fragments. (Rates of decomposition, kinetic shifts

and metastable ions will not concern us here.) From this fragmentation threshold, a value for the heat of formation of the fragment ion can usually be deduced. If the ionization potential of the fragment is known, a value for the energy necessary to dissociate the neutral parent molecule into neutral fragments can be inferred. One of the purposes for investigating the photoionization of free radicals is to obtain ionization potentials of these fragments.

c) In some cases ion-pair formation can be observed, i.e.

$$AB + h\nu \rightarrow A^+ + B^- \tag{1}$$

In such cases, when sufficient auxiliary thermochemical information is known, and there is good reason to believe that the threshold for process (1) is not a delayed one, it is possible to obtain a value for the electron affinity of species B.

d) Abundant evidence has accumulated in the past 16 years to indicate that, in general, ionization not only occurs by a direct mode, i.e.

$$AB + h\nu \rightarrow AB^+ + \bar{e} \tag{2}$$

but also an indirect mode, termed autoionization, and described by

$$AB + h\nu \rightarrow AB^* \tag{3a}$$

$$AB^* \rightarrow AB^+ + \bar{e} \tag{3b}$$

For atoms and small, simple molecules, process (3) usually manifests itself as prominent peaked structure, since process (3a) involves absorption to a specific, quasi-discrete state having the characteristics of a resonance absorption. By contrast, process (2) would be expected to yield a smooth dependence of ionization probability with photon energy.

We note here that for diatomic or polyatomic molecules, the formation of AB^* by process (3a) need not lead to autoionization since (in general) such a state will be able to decay by a competitive mode into neutral fragments, the process called predissociation.

Now with regard to the term "free radicals", there are two commonly understood definitions. Many authors, particularly organic chemists, define a free radical as a system with non-zero spin. Others have come to accept this term to refer to a system which is chemically unstable (i.e. chemically very reactive) but physically stable (i.e. it will not spontaneously decompose). Professor Herzberg has discussed these two views in the introduction to his book "The Spectra and Structure of Simple Free Radicals",[1] and points out some of the inconsistencies. For example, O_2, NO, NO_2, ClO_2 are chemically stable molecules, but have unpaired spins. While the ground state of CH_2 seems well established as a triplet state, its first excited state, not very much higher in energy, is a singlet. For ozone, the ground state appears to be a singlet, although it is fairly reactive.

We shall adopt here a working definition similar to that used by Professor Herzberg — a free radical is any transient species (atom or molecule) that has a short lifetime in the gaseous phase under ordinary laboratory conditions. For the purposes of this lecture, I wish to exclude vibrationally excited molecules having a stable electronic ground state, and high temperature species, which would require another lecture.

With the scope of this lecture more-or-less defined, let me turn to the methods of production of free radicals, the kinds of free radicals that have been studied, and the results currently available.

Methods of Production

1. Electrical discharges and flames
   These are rather uncontrolled ways of tearing apart stable molecules. Spectroscopic absorption or emission studies can be performed on species directly in the discharge or flame. For photoionization measurements, it is necessary to deflect the charged species, and permit the desired radical to flow into an ionization chamber. Many atoms, such as H, N and O, can survive a number of wall collisions and be studied in this way. The afterglow in nitrogen is caused by the recombination of N atoms after flowing downstream from the discharge. I shall report here recent photoionization studies of O and N atoms in our laboratory, produced in this way.

2. Pyrolysis
   Among the earliest investigations of free radicals were those of Paneth and Hofeditz, [2] who thermally decomposed the relatively unstable organometallic gases, such as lead tetramethyl and lead tetraethyl. Mass spectrometric electron impact studies of radicals produced in this way (mostly hydrocarbon free radicals) have been performed by a number of workers, including Eltenton,[3] Langer, Hipple and Stevenson[4] and most notably, Lossing and collaborators.[5] A number of such studies have subsequently been carried out by photoionization tech-

niques, and will be summarized later.

3. Photolysis
   This has been a favorite means of producing reactive intermediates in kinetic studies, but heretofore has not been very
   successful in producing sufficiently high number densities of
   free radicals to enable subsequent photoionization to be studied.
   At the present time, with intense laser beams available at a
   variety of wavelengths, this approach should not be overlooked.

4. Chemical reaction
   Such techniques have been used to produce reactive intermediates
   in kinetic studies, and to measure their number densities by
   titration techniques. For example, OH has been produced by the
   reaction of $H + O_3$ and $H + NO_2$. Nitrogen atom concentrations
   are frequently measured by reacting $N + NO \rightarrow N_2 + O$, and noting
   the change of color of the gas stream. I shall report here
   the recent use in our laboratory of the reaction

$$H + NO_2 \rightarrow OH + NO$$

   to produce a clean source of OH in its vibrational ground state,
   and the photoionization spectrum of OH obtained in this way.

5. Flash photolysis

6. Flash discharge

7. Flash radiolysis

   These pulse techniques have been very successfully employed
by the eminent scientists at the National Research Council of
Canada in recent years to obtain absorption and emission spectra
of a number of free radicals. The techniques and the free radicals studied in this way have been thoroughly described by Dr.
Herzberg in his monograph[1] and in his Nobel laureate address.[6]
The pulse techniques produce a very high instantaneous concentration of radicals, which could well be exploited by both photoionization and photoelectron spectroscopic methods, if the electrical
noise problems could be overcome.

Available Data

   Within the scope that I have outlined above, the photoionization studies that I am aware of can be categorized as experiments
with hydrocarbon (or fluorocarbon) free radicals and radicals of
atmospheric significance. The former include CH_3, CF_3, C_2H_5,

$n-C_3H_7$, $i-C_3H_7$ and C_6H_5; the latter include $O(^3P)$, $N(^4S)$, OH, O_3 and HNO_3.

A. Hydrocarbon and fluorocarbon radicals

In 1962, Elder, Giese, Steiner and Inghram[7] produced various hydrocarbon free radicals in their photoionization apparatus by permitting thermally unstable molecules such as $Pb(CH_3)_4$ and $Hg(C_2H_5)_2$ to impinge upon a hot filament in close proximity to the photon beam. In these experiments, the light source was the hydrogen many-line pseudo-continuum, and measurements did not extend beyond the LiF cut-off at $\sim$ 11 eV. Figures 1-6 in their paper display their various experimental results on CH_3 (from $Pb(CH_3)_4$), C_2H_5 (from sec-butylnitrite), $n-C_3H_7$ (from n-propyl mercury and n-butylnitrite) and $i-C_3H_7$ (from di-isopropyl mercury and isobutylnitrite).

Only a short region beyond the threshold is covered in each instance. The adiabatic ionization potentials and the general shape of the relative ionization cross section curve are in good agreement when the radicals are produced from different sources. The n-propyl and isopropyl radicals clearly give different results, indicating that these isomers retain their identity throughout the thermal decomposition.

It is noteworthy that the adiabatic first ionization potential for CH_3 deduced from these experiments (9.82 ± 0.04 eV) was slightly different from the Rydberg series limit deduced earlier by Herzberg and Shoosmith[8] (9.843 ± 0.001 eV). For the other systems, Elder et al. were reluctant to quote anything more than upper limits, since the ionization cross sections appeared to be rising very slowly from threshold, and the true adiabatic value might be approached only asymptotically. More recently, Lossing and Semeluk re-examined these systems in an electron-impact study using a monochromatized electron beam. Their results (Table 1) for C_2H_5, $n-C_3H_7$ and $i-C_3H_7$ are in excellent agreement with the upper limits given by Elder et al. and suggest that reasonably good adiabatic ionization potentials can be obtained from their data.

In 1967, Lifshitz and Chupka[10] studied CF_3 and re-examined CH_3. For CF_3 the ionization cross section rises very slowly from threshold, making an identification of the adiabatic ionization potential difficult. It is very likely due to an unfavorable Franck-Condon factor, since CF_3 neutral is non-planar, and CF_3^+ is most likely planar. An upper limit of 9.25 ± 0.04 eV was obtained for the adiabatic first ionization potential.

TABLE 1

First ionization potentials of some alkyl free radicals
by photoionization and high resolution electron impact
ionization

Radical	Photoionization[a]	Electron impact[b]
CH_3	9.82	9.84 ± 0.03
C_2H_5	< 8.4	8.38 ± 0.05
$n-C_3H_7$	< 8.1	8.10 ± 0.05
$s-C_3H_7$	< 7.5	7.55 ± 0.05

(a) F. A. Elder, C. F. Giese, B. Steiner and M. G. Inghram,
    J. Chem. Phys. 36, 3292 (1962).

(b) F. P. Lossing and G. P. Semeluk, Can. J. Chem. 48, 955 (1970)

For CH_3 the onset is much more abrupt, but the best value obtained (9.825 ± 0.010 eV) is again slightly lower than that of Herzberg and Shoosmith. The interpretation of the precise shape of the threshold region seems to be responsible for this discrepancy.

The higher energy region of this spectrum shows a distinct rise at ~12 eV. Ab initio calculations of the ground state of CH_3 by Kari and Csizmadia[11] and Whitten and Hackmeyer[12] provide orbital energies which (upon invoking Koopmans' theorem) imply a second ionization potential about 6 eV higher than the first, or at about 16 eV. The strong increase observed in the photoionization spectrum at ~12 eV could very well be the first prominent Rydberg member of the series converging to this second ionization potential, the Rydberg member autoionizing to form CH_3^+ in its ground state. A good photoelectron spectrum of CH_3 is required to further clarify this behavior. An attempt along this line has been reported by Jonathan, Morris, Okuda and Ross[13], but only the threshold region in the vicinity of the 1st ionization potential has been identified.

Recently, Sergeev, Akopyan and Vilesov[13] have reported data on the photoionization of C_6H_5 produced by the pyrolysis of azobenzene. Their data yield an adiabatic first ionization potential of about 8.1 eV. This is of more than passing interest, because other means of deducing this ionization potential have resulted in much higher values. A commonly used scheme is to compare two thresholds, such as

$$AB + h\nu \rightarrow A^+ + B + \bar{e} \qquad (4a)$$

$$AB + h\nu \rightarrow A + B^+ + \bar{e} \qquad (4b)$$

If the ionization potential of A is known, and the difference between the thresholds of processes (4a) and (4b) can be established, one can infer the ionization potential of B.

Such an approach has been used by us at Argonne, for example, to estimate the first ionization potential of the radical C_2H from the reactions

$$C_2HBr + h\nu \rightarrow C_2H^+ + Br + \bar{e} \qquad (5a)$$

and

$$C_2HBr + h\nu \rightarrow C_2H + Br^+ + \bar{e} \qquad (5b)$$

However, when applied to C_6H_5X, the experiments have led to

higher values for the ionization potential of C_6H_5, probably due
to the larger kinetic shifts that occur with larger molecules.
Since the kinetics of decomposition of large molecular ions is
not yet a quantitative science, it is important to establish by
independent means the thermochemistry of the process, which
necessitates measuring the ionization potentials of fragments such
as C_6H_5 in separate experiments.

B. Atmospheric radicals
      The work to be described in this section involves unpublished
studies performed at Argonne on $O(^3P)$, NH_3, OH, O_3 and HNO_3.

1. $O(^3P)$

      Dr. Patricia Dehmer has used the photoionization mass
spectrometric apparatus at Argonne, together with an A.C. discharge through molecular oxygen (sometimes containing argon) to
produce atomic oxygen, which then was permitted to effuse into the
ionization region. A set of grids was employed to repel or deflect
charged species from the discharge. The apparatus for producing
free radicals by this technique is shown in fig. 1. Note that the
radicals are sampled from the center of the discharge tube, and
enter a buffer zone, where they can be thermally equilibrated or
permitted to react with an added gas. This buffer region was used
to produce OH by the reaction of H atoms with added NO_2, in an
experiment to be described later.

      The overall photoionization spectrum of atomic oxygen, ob-
tained with a resolution of 0.42 Å, is shown in Fig. 2. Portions
of this spectrum were obtained with 0.16 Å resolution width
(FWHM). All the results are consistent with an initial atomic
oxygen beam in the ground ^3P state, with the $^3P_{2,1,0}$ components
populated in ratios corresponding to a temperature of $\sim375°K$.

      In brief, the interpretation of this spectrum proceeds as
follows:

      Atomic oxygen in its ground state has the configuration
$(1s)^2(2s)^2(2p)^4\ ^3P$. Removal of an electron from the outermost
orbital results in a p^3 configuration which gives rise to ^4S, ^2D
and ^2P states. The ground state of O^+ is the ^4S state. There
are four optically allowed Rydberg series converging to the first
excited ^2D state (split into $^2D_{5/2}$ and $^2D_{3/2}$ by spin orbit
coupling) and three optically allowed Rydberg series converging
to the second excited ^2P state (split into $^2P_{3/2}$ and $^2P_{1/2}$).
Huffman, Larrabee and Tanaka[14] observed and identified each of
these series in absorption. However, not all of these series are
permitted to autoionize, according to the selection rules for this
process.

      The autoionization selection rules require that the quasi-

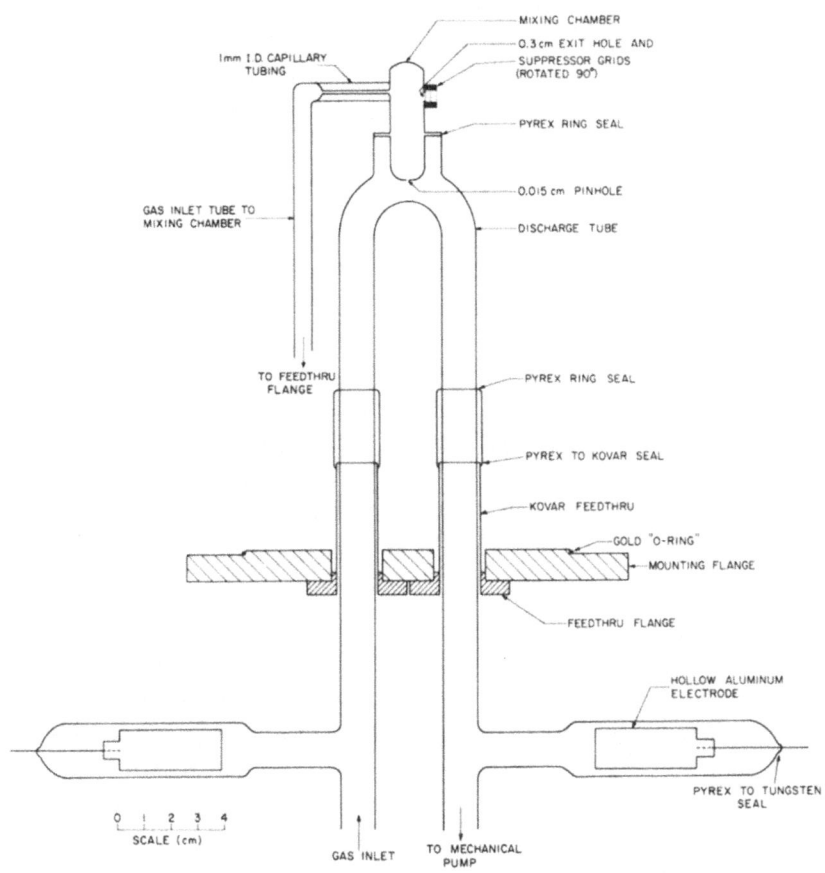

Fig. 1.

Schematic drawing of discharge tube for production of
O and N atoms. The mixing chamber is used for production of OH radicals by reacting H atoms (from discharge) with NO_2.

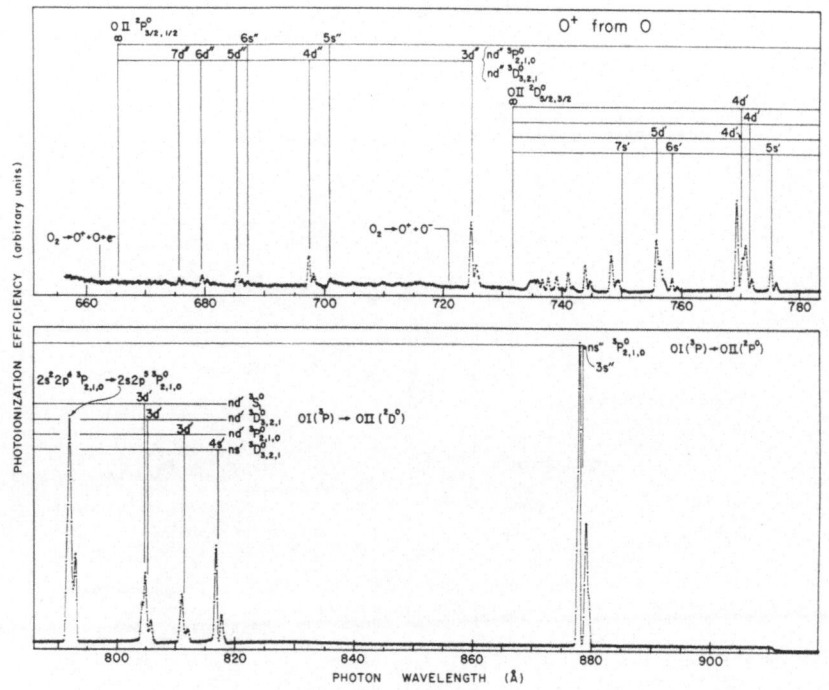

Fig. 2.

Relative photoionization efficiency of atomic oxygen from
ionization threshold (~ 911 Å) to ~ 660 Å, at a resolution
width (FWHM) of 0.42 Å.

discrete state have the same parity and total angular momentum as the underlying continuum. In the limit of pure Russell-Saunders coupling, the values of L and S must also be the same.

For the process

$$O(^3P) + h\nu \rightarrow O^+(^4S) + \bar{e} \tag{6}$$

The electron may depart in an s- or d-wave, according to dipole selection rules. The coupling of the departing electron's angular momentum with that of the ion core gives rise to S or D continua, the optically allowed ones being $^3S_1$ and $^3D_{3,2,1}$. Therefore, in pure Russell-Saunders coupling, ^3P states which are below the second ($^2D_{3/2,1/2}$) ionization limit should not be allowed to autoionize. In addition, $^3P_0$ states should be more rigorously forbidden to autoionize, since they violate the total angular momentum selection rule. There are several examples of ^3P states covered in this study, and all are observed to autoionize in the forbidden region.

The two most intense peaks in the spectrum, one at 878-879 Å being a $3s$ member converging to ^2P, the other an inner shell transition $2s(2p)^5$ at 791-793 Å, are both ^3P states. These (and other forbidden lines) have been observed in emission by Huffman, et al.[14] and by others[15], which means that for these cases autoionization and emission are competitive. Hence, the effect of the selection rules has been to slow down the autoionization rate by several orders of magnitude, but not to prevent it from occurring. The breakdown of these selection rules can be rationalized in terms of spin-orbit coupling.[16]

It is noteworthy that the photoionization cross section of atomic oxygen in the region a few volts beyond threshold is dominated by autoionization. The oscillator strengths for these various transitions are currently being computed, and appear to be in fair agreement with values obtained by other techniques.

2. $N(^4S)$

This spectrum is much simpler, and only a preliminary analysis is currently available. Atomic nitrogen has the ground state configuration $(1s)^2 (2s)^2 (2p)^3 \, ^4S$. Removal of an uppermost electron gives rise to the p^2 set ^3P, ^1D, ^1S. The departing electron can have $\ell = 0$ or $\ell = 2$, and $s = \pm 1/2$. The optically allowed transitions from the neutral ground state must conserve spin, and hence such transitions are limited to the ^3P ionic ground state. As a result, we cannot expect the complex autoionizing structure observed in atomic oxygen, since Rydberg series converging to the ^1D and ^1S excited states are optically forbidden.

The inner-shell excitation 2s → np is permitted, however, and the resulting states can fulfill all the requirements for autoionization. This series, converging to the ^5S state of N^+, accounts for all of the observed autoionization structure in this system. These autoionizations exhibit a characteristic Beutler-Fano profile typical of a small Fano parameter q, whereas the oxygen system displayed peaks above a continuum, characteristic of a large q value. The former implies stronger interaction of the quasi-discrete state with the underlying continuum than the latter.

3. OH

This molecule was produced by the reaction of H atoms (produced in the discharge) with NO_2, in the buffer zone. The relative photoionization cross section data will be reported in a forthcoming publication.

The spectrum is quite complex, and the analysis is still in its preliminary stage. This study does present a good example of the inter-relationship and complementarity of photoionization, photoelectron spectroscopy and the spectra of molecular ions.

We have seen how a knowledge of the excited ionic states can enable us to predict which Rydberg series should be present, and their capability of autoionization. For OH, the ground state configuration is

$$(1s\sigma)^2 (2s\sigma)^2 (2p\sigma)^2 (1\pi)^3 X^2\Pi_i.$$

Removal of an electron from the uppermost occupied orbital leaves a π^2 configuration, which gives rise to $^3\Sigma^-$, $^1\Delta$ and $^1\Sigma^+$. Removal of an electron from the next inner $2p\sigma$ orbital results in a $\sigma\pi^3$ configuration, which gives rise to $^3\Pi_i$ and $^1\Pi$. All of these states have been observed in the isoelectronic NH molecule, although the relative positions of singlet and triplet systems are not firmly established. Only the $A^3\Pi_i \rightarrow X^3\Sigma^-$ transition has been observed in OH^+, although the location of b $^1\Sigma^+$ has been inferred from a predissociation. A composite of high quality calculations[17] and experimental observations[18] enables us to construct a pseudo-photoelectron spectrum, shown in Fig. 3. Thus far, we can only surmise that the rise in the continuum level which we observe at ~15.2 eV is indicative of the new continuum due to the $a^1\Delta$ state manifesting itself at this energy, and the convergence of the overall spectrum at ~16.6—16.7 eV corresponds to the $A^3\Pi_i$ and $b^1\Sigma^+$ convergence limit. A higher resolution photo-

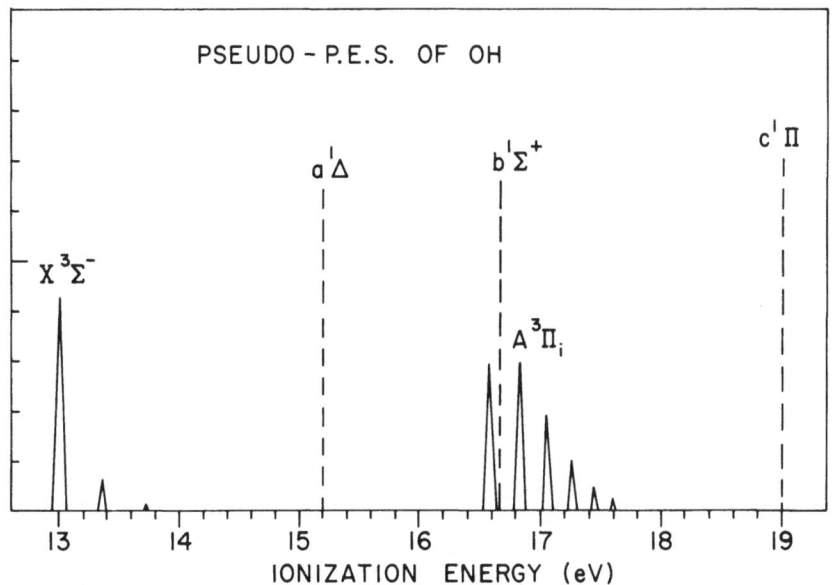

Fig. 3.

Synthetic photoelectron spectrum of OH. The text
describes details of its construction.

ionization spectrum and a real photoelectron spectrum would clearly
help in subsequent analysis. In addition, either a real photoelectron spectrum or an analysis of the present data would
establish the relative energy levels of triplets and singlets for
OH^+. This is an area where PES has an advantage over optical
spectroscopy.

Our best current analysis of the threshold region yields an
adiabatic first ionization potential for OH of 12.99 eV. Foner
and Hudson[19] reported 13.18 eV from a direct electron impact study
of OH. Dibeler, Walker and Rosenstock[20] studied the reaction

$$H_2O + h\nu \rightarrow OH^+ + H + e$$

and reported a threshold value of 18.05 eV for this process.
When combined with well-established values for $\Delta H_f(H_2O)$, $\Delta H_f(OH)$
and $\Delta H_f(H)$, they were able to deduce 12.94 eV for the ionization
potential of OH.

An examination of their data suggests that they selected the
initial rise in the OH^+ spectrum. If allowances for their photon
half-width of ~ 0.04 eV were made, the discrepancy between the two
results would be further reduced.

This example points out the complementarity of photoionization studies on stable and transient species. In one sense, free
radicals are just another kind of molecule, and they display the
same photoionization features as stable molecules. It is difficult, however, to measure fragmentation spectra of free radicals,
because fragments of the same mass from parent molecules or other
species in the system frequently appear at higher photon energy.
On the other hand, a more complete description of the fragmentation process of a stable molecule frequently requires knowledge
of the ionization potential of the fragment molecule, and this is
a major goal in the investigation of free radical photoionization.

4. O_3 and HNO_3

Figures 4 and 5 present preliminary results on these
molecules, in the near threshold region, providing fairly
accurate first ionization potentials and indications of autoionization structure. They are marginal examples of free radicals,
but nevertheless were considered to be of possible current interest to this group and to the scientific community, even in
their incomplete state.

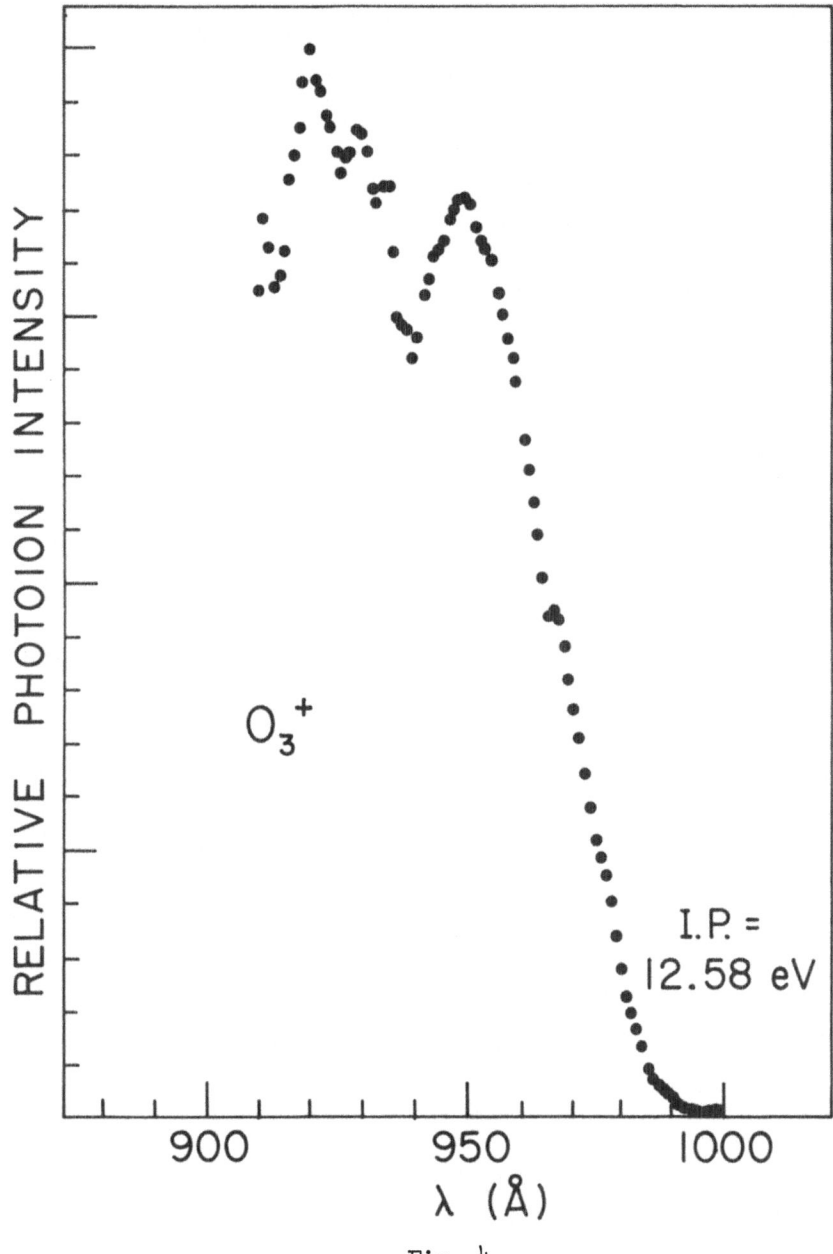

Fig. 4.

Relative photoionization efficiency of O_3 in the threshold region.

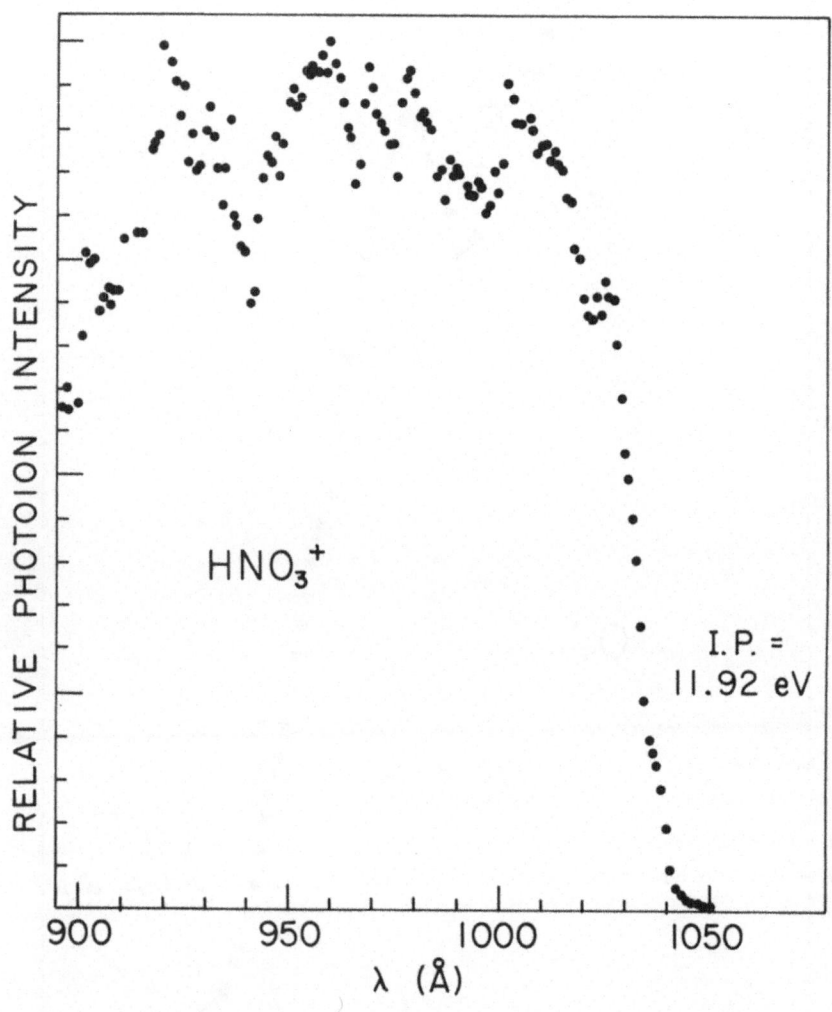

Fig. 5.

Relative photoionization efficiency of HNO_3 in the
threshold region.

References

1. G. Herzberg, "The Spectra and Structures of Simple Free Radicals", Cornell Univ. Press, Ithaca (1971).
2. F. Paneth and W. Hofeditz, Ber. d.d. chem. Ges. 62B, 1335 (1929).
3. G. C. Eltenton, J. Chem. Phys. 10, 403 (1942).
4. A. Langer, J. A. Hipple and D. P. Stevenson, J. Chem. Phys. 22, 1836 (1954).
5. See, for example, F. P. Lossing, "Mass Spectrometry of Free Radicals", Ann. N.Y. Acad. Sci. 67, 499-517 (1957).
6. G. Herzberg, "Spectroscopic Studies of Molecular Structure", Nobel Lecture, Dec. 11, 1971, Publ. by the Nobel Foundation, 1972.
7. F. A. Elder, C. Giese, B. Steiner and M. G. Inghram, J. Chem. Phys. 36, 3292 (1962).
8. G. Herzberg and J. Shoosmith, Can. J. Phys. 34, 523 (1956).
9. F. P. Lossing and G. P. Semeluk, Can. J. Chem. 48, 955 (1970).
10. C. Lifshitz and W. A. Chupka, J. Chem. Phys. 47, 3439 (1967). Chupka & Lifshitz, J. Chem. Phys. 48, 1109 (1968).
11. R. E. Kari and I. G. Csizmadia, J. Chem. Phys. 46, 1817(1967).
12. M. Hackmeyer and J. L. Whitten, unpublished results quoted by M. Krauss, NBS Technical Note 438, "Compendium of ab initio Calculations of Molecular Energies and Properties, Dec. 1967.
13. Yu. L. Sergeev, M. E. Akopyan and F. I. Vilesov, Opt. and Spektr. (Russian) 30, 230 (1971).
14. R. E. Huffman, J. C. Larrabee and Y. Tanaka, J. Chem. Phys. 46, 2213 (1966).
15. B. Edlen, Kgl. Svenska Vetenskapsakad. Handl. 20, 3 (1943).
16. P. M. Dehmer, J. Berkowitz and W. A. Chupka, "Photoionization of Atomic Oxygen from 920 Å to 650 Å", J. Chem. Phys. (accepted for publication).
17. a) M. Horani, J. Rostas and H. Lefebvre-Brion, Can. J. Phys. 45, 3319 (1967).
    b) H. P. D. Liu and G. Verhaegen, Intl. J. Quantum Chem. 5, 103 (1971).
    c) P. E. Cade, Can. J. Phys. 46, 1989 (1968).
18. a) G. Herzberg, "Molecular Spectra and Molecular Structure. I. Spectra of Diatomic Molecules", D. Van Nostrand Co., Inc., Princeton, N.J. (1950).
    b) A. J. Merer, E. K. Achter, M. Horani and J. Rostas, Can. J. Phys. 47, 1723 (1969).
    c) D. Rakotoarijimy, Physica 49, 360 (1970).
19. S. F. Foner and R. L. Hudson, J. Chem. Phys. 25, 602 (1956).
20. V. H. Dibeler, J. A. Walker and H. M. Rosenstock, J. Res. Natl. Bur. Stds. A70, 459 (1966).

PHOTOELECTRON SPECTROSCOPY OF MOLECULAR BEAMS*

J. Berkowitz

Argonne National Laboratory
Argonne, Illinois 60439

I. INTRODUCTION AND CHEMICAL STUDIES

The history of physical science in this century is replete with examples of phenomena initially discovered and investigated by physicists, which have subsequently become tools of the chemists. Visible and infrared spectroscopy, mass spectroscopy, nuclear magnetic resonance and Mössbauer experiments are but a few examples of this trend. I shall try to demonstrate in this paper that the field of photoelectron spectroscopy may develop in a reverse fashion.

Somewhat more than a decade ago, Vilesov, Kurbatov and Terenin (1) reported on the kinetic energy spectrum of electrons observed when monochromatized electromagnetic radiation with energy between 6 and 11.7 eV was incident upon aromatic amines and methyl derivatives of benzene. This early study was limited in its upper energy by the absorption cutoff of LiF windows. About a year later, Turner and Al-Joboury (2) presented the first in a series of papers in which the resonance line of helium (21.217 eV) was used as incident radiation, and analysis of photoelectrons was performed with a concentric cylindrical retarding field.

The introduction of the He(I) line turned out to be highly advantageous, not only because it can be generated as a sharp line with high intensity, but also because the photoionization cross sections of many common gases are near their maximum (3) at this wavelength. In addition, with 21.2 eV one has sufficient energy to eject electrons from the valence orbitals of most compounds. This fortuitous confluence of circumstances strongly influenced the development of molecular photoelectron spectroscopy in the succeeding decade. The energy analyzers have improved during this period, but most studies still emphasize the measurement of ionization potentials with the He(I) line, and interpret their results in terms of the bonding (or antibonding) nature of the orbitals from which the corresponding electrons have been removed. A convenient summary of such data is the handbook by Turner, Baker, Baker and Brundle (4). Other laboratories (5) that have contributed prominently to the measurement of 584 Å photoelectron spectra of volatile molecules include those of Price, Orchard, Heilbronner, Schweig and Schaefer, Bock, Beck, Carlson, Frost and McDowell, Lindholm and others. Today the apparatus for such studies is commercially available and more exotic and complex molecules are continually being added to our store of 584 Å photoelectron spectra.

Some molecules cannot readily be prepared as ambient gases in the steady state, and must be prepared as molecular beams. This class of systems includes free radicals, excited states of molecules, atoms (besides the rare gases and mercury) and in general, materials of low volatility.

The study of free radicals in beams has been pursued in the past few years at Southampton by Jonathan (6) and co-workers, who have detected O, N, H, $O_2{}^1\Delta_g$, CS; at Vancouver by Frost, McDowell (7) and collaborators, who have reported on NF_2, ClO_2, S_2O, O_3 and also CS; and recently at Sussex by Kroto (8) and co-workers.

Relatively involatile systems had been investigated at Argonne by Dr. J. L. Dehmer and the author (9). For this purpose, an apparatus of high collection efficiency was constructed. In initial studies on the thallium halides, it was found that electrons localized on the halogen, which acted as nonbonding lone pairs in the case of the relatively covalent hydrogen halides and methyl halides, were the bonding electrons in the ionic thallium halides. Thus, if the thallium halide in its ground state is closely approximated by the representation Tl^+X^-, then

$$Tl^+X^- + h\nu \rightarrow Tl^+X^o + e$$

occurs when the electron is removed from the halogen-like orbital. If the initial bonding had strong contribution from the coulombic e^2/r potential, which is of the order of ~5 eV, this is drastically reduced in the ion to ~$\alpha e^2/2r^4$, which is of order 0.2 eV. By contrast, an electron removed from a metal-like orbital can be represented by

$$Tl^+X^- + h\nu \rightarrow Tl^{++}X^- + e^-,$$

resulting in an ionic state more strongly bound than the neutral ground state. This behavior has since been verified in the indium halides, and by extensive ab initio calculations as well as semiempirical calculations on the aluminum and gallium halides. Hence, the extension of photoelectron spectroscopy to the study of high temperature species has proved rewarding by providing a dramatic illustration of the different bonding characteristics of ionic molecules (which tend to be involatile) and covalent molecules (which tend to be volatile).

We have recently extended these studies to the very ionic alkali halides. There is, of course, a large transfer of charge occurring in the formation of M^+X^- from M^0 and X^0. The consequence is a filled halogen p-like molecular orbital, while the first metal-like orbital is the next inner p shell (for all cases except lithium). Each of these p-type orbitals can be expected to be split by the cylindrical molecular field into a $p\pi$ and $p\sigma$ orbital, the $p\pi$ being the less bound one on the halogen end, the $p\sigma$ on the metal end since the splitting is caused by potentials of opposite sign. The ionization potential of the halogen-like orbital should be roughly the sum of the electron affinity of the halogen negative ion and its binding energy in the field of the positive ion, or ca. 3 eV + 5 eV $\cong$ 8 eV. As in the thallium, indium and gallium monohalides, we anticipate that removal of an electron from an orbital localized around the halogen will severely disrupt the ionic molecular bond.

The ionization of the metal p-like orbital should occur roughly at the same energy as it does in the free atom, modified by a chemical shift due to charge transfer of the valence s-electron, not unlike the chemical shift observed in ESCA studies. This metal p ionization energy is ~38 eV for sodium, ~25 eV for potassium, ~20-22 eV for rubidium and ~17-19 eV for cesium. Hence, with an ionizing photon energy of 21.2 eV, the only metal p-type orbitals likely to be observed would be in the cesium halides.

The 584 Å photoelectron spectrum of cesium iodide is shown in Fig. 1. We note the two regions of ionic states, the one around 8 eV corresponding to ionization from the iodide 5 p-like orbital, the other around 19 eV corresponding to ionization from the cesium 5p-like orbital. Walker (10), Dehmer and I have examined the splitting of each of these orbitals in some detail. In addition to the aforementioned π-σ splitting, which seems to scale inversely with the ionicity of the molecule, there exists a spin-orbit splitting which has very nearly the same spin-orbit coupling parameter as the corresponding atom. That is, the iodide 5p-like orbital has a spin-orbit coupling parameter close to that of atomic iodine 5p. From ab initio calculations available for 9 of the 20 alkali halides and extrapolations of these results, we have been able to construct a table of π-σ splittings for the outermost, halogen-like orbitals of all of the alkali halides, and also of their spin-orbit splittings. These two types of splitting are of

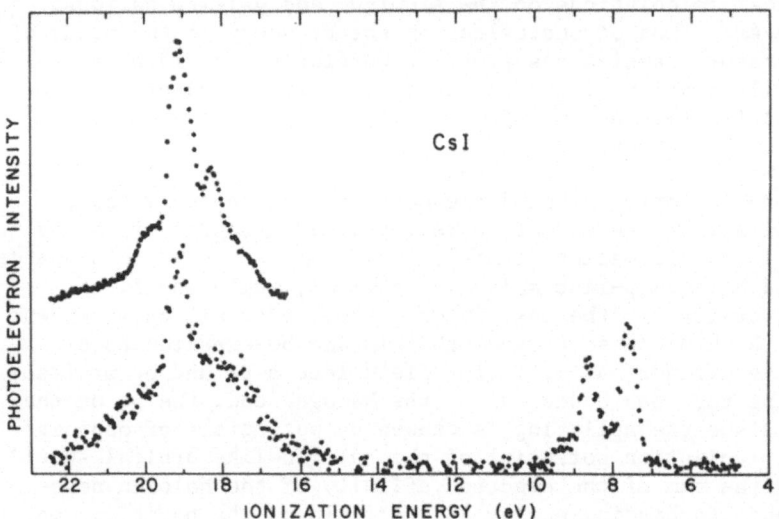

Fig. 1.   The He(I) photoelectron spectrum of CsI.

comparable magnitude, and ultimately it is necessary to construct
and solve a 2 x 2 secular equation to deduce the resulting states.
For example, the iodide 5p-like orbital of cesium iodide has a π-σ
splitting of ~0.12 eV and a spin orbit splitting of 0.63 eV. The
initial splitting diagram is shown in Fig. 2. The $^2\Pi_{1/2}$ and $^2\Sigma_{1/2}$
state, having the same Ω value, then interact, and the magnitude of

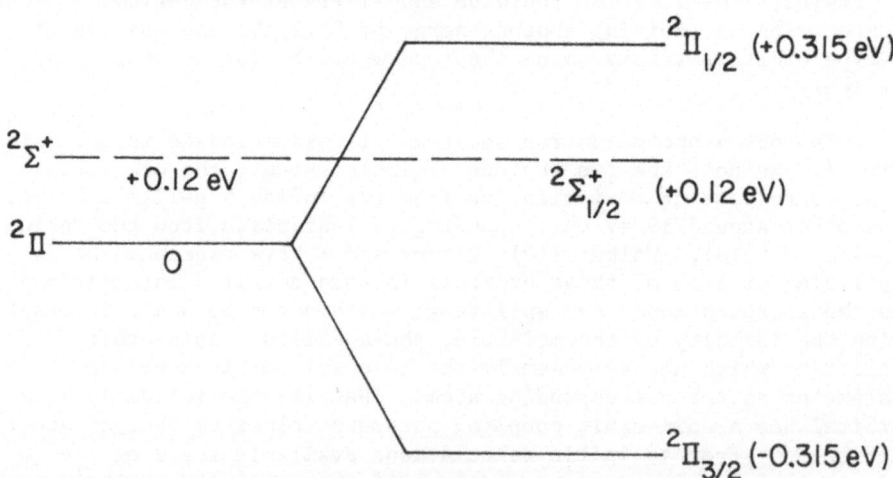

Fig. 2.   Partial splitting diagram of the CsI (iodide orbital)
          ionization, including π-σ and spin-orbit effects.

their repulsion can be obtained from the solution to the secular
equation. The details of this calculation can be found in the
original manuscript (10). The resulting juxtaposition of states
is shown in Fig. 3. The $^2\Pi_{3/2}$ and $^2\Sigma_{1/2}^+$ states are separated by
only ~0.08 eV, while $^2\Pi_{1/2}$ is separated from the mean of the other
states by ~0.95 eV. The experimental data (Fig. 1) reveal two
peaks, the lower one more intense, and separated from the upper
one by about 0.92 eV. This splitting is very nearly that of the
iodine atom, not 2/3 of this value as in HI. This is because the
iodine in CsI is almost spherically symmetric (i.e. the π–σ splitting is small compared to the spin–orbit splitting). The more
ionic alkali halides, which have the smallest π–σ splittings, have
resultant splittings characteristic of individual, spherically
symmetric ions only slightly perturbed by the neighboring ion.

   Taking into account the peak widths, the corresponding calculation for CsBr predicts a partial splitting into two peaks, with
a separation of ~0.4 eV, and for CsCl and CsF a single peak, in
good agreement with the experimental results (10). Predictions
regarding this fine structure for the other alkali halides can be
found in the original article (10).

   A rewarding aspect of photoelectron spectroscopic investigations is that a spectrum of a single molecular system, when finally
understood, enables one to predict the behavior of the entire
homologous series. This is, of course, a consequence of the
molecular orbital structure of matter. One can observe this in the

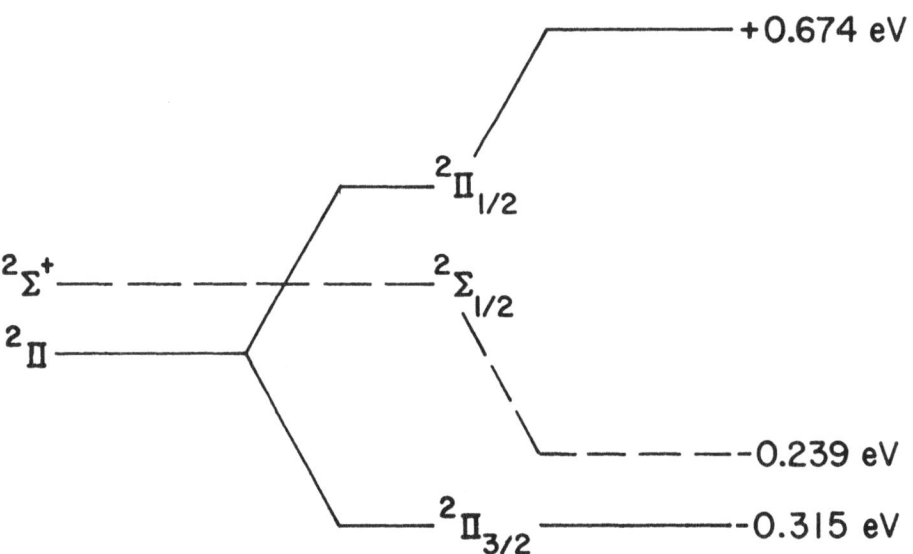

Fig. 3. Final splitting diagram of CsI, including Ω–Ω interaction.

existing spectra of HF, HCl, HBr and HI; the corresponding diatomic halogens; the Group III monohalides of Tl, In, Ga, Al and B; the Group II dihalides, and also the alkali halides. As our knowledge of the ionic states of small molecules increases, it becomes attractive to predict their mass spectra. Table 1 is a composite of parent to fragment (MX^+/M^+) ratios of the diatomic alkali halides, primarily obtained from electron impact data. One can see a strong correlation with the ionicity of the molecule, the most ionic (CsF) having an undetectable parent ion, whereas the least ionic (LiI) has almost an order of magnitude more parent than fragment. We can rationalize these observations from the photoelectron spectroscopic analysis by first noting that 70 eV electron impact ionization will be dominated by ionization from the lower lying halogen-like orbitals. To first order, the magnitude of the $\pi-\sigma$ splitting will determine whether at least one of the fine structure states of this p-like orbital is bound. When the $\pi-\sigma$ splitting is small, both the $^2\Pi$ and $^2\Sigma$ ionic states tend to be repulsive, and hence strongly favor the formation of fragment ions. When the $\pi-\sigma$ splitting is relatively large, as in the cases of LiCl, LiBr and LiI, the $^2\Pi$ state becomes significantly bound, and provides a

TABLE 1

Parent to fragment ratio (MX^+/M^+) in the ionization of alkali halides

	F	Cl	Br	I	Refs.
Li	0.11–0.31	3.39	2.18	7.46	(a)
Na	0.026	0.59	0.71	1.17	(b)
K	0.002	0.17	0.29	0.56	(b)
Rb		0.07			(b)
		0.06	0.155	0.24	(c)
Cs		0.012			(b)
	0.00	0.014	0.07	0.26	(d)
	<.005	0.015	0.028	0.14	(c)

(a) (Electron impact), J. Berkowitz, H. A. Tasman and W. A. Chupka, J. Chem. Phys. 36, 2170 (1962).

(b) (Electron impact), J. Berkowitz and W. A. Chupka, J. Chem. Phys. 29, 653 (1958).

(c) (Photon impact), J. Berkowitz, J. Chem. Phys. 50, 3503 (1969); J. Berkowitz, Adv. High Temp. Chem. 3, 123 (1971).

(d) (Electron impact), P. A. Akishin, L. N. Gorokhov and L. N. Sidorov, Dokl. Akad. Nauk SSSR 135, 113 (1960).

source of parent ion. By including the effects of spin-orbit splitting, it is even possible to rationalize the nonmonotonicity in the fragmentation behavior of LiCl, LiBr and LiI.

So far, we have given examples of other types of molecules which can be examined by 584 Å photoelectron spectroscopy. The information obtained is still primarily of a chemical nature. We have not yet given any evidence to support our thesis that future studies will prove rewarding to atomic and chemical physicists. We shall postpone this demonstration a bit longer to consider briefly the parallel development of photoelectron spectroscopy in the x-ray region that has been pioneered by Prof. Siegbahn and his colleagues at Uppsala (11).

The photoionization cross sections of molecules, which are near their maximum at 584 Å, are typically ~two orders of magnitude smaller in the 8-10 Å region (MgKα, AlKα). Hence, the corresponding photoelectron spectroscopic measurements on free molecules are considerably more difficult at this photon energy. The partial cross sections for ionization from the valence shells decrease even more drastically, and represent a small fraction of the total ionization in the x-ray region. By comparison with the wealth of data accumulated at 584 Å, relatively fewer molecular systems have been studied at 9.9 Å, and most of these studies have been confined to the larger partial cross sections of the inner shells. These inner shells are localized strongly in the vicinity of the atoms that make up a particular molecule. The photoelectron spectra of these inner shells are, to zero[th] order, atomic photoelectron spectra. However, these atomic lines are shifted from their free atom energy values by the <u>chemical</u> environment in which they reside. This chemical shift, typically a few volts, can be measured accurately and has been extensively correlated with the charge distribution within the molecule, and the valence state of particular atoms in the molecule. The only point we wish to make here is that this branch of the photoelectron spectroscopy of molecules has also been strongly oriented toward the determination of chemical information. Indeed, the acronym ESCA (Electron Spectroscopy for <u>Chemical</u> Analysis) is an indication of the primary thrust of this research. X-ray photoelectron spectroscopy has also been extended recently to the study of high temperature species (12), where the combined requirements of working with molecular beams and with low photoionization cross sections makes the experiment very difficult.

II. PHYSICAL STUDIES

The properties that I wish to characterize here as physical ones are intensities and angular distributions, from which one can infer transition probabilities and phase shifts. These quantities are determined by the nature of the continuum wave function, as well as that of the neutral ground state, whereas by and large, chemical properties are discussed in terms of the initial state's molecular orbitals.

In the dipole approximation, which is excellent for low-energy photons, the differential photoionization cross section for un-polarized incident radiation is

$$\frac{d\sigma(\epsilon)}{d\Omega} = \frac{\sigma_t(\epsilon)}{4\pi}\left[1 - \frac{\beta(\epsilon)}{2}\, P_2(\cos\,\theta)\right] \tag{1}$$

where ϵ is the energy of the ejected photoelectron, $\sigma_t(\epsilon)$ is the total photoionization cross section, θ is the angle between the incoming photon direction and the outgoing electron direction, $P_2(\cos\,\theta) = 1/2(3\,\cos^2\theta - 1)$, and $\beta(\epsilon)$ is an asymmetry parameter which, in the approximation of Russell-Saunders coupling and one-electron wave functions can be shown to take the form

$$\beta(\epsilon) = \frac{\ell(\ell-1)\,R_{\ell-1}^2(\epsilon) + (\ell+1)(\ell+2)\,R_{\ell+1}^2(\epsilon) - 6\ell(\ell+1)\,R_{\ell-1}R_{\ell+1}\,\cos\left[\delta_{\ell+1}(\epsilon)-\delta_{\ell-1}(\epsilon)\right]}{(2\ell+1)\left[\ell R_{\ell-1}^2(\epsilon) + (\ell+1)\,R_{\ell+1}^2(\epsilon)\right]} \tag{2}$$

In this latter expression, $R_{\ell\pm1}(\epsilon)$ is the dipole matrix element, defined by

$$R_{\ell\pm1}(\epsilon) = \int_0^\infty P_{n\ell}(r)\, r\, P_{\epsilon,\ell\pm1}(r)\, dr,$$

and $\delta_{\ell\pm1}(\epsilon)$ are the phase shifts of the continuum $\ell \pm 1$ partial waves with respect to free waves.

Even without detailed calculations, these expressions can teach us something about experimental design and the nature of the photoionization process. From eq. (1) we note that when $P_2(\cos\,\theta) = 0$, the differential cross section will be independent of β. This occurs for $\theta = 54^\circ44'$. From the wealth of photoelec-tron spectra currently in the literature, extremely few measure-ments have been made at this angle. Hence, in general, the relative intensities of the various bands, which have been pri-marily measured at $\theta = 90^\circ$, are not meaningful unless one has an independent measurement of β at that energy. However, if one can perform measurements at the $54^\circ44'$ angle, relative intensities can be directly related to calculated intensities without requiring a knowledge of β.

The electron energy analyzer which was previously mentioned as a high transmission device for studies of molecular beams, can conveniently be constructed with θ near the $54°44'$ value. The design parameters of the particular analyzer constructed at Argonne (9) happen to optimize at $\theta = 60°$. In addition, it has no retarding, accelerating or focusing lenses. Taking into account the $\Delta E/E$ effect (13), this analyzer appears to have a flat transmission function for electrons with kinetic energy greater than ~1.5 eV, falling off to lower energy.

I shall describe below three separate experiments which involved accurate relative intensity measurements with this instrument, and show how an interpretation of the results by appropriate theory has given us new insight into the photoionization process.

A. Comparison of photoelectron intensities and Franck–Condon factors in the photoionization of H_2, HD and D_2.

The relevant transitions in the 584 Å photoelectron spectrum of H_2 are

$$H_2(X^1\Sigma_g^+, v'' = o) + h\nu \rightarrow H_2^+(X^2\Sigma_g^+, v' = 0,1,2...) + e^-$$

Designating the initial state as a, and the final state as b, we may formally write the cross section for this process as

$$\sigma(a \rightarrow b) = \frac{8\pi^3 e^2 \nu}{c} \sum_b \sum_\alpha |<b|M_\alpha|a>|^2 \tag{3}$$

where ν is the frequency of incident radiation, e and c have their usual meaning and α is the index for electron coordinates x, y and z. The first summation is over the degenerate final states.

In the Born–Oppenheimer approximation, both initial and final states are written as products of electronic functions ψ, vibrational functions P and rotational functions Y, i.e.

$$|a> = \psi_a(R, \underline{r}_1, \underline{r}_2)\ R^{-1}P_a(v_a, R)Y(J_a, m_a) \tag{4}$$

$$<b| = \psi_b^*(R, \underline{r}_1, \underline{r}_2, E_e)R^{-1}P_b(v_b, R)Y^*(J_b, m_b) \tag{5}$$

where R is the internuclear distance, $\underline{r}_1$ and $\underline{r}_2$ are the coordinates of the two electrons with respect to axes fixed in the molecule, v is the vibrational quantum number and J and M are the rotational quantum number and its projection, and E_e is the electron kinetic energy. If for the moment we assume the rotational functions to be uncoupled and integrate over the rotational coordinates, we can write the transition matrix element as follows:

$$<b|M_{\alpha}|a> = \int_0^{\infty} P_b^* M_{\alpha} P_a dr \tag{6}$$

where

$$M_{\alpha} = \iint \psi_b^* (r_{1\alpha} + r_{2\alpha}) \psi_a \underline{dr_1} \underline{dr_2} \tag{7}$$

Since ψ_b and ψ_a are functions not only of the electron coordinates but also of the internuclear distance R, it follows that M_{α} will also be a function of R. In addition, ψ_b is a function of E_e, and hence M_{α} must also be dependent on E_e. In calculating Franck-Condon factors and identifying them with vibrational intensities, one is explicitly ignoring this dependence, and is writing eq. 6 in the form

$$<b|M_{\alpha}|a> = M_{\alpha}^{(o)} \int_0^{\infty} P_b^* P_a dR \tag{8}$$

where the integral corresponds to the Franck-Condon factor, and $M_{\alpha}^{(o)}$ is the value of M_{α} at a typical R - customarily chosen to be the equilibrium distance in the electronic ground state a. Flannery and Öpik (14) have shown how M_{α} varies with R and E_e in the ionization of H_2, choosing one reasonable model for ψ_b and ψ_a. Itikawa (15) has expanded both upon this model and upon a more sophisticated one for ψ_b. In each case, he has expressed the resulting M_{α} function as a power series

$$M_{\alpha} = \gamma + \delta (R - R_e) + \varepsilon (R - R_e)^2$$

(in which γ, δ and ε are constant), inserted it into eq. 6, and calculated the vibrational intensities.

Experimentally, we (16) have carefully integrated the areas under the vibrational bands obtained in the 584 Å photoelectron spectrum for H_2, HD and D_2. The experiments were performed with a resolution width of 20 meV, where rotational structure was beginning to influence the peak shape. The relative intensity was taken as the area divided by the corresponding electron energy. They were plotted as ratios of the corresponding Franck-Condon factors vs. vibrational number, where the Franck-Condon factors are very precise calculations by Peek (17). The results for D_2 are shown in Fig. 4, together with data by Frost et al. (18) and Itikawa's calculations. If the approximations involved in eq. (8) were valid, we would expect perfect agreement with Franck-Condon factors, in which case the data points should describe a straight horizontal line. Clearly, the data points have a positive slope, indicating a substantial deviation from purely Franck-Condon behavior. The agreement with Itikawa's case (C) is particularly

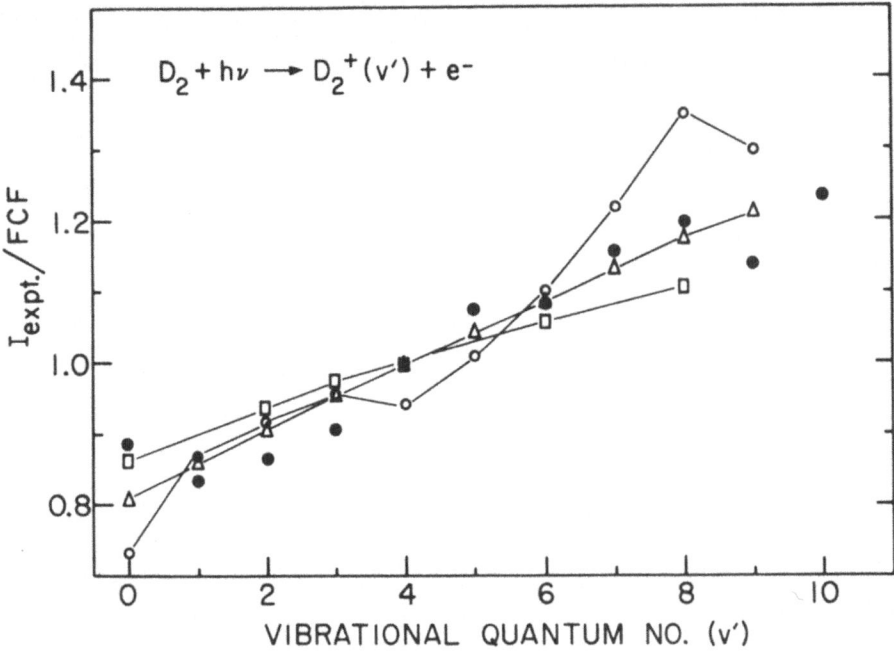

Fig. 4. The ratio of the experimental (or calculated) vibrational
         intensity to the Franck-Condon factor for D_2, plotted as
         a function of the vibrational quantum number v'.
         O O Frost, McDowell and Vroom;   □ □ Itikawa, case (B);
         △ △ Itikawa, case (C); and   ● ● Berkowitz and Spohr.

good, indicating that this particular choice of continuum function
is a reasonable one, and also providing support for the view that
the essence of the physics has been included in the calculations,
despite several approximations in Itikawa's calculations.

B. Rotational band shapes in the photoelectron spectra of HF
    and DF

    It has long been realized that the peak shapes observed in
photoelectron spectroscopy contain useful information. When individual vibrational levels are not resolved, it is assumed that a
broad peak corresponds to a considerable change of geometry on
ionization, while a narrow peak suggests that the ion has a similar
geometry to that of the neutral molecule. However, much less work
has been concerned with the effects of rotational broadening which
may be investigated when individual vibrational levels of the ion
are resolved.

The photoionization process in HF corresponding to

$$HF(^1\Sigma^+, \text{ v''} = o) + h\nu \rightarrow HF^+(^2\Pi_{1/2}, \text{ } ^2\Pi_{3/2}, \text{ v'} = 0, 1, 2, \ldots) + e^-$$

is of particular interest in this context, because the $^2\Pi_{1/2}$ and $^2\Pi_{3/2}$ states are incompletely resolved, and the shape of an individual vibrational band is determined not only by the nature of the rotational transitions, but also by the spin-orbit coupling. By considering the hamiltonian of the entire system, calculating the rotational line strength of each rotational transition and using an experimental line width derived from a nearby atomic line obtained in the same experiment, a theoretical band contour can be constructed. In this instance, the theoretical band contour was calculated for several assumed values of the spin-orbit coupling constant. The best fit enabled us to extract a rather accurate value of the spin-orbit coupling constant from the band contour.

We shall only sketch the outlines of the calculation here. For details see ref. (19).

If we only consider ionization of the outermost $(\pi)^4$ system of HF into the s partial wave, the final state of the system ion plus electron has the configuration $\pi^3\sigma$, which gives rise to $^3\Pi$ and $^1\Pi$ states. These two states are degenerate, i.e. the triplet and singlet Rydberg series approach the same ionic limit. The $^3\Pi$ state consists of three components with Ω = 2, 1 and 0. The $^1\Pi$ state has Ω = 1. In the nonrotating molecule, electric dipole transitions from the ground state of the neutral molecule ($^1\Sigma^+$) are only allowed to the $^1\Pi_1$ state. However, spin-orbit coupling mixes the $^1\Pi_1$ and $^3\Pi_1$ states, and the rotation of the molecule couples the three components of the $^3\Pi$ state by means of spin-uncoupling (20). Hence, transitions become allowed to all four final states. The hamiltonian describing the interactions is of the form

$$H = B[(J_x - L_x - S_x)^2 + (J_y - L_y - S_y)^2] + H^{SO} = H^R + H^{SO} \qquad (9)$$

where J is the total angular momentum, L and S are the electronic orbital and spin angular momenta, and H^{SO} is the spin-orbit contribution. This is written in matrix form in Table 2.

This matrix is then diagonalized, and the rotational line strengths are calculated by using the Hönl-London factors for a $^1\Sigma \rightarrow {}^1\Pi$ transition (21), which are

R branch intensity $\propto$ (J + 2)

Q branch intensity $\propto$ (2J + 1)

P branch intensity $\propto$ (J - 1)

TABLE 2

Matrix elements of $H = H^R + H^{SO}$ for $^3\Pi$ and $^1\Pi$ states in Hund's case (a)[1]

	$^3\Pi_2$	$^3\Pi_1$	$^3\Pi_0$	$^1\Pi_1$
$^3\Pi_2$	$A/2 + B[J^2+J-4]$	$B\sqrt{2(J+2)(J-1)}$	0	0
$^3\Pi_1$	$B\sqrt{2(J+2)(J-1)}$	$BJ(J+1)$	$B\sqrt{2J(J+1)}$	$A/2$
$^3\Pi_0$	0	$B\sqrt{2J(+1)}$	$-A/2 + BJ(J+1)$	0
$^1\Pi_1$	0	$A/2$	0	$B[J^2+J-2]$

[1]A is the spin-orbit coupling for the ion $^2\Pi$ state.

The vibrational band contour is then constructed from the sum of rotational lines, each one broadened by an experimental line width and weighted with the appropriate intensity and Boltzmann factor. The only variable parameters in this approach are the spin-orbit coupling constant and the ionization potential, which we define as the difference in energy between the J = 0 rotational level of the molecule and the J = 3/2 rotational level of the $^2\Pi_{3/2}$ state of the ion. Only the spin-orbit coupling constant affects the band shape.

Figure 5 shows the experimental results for DF$^+$, v' = 0, together with the Ar$^+$ $^2P_{1/2}$ line and the calculated contour that gives the best fit. From this fitting procedure, we have deduced a spin-orbit coupling constant of 290 ± 10 cm^{-1}. This is significantly different from the value 240 cm^{-1} that had been deduced by measuring the partially resolved splitting in the photoelectron spectrum. It is also in line with the ratio of doublet splittings in HX$^+$ and the corresponding X, given in Table 3. For the determination of ionization potentials, rapid scans are preferable since they minimize the likelihood of any change of surface potential occurring between the argon peaks (used for calibration) and the HF$^+$ or DF$^+$ peaks. The values of the ionization potentials obtained from the best fit between calculation and experiment are

HF: 16.044 ± 0.003 eV

DF: 16.058 ± 0.003 eV

These errors are an order of magnitude smaller than the experimental line width. After correcting for molecular and ionic zero point energies, there is a residual discrepancy which may be due to an electronic isotope effect (19, 22).

The ionization potentials given above are slightly different from those obtained previously by photoionization mass spectrometry (23). This is due to the difficulty of choosing a proper

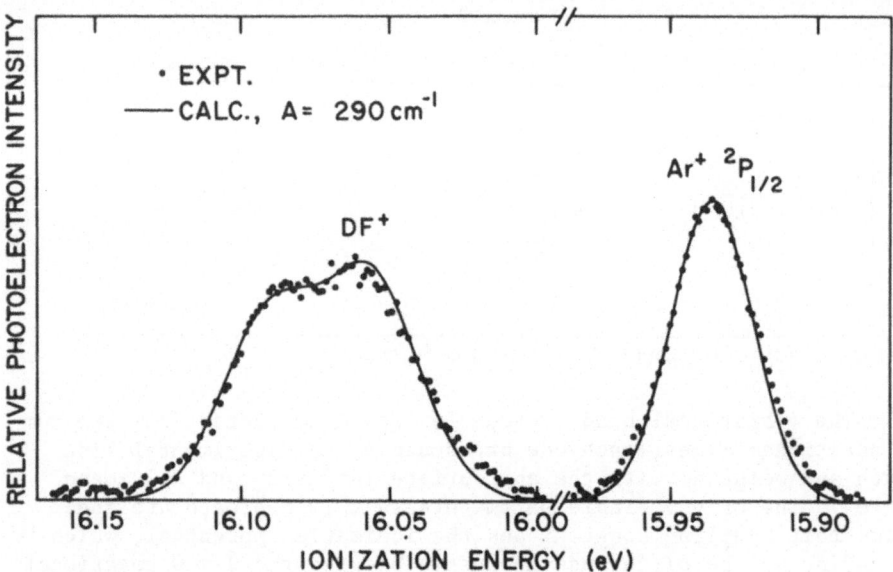

Fig. 5.   Slow scan spectrum of DF⁺, $X^2\Pi_{3/2,1/2}$, v' = 0. Solid
line: calculated contour based on A = 290 cm⁻¹.
•   experimental points.

TABLE 3

Ratio of molecular to atomic spin-orbit splitting

Ion	$^2\Pi_{1/2} - ^2\Pi_{3/2}$, cm⁻¹	$^2P_{1/2} - ^2P_{3/2}$, cm⁻¹	Ratio
HI⁺	5650	7603	0.743
HBr⁺	2653	3685	0.720
HCl⁺	643	881	0.730
HF⁺	290	404	0.72

threshold function for the mass spectrometric determination, as
well as the less detailed treatment of rotation. The results given
here must be considered more precise.

Although only the s wave has been considered in the theory
given here, we have reason to believe (19) that the inclusion of
the dominant d wave will not seriously affect our conclusions. The
d wave will allow larger angular momentum transfers to take place

than for an s wave and transitions become allowed from HF (J) to
HF$^+$ (J ± 1/2, J ± 3/2, J ± 5/2). The effect of these extra
transitions will be to broaden the overall peaks with respect to
that calculated for s waves only, and to reduce any dip between
the two halves of the doublet.

It should be noted that this analysis has presupposed an experiment conducted at θ = 54°44'. If the experiment were conducted
at a very different θ, such as 90°, the angular distribution of
each rotational transition at the energy of the experiment would
have to be known.

C. Spin-orbit coupling and the ratio of photoionization cross
    sections in some atoms.

One of the earliest experiments reported in photoelectron
spectroscopy corresponded to the reaction

$$
\left.
\begin{array}{c}
Ar \\[2mm]
Kr \\[2mm]
Xe
\end{array}
\right\} + h\nu \rightarrow
\begin{array}{l}
Ar^+(^2P_{3/2}, {}^2P_{1/2}) + e^- \\[2mm]
Kr^+(^2P_{3/2}, {}^2P_{1/2}) + e^- \\[2mm]
Xe^+(^2P_{3/2}, {}^2P_{1/2}) + e^-
\end{array}
$$

When measurements become a bit more refined, it was found that the
ratios of intensities of the $^2P_{3/2}$ to $^2P_{1/2}$ were less than 2:1
(their ratio of statistical weights) in every instance. In 1968,
Samson and Cairns (24) reported this ratio to be 1.98, 1.79 and
1.60 for Ar, Kr and Xe, respectively. We have recently remeasured
these ratios with our cylindrical analyzer, and find fair agreement.

In addition, we (25) have measured the ratio of spin-orbit
split states formed in the ionization of the set Zn, Cd and Hg.

Thus

$$
\begin{array}{ll}
Zn \ldots (3d)^{10}(4s)^2 & \\
Cd \ldots (4d)^{10}(5s)^2 \quad + h\nu \rightarrow & \\
Hg \ldots (5d)^{10}(6s)^2 &
\end{array}
\qquad
\begin{array}{l}
Zn^+(^2D_{5/2}, {}^2D_{3/2}) + e^- \\
Cd^+(^2D_{5/2}, {}^2D_{3/2}) + e^- \\
Hg^+(^2D_{5/2}, {}^2D_{3/2}) + e^-
\end{array}
$$

The ratio of statistical weights in this case should be 1.50.
For Zn$^+$ it appears to be close to this value, but for Cd$^+$ it is
1.79, and for Hg, 2.18, i.e. _larger_ than statistical.

In an attempt to rationalize these observations,
Dr. Walker (25) has recently performed some calculations using
Dirac-Slater relativistic wave functions for both the bound and
continuum states. From these calculations has emerged a general

understanding of branching ratios and their departure from statistical weight predictions, which is a generalization of Cooper's
calculations (26). Cooper had previously noted that, although the
dipole selection rules permitted $\Delta \ell = \pm 1$, the $\ell + 1$ channel
usually dominates the cross section.

At very low photoelectron energies, the first major maximum
of the continuum orbital of the $\ell + 1$ channel will lie outside the
region of space occupied by the bound orbital. As the photoelectron energy increases, this maximum moves nearer the nucleus, and
the dipole matrix element increases. When the photoelectron
energy increases still further, the first major maximum of the
continuum orbital begins to overlap the nodes of the bound orbital
(if any) and the dipole matrix element decreases and finally
changes sign, giving rise to the Cooper minimum in the partial
cross section.

Now consider what happens when spin-orbit coupling is introduced. This causes the $j = \ell - 1/2$ component of the bound orbital,
for which the spin-orbit potential is attractive, to be drawn
slightly closer to the nucleus than the $j = \ell + 1/2$. As a result,
when the photoelectron energy increases from zero, the continuum
orbital will have a greater overlap with the $j = \ell + 1/2$ component
of the bound state, and the ratio $\sigma(\ell+1/2)/\sigma(\ell-1/2)$ will be greater
than the statistical value of $(\ell+1)/\ell$. Similarly, the continuum
orbital will overlap the first node of the $j = \ell + 1/2$ before that
of the $j = \ell - 1/2$, and in this part of the spectrum $\sigma(\ell+1/2)/$
$\sigma(\ell-1/2) < (\ell+1)/\ell$. Figure 6 illustrates this. The same considerations hold after the Cooper minimum is passed. Thus we can make
the generalization that if the partial cross section is rising, the
ratio of cross sections is greater than statistical, while if the
partial cross section is falling, the ratio will be less than
statistical. For the rare gases Ar, Kr and Xe, the cross section
is falling at 21.2 eV, while for Zn, Cd and Hg it is rising.

III. SUMMARY AND CONCLUSIONS

There are other physical aspects of photoelectron spectroscopy,
such as the measurement of the angular distribution asymmetry
parameter β as a function of photoelectron energy, and the determination of branching ratios and β's across autoionizing peaks for
which there is neither time nor space at this conference. However,
I hope that I have given enough examples to support the view that
the branching ratio of photoelectron spectroscopy between chemistry
and physics will move more strongly to the latter in the next few
years.

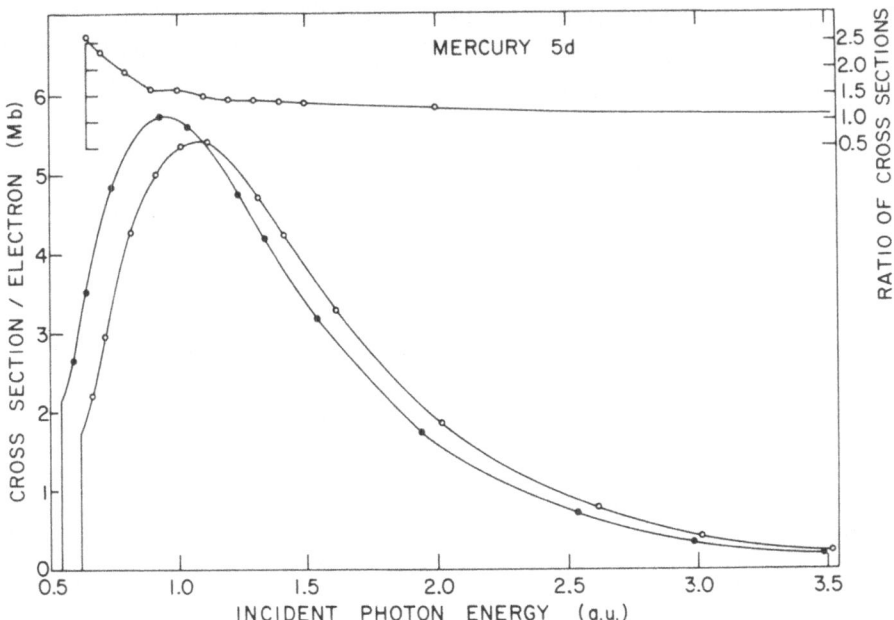

Fig. 6. Cross section per electron for photoionization of the $5d_{5/2,3/2}$ orbital of mercury as a function of incident photon energy. $5d_{3/2}$ – open circles and $5d_{5/2}$ – closed circles. The topmost curve shows $\sigma_{5/2}/\sigma_{3/2}$ as a function of incident photon energy.

REFERENCES

*Work performed under the auspices of the U.S. Atomic Energy Commission.

1. F. I. Vilesov, B. L. Kurbatov and A. N. Terenin, Dokl. Akad. Nauk SSSR 138, 1329 (1961); Soviet Physics Doklady 6, 490 (1961).

2. D. W. Turner and M. I. Al Jobory, J. Chem. Phys. 37, 3007 (1962).

3. J. Berkowitz, to be published.

4. D. W. Turner, C. Baker, A. D. Baker and C. R. Brundle, "Molecular Photoelectron Spectroscopy," Wiley-Interscience, London (1970).

5. As an aid to the reader in locating some of the voluminous
    literature on photoelectron spectroscopy of molecules at
    584 Å, the journals in which recent contributions by the
    aforementioned authors appear are listed below:

    W. Beck, Z. Naturf.; H. Bock, A. Schweig, W. Schaeffer,
    Angew. Chem. Int. Ed.; T. A. Carlson, J. Collin,
    J. Elec. Spectr.; D. C. Frost and C. A. McDowell, Chem.
    Phys. Letters; E. Heilbronner, Helv. Chim. Acta;
    D. R. Loyd, A. F. Orchard, J. Chem. Soc. Far. Trans. 2;
    E. Lindholm, Physca Scripta; W. C. Price, Proc. Roy.
    Soc. and J. Phys. B.

6. N. Jonathan, A. Morris, D. J. Smith and K. J. Ross, Chem.
    Phys. Letters 7, 497 (1970); N. Jonathan, D. J. Smith and
    K. J. Ross, J. Chem. Phys. 53, 3758 (1970); N. Jonathan,
    A. Morris, K. J. Ross and D. J. Smith, J. Chem. Phys. 54,
    4954 (1971); L. Golob, N. Jonathan, A. Morris, M. Okuda and
    K. J. Ross, J. Elec. Spec. 1, 506 (1973).

7. D. C. Frost, S. T. Lee and C. A. McDowell, Chem. Phys.
    Letters 17, 153 (1972); A. B. Cornford, D. C. Frost,
    F. G. Herring and C. A. McDowell, J. Chem. Phys. 54, 1872
    (1971); D. C. Frost, S. T. Lee and C. A. McDowell,
    "Ionization Potentials of Short-lived Species," presented at
    21st Annual Conf. on Mass Spectrometry and Allied Topics,
    San Francisco, Calif., May 20-25, 1973.

8. H. W. Kroto and R. J. Suffolk, Chem. Phys. Letters 17, 213
    (1972).

9. J. Berkowitz, J. Chem. Phys. 56, 2766 (1972); J. Berkowitz
    and J. L. Dehmer, J. Chem. Phys. 57, 3194 (1972).

10. J. Berkowitz, J. L. Dehmer and T. E. H. Walker, "PES of High
    Temperature Vapors IV. The Cesium Halides," J. Chem. Phys.
    (accepted for publication).

11. See, for example, "ESCA - Atomic, Molecular and Solid State
    Structure Studied by Means of Electron Spectroscopy,"
    K. Siegbahn and colleagues, Almquist and Wiksells,
    publishers, Uppsala (1967); "ESCA Applied to Free Molecules,"
    K. Siegbahn and colleagues, N. Holland Publ. Co., Amsterdam
    (1969).

12. Y. S. Khodeev, H. Siegbahn, K. Hamrin, K. Siegbahn, Chem.
    Phys. Lett. 19, 16 (1973).

13. J. Berkowitz and P. M. Guyon, Int. J. Mass Spec. Ion Phys. 6,
    301 (1971).

14. M. R. Flannery and U. Öpik, Proc. Phys. Soc. (London) 86, 491
    (1965).

15. Y. Itikawa, J. Electr. Spectr. and Related Phen. 2, 125 (1973)

16. J. Berkowitz and R. Spohr, J. Electr. Spectr. and Related Phen. 2, 143 (1973).

17. J. M. Peek, private communication.

18. D. C. Frost, C. A. McDowell and D. A. Vroom, Proc. Roy. Soc. (London) A296, 566 (1967).

19. T. E. H. Walker, P. M. Dehmer and J. Berkowitz, "Rotational Band Shapes in Photoelectron Spectroscopy: HF and DF," J. Chem. Phys. (submitted for publication).

20. G. Herzberg, "Molecular Spectra and Molecular Structure. II. Spectra of Diatomic Molecules," D. Van Nostrand, Princeton, N.J. (1950).

21. J. T. Hougen, "The Calculation of Rotational Energy Levels and Rotational Line Intensities in Diatomic Molecules," NBS Monograph 115 (1970).

22. P. R. Bunker, J. Mol. Spec. 28, 422 (1968).

23. J. Berkowitz, W. A. Chupka, P. M. Guyon, J. M. Holloway and R. Spohr, J. Chem. Phys. 54, 5165 (1971).

24. J. A. R. Samson and R. B. Cairns, Phys. Rev. 173, 80 (1968).

25. T. E. H. Walker, J. Berkowitz, J. L. Dehmer and J. T. Waber, "Ratios of Photoionization Cross Sections of Spin-Orbit Split States," Phys. Rev. Lett. 31, 678 (1973).

26. J. W. Cooper, Phys. Rev. 128, 681 (1962).

HIGH RESOLUTION VACUUM ULTRAVIOLET SPECTROSCOPY OF DIATOMIC MOLECULES

A.E. Douglas

Division of Physics
National Research Council, Ottawa

In this short discussion of the high resolution spectroscopy of diatomic molecules in the vacuum ultraviolet, it is impossible to give a review of the instrumentation and the state of knowledge of the spectra. This paper is therefore limited to a discussion of some of the experimental techniques and some of the molecules which are closely associated with the research programs carried out in Ottawa.

INSTRUMENTATION

A very satisfactory review of instrumentation in the vacuum ultraviolet region was written in 1967 by J.A.R. Samson and no attempt will be made to review the many important aspects of the subject covered in that book. This paper deals only with a few special aspects of the subject. Even with this limited objective there are some difficulties in giving a coherent paper since the vacuum ultraviolet is not a single spectral region from the point of view of an instrument maker. The difference between spectroscopy at 1800 and 2800 Å is small. At 1800 Å, oxygen must be removed from the light path but the prisms, lenses, windows, mirrors, light sources and spectrometers are very similar to those used at 2800 Å. The difference between spectroscopy at 800 Å and 1800 Å is large. At 800 Å there is no material useful for prisms, lenses and windows; the reflectivity of all mirrors is poor and sources and spectrometers have to be modified to overcome these difficulties. Below 500 Å where helium is no longer transparent the techniques again have to be modified substantially.

In the spectroscopy laboratory of the Division of Physics we have been concerned with the measurement of the wavelengths of well resolved features of molecular spectra which lie between 600

and 2000 Å. Intensity measurements have played a small role in
our work. Much of the research has involved the spectra of short
lived species. In almost all of this work, photographic recording
of spectra has proved superior to all the various electronic
methods. The following discussion will consider only some aspects
of this limited range of spectroscopy.

Spectrographs

The words "Vacuum Ultraviolet Spectroscopy" in the title
might lead one to think that the vacuum is the central element of
the problem. Although in the early history of this subject the
vacuum presented many difficulties, at present vacuum techniques
are sufficiently well developed to reduce this part of the problem
to a routine engineering operation. All the necessary techniques
are described in a number of books (Van Atta 1965) and will not
be discussed here. Although the need for a vacuum does not add
appreciably to the scientific problems of working at wavelengths
shorter than 2000 Å, it does add considerably to the cost.

Many types of spectrometers have been devised for use in the
vacuum ultraviolet, but for high resolution work down to 600 Å
the concave grating spectrograph, usually in an Eagle mounting,
still remains the most useful instrument. Czerny-Turner and Ebert
type instruments have been used in the region near 2000 Å but the
difficulty in maintaining the high reflectivity of the mirrors
makes them less useful at shorter wavelengths. Below 1200 Å,
grazing incidence concave-grating instruments were used for many
years but the difficulty in maintaining the required accuracy of
alignment of these spectrographs combined with the increased reflectivity of normal incidence gratings have more recently given
the Eagle type spectrograph a substantial place in the spectroscopy
of this region. All work in Ottawa has been carried out on either
a three or a ten metre Eagle spectrograph.

The heart of an Eagle type spectrograph is the concave
grating. Very little information is available on the quality of
gratings measured in terms of the requirement for 600 Å radiation.
It is obvious that required precision of ruling and of the figure
of the blank are ten times more stringent for 600 Å radiation than
for 6000 Å and with present techniques gratings fall short of the
perfection required for the shorter wavelength. It is also clear
however that only a few spectroscopists have pushed existing
gratings to the limit of their ability. Much photographic work is
done using first order spectra of moderate size spectrographs and
the dispersion of the spectrum is insufficient to allow high
resolution to be achieved. For example, a resolving power of
100,000 in the first order at 800 Å, with a 1200 lines/mm grating
in a three metre spectrograph, requires a photographic emulsion
which can resolve 300 lines per mm. Normal commercial fast plates
for the vacuum ultraviolet have a resolution of only 40 - 50 lines
per mm. High resolution spectroscopy therefore requires a large

spectrograph and, since high quality very finely ruled gratings cannot be obtained, it requires the use of the higher orders of coarse gratings.

The spectrograph of highest resolution in the Ottawa laboratory is the 10.6 metre Eagle spectrograph (Douglas and Potter, 1962) for which a number of gratings are available. The four gratings most commonly used are shown in Table 1. The four gratings are arranged in two sets each consisting of two gratings mounted back-to-back so they can be interchanged without breaking the vacuum or disturbing the source. Generally a grating blazed at 1200 Å is used for preliminary work and once suitable conditions have been established in the source the grating with a higher blaze angle is used to photograph the spectrum at high dispersion.

Table 1. Characteristics of gratings used with the 10.6 metre concave grating spectrograph

	blaze Å	coating	lines/mm	disp. Å/mm	region
Set #1	1200	Al/MgF$_2$	1200	0.78	1600 Å
	11000	Al/MgF$_2$	600	0.21	
Set #2	1200	Pt	1200	0.78	800 Å
	3000	Pt	1200	0.19	

The use of a grating in an order higher than the first usually requires some type of predisperser to separate the orders. In the region above 1200 Å we have used a LiF foreprism for this purpose. The prism, usually with a cylinder ground on one face so that it also serves to focus an image of the source on the slit, is mounted about 10 cm in front of the slit. A table, which carries the source, rotates about a point directly under the prism so that the source, the prism-lens combination and the entrance slit of the spectrograph form a crude spectrometer which allows a limit range of wavelengths to enter the spectrograph. This simple arrangement has made it possible to use the high orders of the grating in the 1200 - 2000 Å range and over the past dozen years most of the spectra in this region have been photographed at dispersions between 0.15 and 0.30 Å/mm.

Below 1200 Å this foreprism arrangement is no longer useful since the transmission of LiF is insufficient. While it is true that thin clean LiF windows will transmit radiation well below 1100 Å, we have found that with the thick lenses and prisms which we use it is difficult to work much below 1200 Å.

At wavelengths less than 1200 Å we have used the grazing incidence foregrating arrangement described by Douglas and Herzberg (1957) to separate the orders for large spectrographs and

more recently we have devised a normal incidence foregrating
mounted in such a way that the astigmatism of the foregrating and
grating in the spectrograph cancel. Each of these foregratings
has proved useful but each has its limitations. In the grazing
foregrating arrangement, the distance between the grating and the
source is small and since the slit between them is quite wide,
only clean sources such as the helium or argon continuum can be
used. In the normal incidence arrangement, the source is much
further from the grating and it is possible to protect the grating
by low pressure helium in the light path. The reflectivity of the
normal incidence foregrating is low, however, and much light is
lost.

Over most of the range between 600 and 2000 Å no careful
measurement of the resolving power of the spectrographs has been
made. It is estimated from rough measurements that the resolving
power decreases from 250,000 at 2000 Å to 150,000 at 800 Å. With
most of the photographic plates now in use, it will not be
possible to increase the resolving power significantly without
first increasing the dispersion.

For work below 1200 Å, where no windows are possible,
differential pumping through the slits is often necessary. This
adds to the cost of the spectrograph but presents no technical
problems. Fortunately, in the region above 600 Å which is still
the region of greatest interest to molecular spectroscopists,
helium gas is transparent. Usually the source or the gas under
investigation in an absorption cell can be separated from the
spectrograph by a stream of helium flowing away from the slit.
This helium protects the slit from corrosive emissions from the
source and also reduces the demands on the differential pumping
system since only helium enters the spectrograph through the slit.

It appears unlikely that the resolving power of spectrographs
will improve greatly in the near future. Holographic gratings
(Namioka, Noda and Seya, 1973) have been proposed as a means of
obtaining the necessary precision but they are unlikely to have
much impact on the vacuum region in the near future. A large
ruled grating with 2400 or 3600 lines/mm is required but the
problem of ruling half a million carefully shaped grooves with the
required accuracy is a formidable one. Perhaps there is more hope
in meeting the problem of the low reflectivity of grating in the
region below 1200 Å. Much has been done to improve the reflectivity of gratings during the past fifteen years and it appears
that further improvements can be achieved.

Sources
     In dealing with molecular emission and absorption spectra in
the visible and near ultraviolet, many problems arise in finding
suitable light sources. When dealing with short lived species,
there is the problem of devising suitable intense pulsed sources

and finding means of generating a high concentration of the
unstable species under conditions suitable for spectroscopic
observation. In the vacuum region all of these difficulties are
increased. Although a few diatomic molecules such as CO, N_2 and
H_2 give very strong emission bands in the vacuum region, most
species either give feeble emissions or none at all. In absorption
studies of short lived species, we often find that the desired
species can only be produced in the presence of a large amount of
some other species which has strong continuous absorptions in the
vacuum region. Since the wavenumber dispersion in the vacuum
region is poor even under the best of conditions, useful spectra
have to be obtained with very narrow slits and the exposure times
are usually long. Combined with these difficulties there are
numerous minor problems, such as stray light and deposits on the
windows, all of which are more difficult in the vacuum region than
at longer wavelengths.

    The easiest type of spectra to obtain are the absorption
spectra of stable gases where the light source can consist of
almost any source of continuous radiation. The H_2 continuum has
been known for many years and the continua generated by discharges
through He, Ar, Kr and Xe have been described by Wilkinson and by
the group at the Air Force Cambridge Research Laboratories
(Huffman, Larrabee and Tanaka 1965). In our work with the high
resolution spectrograph, we have found the He and Ar continua very
useful since each covers a sufficiently short wavelength range
that they can be used in the third or fourth orders of a grating
without difficulty from overlapping orders. The Kr and Xe discharges excited in a microwave cavity have proved to be rather
weak for most purposes. In the region between 2000 and 1650 Å we
have found the continuum from commercial high pressure Xe arcs to
be very satisfactory. Throughout the whole region above 900 Å the
Lyman type lamp has been the most widely used source in obtaining
the absorption spectra of both short lived and stable molecules.

    Lyman sources, by which I mean any source in which the
continuum is generated by the discharge of a capacitor in such a
way that a high voltage, high current density discharge passes
through a low pressure gas, have been made in many different
configurations, each having its own particular advantages. When
faced with the long exposures required for a large spectrograph
an important characteristic is the amount of light per flash. The
maximum light per discharge still appears to come from a source
not greatly different from that devised by Lyman. We have usually
employed a source consisting of a ceramic tube about 4 cm long
through which argon at a pressure of 80 torr is slowly pumped.
When the tube is new it has an inner diameter of about 2 mm and an
outer diameter of 14 mm. It is fastened in the middle of a piece
of pyrex glass tubing and co-axial electrodes are mounted on the
ends of the glass tube. A previously unused tube is excited by a
2 uf, 6000 volt discharge and, as the tube diameter increases due

to the discharge, the voltage is increased up to ∿ 12000 volts at the end of the useful life of the tube. The tubes have a life of about 40,000 flashes before the bore becomes too great to be useful. This simple, cheap and old-fashioned source has proved to be the most useful source of continuum in our laboratory.

A high proportion of the short lived species investigated in the laboratory have been produced in either a flash photolysis apparatus (Ramsay 1952) or a flash discharge apparatus (Balfour and Douglas 1968) and this equipment has been used with minor modifications in the vacuum ultraviolet. Although the flash photolysis apparatus has yielded very significant results when used with the three metre spectrograph, even at this dispersion the number of flashes necessary to obtain an adequate exposure is inconveniently large at shorter wavelengths. It has not been possible to use the flash photolysis apparatus effectively with the 10.6 metre spectrograph where longer exposures are required. With the flash discharge apparatus the power input is low and it is possible to discharge the capacitors once or twice a second. With this apparatus the 10.6 metre spectrograph can be used to advantage.

In spite of the large number of different sources of continuous radiation, there is still a need for new sources with a variety of desirable characteristics. Among the newer sources likely to play a role in the spectroscopy of the future, electron storage rings, rare gas emissions and laser sparks seem likely to be important. Synchrotron radiation from electron storage rings designed to give the optimum emission at chosen wavelengths will have considerable impact on spectroscopy in the $500 - 1000$ Å range and will completely revolutionize work below 500 Å. Emission from high pressure Xe and Kr which has been subjected to an intense pulse of high energy electrons also appears promising for the near future. Both of these sources will be described in detail in papers given at this meeting. Finally it is my opinion that the potential of laser induced plasmas has not yet been fully investigated. With the newly developed CO_2-TEA lasers it is quite easy to obtain repeated production of hot plasmas under a wide variety of different conditions and some of these plasmas may be useful as spectroscopic sources.

SPECTRA

In this section I will deal with a few simple principles and a few spectra we have observed in the vacuum ultraviolet, many of which are far from simple. In particular I will attempt to emphasize a few facts learned from the analysis of the spectra of diatomic molecules which may be of some significance in dealing with larger molecules for which little detailed information is available.

Spectra of Valence States

Although there is a tendency to think in terms of Rydberg states when considering the vacuum ultraviolet, it must be remembered that many valence states lie up to and beyond the first ionization potential. A few examples of work now in progress involving valence states may be of interest. The following examples are all rather unusual in that emissions from valence states in the vacuum ultraviolet are used to determine ground state constants.

In 1962 Rao and Venkateswarlu found that, in a discharge through Cl_2, a line of atomic chlorine at 73983 cm^{-1} is excited and that this line is absorbed by molecular chlorine in the same discharge tube. In this way a single rotational level of the upper state of Cl_2 is excited and this level then emits giving a long series of doublets (one P and one R line) involving many vibrational levels of the ground state. Unfortunately the dispersion used by Rao and Venkateswarlu was insufficient to assign the resonance series correctly and this gave rise to some difficulties in understanding the potential curves of Cl_2. We have recently photographed this resonance series at much higher resolution. The new spectra which show the series extending from v=0 to v=59 of the ground state establish unambiguously that the upper state is a J=20 level and the measurements appear to eliminate the difficulties in understanding the potential curves. Furthermore, by placing the discharge tube in a magnetic field, it has been possible to excite three other rotational levels with Zeeman components of the atomic lines and thus obtain three additional resonance series. This work, which is not yet complete, should give data from which an accurate ground state potential curve of Cl_2 can be determined.

Little is known about the ground state of F_2 since the only experimental information comes from Raman spectra. Recently we have found that F_2 gives an extensive band system in emission in the 1000 Å region. This emission band system has been photographed at a dispersion of 0.25 Å/mm and a rotational analysis of the bands is now in progress. The upper state is a $^1\Sigma_u^+$ state with a long bond distance probably associated with the electron configuration $KK(\sigma_g 2s)^2(\sigma_u 2s)^2(\sigma_g 2p)^1(\pi_u 2p)^4(\pi_g 2p)^4(\sigma_u 2p)$. This upper state is highly perturbed probably by a $^3\Sigma$ state. In spite of the difficulties in analysing perturbed bands in the 1000 Å region, the lowest 12 vibrational levels of the ground state have now been observed.

In HF we have a similar example of the ground state being determined from an emission spectrum in the vacuum ultraviolet. The emission, which comes from the $B^1\Sigma^+$ excited state, extends from 1450 to 2670 Å and involves vibrational levels of the ground state from v=7 to v=19. This spectrum which was first analysed by Johns and Barrow (1959) has recently been investigated in more

detail by DiLonardo and Douglas (1973) and gives an accurate
ground state potential curve of HF.

Spectra of Rydberg States
    Time will not permit a full discussion of the nature of
Rydberg states since it is a subject of some hundreds of papers
and presumably one of the major topics of this conference. I will,
however, present a few text-book ideas on the nature of Rydberg
states and make some comments on the limitation of these ideas.

    Let me first remind you that in thinking about a Rydberg
state, it is useful to think of the molecular ion, which will be
called the core, to which is added an electron in a spacially
large orbital. If the electron is sufficiently far away from the
core, the core will appear to be rather similar to a unit positive
point charge and the term values of the system will be similar to
those of atomic hydrogen;

$$T = T_\infty - \frac{R}{n^2} \qquad (1)$$

where T_∞ is the ionization potential, R is the Rydberg constant
and n is the principal quantum number. To a higher degree of
approximation we must expect core polarization and core penetration
by the Rydberg electron to have effects similar to those found in
multi-electron atoms and thus expect the term values to be represented by

$$T = T_\infty - \frac{R}{(n-\delta_\ell)^2} \qquad (2)$$

In this expression δ is a constant called the quantum defect,
which depends on the orbital angular momentum, ℓ, of the electron.
Even at this stage of approximation there are difficulties.
Whereas for atoms n is readily defined even for a diatomic molecule
n is not readily established. In spite of this difficulty,
equation (2) may give a reasonable fit to the term values since it
is only $n-\delta$ and not n and δ separately which is determined
experimentally.

    The term values of equation (2) are further modified in a
linear molecule by the axial (rather than central) field of the
molecule and the terms of a given n and ℓ value are split into
$\ell+1$ components according to the value of λ, the component of ℓ
along the molecular axis. The term values can then be represented
by

$$T_\infty = T - \frac{R}{(n-\delta_{\ell\lambda})^2} \qquad (3)$$

where the value of δ depends upon both ℓ and λ. These three
degrees of approximation are shown in Fig. 1 for the n=3 and n=4
levels.

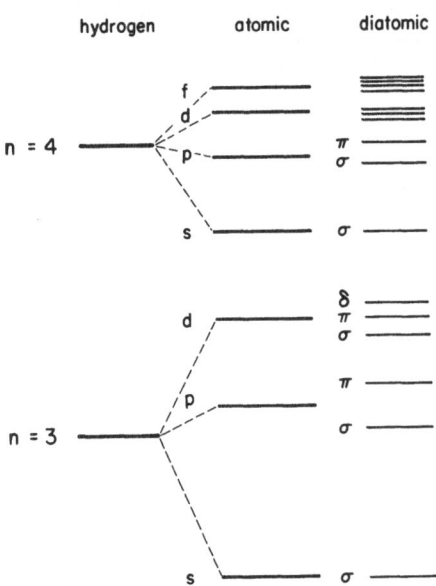

Fig. 1.   Classification of molecular Rydberg orbitals in three
orders of approximation.

In general the splitting between the λ components of a given
n,ℓ term will decrease with increasing n. The rotational constant
which determines the rotational spacing depends on the moment of
inertia of the core and does not change appreciably with the
quantum state of the Rydberg electron. Thus at some value of n we
expect the spacing of the rotational levels to equal and eventually
exceed the spacing between the various λ values of a state of given
n and ℓ. These are the conditions which give rise to case d
coupling (Herzberg, 1950, P. 229) where the rotational angular
momentum R of the core is defined and this couples with the
electron orbital angular momentum to define N. Thus, at high n
values, we expect the pσ and pπ states to become one state which
is usually called a p complex and similarly the dσ, dπ and dδ
states coalesce into a d complex. Expressions for the energies of
the rotational levels in intermediate coupling cases depend upon
the value of ℓ and the fitting of the observed levels to the
theoretical expressions can lead to an unambiguous assignment of a
state to a given ℓ value.

At this stage it appears that Rydberg states are quite simple well-behaved states which start from case b (or a) coupling at low n values and progress to ℓ complexes at high n values. Unfortunately this simple behaviour is the exception rather than the rule and listed below are some of the reasons for complexities in Rydberg states.

(1) There are often strong interactions between Rydberg states of different n and ℓ values such that neither n nor ℓ is defined. Both calculations (Lefebvre-Brion and Moser, 1965, for example) and experimental observations show this mixing to occur for many molecules. The coupling conditions and the rotational structures of bands which result from this mixing have not been fully investigated.

(2) The simple expression given in equation (3) is based on the assumption that the core is a small axial distribution of charge which is polarizable and penetrable but has no other distinguishing characteristics. Actually the core consists of an ion in which electrons occupy orbitals of various type. Exchange interactions between Rydberg and core electrons influence the energies of the Rydberg states.

(3) A particular vibrational level of a Rydberg state may interact with the higher vibrational levels of some lower Rydberg states.

(4) The core may possess orbital or spin angular momentum which will couple with that of the Rydberg electron to give a large number of states. A Rydberg electron in a π orbital, for example, will interact with a $^2\Pi$ core to give Σ^+, Σ^- and Δ states. This type of interaction has been studied in the spectra of the halogen hydrides by Tilford and Ginter (1971).

(5) The core may have low lying excited states. Rydberg series built upon the excited core may interact strongly with those built upon the ground state core.

(6) Valence states may lie in the same region as the Rydberg states and interact strongly with them. Even the distinction between Rydberg and valence states is blurred since a state which is apparently a Rydberg state of a molecule may become a valence state at either large or small internuclear distances.

(7) As is well known from the study of atomic spectra, electron spin coupling conditions may change completely from the low to the high members of a given series. Low lying members of a Rydberg series built on a multiplet core may follow Russell-Saunders coupling but the higher members will almost certainly not do so. This change in coupling conditions will affect the wavelengths, the intensities and the rotational structures of the observed bands in a Rydberg series.

A substantial number of Rydberg series have been observed in the spectra of diatomic molecules but few have been analysed in detail. The Rydberg states of the hydrogen molecule have been investigated in more detail than those of any other molecule (Herzberg and Jungen 1972) but in several respect this spectrum is unique and it will not be discussed here. In the following sections a few comments will be made on some of the molecules in which Rydberg states have been observed in detail.

BH

Probably one of the best examples of relatively simple Rydberg states is found in the absorption spectrum of BH which has been studied by Bauer, Herzberg and Johns (1964). The core has a $^2\Sigma$ ground state and the ionization potential is only 9.77 eV so that there are few complications from the core and the spectrum lies in a convenient region for observation. Near 65000 cm^{-1} a $^1\Sigma$ and $^1\Pi$ state separated by only 767 cm^{-1} were shown to be the two states derived from the 3p Rydberg orbital. The interaction of the two states follow the predictions of theory reasonably well.

Near 66000 cm^{-1} three states are observed; a Σ state at 66079 cm^{-1}, a $^1\Pi$ state 320 cm^{-1} higher and a $^1\Delta$ state 20 cm^{-1} higher than the $^1\Pi$. These three states have been assigned as the states arising from the 3d orbital. In certain respects the three states follow the predictions of theory reasonably well and the rotational energy levels have been fitted to the expressions for the levels of a 3d complex. The relative positions of the three components were however entered into the expressions as parameters to achieve the best fit. The 320 cm^{-1} separation between the Σ and Π components with only a 20 cm^{-1} separation between the Π and Δ components of the complex is not to be expected from simple theory. More recent calculations by Johns and Lepard indicate that the $(4d\sigma, {}^1\Sigma)$ state interacts with the $(4s\sigma, {}^1\Sigma)$ state which lies only 1100 cm^{-1} higher and this mixing is responsible for the large Π-Σ separation. Thus even in this simple molecule there are indications of considerable mixing of Rydberg states.

Nitric Oxide

Aside from the spectrum of hydrogen that of NO is probably the most completely explored. Over a period of twenty years E. Miescher has guided an investigation of this complex spectrum. It is not possible even to attempt a summary of the many problems considered by Prof. Miescher and his collaborators but here one or two aspects of the spectrum will be mentioned.

Clearly the reason for first attempting a complete analysis of the NO spectrum is that it should be relatively simple. The NO$^+$ core has a $^1\Sigma$ ground state and no low lying excited states. Also the ionization potential is low (9.26 eV), the gas may be cooled to liquid oxygen temperatures and several separated isotopes are available for investigation. However, in spite of the expectation of a simple spectrum, it proved to be very complex and it was only through the use of many experimental techniques and several extentions of theory that an almost complete analysis was achieved. Having overcome the many difficulties in the analysis, Miescher's work now stands as one of the basic studies of Rydberg states.

One of the dominant features of the NO spectrum is the strong interaction between Rydberg and valence states. The internuclear distances of the potential minima of the valence states is much greater than that of the ground state while those of the Rydberg states is smaller and the two sets of potential curves intersect. At each intersection large perturbation are observed and small perturbations are found quite far from the points of intersection. Both $\Delta\Lambda=0$ and $\Delta\Lambda=1$ perturbations are observed and in certain regions there are more than two interacting states.

Although many perturbations are observed in the spectrum, there are many bands which are relatively unperturbed and many others where a complete analysis of the perturbation has allowed an accurate determination of the constants which the Rydberg state would have had if it had not been perturbed. Thus in spite of the perturbations (and sometimes because of them) a rather complete picture has been obtained of the Rydberg states of NO. The analysis has identified p complexes which, presumably because of mixing of states, behave only approximately like isolated complexes. The most complete analysis has however been carried out on d and f complexes.

The 4f and 5f complexes have been analysed and treated theoretically by Jungen and Miescher (1969). In the 4f complex, the four components are separated by only ~ 20 cm^{-1} and the complex approaches very close to case d coupling. It was shown that the magnitude of the separation between the λ components and the magnitude of the quantum defect could be treated by assuming that the Rydberg electron is hydrogen-like and is perturbed only by the electric moment and the polarizability of the core. When the d complex was treated in the same way (Jungen 1970) there was little agreement between the calculated levels and those observed. It was shown that, in the 3d complex, there is a large mixing between the 3dσ and the 4sσ orbitals and that the 3dπ orbital is mixed with core π orbitals. Thus even for the d Rydberg complexes of a simple molecule there is no easy way of predicting the term values.

Nitrogen

The spectrum of N_2 in the vacuum region has much in common with that of NO in that the Rydberg states are strongly perturbed by valence states. Unlike NO, near the ionization limit, N_2 shows a very well defined Rydberg series converging to an ionization continuum. Carroll (1973) has recently published spectra of some of the p complexes in which both the term values and the intensities of the rotational lines agree well with theoretical predictions. This spectrum provides a good example of the changes which might be expected in the rotational structures of bands as one progress along a single Rydberg series. From Carroll's paper and the references therein the long and complex history of the Rydberg states of N_2 can be obtained.

Fluorine

The spectrum of fluorine has been studied several times at low dispersion (Gole and Margrave 1972) but the resolution has been insufficient to show the rotational structures of the bands. We have recently investigated the spectrum at high dispersion and have analysed a number of bands. The general description of the spectrum can be obtained from the work of Gole and Margrave and earlier work and here I will only make some comment on the high resolution spectra.

No absorption is observed in the vacuum region at wavelengths longer than 1010 Å and in the region between 1010 and 860 Å three band systems are observed. The bands in the region show resolved rotational structures and in spite of difficulties caused by perturbations an almost complete rotational analysis is possible. One band system clearly involves a valence state and the other two Rydberg states. From 860 to the ionization limit at 787 Å, the spectrum is extremely complex and consists of many diffuse and perturbed bands. Very little detailed analysis will be possible in this region. No Rydberg series have been identified and from the spectrum it would be difficult to determine even a rough value of the ionization potential.

The complex Rydberg spectrum of F_2 is not unexpected. The F_2^+ ion has a $^2\Pi$ ground state with a spin splitting which lies in a range which will cause the maximum difficulty. Also the first excited state of the core is only 21200 cm^{-1} above the ground state and two overlapping sets of Rydberg series are expected. One set corresponds to the removal of a bonding electron and the other to an antibonding electron from the core and the presence of both red and violet shaded bands in the spectrum seems to indicate that both sets are present. Thus in F_2 we expect many overlapping Rydberg series and it appears that there are strong interactions between them. It should however be emphasised that there is agreement between theoretical predictions and observations only in that a very complex spectrum is expected and a complex spectrum is observed.

Hydrogen Fluoride

The absorption spectrum of HF has been photographed at high dispersion in Ottawa with the 10.6 metre concave grating. The spectrum had not been studied previously. The spectrum consists of many thousands of sharp rotational lines in the range from

1040 Å to the ionization limit at 772 Å and even a few lines well
beyond this limit. Although the spectrum can be investigated in
great detail because the lines are sharp and readily measured, it
shows few readily recognized bands. A more careful examination
shows that the long wavelength portion of the spectrum is dominated
by many interlacing bands of the $B^1\Sigma$-$X^1\Sigma$ system, the upper state
being the same one that had been observed in the emission system
mentioned in the earlier section. The absorption bands of the
$B^1\Sigma$-$X^1\Sigma$ system run from the 14-0 band at 96299 cm^{-1} to the 73-0
band at 118338 cm^{-1}. A $^1\Pi$ and $^3\Pi$ state have also been identified
but the spectrum between 110000 and the ionization limit is
extremely complex and it has not been possible to obtain an analysis.

The $B^1\Sigma$ state of HF is a very interesting one. If we use the
united atom approximation for the electron configurations of HF
(since it is known that this approximation gives a good representation of the low lying state of diatomic hydrides) the ground state
has the configuration $(1s\sigma)^2(2s\sigma)^2(2p\sigma)^2(2p\pi)^4$. All excited states
from this neon-like configuration are expected to be Rydberg states
with one electron promoted to the n=3 or higher orbitals. The
lowest of these Rydberg states might be expected to be the $^1\Pi$ and
$^3\Pi$ states arising from the configuration $\ldots(2p\pi)^3(3s\sigma)$ to be
followed by a similar pair of states from $\ldots(2p\pi)^3(3p\sigma)$. A strong
continuum in the 65000 cm^{-1} region appears to involve the
$\ldots(2p\pi)^3(3s\sigma)$ configuration while a strong $^1\Pi$-$X^1\Sigma$ band system
105091, with a weak $^3\Pi$-$X^1\Sigma$ system a little to the red, appears to
account for the $\ldots(2p\pi)^3(3p\sigma)$ configuration. The $B^1\Sigma$ state which
has its minimum at 83305 cm^{-1} does not appear to fit into this
predicted pattern of Rydberg states. This apparent anomaly can be
understood if it is remembered that the bond distance for the
minimum of the $B^1\Sigma$ state is 2.09 Å compared with 0.92 Å for the
ground state and about 1.0 Å for HF$^+$.

The $B^1\Sigma$ state is a well characterized example of a type of
state which occurs for many molecules and which may be described
as either an ionic or a Rydberg state depending on the portion of
the potential curve being considered. It has been shown (DiLonardo
and Douglas 1973) that at large internuclear distances the HF
molecule in the $B^1\Sigma$ state dissociates into H$^+$ and F$^-$ ions and the
potential curve at large internuclear distances is well described
as the Coulomb potential between an F$^-$ ion and a proton. At a
bond distance corresponding to that of the ground state of the
molecule the potential curve of the $B^1\Sigma$ state rises above that of
the $^1\Pi$ state and it appears that it must be assigned to the
Rydberg-like configuration $\ldots(2p\sigma)(2p\pi)^4 3s\sigma$. Thus at small internuclear distances the B state is the first member of a Rydberg
series which converges to an excited state of the HF ion whereas
at large internuclear distances it can not be considered to be a
Rydberg state. In the most readily observed portion of the $B^1\Sigma$
potential curve, the large oscillations of the nucleii carry the
molecule from the Rydberg to the ionic portion of the potential

curve during each vibration.

The complex nature of the absorption spectrum of HF can be explained in the same way as that of F_2. The HF^+ core has a $^2\Pi$ ground state with a doublet splitting of ~ 292 cm^{-1} which is neither large nor small compared to the spacing of the rotational levels. The first excited state of the core is only 24000 cm^{-1} above the ground state. If the $B^1\Sigma$ state is considered to be a member of the Rydberg series leading to the excited state of the ion then in the $B^1\Sigma-X^1\Sigma$ bands which have been analysed there is experimental evidence for perturbation arising from the interaction between the two sets of Rydberg series. Because the spectrum of HF is sharp it may be possible to make some progress in the analysis of the absorption spectrum but it clearly will involve a theoretical treatment which is beyond that readily available at present.

One interesting point in the spectrum of HF, is the observation of a progression of bands beyond the ionization limit. The vibrational spacing indicates that these bands are associated with a state of the Rydberg series leading to the second ionization limit. In spite of the fact that these bands lie considerably beyond the ionization limit, the rotational structure in the bands is sharp.

CONCLUSION

With the highest resolution presently available it is possible to analyse the rotational structures of many band of diatomic molecules in the vacuum ultraviolet. A few examples have been given to show the complexity of the spectra associated with Rydberg series. Although this paper has been limited to a discussion of diatomic molecules, it is clear that similar, and perhaps greater difficulties will exist in analysing the spectra of polyatomic molecules. Johns (1970), for example, has analysed the 3p complex of water and has shown how the interaction between the states of the complex influences the rotational structure. A detailed analysis of additional bands of water which is now in progress shows the complications resulting from predissociation, from interactions in the d complex and from overlapping Rydberg series. Although it has been clear since the early work by Price that Rydberg series can be observed in molecular spectra it is now also very clear that great care must be exercised in assigning un-resolved bands to Rydberg series.

REFERENCES
Balfour, W.J. and Douglas, A.E. 1968. Can. J. Phys. 46, 2277.

Bauer, S.H., Herzberg, G. and Johns, J.W.C. 1964. J. Mol. Spectrosc. 13, 256.

Carroll, P.K. 1973. J. Chem. Phys. 58, 3597.

DiLonardo, G. and Douglas, A.E. 1973. Can. J. Phys. 51, 434.

Douglas, A.E. and Herzberg, G. 1957. J. Opt. Soc. Am. 47, 625.

Douglas, A.E. and Potter, J.G. 1962. Applied Optics 1, 727.

Gole, J.L. and Margrave, J.L. 1972. J. Mol. Spectrosc. 43, 65.

Herzberg, G. 1950. Spectra of Diatomic Molecules (D. Van Nostrand
    Co. Inc.).

Herzberg, G. and Jungen, C. 1972. J. Mol. Spectrosc. 41, 425.

Huffman, R.E., Larrabee, J.C. and Tanaka, Y. 1965. Applied Optics
    4, 1581.

Johns, J.W.C. 1971. Can. J. Phys. 49, 944.

Johns, J.W.C. and Barrow, R.F. 1959. Proc. Roy. Soc. (London)
    A251, 504.

Jungen, C. and Miescher, E. 1969. Can. J. Phys. 47, 1769.

Jungen, C. 1970. J. Chem. Phys. 53, 4168.

Lefebvre-Brion, H. and Moser, C.M. 1965. J. Mol. Spectrosc. 15,
    211.

Namioka, T., Noda, H. and Seya, M. 1973. Sci. Light 22, 77.

Ramsay, D.A. 1952. J. Chem. Phys. 20, 1920.

Rao, Y.V. and Venkateswarlu, P. 1962. J. Mol. Spectrosc. 9, 173.

Samson, J.A.R. 1967. Techniques of Vacuum Ultraviolet Spectroscopy
    (John Wiley & Sons Inc.)

Tilford, S.G. and Ginter, M.L. 1971. J. Mol. Spectrosc. 40, 568.

Van Atta, C.M. 1965. Vacuum Science and Engineering (McGraw-Hill
    Book Company Inc.).

OPTICAL PROPERTIES OF LIQUIDS IN THE VACUUM UV[*]

R. D. Birkhoff,[†] R. N. Hamm, M. W. Williams, and
E. T. Arakawa
Oak Ridge National Laboratory
Oak Ridge, Tennessee
and
L. R. Painter
University of Tennessee
Knoxville, Tennessee

INTRODUCTION

One of the most rapidly developing fields of physics in the last decade has been the study of solids in the vacuum ultraviolet spectral region, that is, the wavelength region between 400 angstroms and 2,000 angstroms or between 30 eV and 6 eV photon energies. Reflection and transmission studies have revealed the properties of many metals, semi-conductors, and insulators and in some cases rather complete band-structure models have been invoked to explain these properties.

In 1965, it occurred to the authors that these same techniques and interpretations might be applied to the study of liquids and in June of 1967 the first paper appeared giving some data on the vacuum ultraviolet reflectance of liquid water. Since this time optical and dielectric data on a dozen liquids have been reported by the ORNL group and its companion program at the University of Tennessee. No detailed theoretical models have been developed for liquids as yet, but it is clear that (1) the optical and dielectric properties of liquids frequently resemble those of their corresponding gases in certain

---
[*]Research sponsored by the U.S. Atomic Energy Commission under contract with Union Carbide Corp. and the University of Tennessee.
[†]Also, Professor of Physics, University of Tennessee, Knoxville, TN.

wavelength regions; (2) liquids containing π electrons show a distinctive absorption at around 7 eV; (3) water and several other liquids have strong <u>collective</u> electron resonances as do many solids.

EXPERIMENTAL BACKGROUND

The possibility of studying liquids in the vacuum ultraviolet did not exist until certain basic experimental developments had taken place. These include (1) the production[1] of relatively inexpensive high efficiency concave diffraction gratings; (2) the development of a simple inexpensive optical monochromator such as that by Seya[2] and Namioka[3] involving only one optical element, the concave grating; (3) the development of a grating calibrator[4] which permitted absolute measurements of grating efficiency in the Seya-Namioka geometry; and (4) the development of reflection type polarizers[5][6] for use in the vacuum ultraviolet. The electron synchrotron[7] as a light source has not as yet been used directly for liquids but will undoubtedly come into use in the future.

Three methods have been developed for studying liquid samples. In the <u>open-dish</u>[8] method, the reflectance of a liquid is measured while the liquid is in equilibrium with its vapor. Although the partial absorption of the light by the vapor is a problem, this is the only method capable of being used at energies beyond 10 eV. The second method is the so-called <u>semi-cylinder</u>[9] method in which the liquid sample is placed in contact with the flat side of a semi-cylinder. Light is incident normally on the curved surface, is reflected specularly at the cylinder-liquid interface, and exits normally through the semi-cylindrical surface again. This method, using semi-cylinders of magnesium fluoride, calcium fluoride, quartz and sapphire, features the possibility of employing reflection at angles around the critical angle, and of avoiding optical absorption in the vapor, but as noted above it cannot be used beyond 10 eV photon energy. The third method is the <u>transmission</u>[10] method using a cell formed of thick planar slabs separated by as little as 500 angstroms of liquid. The transmission method yields the optical absorption directly but as in the semi-cylinder method the limitation of slab transparency to the region below 10 eV is a drawback, and difficulties are sometimes experienced in measuring liquid film thickness.

THEORY

The method by which reflectances as a function of angle and wavelength and transmittances as a function of wavelength are converted into the real and imaginary parts of the index of refraction, n and k, respectively, through the use of Fresnel's equations is well known in theory but sometimes difficult to carry out in practice even with the extensive use of modern computers. Errors in experimental reflectances combined with the insensitivity of reflectance to n or k for some values of these parameters can make this evaluation as much an art as a science.

The square of the complex index of refraction is equal to the complex dielectric constant $\tilde{\epsilon} = \epsilon_1 + i\epsilon_2$. This equation leads to the relation $\epsilon_1 = n^2 - k^2$ and $\epsilon_2 = 2nk$. The energy loss function, $-\text{Im}(1/\tilde{\epsilon})$, obtained optically as above, is proportional to the energy losses an energetic electron would suffer in traversing thin layers of the material. That is, optical measurements in the vacuum UV give information similar to that obtained by measuring the distribution of energy losses experienced by a fast electron in going through a thin layer of matter.

Powell[11] has discussed the general problem of inferring electronic processes and oscillator strengths from curves of ϵ_1, ϵ_2 and $-\text{Im}(1/\tilde{\epsilon})$. For valence and outer shell electrons, ϵ_1 and ϵ_2 may be represented by

$$\epsilon_1(\omega) = 1 - \frac{f_o \omega_p^2}{(\omega^2 + g_o^2)} - \sum_{j=1}^{\bar{m}} \frac{f_j \omega_p^2 (\omega^2 - \omega_j^2)}{(\omega^2 - \omega_j^2)^2 + g_j^2 \omega^2} \tag{1}$$

$$\epsilon_2(\omega) = \frac{f_o \omega_p^2 g_o}{\omega(\omega^2 + g_o^2)} + \sum_{j=1}^{\bar{m}} \frac{f_j \omega_p^2 \omega g_j}{(\omega^2 - \omega_j^2)^2 + g_j^2 \omega^2} \tag{2}$$

where the f_i and g_i are oscillator strengths and damping constants, respectively, and ω_p is the "plasma" frequency. The terms under the summation represent one electron transitions whereas the remainder is the response of a free electron gas. It is clear that if ω_p is quite

different from any of the ω_j's, that the collective resonance will be
identifiable and separable from the one electron resonances but that
where the ω's are of similar value, a complicated distortion of the
dielectric response will result. Little fitting of these functions has
been reported as yet for liquids but will undoubtedly become more
popular in the future. Until then many interpretations of dielectric
data will have to be viewed as conjectural.

Where the ω_j's are well separated, the following features are
noted. An "s-shaped" pattern in ϵ_1 coincident in energy with a
maximum in ϵ_2 represents a one electron resonance. If ϵ_1 and ϵ_2 are
"small", ϵ_1 is increasing, ϵ_2 is decreasing with energy, and $-Im(1/\tilde{\epsilon})$
has a maximum, all at a particular energy, then this is interpreted[12]
as a collective electronic oscillation or plasmon response.

CARBON RING LIQUIDS

Optical and dielectric data have been obtained on several carbon
ring liquids out to 10.5 eV. In Fig. 1 the dielectric constants ϵ_1 and ϵ_2
are plotted for benzene.[13] The strong peak in ϵ_2 at the same energy
at which ϵ_1 has an "s-shaped" characteristic is thought to represent
absorption by the π electrons at this energy. Both ϵ_1 and ϵ_2 are
reasonably small at around 7 eV, and this combined with the observation that ϵ_1 is increasing and ϵ_2 decreasing with energy are the
conditions for the existence of a collective electronic excitation.
The figure also shows the energy loss function for this liquid and
indicates a strong absorption of energy to collective excitation for an
electron beam in liquid benzene at about 7.1 eV. Evidence[14] that the
resonance absorption at 7 eV is really due to π electrons may be obtained by looking at Figs. 2, 3, and 4. The liquids 1,3 cyclohexadiene,
cyclohexene, and cyclohexane have 2, 1, and no π electrons, respectively. The s-shaped characteristics of ϵ_1 and the peaks in ϵ_2 at around
7 eV are seen to increase to no structure and zero, respectively, as
the number of double bonds declines.

WATER

Perhaps half of the total effort on optical measurements of
liquids has been expended on water. Data have now been obtained out
to 26 electron volts. Figure 5[15] shows the reflectance of water
and its complex index of refraction. Of greater help in interpretation
are the data of Fig. 6[15] showing the dielectric constants and the

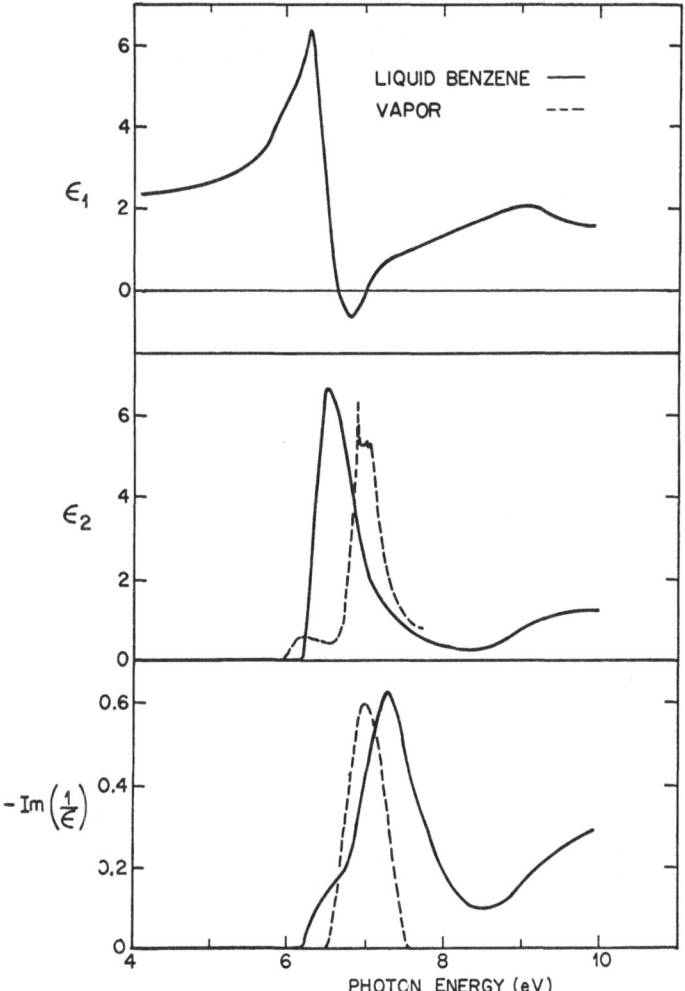

Figure 1. Dielectric constants ϵ_1 and ϵ_2 and energy-loss function $-\mathrm{Im}\,\epsilon^{-1}$ for pure liquid benzene (solid line). Vapor data is represented by dashed lines. (Refer to Ref. 13 for full explanation.)

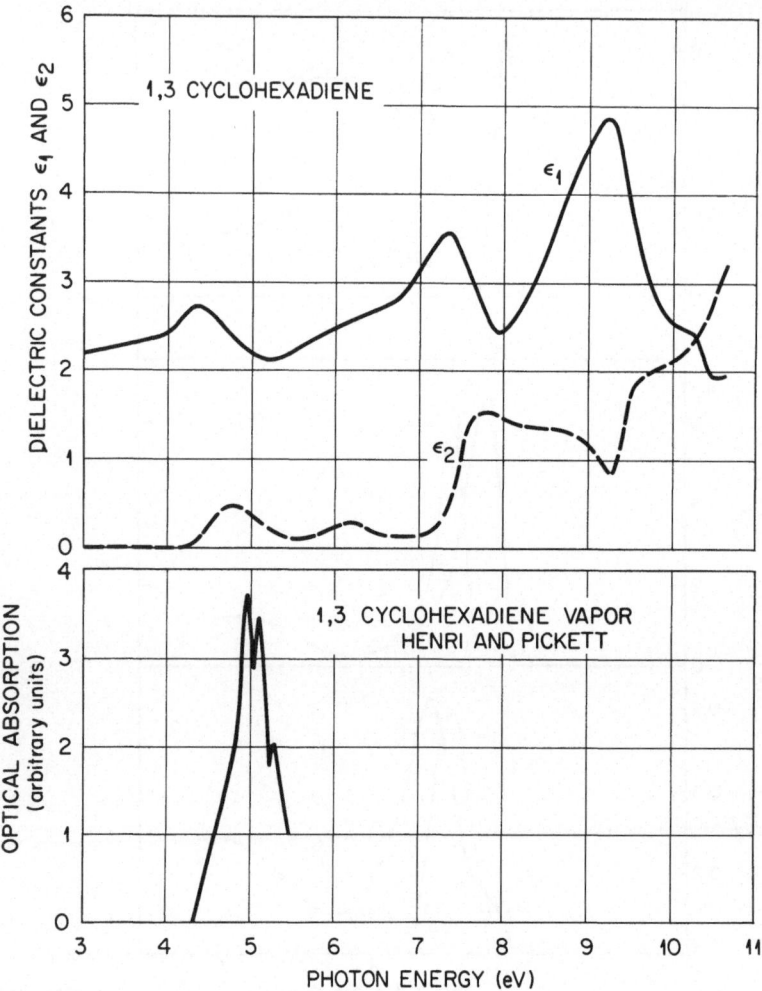

Figure 2. Dielectric constant of liquid 1,3-cyclohexadiene and the absorption coefficient of 1,3-cyclohexadiene vapor as a function of incident photon energy. (Refer to Ref. 14 for full explanation.)

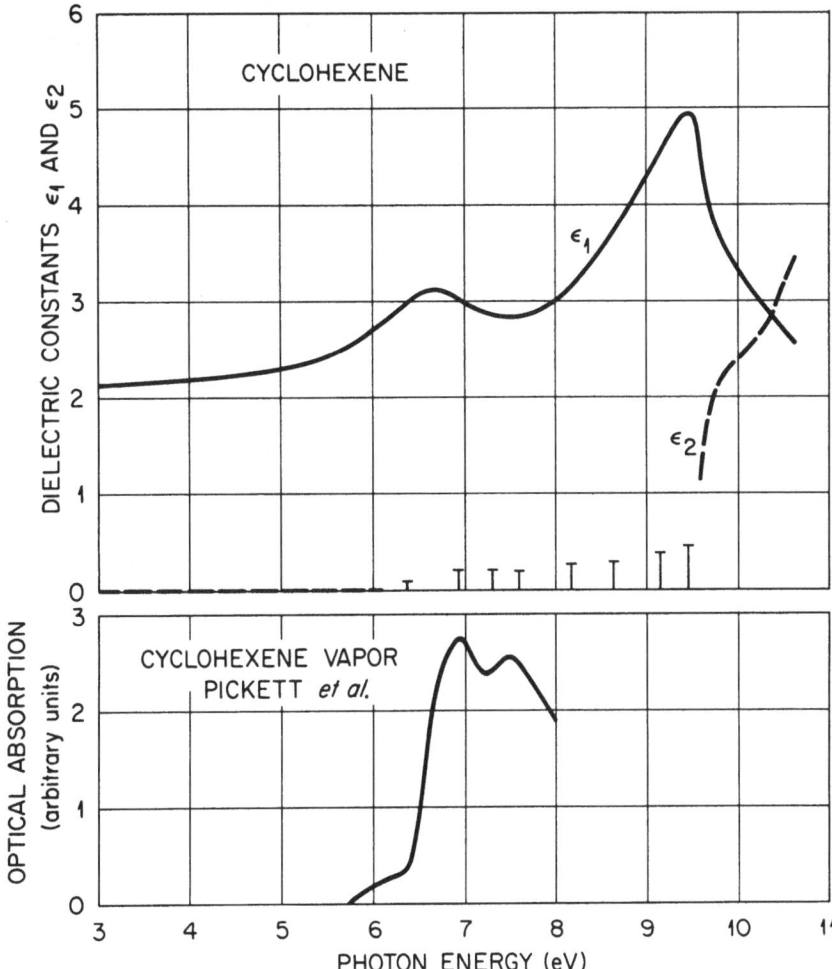

Figure 3. Dielectric constants of liquid cyclohexene and the absorption coefficient of cyclohexene vapor as a function of incident photon energy. (Refer to Ref. 14 for full explanation.)

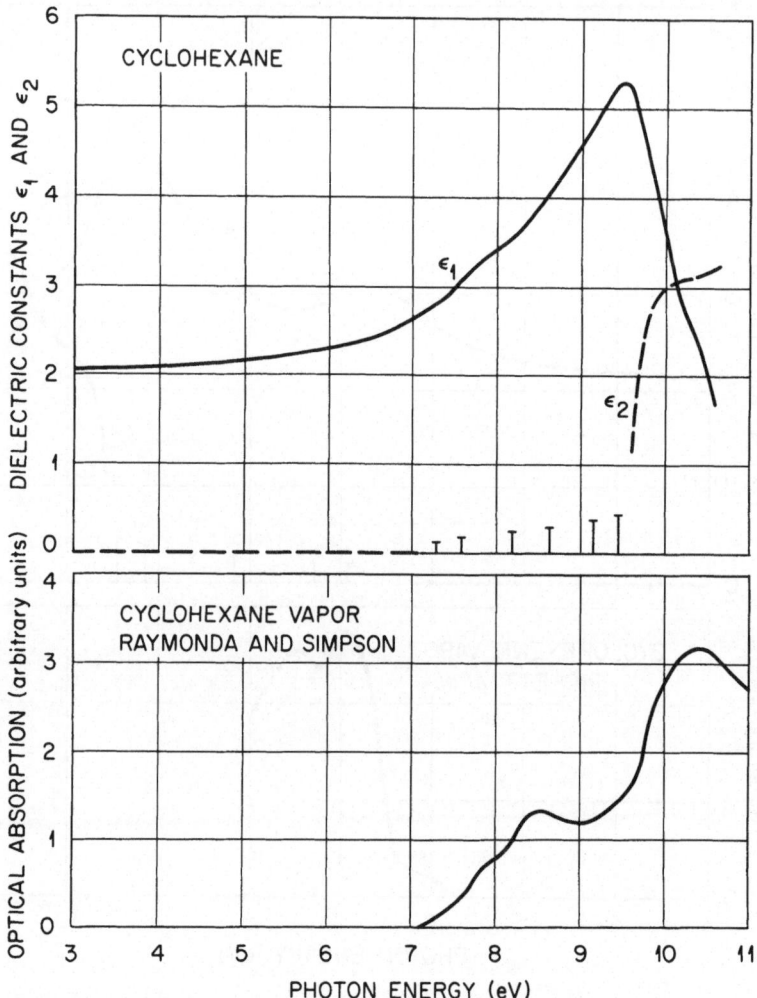

Figure 4. Dielectric constants of liquid cyclohexane and the
absorption coefficient of cyclohexane vapor as a function of
incident photon energy. (Refer to Ref. 14 for full explanation.)

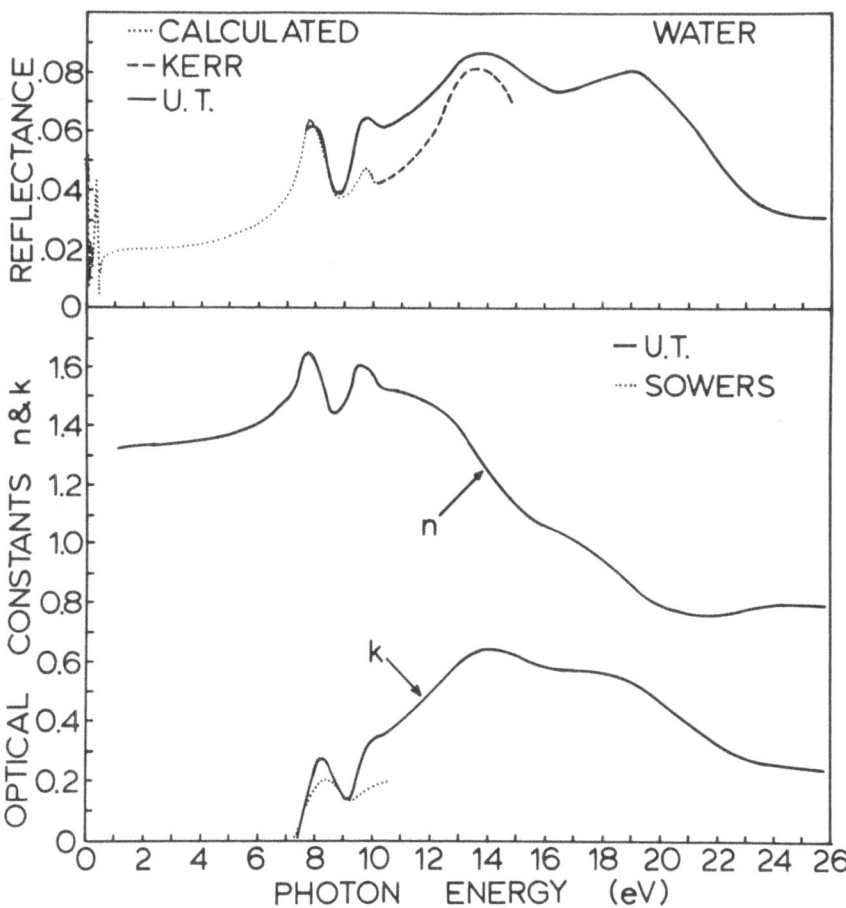

Figure 5. Near-normal reflectance and optical functions of liquid water. (Refer to Ref. 14 for full explanation.)

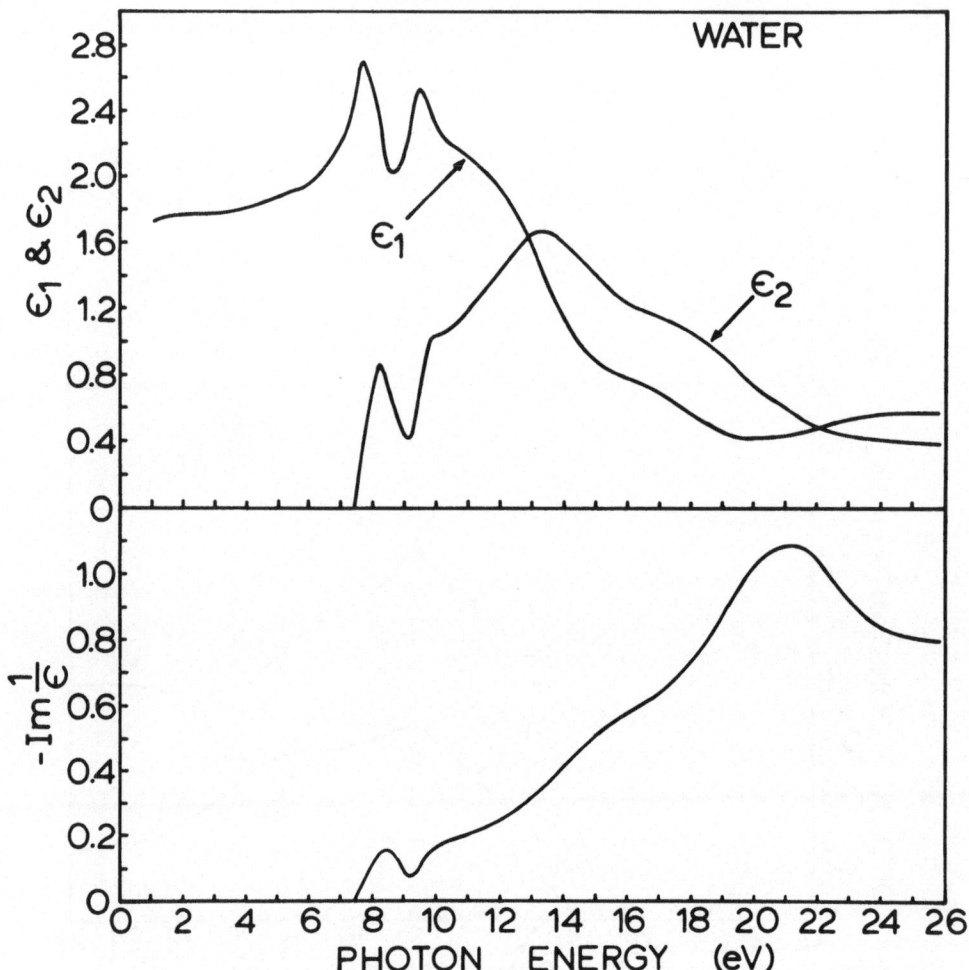

Figure 6.   Dielectric functions and energy loss function for liquid water.

energy loss function. Structure around 8 eV is attributed to dissociation processes. Enhanced absorption above 12.5 eV may be associated with the ionization phenomenon in the liquid, but the most distinguishing characteristic is the very large peak in the energy loss function at 21 eV. The collective electron resonance at 21 eV is a major source of energy absorption for energetic electrons in water. The effective number of electrons participating in the excitation process is calculated by

$$n_{eff}(E_o) = \frac{m_{eff}}{2\pi^2 \hbar^2 N e^2} \int_0^{26} E[-Im(1/\tilde{\epsilon})]\, dE \tag{3}$$

and gives n_{eff} = 2.9 at 26 eV so that about half of the electrons in hydrogen and above the 2-s level in oxygen take part in this collective electron resonance. The processes involved in the deexcitation of this collective resonance can only be speculated upon at this time, but they will provide an interesting study area for the future.

OTHER LIQUIDS

    The optical properties of only six other liquids have been measured in the vacuum ultraviolet. Liquid glycerol[16] with no π electrons shows no absorption until nearly 8 eV as shown in Fig. 7. Of most interest for this molecule is the large peak in the energy loss function at about 20.4 eV, which is attributed to a collective oscillation of the σ electrons. Using a sum rule similar to Eqn. (3) only with ϵ_2 substituted for $(-Im\, 1/\epsilon)$ shows[16] that out of the 38 valence electrons (26 σ and 12 non-bonding), an effective number of about 17 electrons participate in optical absorption energies up to 22 eV.

    Studies on silicon vacuum pump oils,[17][18] Dow Corning 200, 704, and 705 (dimethyl polysiloxane, tetramethyltetraphenyltrisiloxane, and trimethylpentaphenyltrisiloxane) show structure in the optical and dielectric constants at around 6 eV for the second two (see Fig. 8) but no structure for the first. This is consistent with the explanation that this structure is due to the π electrons, since DC 200 has no π electrons and the others do in the phenyl groups. The shapes of the curves for ϵ_1 and ϵ_2 in this region produce large peaks in the energy loss functions with the same characteristics as seen in liquid benzene, and so we can attribute at least a portion of this absorption characteristic to a collective oscillation of the π electrons. At much higher

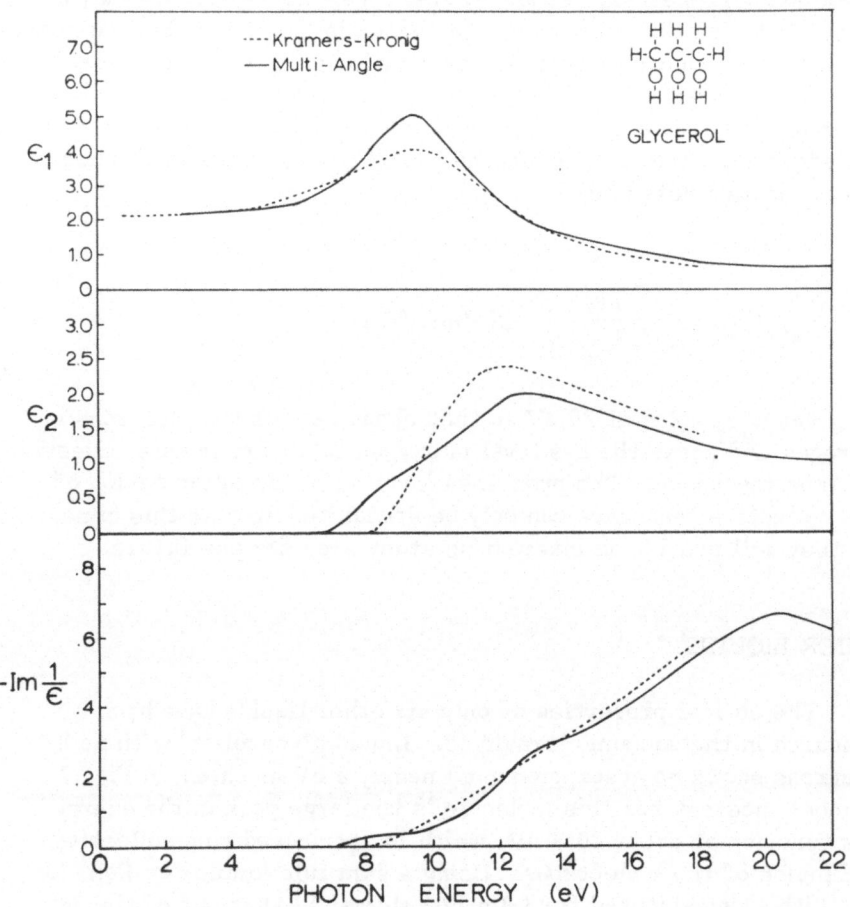

Figure 7.   Dielectric functions and energy loss function for liquid
glycerol.   (Refer to Ref. 16 for full explanation.)

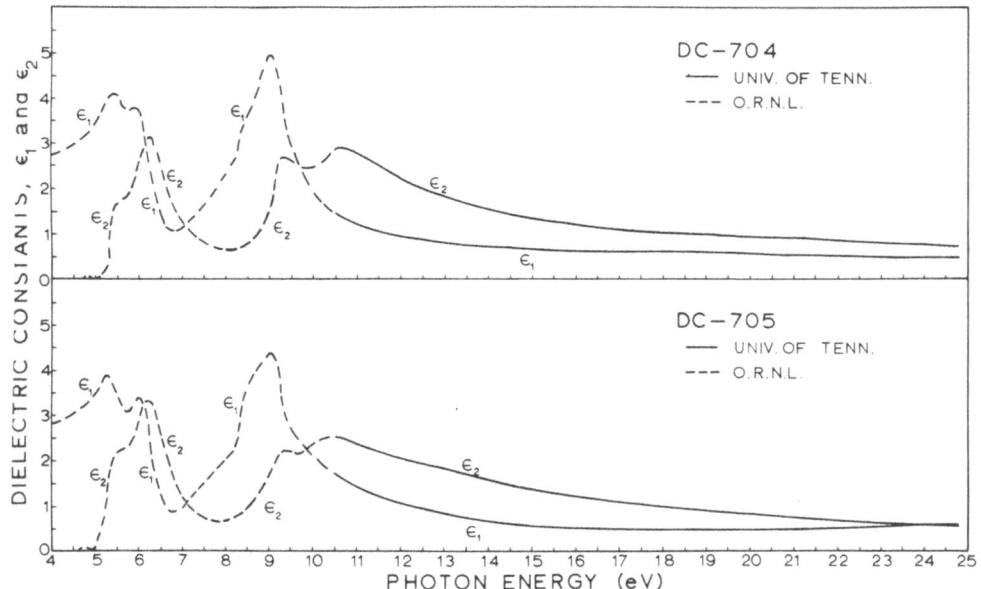

Figure 8. Dielectric functions for silicon pump oils DC-704 and DC-705 in liquid phase (Ref. 18).

energies the dielectric constants yield another peak in the energy loss function as shown in Fig. 9 which is much broader and is attributed to a collective oscillation of the π and σ electrons. This behavior is similar to that found for graphite.[19] When we make a sum rule calculation[18] for this optical data, we find that 127 out of 168 valence electrons are involved in optical absorption for DC-704, and 133 out of 190 for DC-705, both up to an energy of 25 eV.

Optical data has been obtained also on the linear hydrocarbon hexane and this is shown[10][20] in Fig. 10. No absorption occurs until nearly 8 eV because of the lack of π electrons. Than a single electron resonance gives an "s-shape" to the ϵ_1 curve and a peak to the ϵ_2 curve.

One way of comparing a liquid and a vapor is to compute an optical absorption cross-section per molecule for both. For the molecule carbon tetrachloride this is shown[10] in Fig. 11 which indicates the similarity in absorption at around 7 eV for the liquid and the vapor. This peak has been attributed to an $n \rightarrow \sigma^*$ transition of the non-bonding chlorine electrons. The structures at 9 and 9.7 are probably the result of $\sigma \rightarrow \sigma^*$ transitions. For this molecule as for most others no optical data exist for vapor phase absorption in the far vacuum ultraviolet.

DISCUSSION

There is no theory to explain quantitatively the optical data on liquids given above. However, two of the authors' colleagues, R. H. Ritchie and V. N. Neelavathi, have begun theoretical studies on the dielectric properties of liquids which bear on the collective oscillations in the liquids. Since collective oscillations play such a prominent role in shaping the dielectric response, their studies mark an important first step toward developing a theoretical understanding of the electronic structure of liquids.

Perhaps the most interesting aspect of the experimental work described above is in its application to the study of biological systems. It is clear that the same techniques which were used to study liquid water can also be used to study biologically important molecules in solution. The possibility of studying these molecules in a near normal life environment is a challenging one and will undoubtedly determine the course of our research over the next few years.

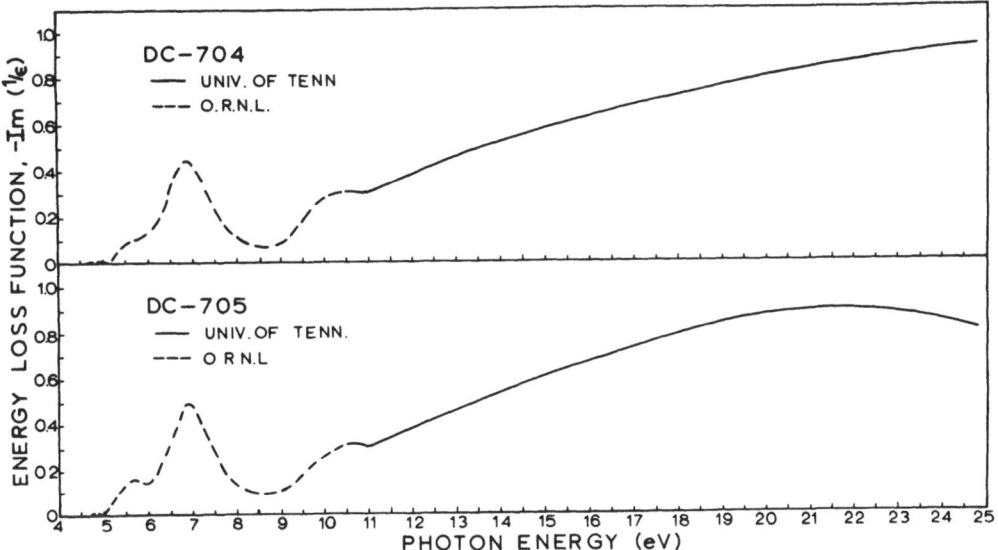

Figure 9. Energy loss functions for liquid silicon pump oils.

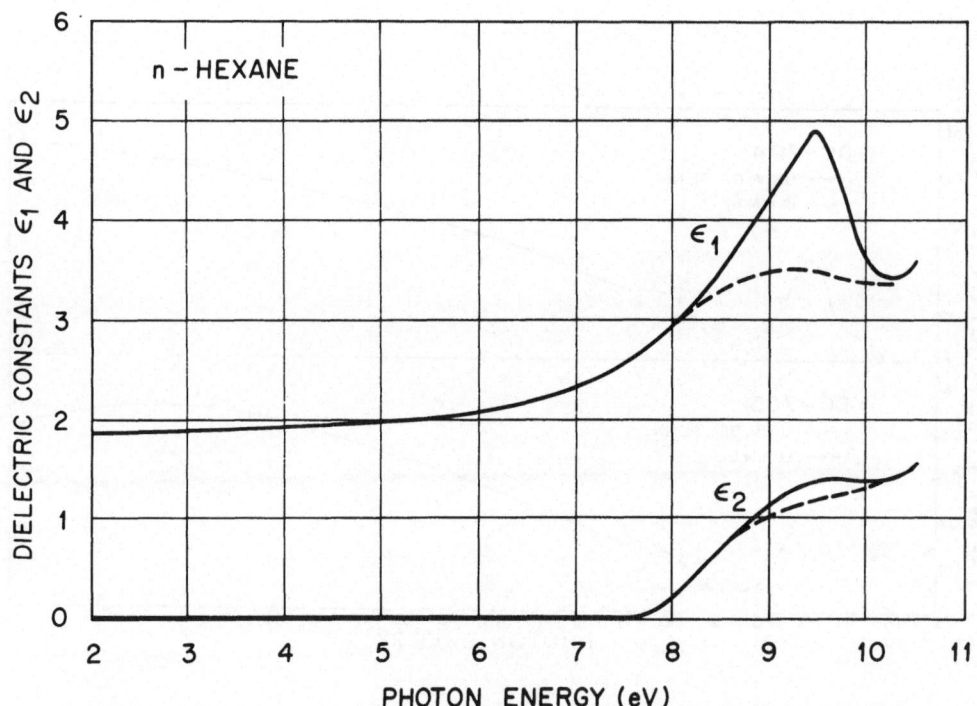

Figure 10.   Dielectric functions of liquid hexane taken from Ref. 20.

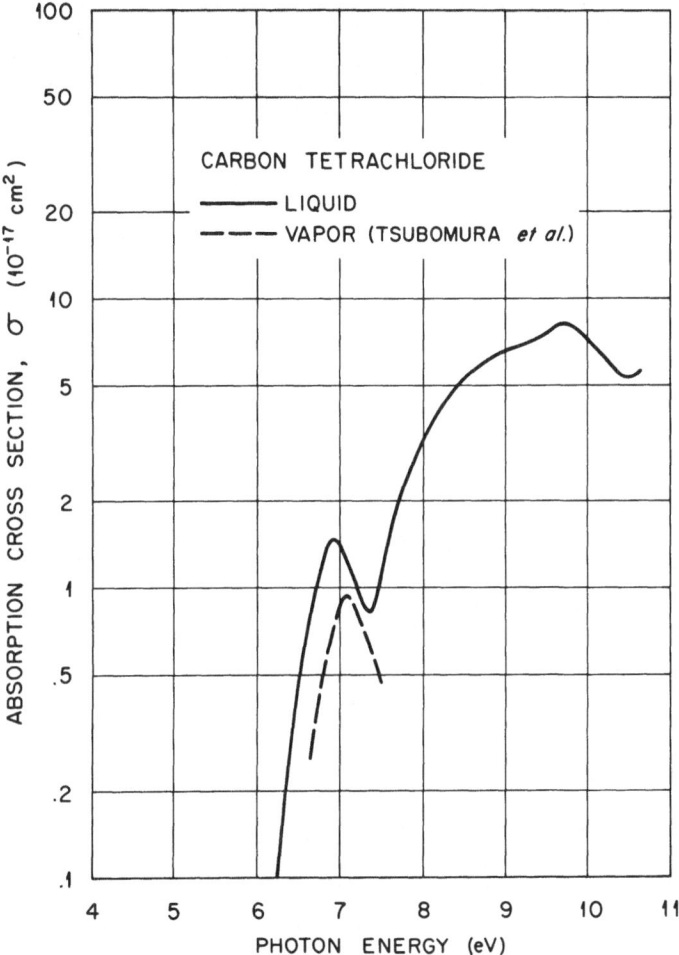

Figure 11. Absorption cross section σ of carbon tetrachloride liquid and vapor as a function of photon energy. (See Ref. 10.)

ACKNOWLEDGEMENTS

The authors wish to acknowledge the active participation in the above research of their three graduate students, James M. Heller and Drs. G. D. Kerr and B. L. Sowers.

BIBLIOGRAPHY

(1) Diffraction Grating Handbook, Bausch and Lomb, Inc., Rochester, New York (1970).

(2) M. Seya, Sci. of Light, 2, 8 (1952).

(3) T. Namioka, Sci. of Light, 3, 15 (1954).

(4) D. C. Hammer, E. T. Arakawa, and R. D. Birkhoff, Applied Optics, 3, 79 (1964).

(5) R. N. Hamm, R. A. MacRae, and E. T. Arakawa, J. Opt. Soc. Am., 55, 1460 (1965).

(6) V. G. Horton, E. T. Arakawa, R. N. Hamm, and M. W. Williams, Applied Optics, 8, 667 (1967).

(7) R. P. Godwin, Springer Tracts in Modern Physics, Vol. 51, Springer-Verlag, Berlin (1969).

(8) L. L. Robinson, L. C. Emerson, J. G. Carter, and R. D. Birkhoff, J. Chem. Phys., 46, 4548 (1967).

(9) L. Robinson Painter, R. D. Birkhoff, and E. T. Arakawa, J. Chem. Phys., 51, 243 (1969).

(10) B. L. Sowers, M. W. Williams, R. N. Hamm, and E. T. Arakawa, J. Chem. Phys., 57, 167 (1972).

(11) C. J. Powell, J. Opt. Soc. Am., 59, 6, 738 (1969).

(12) H. Ehrenreich and H. R. Philipp, in Proc. of the International Conference on the Physics of Semiconductors, ed. A. C. Strickland (Bartholomew Press, Dorking, England, 1962) pp. 367-374.

(13) M. W. Williams, R. A. MacRae, R. N. Hamm, and E. T. Arakawa, Phys. Rev. Letters, 22, 21, 1088 (1969).

(14) B. L. Sowers, E. T. Arakawa, and R. D. Birkhoff, J. Chem. Phys., 54, 6, 2319 (1971).

(15) J. M. Heller, R. N. Hamm, R. D. Birkhoff, and L. R. Painter, in press.

(16) J. M. Heller, Jr., R. D. Birkhoff, M. W. Williams, and L. R. Painter, Rad. Res., 52, 1, 25 (1972).

(17) B. L. Sowers, M. W. Williams, R. N. Hamm, and E. T. Arakawa, J. Appl. Phys., 42, 11, 4252 (1971).

(18) G. D. Kerr, M. W. Williams, R. D. Birkhoff, and L. R. Painter, J. Appl. Phys. 42, 11, 4258 (1971).

(19) E. A. Taft and H. R. Philipp, Phys. Rev., 138, A197 (1965).

(20) B. L. Sowers, R. D. Birkhoff, and E. T. Arakawa, "Optical Properties of Liquid CCl_4, C_6H_{14}, C_6H_{12}, C_6H_{10}, C_6H_8, and C_6H_6 in the Vacuum Ultraviolet," Oak Ridge National Laboratory Report ORNL-TM-3665, March 1972.

THE FAR ULTRA-VIOLET ABSORPTION SPECTRA OF ORGANIC MOLECULES: LONE PAIRS AND DOUBLE BONDS

C. SANDORFY

Université de Montréal, Département de Chimie
Montréal, Québec, Canada

INTRODUCTION

The far-ultraviolet spectra of organic molecules can be conveniently discussed in terms of the following categories: saturated molecules containing bonding σ-electrons only; saturated heteroatomic molecules with both bonding σ-electrons and lone pairs of electrons; unsaturated molecules containing bonding σ and π-electrons; unsaturated molecules with bonding σ and π electrons and lone pairs of electrons. In a previous lecture at this Institute Dr. M.B. Robin stressed the principal characteristics of the far-ultraviolet spectroscopy of saturated molecules. In one simple sentence one might say that electrons in bonding σ-orbitals start absorbing at 170 or 160 nm and when the molecule contains heteroatoms then, except for fluorine, the bands due to the transitions of their lone pair electrons dominate the spectrum up to, at least 120 nm and begin at longer wavelengths. Almost all these bands can be interpreted as Rydberg bands converging to the first (lowest) ionization potential of these molecules. The Rydberg nature of most of the bands encountered in the far-UV spectra of organic molecules establishes a close connection between optical vacuum ultraviolet (VUV) and photoelectron (PE) spectroscopy.

The present lecture is devoted to spectra due to the lone pairs of heavier atoms like $C\ell$, Br, I and S and to double bonds in olefins and in benzene derivatives.

Halogen derivatives of methane

Whatever the limitations of the orbital concept it enables us to achieve an overall understanding of the spectra of these compounds.

In methane itself the "orbital structure" of the ground state is well established. The highest filled ("frontier") orbital is a triply degenerate f_2 orbital under T_d symmetry, with an IP of about 13 eV (the PE band is broad covering the range from 12.7 to 16.0 eV) while the next orbital is a_1 with about 23 eV rather far from the first IP (1). The halides might have C_{3v}, C_{2v} or C_s symmetry and Table 1 shows the correlations between the orbitals of the united atom and methane and what they become under reduced symmetry.

Neither the first IP nor the absorption bands, at least not up to 120 nm, will be due to transitions departing from these orbitals, however. In the presence of one or more $C\ell$, Br or I atoms they will be preceded by molecular orbitals formed essentially from halogen lone pair atomic orbitals. The latter yield MO's having symmetries given in Table 2.

United atom	T_d	C_{3v}	C_{2v}	C_s
s_g	a_1	a_1	a_1	a'
p_u	f_2	$a_1 + e$	$a_1 + b_1 + b_2$	$2a' + a''$

Table 1

Correlations for the molecular orbitals of methane derivatives formed from bonding σ atomic orbitals.

CH_3X	C_{3v}	e
CH_2X_2	C_{2v}	$a_1 + a_2 + b_1 + b_2$
CHX_3	C_{3v}	$a_1 + a_2 + e + e$
CX_4	T_d	$e + f_1 + f_2$

Table 2

The symmetries of molecular orbitals based on lone pair atomic orbitals for methane halides. $X = C\ell$, Br or I.

The PE spectra of methylhalides were thoroughly discussed by Turner et al. (1)

The UV spectra, up to at least 120 nm, are due to transitions departing from such lone pair molecular orbitals. (Naturally, if there is a bonding molecular orbital of the same symmetry there will be mixing). In what follows we shall compare some spectra in the series chlorine, bromine, iodine and, at the same time, we shall study the effect of substitution of the hydrogens by fluorines in these methane derivatives. Table 3 compares the lowest three or four IP's for $CH_3C\ell$, CHF_3 and $CF_3C\ell$.

The first IP in $CH_3C\ell$ which is related to the doubly degenerate lone pair orbital moves from 11.3 eV to 13.0 eV when the hydrogens are replaced by fluorine. This very significant stabilizing effect or decreased availability of the chlorine lone pair electrons illustrates one of the basic facts of fluorine chemistry.

The next orbital is of mainly $C-C\ell$ bonding character in both $CH_3C\ell$ and $CF_3C\ell$ as can be inferred from the regular trend occurred upon replacing $C\ell$ by Br and by I (4). The stabilizing effect of the fluorines is somewhat less pronounced in this orbital. Comparing CHF_3 and $CF_3C\ell$ we see that the higher IP's whose assignments are based on calculations by Brundle, Robin and Basch (2) on CHF_3 do not change much when H is replaced by $C\ell$. It is also seen that the mainly C-H orbital in CHF_3 has about the same energy as the mainly $C-C\ell$ orbital in $CF_3C\ell$.

$CH_3C\ell$		CHF_3		$CF_3C\ell$	
$\overline{C\ell}$	11.3 e	C-H	14.8 a_1	$\overline{C\ell}$	13.0 e
$C-C\ell$	14.4 a_1	$\overline{F}$	15.5 a_2	$C-C\ell$	15.0 a_1
		$CF_3-\overline{F}$	16.2 e	$\overline{F}$	15.55 a_2
CH_3	15.4 e	$\overline{F}$	17.2 e	$CF_3-\overline{F}$	16.5 e
				$\overline{F}$	17.4 e

Table 3

Comparison of the lowest IP's (in eV) of $CH_3C\ell(1)$, $CHF_3(2)$ and $CF_3C\ell(3)$.
$\overline{C\ell}, \overline{F}$: orbital of mainly lone pair character.
$C-C\ell$, CH_3 : orbital mainly populated in the respective group or bond.

The spectrum of CH_3Cl was measured a long time ago by Price (5) and by Zobel and Duncan (6). In the 200–120 nm region of the spectrum there is a weak band having its maximum near 57900 cm^{-1} and a much stronger band at about 62500 cm^{-1} followed by a very diffuse band near 71100 cm^{-1}. The weak band (ϵ about 50) has a term value of about 32000 cm^{-1} higher than would be expected for any Rydberg series. It is probably better described as a valence shell transition. Duncan (6) suggested that the orbital of the excited electron is a MO antibonding in the C–Cl link. Since the highest occupied orbital in the ground state is mainly populated in that bond this assignment seems to be acceptable. Some computational help in the assignment of this photochemically very important band would be welcome.

The strong band centered at 62500 cm^{-1} is readily assigned to the first member of an s type Rydberg series converging to the first IP ($a_1 \leftarrow e$; $^1E \leftarrow {}^1A_1$). Its term value is about 28600 cm^{-1}. The diffuse bands at 71100 cm^{-1} can be taken for the first members of p type series.

If we replace Cl by Br and I these band move to lower energies. (See Table 4) ((7) and references given in it).

They clearly follow the trends in the IP's; 11.3 → 10.5 → 9.5 eV. (The dissociation energies of the C–X bond vary in the same way : 3.5 → 2.9 → 2.3 eV) (7). At higher frequencies p and d type series begin.

The symmetry of the excited state relating to the weak $\tilde{A}$ band is not known.

	CH_3Cl	CH_3Br	CH_3I
$\tilde{A}$	57900 cm^{-1}	49000	38600
$\tilde{C}$	62500	56000	49700
$\tilde{D}$		59100	54600

Table 4

Approximate location of the lowest frequency bands of methylhalides (7).

Spin-orbital splittings that might occur in the spectra of these molecules are discussed in detail in Herzberg's Volume III (7). For $C\ell$ these are expected to be small (of the order of 0.1 eV) and practically undetectable under moderate resolution while the cases of Br and I can be considered as typical for weak and strong spin-orbital coupling respectively. Actually, bands were found in both PE and UV spectra of methylhalides whose existence is due to these spin-orbital interactions. In the spectra of both CH_3Br and CH_3I a part of these bands (for the 3s ← e band) were identified. (The strongest ones are $\tilde{C}$ and $\tilde{D}$, Table 4).

If we now turn to $CF_3C\ell$ (3) (Fig. 1) we see that the UV spectrum underwent a large hypsochromic shift just as the first IP. The weak band of $CH_3C\ell$ moved from about 57900 cm^{-1} to about 71000 cm^{-1}, the strong 3s Rydberg band from 62500 to 78000 cm^{-1}. This is partly true for CF_3Br as well (Fig. 2). The spectrum is shifted to lower frequencies with respect to $CF_3C\ell$ as does the first IP. (13.0 → 12.0 eV). The strong Rydberg band moved from 56000 to 70500 cm^{-1} upon replacement of the hydrogens by fluorines. (The situation is less clear in relation to the weak

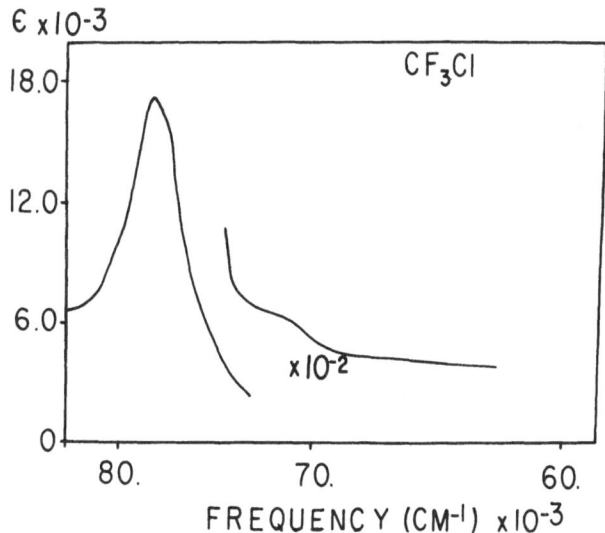

Figure 1

The far-ultraviolet absorption spectrum of $CF_3C\ell$ (200–120 nm)

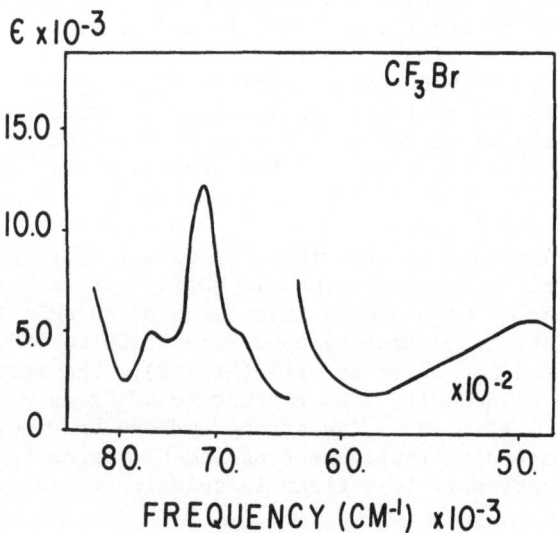

Figure 2

The far ultraviolet absorption spectrum of CF_3Br (200–120 nm)

band at 48000–49000 cm^{-1}). The band at 67900 cm^{-1} is separated by about 0.3 eV from the strong Rydberg band which is of the right order of magnitude for spin–orbital splitting in the case of Br. This band is very probably one of the bands resulting from the splitting. On the high frequency side of the strong band there is another band which can be accounted for as the first member of a p type Rydberg series (3).

What do we find in the UV spectrum beyond the first IP ? Figs. 3 and 4 show the absorption spectra of CF_3Cl and CF_3Br measured in our laboratory (8). For CF_3Cl we find five broad bands between 120 and 60 nm. (cf. the spectra of fluoromethanes (9)(10)).

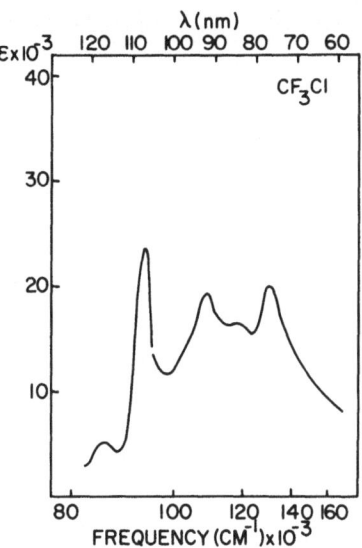

Figure 3

The far ultraviolet absorption spectrum of CF_3Cl (120–60 nm)

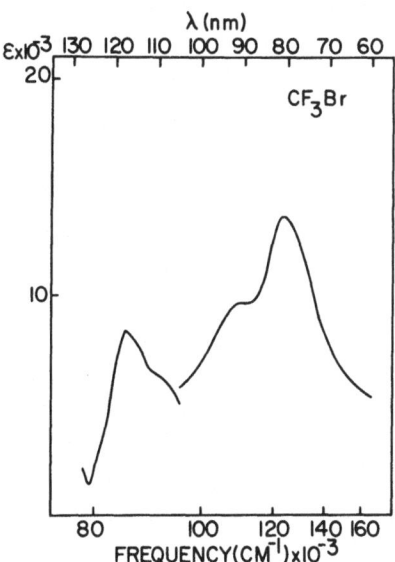

Figure 4

The far ultraviolet absorption spectrum of CF_3Br (120–60 nm)

ν_{max} cm^{-1}	ϵ_{max}	Assignment	Related I.P. (eV)	Term cm^{-1}
85400	5000	3 p ← $\overline{Cl}$	13.0	19500
93400	24000	3 s ← (C–Cl)	15.0	27600
108660		3 s ← $\overline{F}$	17.4	31700
		3 s ← $\overline{F}$–CF$_3$	16.5	24440
		4 s ← C–Cl	15.0	
11900		3 p ← $\overline{F}$	17.4	21360
		3 d ← $\overline{F}$–CF$_3$	16.5	13300
130000		4 s ← $\overline{F}$	17.4	

Table 5

Tentative assignments for the higher electronic transitions of CF$_3$Cl.

In Table 5 we summarized some tentative assignments for these. In some cases more than one assignment would fit a given frequency and these bands probably receive contributions from more than one transition. More important than the actual assignments is the fact that beyond the first IP (about 104000 cm^{-1}) we find broad but discret bands which can be interpreted as Rydberg bands converging toward second and higher IP's and these are not lost in the continuum following the first IP. This is in contrast to the respective parts of the spectrum of paraffin hydrocarbons (11)(12). As can be seen from Table 5 the observed bands are due to Rydberg transitions related to IP's of either bonding or fluorine lone pair electrons.

The comparison between the 120–60 nm part of the spectrum of CF$_3$Cl and CF$_3$Br is interesting from the following point of view. As we have seen the first two IP's ($\overline{Cl}$ and C–Cl) undergo significant shifts from Cl to Br but the following IP's which involve orbitals having fluorine lone-pair or C–F bonding character are hardly affected. Those bands which belong to Rydberg series converging to the two first IP's undergo similar shifts from Cl to Br but the high frequency bands which belong to series converging to the F IP's remain practically at their places.

This similarity of the spectra of CF$_3$Cl and CF$_3$Br beyond the bands which belong to the first two IP's confirms the assignments of these high frequency bands to Rydberg transitions related to the IP's involving fluorine.

If we put two $C\ell$ atoms (3) on the same carbon we expect the
lone pair AO's to interact and since there is no degeneracy
under C_{2v} symmetry to yield four close lying IP's. This is what
Turner et al. (1) found in the case of $CH_2C\ell_2$ and we found the
same in the case of $CF_2C\ell_2$ (Fig. 5) except that they are shifted
to higher frequencies (by about 1 eV.(11.31 and 12.18, two partly
resolved double bands for $CH_2C\ell_2$ and 12.3, 12.6, 13.2 and 13.5 eV
for $CF_2C\ell_2$ (3))). The fifth PE band (14.4 eV, $C-C\ell$) has well
developed vibrational fine structure indicating a stable ion.
The main vibrational interval is about 385 cm^{-1} which is likely
to correspond to a $C-C\ell$ deformation mode.

The UV spectrum of $CF_2C\ell_2$ has been first measured by Zobel
and Duncan (6) and by Stokes and Duncan (13). Fig. 6 shows its
spectrum measured in our laboratory. At about 56460 cm^{-1} we find
a band which is somewhat broader and more intense than the corresponding band of $CF_3C\ell$. This band has been interpreted as a
$(C-C\ell)^* \leftarrow C\ell$ transition. Since now we have two close lying IP's
(12.3 and 12.6 eV) we can assume that the band receives contributions from transitions departing from both. At 65400 cm^{-1} we
find a weaker but broad band with a pronounced shoulder which has

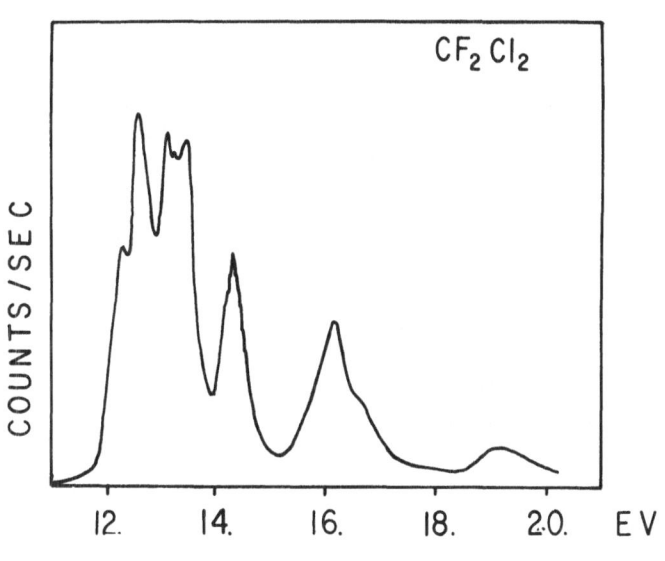

Figure 5

The photoelectron spectrum of $CF_2C\ell_2$

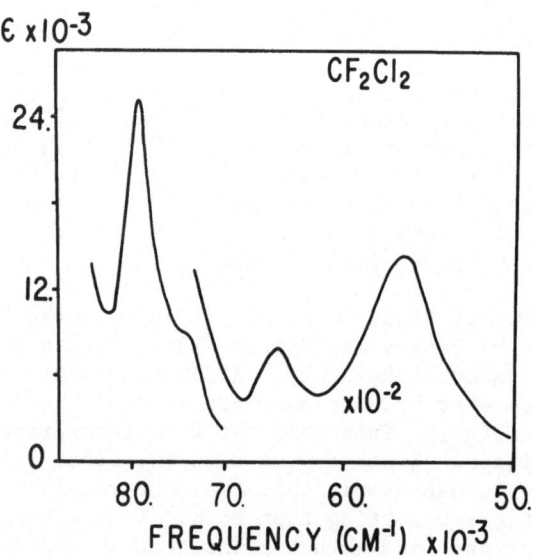

Figure 6

The far ultraviolet absorption spectrum of CF_2Cl_2 (200-120 nm)

no analogue in the spectrum of CF_3Cl. It is logical to assign
it to $(C-Cl)^* \leftarrow Cl$ transitions from the two other lone pair IP's
(13.2 and 13.5 eV). The strong band at about 79100 cm^{-1}, the
shoulder at 74000 and the band at 84400 cm^{-1} have the right term
values to be 3 s $\leftarrow Cl$ bands. The following two bands near 87000
are better fitted by 3 p $\leftarrow Cl$ (13.2 and 13.5 eV). At higher
frequencies and beyond the first IP we again find discret bands
which are readily interpreted as Rydberg transitions related to
higher IP's.

The spectra of all halogenated derivatives of methane which
we examined can be interpreted along similar lines. Let us summarize the main points again:

a) The lowest IP's are due to ionization from orbitals formed
by lone pair type AO's of Cl, Br or I. They are followed toward
higher energies by IP's corresponding the ionization from mainly
C-X or C-H bonding orbitals. The fluorine lone pairs and C-F
bonding orbitals come in usually beyond 15.5 eV.

b) The lowest frequency bands in all the UV spectra seem to be
valence-shell type transitions of the $(C-X)^* \leftarrow \overline{X}$ type.

c) The strong bands at higher frequencies can all be interpreted as Rydberg transitions which are related to the first, $\overline{X}$, IP's up to about 90000 cm^{-1}.

d) Beyond that limit we find discret bands in the fluoroderivatives which are Rydberg transitions converging to higher, C-X, then C-F or F type IP's.

Halogen derivatives of ethane

Ethyl-chloride, bromide and iodide have spectra similar to those of the related methyl-halides (14)(7). This similarity extends to fluorinated ethyl-halides. To illustrate this we show the PE (Figs. 7, 8) and UV (Figs. 9, 10) spectra of CF_3-CF_3 and CF_3-CF_2Cl (15).

In ethane itself of the two uppermost filled orbitals of the ground state one is mainly populated in the C-H bonds the other in the C-C bond and they are close to one other so that the determination of their order constitutes a delicate problem (16-23). In the case of hexafluoroethane, however, there is no doubt that the frontier-orbital is of C-C character (about 14.5 eV, Fig. 9). If we replace one fluorine by a chlorine, the characteristic lone pair IP takes the first place (Fig. 10). The next peak (near 14 eV) can still be assigned to an orbital of mainly C-C character

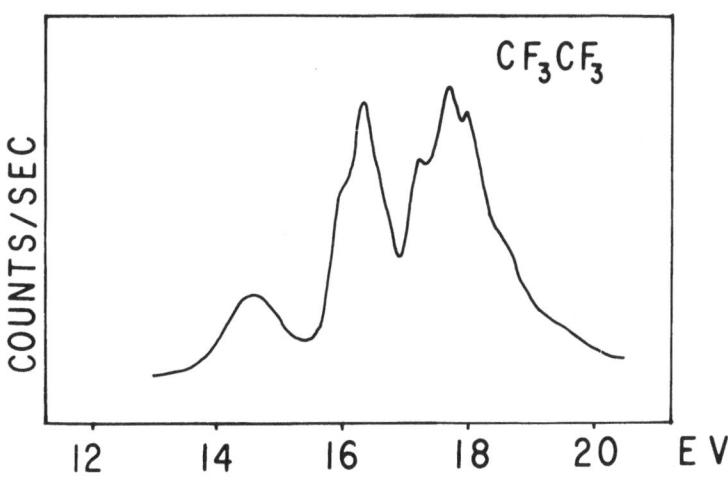

Figure 7

The photoelectron spectrum of C_2F_6

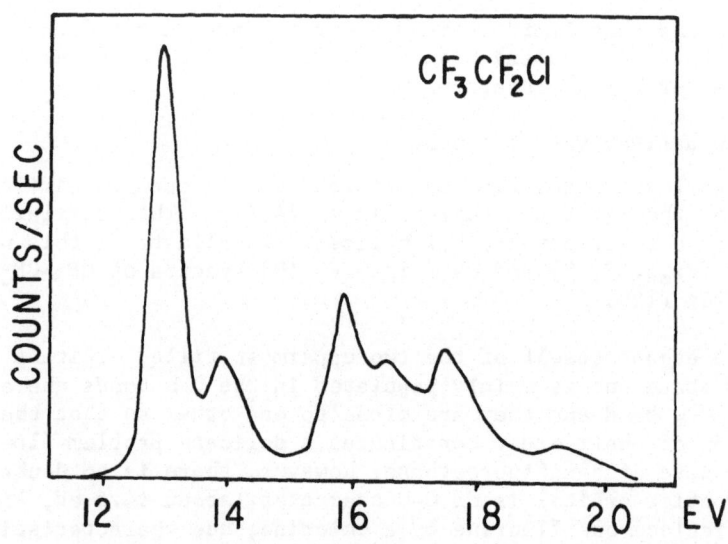

Figure 8

The photoelectron spectrum of C_2F_5Cl

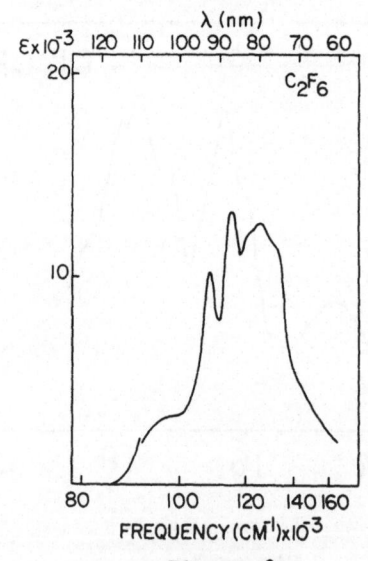

Figure 9

The far ultraviolet spectrum of C_2F_6

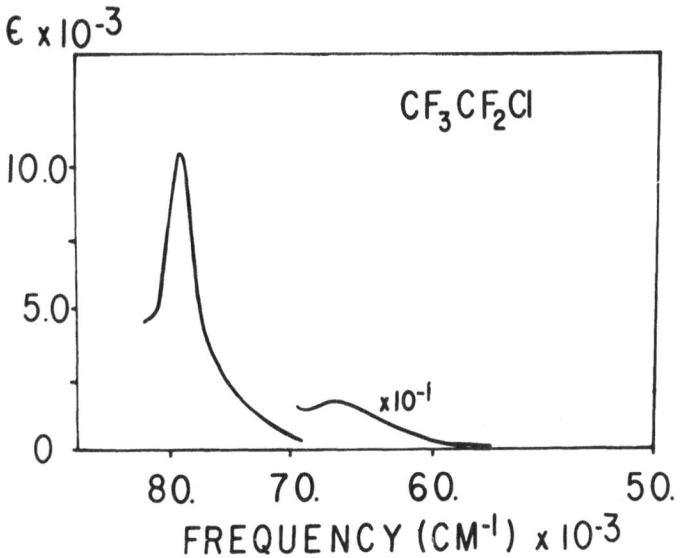

Figure 10

The far ultraviolet spectrum of C_2F_5Cl

although from the very similar C–C and C–Cl bond dissociation
energies (24) we can infer that there will be a great deal of
mixing.

The UV spectrum of CF_3–CF_3 (15) lies beyond 120 nm. The
first band is at 97560 cm^{-1} and with a term value of 22230 cm^{-1}
it is readily interpreted as a 3 p ← a_1 transition departing from
the 14.5 eV C–C IP. It is followed by several other bands which
can be accounted for as Rydberg transitions for the higher IP's
having mainly $\overline{F}$ or C–F character. As expected, the spectrum of
CF_3–CF_2Cl is quite different (Fig. 12) and it is very similar to
the spectrum of CF_3Cl. The first weak band at about 67000 cm^{-1}
is like the $(C-Cl)^* ←$ Cl and the strong band at 79000 cm^{-1} like
the 3 s ← Cl band of CF_3Cl and they are naturally assigned in the
same way. This resemblance confirms the mainly localized character of the first and the Rydberg character of the second: the
size of the molecule is of little importance.

Under the low symmetry of CF_3–CF_2Cl the lone pair AO is no
more degenerate. The split must be slight, however; all we see
is a shoulder (at higher resolution) on the $\overline{Cl}$ band. If there are
two or more chlorines in the molecule we obtain an increased number of bands in both PE and UV spectra, due to the interactions

between the lone pair AO's.

Ethylene derivatives

In the PE spectrum of ethylene the band of lowest energy which is related to ionization from the π orbital is at 10.51 eV (25). The σ-orbitals give peaks at higher energies, 12.38, 14.47, 15.68 and 18.87 eV (all adiabatic).

The electronic spectrum of ethylene was first measured by Price and Tutte (26) in 1940 and was thoroughly studied by Wilkinson and Mulliken (27)(28), Merer and Schoonveld (29) and Merer and Mulliken (30). The $\pi^* \leftarrow \pi$ band has a broad maximum near 162 nm which is preceded by a very long progression whose origin is probably at as low a wavelength as 265 nm. It gives strong support to the assumption of a perpendicular structure for the V state. Prominent vibrations in this fine structure are the $C = C$ stretching mode (ν_2) and double jumps of the twisting vibration (ν_4) whose ground state values are 1623 and 1023 cm^{-1} respectively. They are greatly reduced in the V state. Partly superimposed to this band is the first member of a ns type Rydberg series which converges to the first (π) IP at 10.51 eV. In several methyl and fluoroderivatives of ethylene this band appears at longer wavelengths (210-190 nm) and was the topic of much discussion. Somewhat less adequately the excited orbital might be described as σ^* instead of 3s since the orbital is likely to have dimensions comparable to the molecular core but its membership in a Rydberg series makes the former description preferable. The members of the ns series have vibrational fine structure which is again dominated by ν_2 and ν_4 but ν_2 (about 1300-1400 cm^{-1}) undergoes a lesser decrease with respect of its ground state value than in the V state. Merer and Schoonveld have shown that in the 3s state the molecule deviates from coplanarity by approximately 25° Two other Rydberg series are found, which also converge to the first π IP. Only continuous absorption was observed between 100 and 60 nm.

Fluorinated ethylenes (31) are interesting, since it is a well established fact that fluorine substitution has a large stabilizing effect on the energies of σ-orbitals but it hardly affects the energies of π-orbitals.

The PE spectra of some of the fluoroethylenes were measured by Lake and Thompson (32) and by Bralsford, Harris and Price (33).

The first (π) IP decreases slowly with the number of fluorine atoms, the adiabatic value going from 10.51 for ethylene itself, though 10.37 for monofluoroethylene to 10.11 eV for tetrafluoroethylene.

The value of the IP related to the lowest σ-orbital is of some interest. In ethylene itself it is at about 13.0 (vert.) in monofluoroethylene it moved to 13.79 (vert.) and in 1,1-difluoroethylene to 14.79 eV. (vert.). (Or 12.38 (ad) for ethylene and 14.06 (ad) for 1,1-difluoroethylene). Unfortunately in the highly substituted derivatives the $\overline{F}$ and C-F type IP's come in strongly so that the C-C orbital cannot be identified with any degree of certainty. This is perhaps sufficient to illustrate the σ-stabilizing effect of the fluorines and also the fact that the highest σ-orbital of olefins has a much higher energy than in paraffins (12.38 for ethylene, 11.56 eV for ethane).

The ultraviolet spectra of fluoroethylenes exhibit the same general features as the spectrum of ethylene itself (31). The $\pi^* \leftarrow \pi$ band in nearly at the same place in all derivatives. (Except in tetrafluoroethylene).

The long tail of the $\pi^* \leftarrow \pi$ band which exists in the spectrum of ethylene seems to be absent in the fluoro-derivatives. However, measurements at higher vapor pressures would be needed in order to ascertain this. The three Rydberg series of ethylene (ns, np and nd or series R, R' and R'') are present in all fluoroethylenes and converge to the first (π) IP. (Figs. 11 and 12). The 3s ($\equiv$ 3 R) band is prominent (at about 190 nm) and

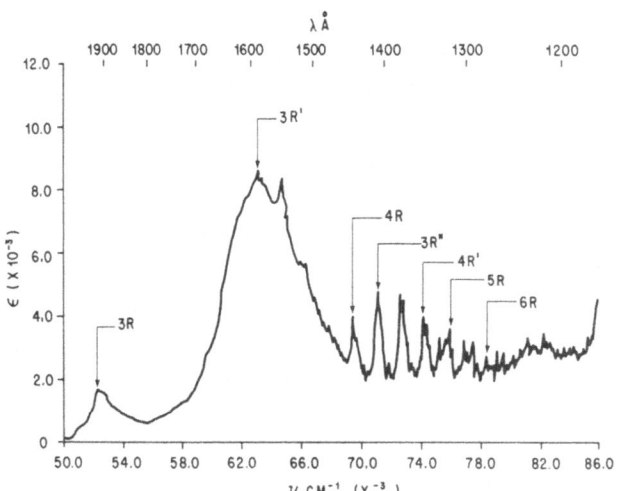

Figure 11

The far ultraviolet spectrum of cis 1,2-difluoroethylene

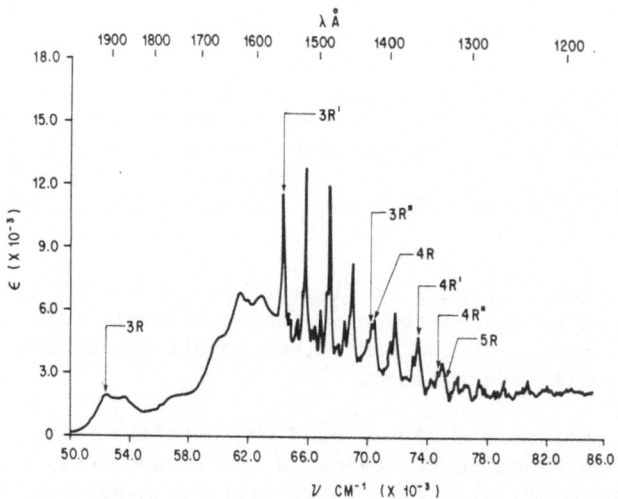

Figure 12

The far-ultraviolet spectrum of trifluoroethylene

broader than the others in several derivatives. This band is
known from the spectra of methylethylenes (30)(34). The Rydberg
bands have vibrational fine structure. It is dominated by quanta
of the C=C stretching vibration and double quanta of a twisting
vibration.

We found no evidence of Rydberg bands related to second and
higher IP's up to the limit of our measurements (120 nm).

If instead of fluorine, we substitute chlorine atoms onto
the double bond the picture changes drastically. As to the PE
spectra one might ask if the lowest IP is π or $\overline{Cl}$ in character ?
The answer is that it is still essentially π. The PE spectrum
of chloroethylene was measured by Lake and Thompson (32). The
lowest IP is at 10.00 eV (adiabatic) or 10.18 eV (vertical), at
somewhat lower energies than in ethylene itself (10.51 eV).
Chlorine has two lone pair orbitals one of which has its axis
parallel with those of the π-orbitals of the carbons the other
is perpendicular to it as well as to the direction of the C-C
bond. We could call it a typical σ lone pair, "n". The
π lone pair interacts with the C – C π bond raising

its energy somewhat (i.e. making the first IP lower); the frontier orbital is now of C–C–Cℓ character, bonding in C–C but antibonding in C–Cℓ. On the other hand the π lone pair will gain
some C–Cℓ bonding character and its energy slightly lowered (its
IP will be heightened). This amounts to a lifting of the degeneracy of the "original" Cℓ lone pair orbitals. Actually we find
two typical, sharp lone pair peaks in the PE spectrum at 11.72
and 11.87 eV respectively.

The ultraviolet spectra of chloroethylenes were measured by
Walsh and his collaborators (35)(36). Although some problems
remain the main points are clear. Outside of the π* ← π band
(180–160 nm) we find ns, np and nd type Rydberg series, all converging to the π IP. The first member of the ns series seems to
be a broad band between 200 and 190 nm. In the higher frequency
part of·the 200–120 nm region there are bands which can be interpreted as Rydberg bands related to the $\overline{Cℓ}$ lone pair IP's. Further
work is needed on this, we think.

Scott and Russell(37) measured the VUV spectra of several
fluorochloroethylenes. They are of the same general type. Neither the π* ← π, nor the three Rydberg series which converge to
the π IP are significantly displaced. This again substantiates
the relative insensitivity of the energy of π orbitals to fluorine substitution.

Benzene derivatives

The energy levels and transitions in benzene were the subject of hundreds of publications. One of the more recent problems is the possible existence of a σ level between the e_{1g} and
a_{2u} levels (π) in the ground state (38–41). The least that we
can say is that energies of the $a_{2u}(\pi)$ and the highest σ orbital
are very close to one other, probably within the validity of
Koopmans'theorem. Open-shell theoretical chemistry must become
a great deal more accurate before we can decide if the static
order of these three orbitals is the same as the order of the
observed IP's. One thing is sure: this delicate order can be
modified by substitution. In particular in fluorinated benzene
derivatives the order is surely $e_{1g}(\pi) > a_{2u}(\pi) > e_{2g}(\sigma)$. This
is another consequence of the strong stabilizing effect of fluorine substitution on σ-orbitals. The PE spectrum of 1,3,5-tri-
fluorobenzene shows this rather clearly: the fine-structured π
bands are followed by the more diffuse first σ band (42) (Fig.13).
In mono – and difluorobenzenes the two latter bands overlap.

There is no need for discussing the near-ultraviolet spectrum of benzene here. As it is well known three π* ← π bands
originate from the $e_{2u} ← e_{1g}$ transition, with excited states

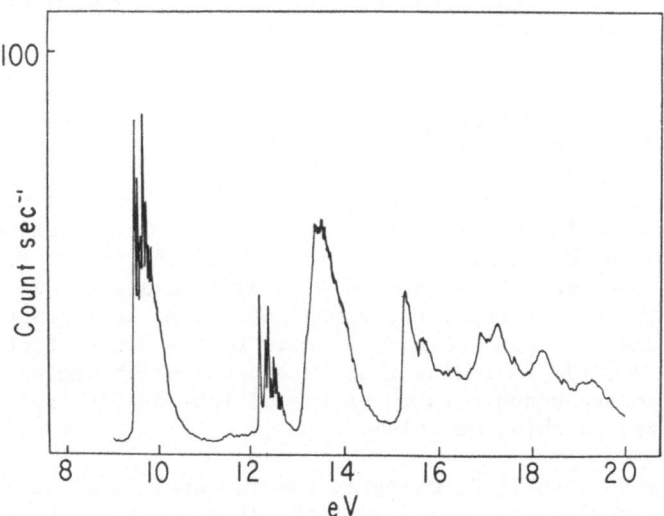

Figure 13

The photoelectron spectrum of 1,3,5-trifluorobenzene

B_{2u}, B_{1u} and E_{1u}. Only the last one is electronically allowed and it is centered at about 182 nm. Are there any other valence-shell transitions ? It is hard to tell. There is a broad band near 65000 cm^{-1}, at the high frequency side of the very strong E_{1u} band which might be $b_{2g} \leftarrow e_{1g}$ that is $^{1}E_{2g} \leftarrow {}^{1}A_{1g}$. It is, of course forbidden in benzene but it is allowed in many derivatives. Its intensity is comparable to that of the B_{1u} band. All other bands are better interpreted as Rydberg transitions. Wilkinson (43) found four Rydberg-series, one of p and three of f or forbidden d type. The latter are probably the three electronically allowed members of the f manifold and are very close to each other for members with the same n. Series of the s and d type are electronically forbidden. This interesting situation is explained in Table 6 in which all the selection rules are contained, for both Rydberg and valence-shell transitions. All these series converge to the lowest π IP.

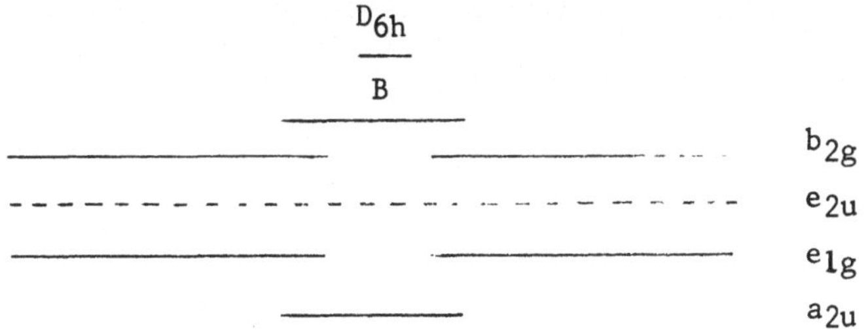

$$D_{6h}$$

B

_____ b_{2g}

- - - - - - - - - - e_{2u}

_____ e_{1g}

_____ a_{2u}

Valence-shell transitions	Resulting excited states	
$e_{2u} \leftarrow e_{1g}$	B_{2u}	forb.
	B_{1u}	forb.
	E_{1u}	x,y
$b_{2g} \leftarrow e_{1g}$	E_{2g}	forb.
$b_{2g} \leftarrow a_{2u}$	B_{1u}	forb.
$e_{2u} \leftarrow a_{2u}$	E_{2g}	forb.

Rydberg transitions	Resulting excited states	
$ns \leftarrow e_{1g}$	E_{1g}	forb.
$np \leftarrow e_{1g}$	E_{1u}	x,y
	$A_{1u} + \underline{A_{2u}} + E_{2u}$	z
$nd \leftarrow e_{1g}$	E_{1g}	forb.
	$E_{2g} + A_{2g} + A_{1g}$	forb.
	$E_{1g} + B_{2g} + B_{1g}$	forb.
$nf \leftarrow e_{1g}$	E_{1u}	x,y
	E_{2u}	forb.
	E_{2u}	forb.
	$A_{1u} + \underline{A_{2u}} + E_{2u}$	z
	$B_{1u} + B_{2u} + E_{1u}$	x,y

Table 6

Summary of the electronic transitions (valence-shell and Rydberg) and selection rules for benzene.

What happens if we perturb the symmetry of benzene ?
Gilbert, Sauvageau and Sandorfy (44)(45) measured the absorption
spectra of a nearly complet set of fluorobenzenes. Some intriguing observations can be made.

a) Even though under the reduced symmetry of many of the derivatives the highly allowed $^1E_{1u} \leftarrow ^1A_{1g}$ band should be split we
never found any clear evidence for this. One of the components
resulting from the split must be much stronger than the others.

b) The same Rydberg series appear in all fluorinated derivatives
regardless their symmetry. (Actually under the given experimental conditions we found only one p series and one series with a
low quantum defect). This shows that fluorine substitution does
not disturb strongly the π levels. Actually it was found that
vibronic interactions are the main source of intensity for D_{6h}
forbidden bands even in derivatives in which these become electronically allowed.

The PE spectra of a number of other substituted benzenes
were thoroughly studied. Halogenated benzenes were examined by
Turner et al. (1)(41), Lindholm et al. (38)(46)(47) and by Rabalais
et al.(48) and toluene, fluorobenzene, phenol and aniline by
Debies and Rabalais (49). We know from these very interesting
works that the $e_{1g}(\pi)$ level of benzene is split by 0.5–1.0 eV in
all the monoderivatives and that these π levels are followed by
what were the $e_{2g}(\sigma)$ and $a_{2u}(\pi)$ levels in benzene. In chloro,
bromo – and iodo benzene the halogens supply two lone pair orbitals, one conjugating with the π AO's of the ring and one
having its axis in the plane of it. This yields two close lying
IP's just us in vinylchloride. The in-plane one yields the
sharper band, the other is stabilized by interaction with the
other π-orbitals, it is somewhat broader and shifted to higher
energies. While the lowest π IP of these molecules is never
significantly higher than about 9 eV, the lone pair levels are
near 13.9 eV with fluorobenzene, 11.8 with chlorobenzene, 11.3
with bromobenzene and 10.4 eV with iodobenzene. In no case are
the orbitals of lone pair character the lowest IP's. The same is
true for the other substituents. According to Debies and Rabalais
(49) the "n" level follows the two split $e_{1g}(\pi)$ levels in aniline,
it comes still before the highest σ level for phenol but are
below the $\sigma(e_{2g})$ and $a_{2u}(\pi)$ levels in fluorobenzene. Murrell and
Suffolk (50) has shown that in pyridine too the n orbital has a
higher energy than the lowest π levels.

The main point is, from our point of view that the frontier
orbital is π in all simple benzene derivatives. Thus we expect
that the VUV spectra will contain mainly Rydberg series converging to the lowest π IP just as in benzene itself. Actually p
and f or d type series are found in all cases known to us. There

might be additional Rydberg series in the less symmetrical derivatives but little is known about these. In 1,3 and 1,4-difluo-
robenzenes we found a series converging to the second π IP which
resulted from the splitting of benzene's e_{1g} level. With chlorobenzene Price and Walsh (51) found bands near 140 nm which they
assigned as Rydberg bands belonging to the $\overline{Cl}$ IP (52).

We believe that the far ultraviolet spectroscopy of benzene
derivatives should be given renewed attention in view of recent
PE results and that this sector is not yet ripe for a review.

Molecules containing divalent sulfur

Organic molecules containing divalent sulfur pose certain
problems related to the extent of the involvement of the sulfur
3d orbitals in the wave functions of the ground - and excited
states and the interpretation of their ultraviolet absorption
spectra. We have carried out, in our laboratory (53) a comparative PE and VUV study of the spectra of tetrahydropyran and
tetrahydrothiopyran on the one hand and methylvinylether and
methylvinylthioether on the other.

As to the two saturated molecules their PE spectra are very
similar except for a shift to lower energies for the sulfur compound. The first PE bands have the typical lone pair appearance.
The vertical IP's are 9.46 and 8.39 eV for the ether and for the
thioether respectively reflecting the better shielding of the S
3p orbital. The second IP is stabilized in the sulfur compound
due partly to the large C-S bonding character of the respective
orbital. The difference between the first two IP's is 1.44 eV
for tetrahydropyran and 2.55 eV for tetrahydrothiopyran (54).
The main point is that there is a fargoing analogy between the
PE spectra of saturated ethers and thioethers. They give absolutely no evidence of the participation of S 3d AO's in the
valence orbitals.

The ultraviolet spectra of the two compounds are very different, however. That of tetrahydropyran was studied in detail
by Hernández (55). The first, highly structured band system
whose 0-0 band is at 51940 cm^{-1} is best described as a 3s $\leftarrow$ n
(O 2p) transition which is analogous to similar transitions
found in water and all simple alcohols and ethers. It is an extended band system dominated by low frequency ring vibrations.
The second band system (54000-60000 cm^{-1}) can be assigned to the
3p $\leftarrow$ n (O 2p) manifold and the strong band centered at 68000 cm^{-1}
probably receives contributions from several Rydberg-transitions
(3d, 4s, etc.). Tetrahydrothiopyran has a rather different spectrum, its low frequency part consisting of a number of fine bands
which cannot belong to the same transition (Fig. 14). McGlynn
et al. (56) predicted on the basis of theoretical calculations

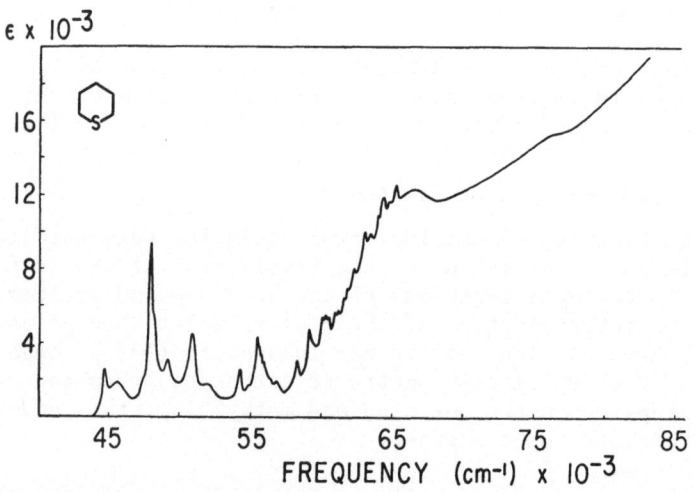

Figure 14

The far-ultraviolet absorption spectrum of tetrahydrothiopyran

that the lowest energy transition should go to an excited orbital largely S 4s in character followed by three transitions to
degenerate S 3d levels and predicted the very small energy difference among these.

A comparison between s, p, d, ... term values found with
O and S compounds is helpful in the assigning these bands. In
both the atoms and in simple molecules the sulfur terms are
somewhat lower than the oxygen terms but in all cases s > p > d.
Now the first sharp peak at 44700 cm^{-1} yields a term value of
23000 with the 8.39 eV IP which is an acceptable value for an s
type Rydberg orbital. The following sharp peaks at 47990, 50930,
54260 and 55480 cm^{-1} give respectively 19700, 16700, 13420 and
12200 cm^{-1}. It is logical to assign the first one to the first
member of a p type Rydberg series while the three other bands
may correspond to the first members of Rydberg series where the
excited orbital is of mixed p-d character containing an increasing amount of d in the above order.

It is believed that this is the essential of the participation of d electrons in the low lying electronic states of organic
sulfur compounds: because of the availability of d AO's of relatively low energy the first members of the d or mixed p-d type
Rydberg series have low frequencies. p-d mixtures are made possible by the low symmetry of these molecules.

The conditions are different with unsaturated ethers and
thioethers. The PE spectra are again very similar. For methylvinylether the first IP (vertical) is 9,14 eV. The lone pair is
conjugated with the double bond and the frontier orbital is essentially a π orbital, bonding in C = C but antibonding in C - O
(57). Actually the vibrational fine structure is dominated by a
frequency of about 1300 cm^{-1} which might be lowered C = C mode
or the heightened C - O mode or a mixture of these.

Methylvinylthioether has its first PE band at 8.44 eV. It
is also a π type IP but the fine structure is now dominated by a
frequency of 600 cm^{-1} showing that the main change occurs in the
C - S bond, not in the C = C bond. Accordingly the PE band is
sharper with the thioether than with the ether.

To our surprise we find that the UV spectra of this unsaturated ether and thioether are also fairly similar (Figs. 15 and
16).

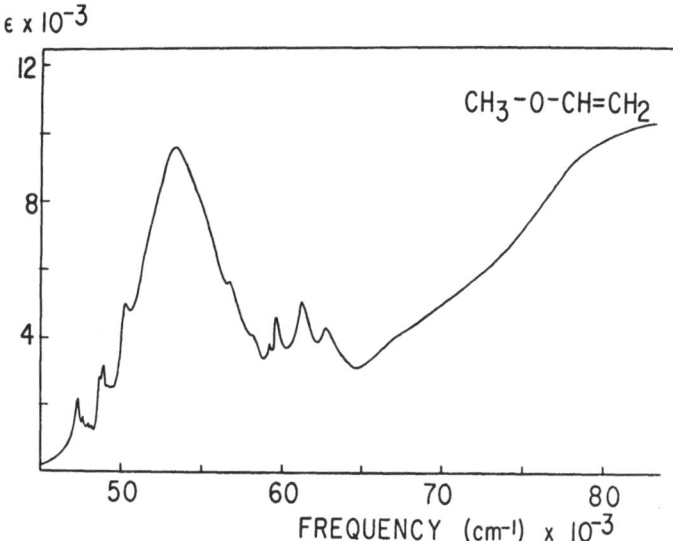

Figure 15

The far ultraviolet absorption spectrum of methylvinylether

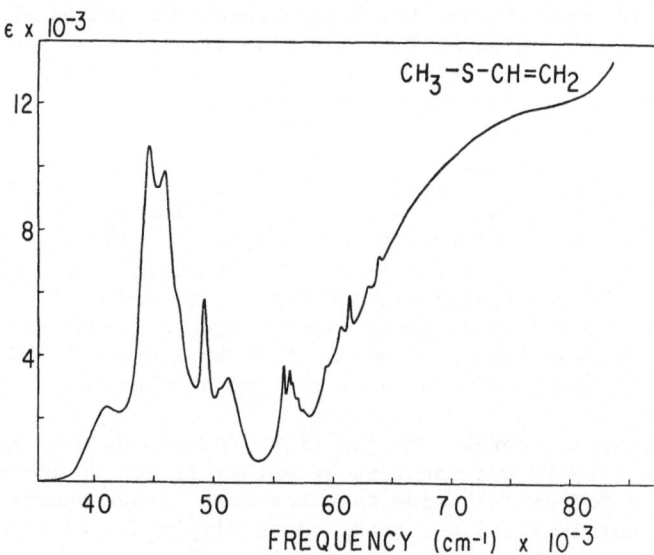

Figure 16

The far ultraviolet absorption spectrum of methylvinylthioether

In the spectrum of methylvinylether the broad $\pi^* \leftarrow \pi$ band is centered at 53250 cm^{-1}. It is in partial coincidence with a 3s type Rydberg band whose fine structure is dominated by a relatively high C = C frequency (about 1450 cm^{-1}). We find p and d type Rydberg bands at higher frequencies. The main point is that except for the expected shifts the UV spectrum of methylvinylthioether can be described in almost the same manner. There is no evidence for the intervention of the S 3d orbitals in the low lying excited states in the case of this unsaturated thioether.

It appears then that this is characteristic of transitions originating from the essentially lone pair orbital of sulfur only. When the lowest IP is essentially $\pi^* \leftarrow \pi$ we find no low lying levels of d character.

A part of the results on which this review is based have been obtained in our own laboratory. The author wishes to express his appreciation to his former and present collaborators: Prof. B.A. Lombos, Mr. P. Sauvageau, Dr. G. Bélanger, Dr. R. Gilbert, Dr. A.A. Planckaert, Mr. J. Doucet, Mr. R. Macaulay and Mr. D. Goutier for their valuable contributions.

(1) D.W. Turner, C. Baker, A.D. Baker, and C.R. Brundle, "Molecular Photoelectron Spectroscopy", Interscience, New York and London. 1970.

(2) C.R. Brundle, M.B. Robin, and H. Basch, J. Chem. Phys., 53, 2196 (1970).

(3) J. Doucet, P. Sauvageau, and C. Sandorfy, J. Chem. Phys., 58, 3708 (1973).

(4) D.C. Frost and C.A. McDowell, Proc. Roy. Soc. London A241, 194 (1957).

(5) W.C. Price, J. Chem. Phys., 4, 539 (1936).

(6) C.R. Zobel and A.B.F. Duncan, J. Am. Chem. Soc., 77, 2611 (1955).

(7) G. Herzberg, "Molecular Spectra and Molecular Structure", Vol. III. Van Nostrand, Princeton, N.J. 1967.

(8) P. Sauvageau, R. Gilbert, and C. Sandorfy, to be published.

(9) W.R.Harshbarger, M.B. Robin and E.N. Lassettre, J. Electron Spectrosc. 1, 319 (1973).

(10) P. Sauvageau, R. Gilbert, P.P. Berlow, and C. Sandorfy, J. Chem. Phys., 59, 762 (1973).

(11) R.J. Schoen, J. Chem. Phys., 37, 2032 (1962).

(12) E.E. Koch and M. Skibowski, Chem. Phys. Lett., 9, 429 (1971).

(13) S. Stokes and A.B.F. Duncan, J. Am. Chem. Soc., 80, 6177 (1958).

(14) W.C. Price, J. Chem. Phys., 4, 547 (1936).

(15) R. Gilbert, P. Sauvageau, and C. Sandorfy, to be published.

(16) B.A. Lombos, P. Sauvageau, and C. Sandorfy, J. Mol. Spectry 24, 253 (1967).

(17) C. Sandorfy, B. Lombos, and P. Sauvageau. XIII Colloquium Spectroscopicum Internationale, A. Hilger, London, 1967, p. 370.

(18) J.W. Raymonda and W.T. Simpson, J. Chem. Phys. 47, 430 (1967).

(19) E.F. Pearson and K.K. Innes, J. Chem. Phys., 30, 232 (1969).

(20) W.A. Lathan, L.A. Curtiss, and J.A. Pople, Mol. Phys., 22 1081 (1971).

(21) E.N. Lassettre, A. Skerbele, and M.A. Dillon, J. Chem. Phys., 49, 2382 (1968).

(22) J.N. Murrell and W. Schmidt, J. Chem. Soc. Faraday Trans. II, 1709 (1972).

(23) B. Narayan, Mol. Phys., 23, 281 (1972).

(24) F.A. Bower, International Conference on the Ecology and Toxicology of Fluorocarbons; Andover, Mass. 1972 (E.I. Du Pont de Nemours and Company, Wilmington, Del.).

(25) A.D. Baker, C. Baker, C.R. Brundle, and D.W. Turner, Int. J. Mass. Spec. Ion Phys., 1, 285 (1968).

(26) W.C. Price and W.T. Tutte, Proc. Roy. Soc. (London), A174, 207 (1940).

(27) P.G. Wilkinson and R.S. Mulliken, J. Chem. Phys., 23, 1895 (1955).

(28) P.G. Wilkinson, Can. J. Phys., $\underline{34}$, 643 (1956).

(29) A.J. Merer and I. Schoonveld, Can. J. Phys., $\underline{47}$, 1731 (1969).

(30) A.J. Merer and R.S. Mulliken, Chem. Rev., $\underline{69}$, 639 (1969).

(31) G. Bélanger and C. Sandorfy, J. Chem. Phys., $\underline{55}$, 2055 (1971).

(32) R.F. Lake and Sir H. Thompson, Proc. Roy. Soc. (London) $\underline{A315}$, 323 (1970).

(33) R. Bralsford, P.V. Harris, and W.C. Price, Proc. Roy. Soc. (London), $\underline{A258}$, 459 (1960).

(34) M.B. Robin, H. Basch, N.A. Kuebler, B.A. Kaplan, and J. Meinwald, J. Chem. Phys., $\underline{48}$, 5037 (1968).

(35) A.D. Walsh and P.A. Warsop, Trans. Faraday Soc., $\underline{64}$, 1418, 1425 (1968).

(36) A.D. Walsh, P.A. Warsop, and J.A.B. Whiteside, Trans. Faraday Soc., $\underline{64}$, 1432 (1968).

(37) J.D. Scott and B.R. Russell, Chem. Phys. Letters $\underline{9}$, 375 (1971).

(38) L. Åsbrink, O. Edquist, E. Lindholm, and L.E. Selin, Chem. Phys. Lett., $\underline{5}$, 192 (1970).

(39) C.R. Brundle, M.B. Robin, and N.A. Kuebler, J. Am. Chem. Soc., $\underline{94}$, 1466 (1972).

(40) B. Narayan and J.N. Murrell, Mol. Phys., $\underline{19}$, 169 (1970).

(41) A.D. Baker, C.R. Brundle, and D.W. Turner, Int. J. Mass. Spectr. Ion Phys., $\underline{1}$, 443 (1968).

(42) R. Gilbert, P. Sauvageau, and C. Sandorfy, Chem. Phys. Lett., $\underline{17}$, 465 (1972).

(43) P.G. Wilkinson, Can. J. Phys., $\underline{34}$, 596 (1956).

(44) R. Gilbert, P. Sauvageau, and C. Sandorfy, Can. J. Chem., $\underline{50}$, 543 (1972).

(45) R. Gilbert and C. Sandorfy, Chem. Phys. Lett., $\underline{9}$, 121 (1971).

(46) L. Åsbrink, E. Lindholm, and O. Edquist, Chem. Phys. Letters $\underline{5}$, 609 (1970).

(47) L. Åsbrink, C. Fridh and E. Lindholm, Chem. Phys. Letters $\underline{15}$, 567 (1972).

(48) J.W. Rabalais, L.O. Werme, T. Bergmark, L. Karlsson, and K. Siegbahn, Int. J. Mass. Spectr. Ion Phys., $\underline{9}$, 185 (1972).

(49) T.P. Debies and J.W. Rabalais, J. Electron Spectrosc., $\underline{1}$, 355 (1972/73).

(50) J.N. Murrell and R.J. Suffolk, J. Electron Spectrosc., $\underline{1}$, 471 (1973).

(51) W.C. Price and A.D. Walsh, Proc. Roy. Soc. (London), $\underline{A191}$, 22 (1947).

(52) A. Quemerais, M. Morlais, and S. Robin, Compt. Rend. Acad. Sci. Paris, $\underline{B265}$, 649 (1967).

(53) A.A. Planckaert, J. Doucet, and C. Sandorfy, to be published.

(54) D.C. Frost, F.G. Herring, A. Katrib, C.A. McDowell, and
      R.A.N. McLean, J. Phys. Chem., 76, 1030 (1972).
(55) G.J. Hernández, J. Chem. Phys., 38, 2233 (1963).
(56) S.D. Thompson, D.G. Carroll, F. Watson, M. O'Donnell, and
      S.P. McGlynn, J. Chem. Phys., 45, 1367 (1966).
(57) D. Goutier and R. Macaulay, to be published.

THE ETHANE PROBLEM

C. SANDORFY

Université de Montréal, Département de Chimie
Montréal, Québec, Canada

I. The ground state

It all goes back to 1935 when Mulliken applied united molecule considerations to ethane (1). If we take ethane for two
semi-united atoms bound together by a σ-bond we have the following configuration:

$$[sa_1]^2 \quad [sa_1]^2 \quad [\pi\ e]^4 \quad [\pi\ e]^4 \quad [\sigma + \sigma]^2 \ {}^1A_1$$

$$CH_3 \qquad CH_3 \qquad CH_3 \qquad CH_3 \qquad C-C$$

where the carbon 1s orbitals are omitted and C_{3v} symmetry is
taken for the methyl groups. Now, the important point is that
if we allow for interaction between the two "sides" the C-C orbital may not come out first but it may be squeezed between the
two degenerate MO's resulting from the interaction of the two
$[\pi\ e]$ orbitals. Before considering the results of modern, highly
computerized quantum chemical calculations let us have a look at
the correlation diagrams.

The united molecule of ethane is the fluorine molecule, F_2.
For F_2 the ground state has the following configuration (2):

$$KK\ (\sigma_g 2s)^2\ (\sigma_u 2s)^2\ (\sigma_g 2p)^2\ (\pi_u 2p)^4\ (\pi_g 2p)^4\ {}^1\Sigma_g^+$$

where the order of the three last levels is to be noted.

The orbital correlation diagram for X_2H_6 molecules under

D_{3d} symmetry is given in Herzberg's Vol. III (3). This time the order is

$$2a_{1g} < 2a_{2u} < 1e_u < 1e_g < 3a_{1g}$$

but at shorter X-X distances the $3a_{1g}$-$1e_g$ order is reversed.

In Katagiri and Sandorfy's Pariser-Parr type treatment(4) which was the first attempt to interpret the spectra of small paraffins by a semi-empirical method the order of the orbitals is again

$$1a_{1g} < 2a_{2u} < 1e_u < 3a_{1g} < 1e_g .$$

In recent years a number of ab initio calculations were performed on ethane (5-12). These were reviewed by Lathan, Curtiss and Pople (13) (LCP). They generally agree that the ground state configuration is

$$(1a_{1g})^2 \; (1a_{2u})^2 \; (2a_{1g})^2 \; (2a_{2u})^2 \; (1e_u)^4 \; (3a_{1g})^2 \; (1e_g)^4 \; {}^1A_{1g} .$$

Ab initio calculations tend to be quite reliable if they apply to the ground state. The above order is very probably the correct one. The least we can say is, in view of all these considerations, that the two highest levels are $3a_{1g}$ and $1e_g$ and that they are probably very close to one other.

LCP (13) pointed out that the order $3a_{1g} < 1e_g$ appears to apply for a wide variety of basis sets. The $3a_{1g}$ orbital is mainly carbon-carbon (σ) bonding while $1e_g$ is carbon-hydrogen bonding but carbon-carbon antibonding.

Murrell and Schmidt (14) obtained $3a_{1g}$ on the top. These authors took care of representing adequately the so-called through space interactions between the methyl groups but used approximate values for three and four center integrals.

Macaulay (15), in this laboratory, performed an even more accurate calculation on ethane using the same geometry as LCP but a 6-31G basis set (16) while the former authors used a 4-31G basis set. He obtained the same order for the molecular orbitals that is,

$$1a_{1g} < 1a_{2u} < 2a_{1g} < 2a_{2u} < 1e_u < 3a_{1g} < 1e_g$$

with as orbital energies -11.2184, -11.2181, -1.0218, -0.8446, -0.5972, -0.5086 and -0.4837 atomic units in this order. This corresponds to -14.95, -12.73 and -12.11 eV for the three upper-most orbitals after multiplication by Robin's empirical 0.92 factor (17). The overlap populations are:

for $1e_g$, -0.385 in C-C and 0.313 in the C-H; (counting both partners)

and for $3a_{1g}$, 0.474 in C-C and 0.034 in the C-H.

Unfortunately, whatever the merits of these calculations they are still taking only insufficient account of correlation and we cannot say that the matter is definitely settled.

There is something more important from our point of view, however, than the static order of the orbitals. What we are really interested in is to identify the orbital of departure which corresponds to the first (lowest) ionization potential and to the ultraviolet transitions of lowest energy. Now, this must not be the highest orbital. Indeed, $1e_g$ and $3a_{1g}$ are so close to one-other that even a moderate departure of Koopmans' theorem could reverse their order.

II. The ion

Lathan, Curtiss and Pople (13) performed an ab initio open-shell calculation on the $C_2H_6^+$ ion. They varied the geometry and obtained the result that the cation (in its ground state) has the symmetry D_{3d} as does the molecule and species $^2A_{1g}$. The species $^2A_{1g}$ would mean that even though $1e_g$ is the frontier orbital in the unperturbed molecule the lowest vertical IP relates to ion-ization from $3a_{1g}$. This ion would have a very long C-C bond (about 1.8-2.0Å) and nearly planar CH_3 groups. The calculated adiabatic ionization energy obtained through these calculations was 10.24 eV while the experimental value is 11.56 eV (18).

The HeI photoelectron spectrum of ethane is known from the works of Turner and his coworkers (18)(19). Between about 11.5 and 14.5 eV there is a triple band with centers at 12.0, 12.7 and 13.4 eV. The first one has vibrational fine structure with the adiabatic IP at 11.56 eV and a vibrational progression of about 1170 cm^{-1}. These three bands must belong to ionization from the $1e_g$ and $3a_{1g}$ levels. This is quite certain. But which is which? In a ^2E_g state the ion is expected to be subjected to the Jahn-Teller effect and the 0.7 eV splitting between the two first maxima is of the expected order of magnitude. (Like in methane or cyclopropane, (18)). This, however, is compatible with either 12.0 and 12.7 or 12.7 and 13.4 for ^2E_g. Could the observed vibrational fine structure help ? As we shall see the same problem will have to be faced in relation with the ultra-violet absorption spectrum.

Ethane in its ground state has five vibrations which might be responsible for the observed frequency of 1170 cm^{-1}. These are:

ν_2	1375 cm^{-1}	CH sym. bending	a_{1g}
ν_6	1379	CH sym. bending	a_{2u}
ν_8	1486	CH antisym. bending	e_u
ν_{11}	1460	CH antisym. bending	e_g
ν_3	993	C-C stretching	a_{1g}

Now, as we have seen, the $3a_{1g}$ orbital is strongly C-C bonding and slighly C-H bonding while the $1e_g$ orbital is C-C antibonding and C-H bonding. Thus at first sight if we take out an electron from the $3a_{1g}$ orbital we should obtain an ion with a lengthened C-C distance; the vibrational structure of the related PE band being dominated by a decreased ν_3. This type of reasoning indirectly implies Koopmans' theorem. LCP say, at this point with some justification that because the $3a_{1g}$ orbital in "... neutral C_2H_6 is mainly localized on the carbon atoms, whereas $1e_g$ orbitals are distributed over both carbon and hydrogen ... removal of an electron from $3a_{1g}$ leads to a more pronounced 'positive hole' and more redistribution occurs". Actually, according to their calculations the HCH angle undergoes a significant change upon ionization from the $3a_{1g}$ level so that the appearance of a vibration of CH bending character in the fine structure of the first PE band would not be surprising. The 1170 cm^{-1} interval that is found could, of course, not be a decreased C-C frequency. If this interpretation is correct the observed vibrational interval can only be ν_2, the totally symmetrical CH bending vibration. It is perhaps strange, however, that despite of the considerable change in the C-C distance ν_3 does not show up prominently. Let us hope that this is due to the relatively low resolution of the PE spectrometer.

If the electron is departing from the $1e_{1g}$ orbital the C-C bond must contract since the orbital is antibonding in the C-C link. Some rearrangement is likely to occur but it is reasonable to expect that the bond will remain relatively short. Lassettre, Skerbele and Dillon (20) made this suggestion about five years ago. In this case we should observe an increased or nearly unchanged ν_3 frequency which would be, however, highly mixed with C-H bending. The increased ν_3 and the decreased ν_2 would have fairly similar frequencies. What we see then might be two close-lying vibrations in the fine structure which we cannot resolve under the given experimental conditions. All we can say is that the observed vibrational fine structure of the first PE band is compatible with both interpretations. In both

cases we should expect a mainly CH bending vibration to be prominent in the fine structure.

Deuteration can provide relative help only. It is enough to
note that in the <u>ground state</u>, ν_2 has a C_2H_6/C_2D_6 ratio of 1.187
while ν_3's ratio is 1.165, almost the same! The C-C stretching
and C-H bending motions are highly mixed in the ground state,
probably more so in C_2D_6 than in C_2H_6.

We measured the PE spectra of a few deuterated ethanes.
Perhaps the most significant result was that with both CH_3-CD_3
and CH_2D-CH_2D we obtained the same vibrational interval, about
1050 cm^{-1}. Since the latter has no degenerate vibrations this
shows that we do not need to involve degenerate vibrations to
explain the observed fine structure. This makes it very likely
that the observed interval belongs to a totally symmetrical vibration that is ν_2 or ν_3.

We were hoping to resolve a CH_3 and a CD_3 bending vibration
for CH_3-CD_3 but we only found one well defined interval, about
1050 cm^{-1}. The shape of the bands is asymmetrical, however, so
this may be just a matter of resolution.

For C_2D_6 the vibrational interval was found to be 920 cm^{-1}
giving an isotopic ratio of 1.27.

This large ratio shows that the vibration is a mainly C-H
bending vibration[*]. As we have stated this is compatible with
either $^2A_{1g}$ or ^2E_g species for the ion. Photoelectron spectroscopy did not supply us with a decisive argument to choose
between these possibilities. Only the general appearance of the
PE spectrum , the first band (11.56 eV) having fine structure,
the two others (12.7 and 13.4) diffuse, seems to support $^2A_{1g}$.

III. The ultraviolet spectra

That ethane possesses electronic absorption bands exhibiting
vibrational fine structure became known in 1967 through the vacuum ultraviolet measurements of Raymonda and Simpson (21) and of
Lombos, Sauvageau and Sandorfy (22)(23) and by the electron impact work of Lassettre, Skerbele and Dillon (20).

Between 70000 and 90000 cm^{-1} there are two, perhaps three
electronic band systems with vibrational structure. (Fig. 1).
We first consider the lower frequency part of this spectral
region.

---

[*] Professor W.C. Price commented at this point that the same vibrational intervals are found for methyl halides; this is another argument in favor of the CH character of this vibration.

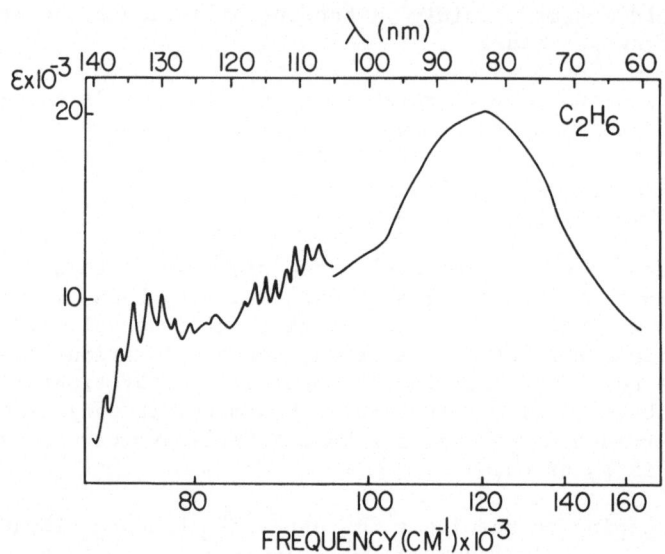

Figure 1

The vacuum ultraviolet spectrum of ethane[*]

A prominent vibrational progression is found with average
interval of about 1150 cm^{-1} although the many shoulders that can
be seen indicate the presence of more than one vibration in the
structure. (Or irregularities due to the Jahn-Teller effect
(22)(24). Higher resolution would be needed to ascertain this).
The first interval is certainly less, about 1080 cm^{-1}. The same
applies to the deuterated ethanes. C_2D_6 has an average interval
of about 920 cm^{-1} with the first frequency about 780 cm^{-1}, in
CH_3-CD_3 the main interval is 1060, the first one 900 cm^{-1}. We
assigned the main interval to a CH bending vibration and the
electronic transition to a $^1E_u \leftarrow {}^1A_{1g}$ transition. The C_2H_6/C_2D_6
isotopic ratio was found to be 1.24 by Lassettre, Skerbele and
Dillon (20), 1.31 by Raymonda and Simpson (21) and 1.29 by us
(22). The electronic assignment was thought to correspond to the
orbital assignment $a_{2u} \leftarrow e_g$ and was based on the CH character of
the vibrational progression and Katagiri and Sandorfy's (4) calculations. It implied <u>perpendicular</u> polarization for the electronic transition. This assignment differed from Raymonda and
Simpson's who on the basis of their independent system model suggested that it is $^1A_{2u} \leftarrow {}^1A_{1g}$ and polarized parallel to the C-C

* This spectrum has been measured by R. Gilbert and P. Sauvageau
in the author's laboratory using a helium discharge tube for
lightsource. A similar spectrum has been obtained by E.E. Koch
and M. Skibowski (Chem. Phys. Lett., 9, 429 (1971)) who used a synchrotron source. cf. also Lassettre's electron impact spectrum(20).

line.

This assignment is important from both the spectroscopic and photochemical point of view. We continue its discussion on the basis of the high resolution work of Pearson and Innes (25) and then put together the available UV and PE results.

Pearson and Innes in 1969 measured the spectra of C_2H_6 and C_2D_6 with a 21 ft. instrument with resolution of the order of 10 cm^{-1} for the sharpest maxima. In the case of C_2D_6 they were able to resolve the rotational contour of the 0-0 band. The latter is at 71,099 for C_2D_6 and 70,368 cm^{-1} for C_2H_6. The difference, 731 cm^{-1}, is quite large and can only be due to a large reduction in a CH frequency (and its zero-point energy) occurred upon excitation. The band is obviously allowed. (The oscillator strength (22) is about f = 0.30). It is very important that Pearson and Innes were able to resolve three vibrational frequencies, all of them totally symmetrical. These are, in cm^{-1} :

		Ground state	Excited state
C_2D_6	ν_1	2083.0	1751
	ν_2	1154.5	781
	ν_3	843.0	616
C_2H_6	ν_1	2953.7	2215
	ν_2	1388.4	1053
	ν_3	994.8	755

These assignments are in excellent agreement with the Teller-Redlich product rule. The main point is that all three frequencies decrease upon excitation. The decrease in the C-H stretching frequency ν_1 is very significant because it shows that the C-H distance decreased considerably in the excited state. (LCP did not predict this for the $^2A_{1g}$ ion).

A further result is that the ν_2(CH) progression is longer than the ν_3(C-C) progression, showing that the more important changes occur in the HCH angles. (We believe that the lower frequency in our above mentioned moderate resolution spectra, 1080 cm^{-1} for C_2H_6, is ν_3 and the higher one, 1150 cm^{-1} is ν_2).

Pearson and Innes carried out a band contour analysis on the 0-0 band of C_2D_6 which was well resolved. Using the accurate rotational constant known for the ground state they there able to show conclusively that the observed rotational contour cannot be matched if the transition is a parallel transition even if allowance is made for the possibility of internal rotation in the excited state. The band shape was matched supposing that the transition is perpendicular, with a vibronic angular momentum around the figure axis. Since the transition is electronically allowed the only possible assignment is $^1E_u \leftarrow ^1A_{1g}$ in agreement with that of Lombos, Sauvageau and Sandorfy (22). We believe that this definite.

We are also interested in the "orbital assignment". How can the ^1E_u state be obtained ? Should departure take place from le_g the excited orbital could be a_{1u}, a_{2u} or e_u. (In the latter case the direct product is $A_{1u} + A_{2u} + E_u$). If the level of departure is $3a_{1g}$ the excited orbital must be e_u.

Now, a very important point is the Rydberg character of the transition which is strongly indicated by the fargoing resemblance between the fine structures of the PE and UV bands. Under D_{3d} symmetry the ns, np, nd Rydberg orbitals have symmetries a_{1g}, $a_{2u} + e_u$ and $a_{1g} + 2e_g$ respectively. If the molecule keeps its center of symmetry in the excited state, ns and nd type transitions are $g \leftrightarrow g$ and forbidden by the Laporte rule from both $3a_{1g}$ and le_g. In view of the high intensity of the band this leaves us with the possibility of an np type Rydberg transition. This is possible from both $3a_{1g}$ and le_g.

From the former we have $2e_u \leftarrow 3a_{1g}$, from the latter $2e_u \leftarrow le_g$ or $3a_{2u} \leftarrow le_g$, all with perpendicular polarization. The term value computed from the first adiabatic IP (11.56 eV = 93.243 cm^{-1}) and the 0-0 band of Pearson and Innes (70,368 cm^{-1}) is 22,875 cm^{-1}. It is an acceptable although somewhat too high a value for a 3p Rydberg term. It would be even higher, about 29000 cm^{-1}, with the next IP, about 12.4 eV (roughly estimated adiabatic value) and as we have seen the transition to a 3s Rydberg orbital would be $g \leftrightarrow g$.

First, let us suppose that the orbital of departure is the same for both the ionization and first electronic transition of lowest energy. The great similarity between the vibrational fine structure of the first PE and first UV bands leads us naturally to this supposition. Then, if the transition is $2e_u \leftarrow 3a_{1g}$ the excited Rydberg state would be related to the $^2A_{1g}$ ion which possesses a very long C-C bond but C-H bonds of near normal length (11). The $2e_u$ Rydberg orbital has some C-C bonding and C-H antibonding character but being of Rydberg character it is unlikely to be very far of being nonbonding. Thus the addition of the $2e_u$ orbital plays about the same role as electronic rearrangement in the case of the ion. Cleavage could be expected in the long C-C bond through its highly anharmonic stretching vibration.

If, on the other hand, the excitation takes place from the le_g orbital the resulting Rydberg state should resemble the

^2E_g ion which has longer C–H bonds and a relatively short C–C bond. The addition of the $3a_{2u}$ orbital would weaken all the bonds while the $2e_u$ orbital would strengthen the C–C and weaken the C–H bonds to some extent. Again, since the Rydberg orbital is large, it is unlikely to create qualitative differences with respect to the ion. We would then expect C–H rather than C–C cleavage.

The above argument is certainly oversimplified. It is based on symmetry requirements and disregards other factors such as the Jahn-Teller effect. We believe, however, that the latter is a much less important factor in the Rydberg states than in the ion. Indeed no sure manifestation of it has been found in the UV spectrum (25).

We shall have to look at the photochemical facts to decide the issue.

Before doing this, however, we should like to consider another interesting possibility. As we said the $^2A_{1g}$ and ^2E_g ions are both compatible with the observed fine structure, so the resemblance between the fine structures of the PE and UV bands is not really a proof for a common orbital of departure in this case. In other terms, if, as Pople suggests there is a reversal in the static order of the orbitals of ethane upon ionization one might ask if the electronic transition is also affected by this reversal. We usually apply, tacitly, a kind of Koopmans' theorem to electronic transitions. In the case of Rydberg transitions this seems to be more natural, since Rydberg bands generally resemble the respective PE band. First members of Rydberg series, however, are the less "Rydberg" of all members. Because of the not quite negligible size of the molecular core they often have some valence-shell character. Thus even if rearrangement reverses the Koopmans' theorem order of the IP's it may not do this with the order of electronic transitions. In the particular case of the first singlet-singlet transition of ethane this would mean that whether or not the first PE band is from $3a_{1g}$ and leads to a $^2A_{1g}$ ion, the UV band could be $3a_{2u}$ $1e_g$ or $2e_u$ $1e_g$. In Rydberg language the transition would be again to a 3p type orbital.

This assignment is compatible with all the main theoretical and experimental observations which are known to us.

a) It does not contradict Pople's suggestion that the 11,56 eV band corresponds to a $^2A_{1g}$ ion.

b) It agrees with Pearson and Innes' result that the polarization of the band is perpendicular to the C–C direction.

c) It would imply that the main vibrational progression is a mainly CH bending mode.

d) It would make it understandable that between 8 and 10 eV the main photochemical primary steps involve the breaking of C-H, not C-C bonds.

e) It would be in conformity with the orbital correlation diagram. (Unlike $e_u \leftarrow 3a_{1g}$).

The disconcerting conclusion is that neither quantum chemistry, nor photoelectron, nor vacuum ultraviolet spectroscopy has enabled us to choose between excitation from $3a_{1g}$ or $1e_g$. We shall see, however, in the next section that in order to explain well established photochemical facts we have to assume that the first UV band departs from the $1e_g$ level whatever might be the case for the ion.

It must be pointed out that the assumption is implied that the molecule conserves its center of symmetry in the excited state. For this we have no other proof than Lathan, Curtiss and Pople's calculations on $C_2H_6^+$. If this turned out to be wrong almost everything would become possible including a $3s \leftarrow e_g$ transition which was Mulliken's original assignment. (Which would still give a perpendicular transition and C-H cleavage).

In the VUV spectrum of ethane there is another structured band occupying the 80000-95000 cm^{-1} region (20)(26)(27). Under the resolution of our 1 m spectrometer it exhibits a vibrational progression with somewhat irregular intervals ranging from 1050 to 1250 cm^{-1}. The first member can be approximately located at 84800 cm^{-1}. (For C_2D_6 the average spacing is 950 cm^{-1} and the average isotopic ratio is 1.21). Between the two band systems, near 80000 cm^{-1} we find an intermediate area which may belong to a separate electronic transition. These bands might well represent the $3p \leftarrow 3a_{1g}$ transitions ($^1E_u \leftarrow {}^1A_{1g}$; $a_{2u} \leftarrow 3a_{1g}$; $e_u \leftarrow 3a_{1g}$). The second member of the first Rydberg series $4p \leftarrow 1e_g$ could also come in this part of the spectrum. With the insufficiently resolved fine structure which is at our disposal we can only speculate. It is remarkable, however, that ethane possesses two or three stable excited states linked to transitions of bonding σ electrons.

IV. Possible weak transitions

All bands which were discussed so far are due to electronically allowed singlet-singlet transitions. The photochemist will be certainly interested by a review of weak transitions which might be hidden under the strong allowed bands. Most of them could appear through vibronic interactions or through a decreased excited state symmetry. We shall use Rydberg language only. For the sake of completeness the allowed bands are also included.

a) From $3a_{1g}$.

$$3s(a_{1g}) \leftarrow 3a_{1g} \; ; \quad {}^{1}A_{1g} \leftarrow {}^{1}A_{1g} \; , \text{ forb.}$$

$$3p(a_{2u}) \leftarrow 3a_{1g} \; ; \quad {}^{1}A_{2u} \leftarrow {}^{1}A_{1g}$$

allowed, polarized parallel to the C-C axis.

$$3p(e_{u}) \leftarrow 3a_{1g} \quad {}^{1}E_{u} \leftarrow {}^{1}A_{1g} \text{ allowed}$$

$$3d(a_{1g}) \leftarrow 3a_{1g} \quad {}^{1}A_{1g} \leftarrow {}^{1}A_{1g}, \text{ forb.}$$

$$\text{twice} \quad 3d(e_{g}) \leftarrow 3a_{1g} \quad {}^{1}E_{g} \leftarrow {}^{1}A_{1g}, \text{ forb.}$$

b) From $1e_{g}$.

$$3s(a_{1g}) \leftarrow 1e_{g} \; ; \quad {}^{1}E_{g} \leftarrow {}^{1}A_{1g}, \text{ forb.}$$

$$3p(a_{2u}+e_{u}) \leftarrow 1e_{g} \quad ({}^{1}E_{u} \leftarrow {}^{1}A_{1g}, \text{ allowed})$$

$${}^{1}A_{1u} \leftarrow {}^{1}A_{1g}, \text{ forb.}$$

$${}^{1}A_{2u} \leftarrow {}^{1}A_{1g}, \text{ allowed}$$
polarized along the figure axis

$$({}^{1}E_{u} \leftarrow {}^{1}A_{1g}, \text{ allowed})$$

$$3d(a_{1g}+2e_{g}) \leftarrow 1e_{g}; \quad {}^{1}E_{g} \leftarrow {}^{1}A_{1g}, \text{ forb.}$$

$$\text{twice} \quad {}^{1}A_{1g} \leftarrow {}^{1}A_{1g}, \text{ forb.}$$

$$\text{twice} \quad {}^{1}A_{2g} \leftarrow {}^{1}A_{1g}, \text{ forb.}$$

$$\text{twice} \quad {}^{1}E_{g} \leftarrow {}^{1}A_{1g}, \text{ forb.}$$

Singlet-triplet transitions are as numerous as singlet-singlet transitions. Only the ones of lowest frequency stand a chance to be detected by absorption spectroscopy. Little is known about Rydberg triplets. Katagiri and Sandorfy predicted the first one to lie at 0.5-1 eV to lower frequencies from the first singlet-singlet band. Brongersma and Oosterhoff (28) found these experimentally by low energy electron impact measurements for several paraffins.

V. The primary steps

The author who cannot pretend to have a wide experience in photochemistry will make no attempt to review the literature relating to the photodecomposition of ethane. From the many good papers published on this subject he mentions only those of Hampson, McNesby, Akimoto and Tanaka (29)(30),Hampson and McNesby (31) and Lias, Collin, Rebbert and Ausloos (32) who examined the photolysis of ethane using the xenon (8.4 eV), krypton (10.0 eV) and argon (11.6–11.8 eV) resonance lines respectively. The following discussion will be based on the conclusions of the last paper and of a review by Lias and Ausloos (33).

Fig. 1 in ref. (32) is most instructive. It shows the yields of primary processes in the photolysis of ethane as a function of the energy of the exciting photons. Four primary steps were proposed:

$$(1) \quad C_2H_6 \;\rightarrow\; CH_4 + CH_2$$
$$(2) \quad C_2H_6 \;\rightarrow\; C_2H_5 + H$$
$$(3) \quad C_2H_6 \;\rightarrow\; C_2H_4 + H_2$$
$$(4) \quad C_2H_6 \;\rightarrow\; CH_3 + CH_3$$

At 8.4 eV 85% of the decomposition is due to (3); an astonishing result. Is it credible that this reaction occurs in a Rydberg state resembling the $^2A_{1g}$ ion which according to LCP has a C–C bond length of 1.8–2.0Å but near normal C–H bond lengths ? To the contrary, the great importance of reaction (3) is readily understood if the excited state is the ^1E_u state obtained through excitation from the le_g orbital as proposed above. Looking only at this orbital the excitation should result in a shortening of the C–C bond and even if subsequent rearrangement or the addition of the Rydberg orbital lengthens it somewhat it will still be a bond well prepared for the formation of excited ethylene with departure of the hydrogens from the weakened C–H bonds.

As to the excited states of ethylene at least two of these have been thoroughly studied. In the V state it has a perpendicular structure with a C–C bond length estimated by Merer and Mulliken (34) to 1.44 Å and in the lowest (3s) Rydberg state which deviates by about 25° from coplanarity (35)(36), 1.41 Å. It seems to us unlikely that the $^2A_{1g}$ ion of ethane with a bond length of over 1.80 Å could be conducive to any of these states.

As to the reasons why the simple C–H cleavage (2) is increasing in importance as the energy of the exciting photons increases we can only refer to Ausloos and coll.: " ... as one goes to higher energies, one would expect the excited molecule to have a shorter dissociative lifetime, and hence to undergo processes requiring rearrangements (such as (1) and (3) with a lower probability". We might add that more and more of the molecules reacting in the first excited state (singlet or triplet) will get there via the higher excited states (above 10 eV). Some molecules might react in these higher states and since at least part of these implies excitation from $3a_{1g}$ the increasing number of molecules decomposing through C-C cleavage becomes at least qualitatively understandable.

The problem is even more complicated. 8.4 eV is less than the energy of the 0-0 band in the spectrum (70368 $cm^{-1} \cong 8.7$ eV). This means that the molecules might decompose in one of the forbidden states which are expected to lie at the low frequency side of the first allowed singlet-singlet absorption band. According to the above reasoning the state in which most molecules react should be an E state; even then.

Any attempt to go further would probably be unwise.

The author is indebted to Mr. P. Sauvageau, Mr. R. Gilbert and Mr. J. Doucet who measured some of the spectra and to Mr. R. Macaulay and Mr. D. Goutier who carried out theoretical calculations on ethane referred to in this lecture. Their results will be published elsewhere.

(1)   R.S. Mulliken, J. Chem. Phys., 3, 517 (1935).
(2)   G. Herzberg, "Molecular Spectra and Molecular Structure",
      Vol. I. D. Van Nostrand, Princeton, New Jersey, 1950
      (p. 390).
(3)   G. Herzberg, "Molecular Spectra and Molecular Structure",
      Vol. III. D. Van Nostrand, Princeton, New Jersey, 1966
      (p. 326).
(4)   S. Katagiri and C. Sandorfy, Theoret. Chim. Acta 4, 203
      (1966).
(5)   W.E. Palke and W.N. Lipscomb, J. Am. Chem. Soc., 88, 2384
      (1966).
(6)   E. Clementi and D.R. Davis, J. Chem. Phys., 45, 2593
      (1966).
(7)   R.J. Buenker, S.D. Peyerimhoff, and J.L. Whitten, J. Chem.
      Phys., 46, 2029 (1967).
(8)   W. Fink and L.C. Allen, J. Chem. Phys., 46, 2261 (1967).
(9)   R.M. Pitzer, J. Chem. Phys., 47, 965 (1967).
(10) A. Veillard, Chem. Phys. Lett., 3, 128, 565 (1969).

(11) W.A. Lathan, W.J. Hehre, and J.A. Pople, J. Am. Chem. Soc.,
      93, 808 (1971).
(12) R.J. Buenker, S.D. Peyerimhoff, L.C. Allen, and J.L. Whitten,
      J. Chem. Phys., 45, 2835 (1966).
(13) W.A. Lathan, L.A. Curtiss, and J.A. Pople, Mol. Phys., 22,
      1081 (1971).
(14) J.N. Murrell and W. Schmidt, J. Chem. Soc. Faraday Trans.
      II, 1709 (1972).
(15) R. Macaulay, to be published.
(16) W.J. Hehre, R. Ditchfield, and J.A. Pople, J. Chem. Phys.,
      56, 2257 (1972).
(17) C.R. Brundle, M.B. Robin, and H. Basch, J. Chem. Phys.,
      53, 2196 (1970).
(18) D.W. Turner, C. Baker, A.D. Baker, and C.R. Brundle,
      "Molecular Photoelectron Spectroscopy", Interscience, New
      York and London, 1970.
(19) A.D. Baker, C. Baker, C.R. Brundle, and D.W. Turner, Int.
      J. Mass. Spectr. Ion Phys., 1, 285 (1968).
(20) E.N. Lassettre, A. Skerbele, and M.A. Dillon, J. Chem.
      Phys., 49, 2382 (1968).
(21) J.W. Raymonda and W.T. Simpson, J. Chem. Phys., 47, 430
      (1967).
(22) B.A. Lombos, P. Sauvageau, and C. Sandorfy, J. Mol. Spectry
      24, 253 (1967).
(23) C. Sandorfy, B. Lombos, and P. Sauvageau, "XIII Colloquium
      Spectroscopicum Internationale", A. Hilger, London 1967
      (p. 370).
(24) H.C. Longuet-Higgins, V. Öpik, M.H.L. Pryce, and R.A. Sack,
      Proc. Roy. Soc., A244, 1 (1958).
(25) E.F. Pearson and K.K. Innes, J. Mol. Spectry 30, 232 (1969).
(26) E.E. Koch and M. Skibowski, Chem. Phys. Lett., 9, 429
      (1971).
(27) R. Gilbert, P. Sauvageau, and C. Sandorfy, to be published.
(28) H.H. Brongersma and L.J. Oosterhoff, Chem. Phys. Lett., 3,
      437 (1969).
(29) R.F. Hampson, J.R. McNesby, H. Akimoto, and I. Tanaka,
      J. Chem. Phys., 40, 1099 (1964).
(30) H. Akimoto, K. Obi, and I. Tanaka, J. Chem. Phys., 42,
      3864 (1965).
(31) R.F. Hampson and J.R. McNesby, J. Chem. Phys., 42, 2200
      (1965).
(32) S.G. Lias, G.J. Collin, R.E. Rebbert and P. Ausloos, J.
      Chem. Phys., 52, 1841 (1970).
(33) P.J. Ausloos and S.G. Lias, Ann. Revs. Phys. Chem., 85
      (1971).
(34) A.J. Merer and R.S. Mulliken, Chem. Revs., 69, 639 (1969).
(35) A.J. Merer and L. Schoonveld, J. Chem. Phys., 48, 522
      (1968).
(36) A.J. Merer and L. Schoonveld, Can. J. Phys., 47, 1931
      (1969).

A POT-POURRI OF UV AND PE SPECTRA OF IODIDES

D.R. Salahub and R.A. Boschi

Department of Applied Mathematics
University of Waterloo, Ontario and
TWP Sandoz AG, Basel, Switzerland.

Firstly, we would like to thank Professor Sandorfy for inviting us to make, as an addendum to his lecture, a few brief remarks concerning the UV and PE spectra of iodine containing molecules (1-4). As the title indicates we will be neither systematic nor complete. Instead, we have chosen, here and there, a few of what we consider to be the most interesting aspects of these spectra.

PE SPECTRUM OF VINYL IODIDE

Sandorfy has raised the question whether the π orbital in the vinyl halides is always the uppermost orbital, as is the case in vinyl chloride, or whether the lone-pair orbital localized mainly on the halogen becomes the least tightly bound for vinyl bromide or vinyl iodide. One can go through a rather long argument based on the series of halobenzenes and vinyl halides (4) to show that in vinyl iodide the order is $n(a'')$, $n(a')$, π, where $n(a'')$ and $n(a')$ are the out-of-plane and in-plane iodine lone-pairs respectively. The same information, however can be derived with reasonable confidence from the spectrum of vinyl iodide alone (Fig. 1). The second band of the spectrum is surely due to the in-plane lone-pair and in fact has the typical sharp, strong profile with limited vibrational structure. The first and third bands are broader and diffuse due to the mixing of the $n(a'')$ and π levels. The first is however still quite sharp and strong and certainly has more of the characteristics of a lone-pair band than a π band while the opposite is true for the third. For vinyl bromide (5) the situation is intermediate

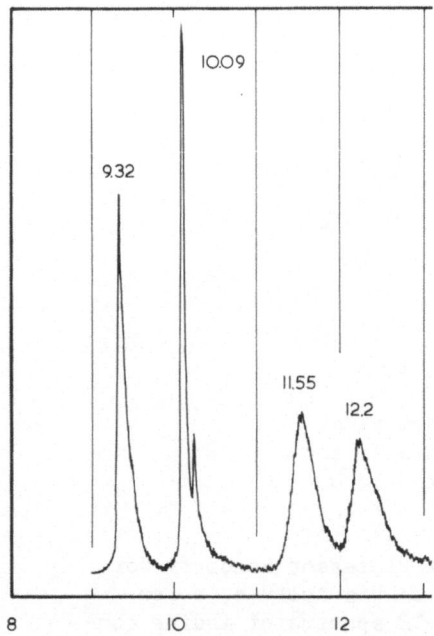

Figure 1.   The PE Spectrum of Vinyl Iodide

between that of vinyl chloride and vinyl iodide, the a'' lone-
pair being highly mixed with the π orbital.

UV SPECTRA OF ALKYL IODIDES

    The UV spectra of the alkyl iodides (1, 2) consist of a
broad structureless band near 39,000 cm^{-1} followed by a large
number of more or less sharp bands. The first band is due to an
σ* ←n valence shell transition. The remaining bands can be fit
into Rydberg series leading to the first two IPs. The presence
of the iodine atom leads to large interactions between spin and
orbital angular momentum so that configurations involving unpaired electrons in degenerate MOs with large coefficients on
the iodine must be treated using double groups (6) and (Ω,ω)
coupling. Of course for the higher alkyl iodides there is,
rigourously, no degeneracy; however, the constancy of the spectral
splittings indicates that the lone-pairs are quite well localized
and are effectively in a potential which has C_{3v} symmetry. Group
theoretical considerations then lead us to expect two Rydberg
s series, six p and ten d series converging to each of the first
two IPs (corresponding to the spin-orbit split $E_{3/2}$ and $E_{1/2}$

n	CH_3I	Xe ns$[1\frac{1}{2}]^O$ J = 2
6	27218	30766
7	14358	12645
8	7736	7029
9	4827	4436
10	3297	3074
11	2393	2255
12	1863	1725

Table 1.   Comparison of term values for Rydberg series of CH_3I and Xe(cm^{-1}).

states of the ion).   In fact we have managed to identify three series converging to each IP (1, 2).

In order to characterize the series we have found it useful to make comparisons with the "semi-united" atom Xenon.   In Table 1 we show the term values for the δ = 4.24 series of CH_3I and the ns $[1\frac{1}{2}]^O$ J = 2 series of Xe(7).   The agreement is quite good and supports the assignments of this series, which is the strongest in the spectrum to an ns series.   Moreover it allows us to overcome the ambiguity in the determination of separate values for n and δ in the Rydberg formula.   The first member of the Xe ns series has n=6 and this leads to a δ value for the methyl iodide series of 4.24.   This also substantiates the assignment of the first weak broad band to a valence transition since 6s is the lowest Rydberg state expected.

One last comment on these spectra concerns the utility of using the vibrational structure in the UV and PE spectra to help in the characterization of a band as a Rydberg transition.   In Table 2 we have gathered the frequencies of the main vibrational progression for the various bands of the UV and PE spectra of some 1-iodoalkanes.   We conclude that the potential surfaces in the B, C, D and E states are not radically different from those of the ion and that these are properly classified as Rydberg states.

PE SPECTRA OF FLUOROIODOETHANES

The final topic we wish to deal with is the order of ionizations due to CC and CH electrons in ethyl iodide.   The order in ethane itself has been the subject of some controversy.   In order to decide the order in C_2H_5I we have studied the PE spectra of C_2H_5I, CF_3CH_2I and C_2F_5I.   As is well known the substitution of fluorine for hydrogen causes a stabilization of most of the

Molecule	Ultra-violet spectrum					Photoelectron Spectrum	
	$\overline{X}$	B	C	D	E	1st band	2nd band
CH_3I	1251		1100	1075	1210	1210	1220
C_2H_5I	1197	1125	1140	1160		1090	1120
C_3H_7I	1185	1100	1125	1115	1165	1190	1200
C_4H_9I	1190	1125	1140	1160	1160	1090	1120
$C_5H_{11}I$		1125	1135	1100	1180	1130	
$C_6H_{13}I$		1150	1150	1150	1190	1100	1070

Table 2. Frequencies of the principal vibrational progression in the ground state, the B, C, D and E bands in the optical spectra and the first two bands of the photo-electron spectra (cm^{-1}).

energy levels. Let us assume that the stabilization in any of the above three molecules is about the same for CI, CC and CH bands. If we then shift the spectra so that the CI bands co-incide we obtain Fig. 2 and we see clearly that the second band is very strong for C_2H_5I, diminishes in intensity for CF_3CH_2I where three hydrogens have been removed and disappears altogether for C_2F_5I. The third band is roughly constant in the three

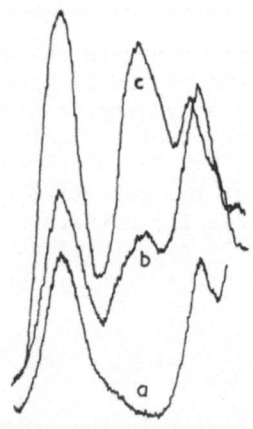

Figure 2. Comparison of the spectra of C_2H_5I(c), CF_3CH_2I(b), and C_2F_5I(a), shifted so that the bands due to ionization of the CI bonding electrons coincide.

spectra except for a diminution of intensity which can be explained on the basis of the high electron attracting power of fluorine. Thus in ethyl iodide ionization from a CH orbital precedes that from the CC orbital.

REFERENCES

(1) R.A. Boschi and D.R. Salahub, Mol.Phys. $\underline{24}$, 289 (1972)
(2) R.A. Boschi and D.R. Salahub, Mol.Phys. $\underline{24}$, 735 (1972)
(3) D.R. Salahub and R.A. Boschi, Chem. Phys. Letters, $\underline{16}$, 320 (1972)
(4) R.A. Boschi and D.R. Salahub, to be published
(5) D. Chadwick, D.C. Frost, A. Katrib, C.A. McDowell and R.A.N. McLean, Can.J.Chem. $\underline{50}$, 2647 (1972)
(6) G. Herzberg, Molecular Structure and Molecular Spectra. III. Electronic Spectra and Electronic Structure of Polyatomic Molecules, Van Nostrand, (1966)
(7) C.E. Moore, Atomic Energy Levels, Vol. III, Circular 467 (National Bureau of Standards) (1958).

BEYOND THE ORBITAL APPROXIMATION

R. Daudel

Université de Paris VI et
Centre de Mécanique Ondulatoire Appliquée
Paris, France

The purpose of this paper is more to present new concepts, to suggest new methods than to present many results. This paper is concerned with a work in progress.

The need for a language adapted to very elaborate wave-functions

The shell-model, in which a molecular state is represented by a single determinant built on a simple set of molecular orbitals has been proved to be very useful. But it has drastic limitations. The near Hartree-Fock value of the electronic dissociation energy D_e of oxygen is 1.43 ev. This value is only a quarter of the experimental one (5.21 ev). Therefore the Hartree-Fock wave-function gives a very bad representation of the chemical bond. A full valence CI on a minimum basis leads to 3.81 ev (1) for that electronic dissociation energy. It is nevertheless 1.4 ev less than the experimental value. A large CI on a large basis is necessary.

It is clearly seen that very elaborate wave-functions are necessary to represent conveniently a chemical bond.

Even for qualitative predictions a large CI is necessary. For F_2, one of the earliest extended basis SCF calculations (2) showed the Hartree-Fock energy of F_2 to lie more than lev above the energy of two F atoms, leading to the wrong conclusion that the F_2 molecule would be unstable. S. Peyerimhoff (3) has recently shown that all cis-butadiene is correctly predicted by CI to be more stable than cyclobutene by 0.34 ev (exp 0.40 ev) in

contrast to the S.C.F. calculation alone which finds butadiene
less stable by 0.2 ev.

The need for very large CI is again more obvious when excited states are concerned. Hosteny et al.(4) have shown for
example that even full π CI is not sufficient to predict the
main vertical excitation energies of butadiene with an error of
less than 10 percent.

Many procedures have been proposed to compute CI wave-functions including the iterative natural orbital method (5), PCILO
(6) and CIPSI (7) procedures.

But it is difficult to understand the physical (or chemical)
meaning of a large CI wave-function. With such a wave-function
it is no longer possible to use the concept of localized orbital
to represent a chemical bond. We need a new language. Furthermore the calculations of large CI wave-functions are computer
time-consuming. We also need a process to reduce the necessary
number of configurations. The loge theory is an attempt to
serve both purposes.

The loge theory

I have not the time to describe the loge theory in detail
here (8). I shall give only one example. The basic idea is to
divide the space of a molecule into volumes in such a way as to obtain the maximum amount of information about the localizability
of the electrons. Let us consider the BH molecule. It contains
6 electrons. Let us consider a partition of the space into three
volumes (Figure 1)

a) a sphere of radius R centered at the boron nucleus (Vol. A)

b) the part of a cone of angle φ having as axis the BH line and
containing the hydrogen nucleus (Vol. B)

c) the remaining part of the space (Vol. C).

To find a certain distribution of electrons in the various
volumes is a certain electronic event. Figure 2 invokes such
events.

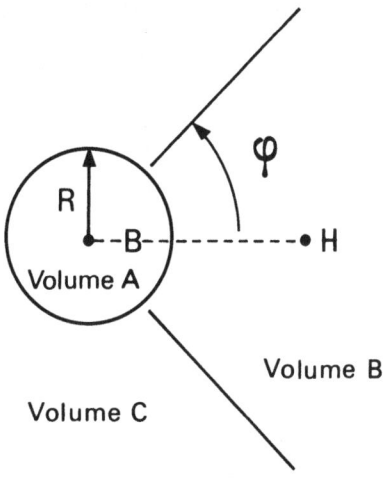

Figure 1

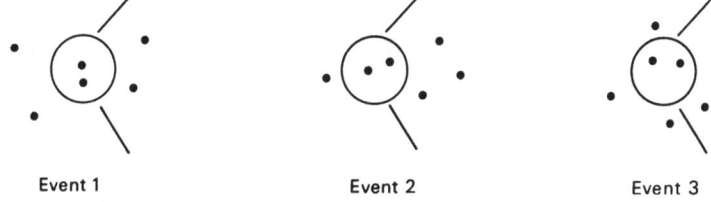

Figure 2

Let us consider any kind of wave-function ψ associated with a given state of the molecule. From that function (which can be a large CI function) it is possible to calculate the probability of each electronic event for that state. For example the probability of event 2 (two electrons in volume A, three in volume B and one in volume C) is :

$$p_2 = N \int_A dv_a dv_b \int_B dv_c dv_d dv_e \int_C dv_f \left| \psi(M_a, M_b \ldots M_f) \right|^2$$

N being a simple factor resulting from the undistinguishability of the electrons.

The <u>relative missing information function</u> associated with the various probabilities p_i is :

$$I_r = \frac{\sum\limits_i p_i \log_2 \frac{1}{p_i}}{\log_2 \nu}$$

if ν denotes the number of electronic events.

To obtain the maximum amount of information we must calculate the value of φ and R which minimize the missing information function I. This calculation has been done (9) for the groundstate of the BH molecule. The function I reaches its minimum value for :

$$R = 0.7 \text{ a.u.}$$
$$\varphi = 73°$$

It can be said that the corresponding partition of the space is for the groundstate of BH the <u>best partition into three loges</u> (possessing a sphere and a cone as frontiers). It would be the absolute best partition into loges if all other shapes of loges were explored, I_r remaining the absolute minimum for the foregoing partition. Obviously we can compare between them the best partitions into two loges, three loges ... and retain the partition corresponding to the smallest minimum value of the relative missing information function.

The mathematical analysis of the function I_r shows that when it reaches a minimum there is usually <u>a leading event</u>, that is to say an event possessing a probability much more important than all other probabilities.

It is the case here: the leading event corresponds to the occurence of two electrons in each loge: the probability of this leading event is about 0.7. This is a very high probability if we take account of the great number of possible events.

Figure 3 shows the position of the frontiers of the three loges on the electronic density map. It is seen that the frontier of the central loge corresponds to a zone in which the effect of the binding begins (the end of the spherical isodensity contours). The central loge can be considered to be the boron core loge.

The frontier between the two other loges is made of steepest descent lines on the density map. The loge containing the hydrogen atom may be called a two-electron BH bond loge. The other loge can be called a lone pair loge. We must underline the fact that even if there is a high probability of finding two electrons (and two only) in the lone pair loge, there is also a small probability of finding in it zero electron or more than two electrons.

It appears that the volume associated with the lone pair is significantly larger than the volume of the bond loge. A lone pair is more "bulky" than a two electron bond. This result is a mathematical justification of the hypothesis which has been used by Gillespie to predict correctly the geometry of a very large number of molecules in their ground state (10). Let us add that by using an idea of Robb et al. (11) it would be possible to precise the "size" of the superficial loges.

Such are the fundamental ideas of the loge theory. As it is a general theory it can be applied in all fields of quantum chemistry and therefore in many topics of interest for our Summer Institute.

There are many ways of using that theory. The most obvious is the analysis of any wave-function in terms of loges as we did in the case of the BH molecule. This analysis gives a very clear understanding of the electronic structure described by the wave-function. By following that way one can obtain very direct informations on the change in the electron localizability accompanying excitation and ionization, including informations on the localizability of excitons[*]. We can also follow the change of localizability of electrons during a photochemical process.

---

* With Professor Sandorfy we are planning to study the change of the size of lone pair with excitation and therefore their effect on the geometry of excited states to extent if possible to excited states the Gillespie problematic.

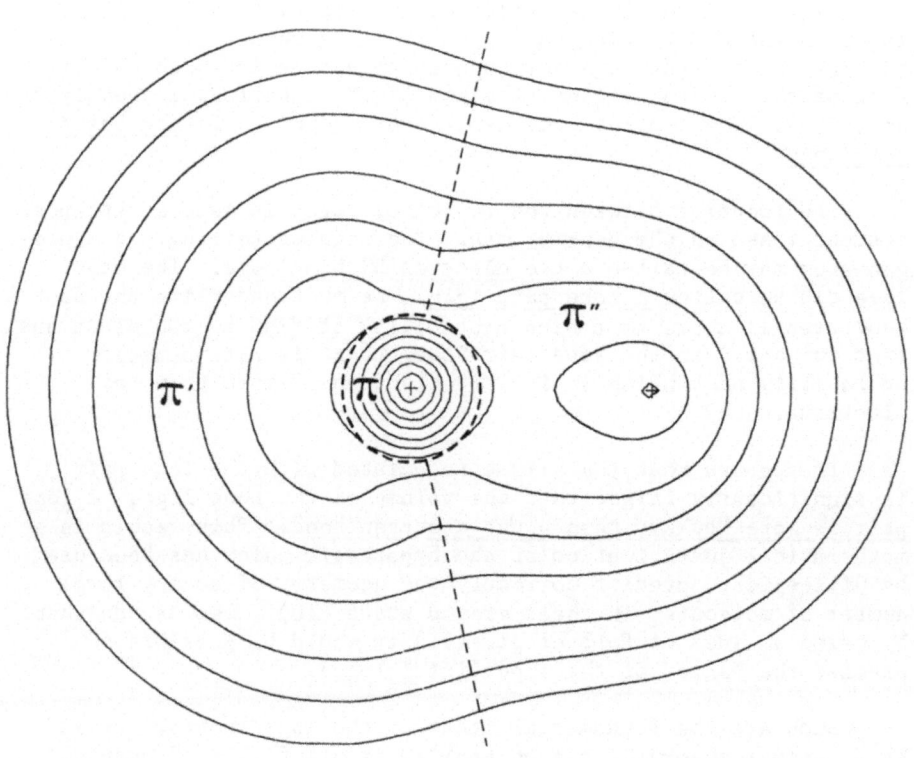

Figure 3

However, the explicit knowledge of the wave-function is not always necessary to discuss a problem: the formalism, the basic
ideas, of the loge theory can be used to solve some problems.

Furthermore the loge theory can be used to suggest new procedures to calculate wave-functions. That theory can also give
a better understanding of the efficiency of some computational
methods already known.

We have not the time to describe in detail all possible applications of loge theory in the framework of this Summer Institute. We shall only give a few examples.

Change in the localizability of electrons during ionization

An analysis of the ionization process in terms of the loge
theory shows easily that ionization can lead to

a) a decrease of the electron localizability

b) a increase of that localizability

c) no significant effect.

All cases are possible depending on the nature of the molecule
and on the states involved.

In various cases the ionization affects particularly one
loge. This happens for molecules possessing a lone pair loge.

The lowest ionization of methylamine produces mainly a hole
in the lone pair loge. In the normal molecule, during the
leading event, there are two electrons in that loge. For the
ion, during the leading event there is only one electron in the
corresponding loge. The change in lozalizability is not very
significant.

But let us now consider diamino methane, that is to say a
molecule in which there are two analogous lone pair loge (Fig. 4).
During the leading event there are two electrons in each loge.

If, as a first guess, we assume that the topology of the
loges is not destroyed by the ionization we can predict that for
the ion the leading event will be represented by one of the two
diagrams of figure 5.

However, on symmetry grounds event I and event II must have
equal probabilities; therefore, each event cannot have a probability higher than 0.5. There is no longer a real leading
event. This situation is very bad from the information theory
viewpoint. Let us assume, for example, that the probability of
each event is exactly 0.5. The missing information function
takes the value :

$$I = 0.5 \log_2 2 + 0.5 \log_2 2 = 1$$

It is readily seen that this is the maximum possible value
for a missing information associated with two events. Therefore
the best partition into loges for the normal molecule becomes
the worst partition into loges of the ion. To improve that partition we can withdraw the frontier between the lone pair loges.
The new partition is now represented by the unique diagram of
figure 6.

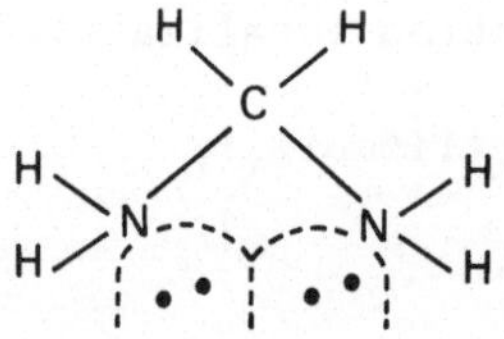

Figure 4

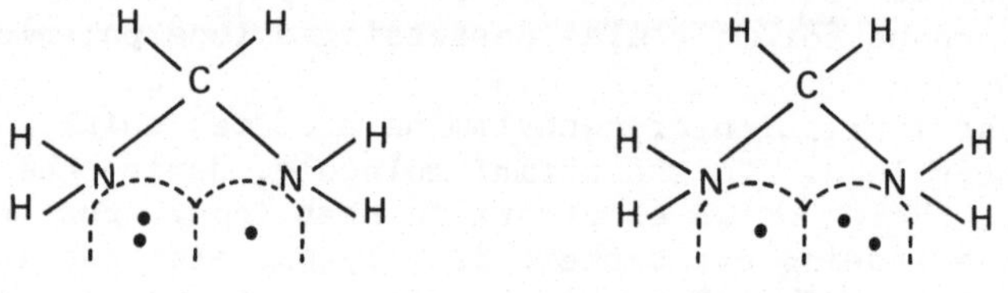

Figure 5

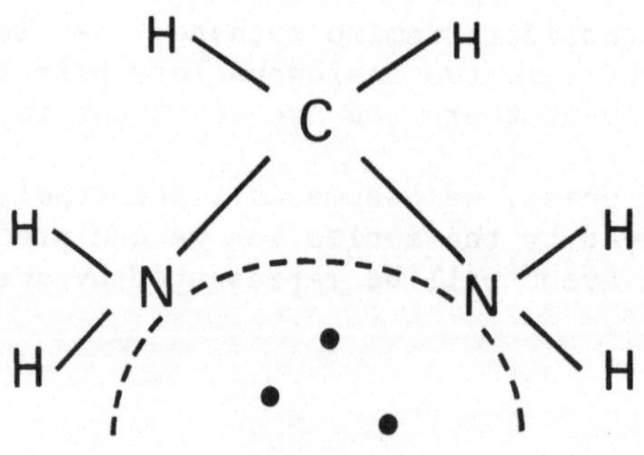

Figure 6

The probability of the corresponding event is one (in the foregoing conditions) and the missing information function becomes:

$$I = \log_2 1 = 0$$

It reaches its smallest possible value. The new partition is
therefore a good one. It shows a three electron loge <u>delocal-
ized</u> between the two nitrogen atoms. Therefore, the ionization
reduces the localizability of the electrons and in good partition
into loges, there is necessarily a large loge delocalized on non
adjacent atoms. As a consequence it is extremely dangerous to
use without caution the photoelectronic spectrum data to measure
the electron localizability of the <u>initial state</u> of a molecule.
The electron localizability of both initial and ionized states
plays an important role in the photo-electron spectrum of a
molecule.

<u>The localizability of the electron as a starting point to calcu-
late an elaborate wave-function</u>:

    We saw that starting from a wave-function the loge theory
permits to obtain informations on the localizability of electrons. But, vice versa any information about the localizability
of electrons in a molecule can be helpful to build a wave-function for that molecule. Such informations can be obtained from
a rough wave-function already calculated or from chemical in-
tuition or from experimental data or on analogy grounds if for
analogous molecules an analysis has been already made. The
loge theory can be useful with that respect because it helps to
rationalize the results obtained on the localizability of electrons.

    Without any calculations we can easily predict that, for a
partition of the space of the He LiH$^+$ ion into three loges, the
diagrams invoked by figure 7 will be the most important events,
the first one being the leading event.

    These diagrams suggest to associate with the ion the trial
function :

$$\psi = \mathcal{Q} \left\{ aK\ (1,2)\quad L\ (3,4)\quad 1'\ (5) \right.$$

$$\left. + b\ K\ (1,2)\quad 1\ (3)\quad L'\ (4,5) \right\} \sigma$$

where $\mathcal{Q}$ is an antisymmetrisor, K the lithium core function,
L and 1 functions associated with the He Li bond loge, L' and
1' functions associated with the LiH bond loge and σ a convenient spin function, a and b are simply variation coefficients.

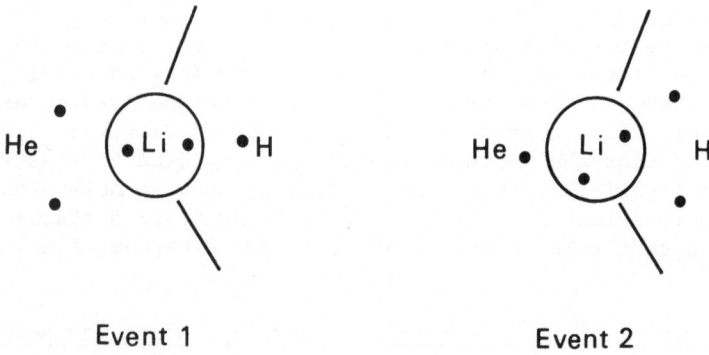

Event 1   Event 2

Figure 7

The best loge functions and the coefficients can be obtained by
solving the usual variation equation :

$$\delta < \psi \, |H - E|\psi \,> = 0$$

In practice it is not necessary to use functions completely
localized in their loges. We can simply extend each loge function on a convenient basis set (gaussian or Slater functions for
example) and permit a certain overlap between the loge functions.
The variation procedure makes the rest. To do the effective
calculations one can follow the group function method proposed
by McWeeny et al. (12). We guess all loge functions except one.
We calculate the best loge function in the field on the others
and by iteration each loge function is improved until con-
vergency. This procedure amounts to do local CI and self consistent treatment between the various local CI. The number of
necessary configurations is much less than it would be if the CI
was done over the whole molecule. Furthermore the loge functions obtained are transferable from a molecule to another
analogous one, at least in a first approximation, as they are

associated with molecular fragments possessing a chemical individuality.

With these ideas in mind, let us discuss a very simple example: Rydberg states. The figure 8 contains the diagrams associated with the most important events of a Rydberg state. Surrounding a central system of loges (more or less complex) there is a superficial loge in which there is a high probability of finding one electron and one only (<u>leading event</u>) and small probabilities of finding zero or two electrons (satellite events)[*] We are led to consider the following wave-function :

$$\Psi = \mathcal{Q}\left\{a\ K\ (1,\ldots N-1)\ L\ (N) + b\ K'\ (1,\ldots N) + \right.$$
$$\left. c\ K''\ (1,\ldots N-2)\ L''\ (N-1,\ N)\right\}\ \sigma$$

the first term in the brackets being the leading one. An approximate representation of the function K' $(1,\ldots N)$ can be the SCF wave-function associated with the ground state of the molecule. Similarly functions K' and K'' can be obtained by omitting from K the last occupied orbital or the two last ones respectively.

Event 1   Event 2   Event 3

Figure 8

---

[*] This situation is rather general. For a loge belonging to a good partition there is usually a leading event and two satellite events, other events being negligible.

As we know during the calculation of the SCF wave-function one can
also compute various virtual orbitals. The functions L and L" can
be built by following the usual CI procedure on this basis set of
virtual orbitals. The variation equation will furnish a set of
solutions which are approximate wave-functions for various Rydberg
states.

The calculation made by H. Lefebvre-Brion et al. (13) for the
CO molecule amounts to use such a procedure, the leading event
being only taken into account. The basis set used in this calculation was made of 1s, 2s, 2p, 3s, 3p, 3d, 4s, 4p and 4d atomic
orbitals on each center where Slater's rules were used to calculate the orbital exponents. Table I makes it possible to compare
theoretical results with experimental data.

Table I

Observed energies (ev)	Assignment proposed by experimentalists	Calculated energies (ev)
10.777	B $^1\Sigma^+$	11
11.397	C $^1\Sigma^+$	11,5
11.521	E $^1\Sigma^+$	12,5
10.393	b $^3\Sigma^+$	10,3
11.415	c $^3\Sigma^+$	11,4

In all cases (the $E^1\Sigma^+$ excepted) the difference between observed and calculated excitation energies is only 0.1 - 0.2 ev,
while for $E^1\Sigma^+$ it is about 1 ev. That discrepancy led to the
conclusion that the assignment of the experimental energy 11.521
to an $E^1\Sigma^+$ level was wrong and that this energy must correspond
yo a Π level. Tilford et al. (14) have done a rotational analysis of that level and confirmed that this theoretical proposal
was right.

Loge theory and exciton localizability

In the framework of the loge theory it is possible (15) to express any average value of an operator Ω associated with a given molecule as a sum of loge contribution $\overline{\Omega}_J$ and of loge pair contribution $\overline{\Omega}_{J,J'}$:

$$\overline{\Omega} = \sum_J \overline{\Omega}_J + \sum_{J,J'} \overline{\Omega}_{J,J'}$$

If we use this formula to express the energy of a Rydberg state in the foregoing treatment three terms appear: the energy associated with the CO core, the energy associated with the external loge and the interaction energy between the core and the external loge. As for all Rydberg states the first term has been made a constant. One can say that the energy of the photon (the exciton) which produces the transition from a Rydberg state to another one is localized in the external loge.

This is a rather general situation. For many molecular excitations we may say that the exciton is localized in one or in a small number of loges.

Let us go back to a molecule containing a lone pair and let us assume a state in which an exciton is localized in the lone pair loge. Let us now consider a molecule in which there are two analogous lone pair loges with the same environment (as in our di-amino methane molecule).

On symmetry grounds it can be shown that for a stationary state the energy of the exciton cannot be localized in one of the two lone pair loges. It must be shared between the two lone pair loges.

But if now the molecule is not isolated, if collisions with other molecules are possible, the situation is different: the collision destroys the symmetry: the partial localization of the exciton in one loge becomes possible and the collisions can produce the transfer of the exciton from a loge to another.

This intramolecular transfer of energy can be followed experimentally. If the molecular state under consideration is a triplet state one can register the E.P.R. signal. Its shape is usually affected by the exciton transfer and as the rate of this transfer depends on the temperature the E.P.R. signal will depend on temperature. Such a temperature effect is an indication on the possibility of an exciton transfer.

This phenomenon has been effectively observed in case of rather complex molecule and has been completely interpreted theoretically (16).

References

(1) See for example : H.F. Schaefer, The Electronic Structure of Atoms and Molecules, Addison Wesley Publishing Co (1972) p. 205.

(2) A.C. Wahl, J. Chem. Phys. 41, 2600 (1964).

(3) S. Peyerimhoff, The Sixth Jerusalem Symposium on Chemical and Biochemical Reactivity (1973).

(4) R.P. Hosteny, T.H. Dunning, R.R. Gilman, A. Pipano, and I. Shavitt (unpublished results quoted by Schaefer loc. cit. p. 355).

(5) C.F. Bender and E.R. Davidson, J. Phys. Chem. 70, 2675 (1966).

(6) S. Diner, J.P. Malrieu, and P. Claverie, Theoret. Chimica Acta, 13, 1 (1969).

(7) S. Huron, J.P. Malrieu, and P. Rancurel, J. Chem. Phys. (in press).

(8) A review of that theory is given in : C. Aslangul, R. Constanciel, R. Daudel, and P. Kottis, Adv. Quantum Chem. 6, 93 (1972).
     See also R. Daudel in Wave Mechanics, Butterworths (1973) p. 61.

(9) R. Bader, R. Daudel, and R. Stevens (Unpublished results).

(10) R.J. Gillespie, Molecular Geometry, Van Nostrand (1972).

(11) M.A. Robb, W.J. Haines, and I.G. Csizmadia,   J. Am. Chem. Soc. 95, 42 (1973).

(12) M. Klessinger and R. McWeeny, J. Chem. Phys. 42, 3343 (1965).

(13) H. Lefebvre-Brion, C.M. Moser, and R.K. Nesbet, J. Mol. Spectroscopy 13, 418 (1964).

(14) S.G. Tilford, J.T. Vanderslice, and P.G. Wilkinson, Can. J. Phys. 43, 450 (1965).

(15) C. Aslangul, R. Constanciel, R. Daudel, L. Esnault, and E. Ludena (in press).

(16). See for example: J.P. Lemaistre and Ph. Kottis, in Electron Spin Relaxation in Liquids, Plenum Press (1972) p. 455 and C. Aslangul et al. (1972) (loc. cit.).

NATURAL AND MAGNETIC CIRCULAR DICHROISM SPECTROSCOPY IN THE VACUUM ULTRAVIOLET

O. Schnepp

Department of Chemistry, University of Southern California
Los Angeles, California 90007

I. INTRODUCTION

1. <u>General.</u> In the field of spectroscopy it is well known
that a well-structured spectrum can be analysed and a great deal
of information obtained from it. However, there are many instances
of spectra which do not contain fine structure. In particular,
continuous spectra are observed in cases of predissoication or
dissociation - i.e. in cases where elementary photochemical reactions occur. Also at higher energies, interactions between different
states and dissociation processes become progressively more important. It is therefore a question of prime concern how excited
electronic states can be characterized in cases of lack of fine
structure.

Consideration of this problem caused us to investigate the
potential of Circular Dichroism (CD) spectroscopy. CD spectroscopy
can be carried out in two ways: we can study the natural CD of
molecules which are optically active and contain the chromophore
of interest or we can study the magnetic field induced circular dichroism (MCD) spectrum of the symmetric chromophore itself. In the
former case, it is assumed that the substituents which produce
optical activity have excited states which are high in energy and
therefore the spectrum observed is, in fact, that of the chromophore
with only minor perturbation.

Once it was decided to carry out CD spectroscopic investigations,
it soon became clear that it would be most desirable to remove the
short wavelength limit to such investigations at about 190 nm imposed by commercially available instrumentation. It was therefore
decided to develop instrumentation (Schnepp, Allen and Pearson, 1970)

for the vacuum ultraviolet for both natural CD as well as MCD
spectroscopy. Another instrument for natural CD measurements was
developed by Johnson (1971).

2. <u>Natural CD</u>. The phenomenon of optical rotation and circular dichroism are related to the difference in behaviour of the
molecule to right and to left-circularly polarized light. The
optical rotation ϕ is related to the difference in refractive index
for these two polarizations.

$$\phi = \frac{\pi}{\lambda} (n_L - n_R) \tag{1}$$

and the circular dichroism θ is similarly related to the absorption
coefficient difference:

$$\theta = \frac{\pi}{\lambda} (k_L - k_R) \tag{2}$$

We shall use the usual decadic molar absorption coefficient ε and
measure CD in terms of $\Delta\varepsilon$:

$$\Delta\varepsilon = \varepsilon_L - \varepsilon_R \tag{3}$$

The above measurables are related to quantum-mechanical transition
moment matrix elements as given below:

$$R = \int \Delta\varepsilon(\nu)/\nu \, d\nu \sim Im \, [<0|\underset{\sim}{\mu}|1> \cdot <1|\underset{\sim}{m}|0>] \tag{4}$$

For comparison we give the well known analogous expression for the
absorption coefficient ε:

$$D = \int \varepsilon(\nu)/\nu \, d\nu \sim <0|\underset{\sim}{\mu}|1> \cdot <1|\underset{\sim}{\mu}|0> \tag{5}$$

In the above $|0>$ and $|1>$ designate ground and excited electronic
states and μ and m are the electric dipole and magnetic dipole
moment operators, respectively.

There are several ways in which a CD spectrum can be helpful
to characterize electronic states. First of all, we note that the
CD spectrum ($\Delta\varepsilon$) can be positive or negative. This property is
often instrumental in resolving close-lying or overlapping electronic transitions, when the two bands have opposite signs. Also, since
the quantities ε and $\Delta\varepsilon$ depend on different matrix elements, the
absorption and CD spectrum contours are quite different; the
absolute quantities of ε and $\Delta\varepsilon$ at the peaks have different ratios
depending on the transition. Examination of eqs. (4,5) indicates
that the ratio $g = \Delta\varepsilon/\varepsilon$ is a measure of the magnitude of the magnetic dipole transition moment relative to the electric dipole
transition moment. Some reflection leads us to expect a relatively
large value of g for transitions which were magnetic dipole allowed
(electric dipole forbidden) in the unsubstituted symmetric

chromophore whereas a small value of g is indicative of the opposite case. The absolute magnitudes of the "anisotropy factor" g to be expected in each case will depend on calibration for the chromophore under study and the relative magnitudes will be of real significance.

In practice the two considerations described above have been found most useful in the interpretation of electronic spectra. In principle, however, there is further information contained in the CD spectrum, i.e. the absolute sign and the rotatory strength R as defined in eq. (4). In order to make efficient use of these data we require more reliable theory than is generally available at this time. It should, however, be pointed out that such calculations have been carried out in some instances and have been useful for characterization of excited states.

3. <u>Magnetic Circular Dichroism (MCD)</u>. In our considerations of MCD we shall be concerned with two types of line shapes which correspond to first and second order Zeeman effects, respectively. The first type is a derivative line shape with the central zero of $\Delta\varepsilon$ coinciding with the absorption peak. Such a line shape is obtained for a degenerate excited state which undergoes a first order Zeeman splitting in the magnetic field. This splitting is usually very small (of the order of a few wavenumber units) compared to the band width. The two components are circularly polarized and of opposite sense, resulting in the symmetric derivative line shape for $\Delta\varepsilon$ vs. ν as described above. When such a line shape is observed, it is definitive evidence for a degenerate electronic state and the magnetic moment of the state can be obtained from the spectrum as will be shown. This type of MCD feature is referred to as an A-term, and it is quantitatively characterized by the first moment of the MCD curve. The A-term is related to the following quantum mechanical expression (Stephens, Mowery and Schatz, 1971).

$$\mathcal{Q}(A \rightarrow J) = -1/2 \sum_{\lambda,\lambda'} \text{Im} \left[\langle A|\underline{\mu}|J_\lambda\rangle \times \langle J_\lambda'|\underline{\mu}|A\rangle \cdot \langle J_\lambda|\underline{m}|J_\lambda'\rangle \right]. \quad (6)$$

In this equation A designates the ground state J and the degenerate excited state having components J_λ. As before, μ and m designate the electric dipole and magnetic dipole moment operator, respectively. We note that the absorption strength for the same transition is given by:

$$D(A \rightarrow J) = \sum_\lambda \langle A|\mu|J_\lambda\rangle \cdot \langle J_\lambda|\underline{\mu}|A\rangle \quad (7)$$

As a result, the ratio $\mathcal{Q}/D$ gives the expectation value of the magnetic moment in state J. Stephens (1970) has shown that the first moment of the MCD band is a valid quantity in the above context for an envelope of overlapping lines.

The second type of line shape of interest here is a regular band shape determined by the same parameters as the absorption band. This is known as a B-term and it is characterized by the zeroeth moment of the MCD curve, given by the expression as below:

$$B = \sum_{\lambda} \text{Im} \; [\sum_{K_K \neq J} \langle A|\underset{\sim}{\mu}|J_\lambda\rangle \; x \; \langle K_K|\underset{\sim}{\mu}|A\rangle \cdot$$

$$\cdot \; \langle J_\lambda|\underset{\sim}{m}|K_K\rangle / \hbar\omega_{KJ}] \tag{8}$$

The B-term is caused by a second order Zeeman effect where state K (components K_K) mixed with state J in the magnetic field. The B-term contribution for state J due to state K is equal but opposite in sign for state K due to state J. Thus, if J and K are the only two interacting states, their MCD spectra will be of equal integrated intensities and of opposite signs.

As before, changes in sign are often helpful to resolve and characterize overlapping transitions, also when these are continuous bands. In addition, as already pointed out, the appearance of an A-term characterizes a degenerate state and allows the measurement of its magnetic dipole moment. Further, the study of B-terms reveals interactions and relationships between different electronic states.

II. INSTRUMENTATION

A schematic of the vacuum ultraviolet CD instrument built in our laboratory (Schnepp, et al., 1970) is given in fig. 1. The light source is a Hinteregger hydrogen lamp which has been modified in various ways and is operated at 1.5 KW and a H_2 pressure of 5 mm. The monochromator is a McPhearson 1 m vacuum spectrometer. The monochromatic beam emerging from the exit slit is collimated by a LiF lens and is split into two oppositely linearly polarized beams by a MgF_2 Wollaston prism. The prism is optically contacted. One of these beams is eventually discarded. The other linearly polarized beam passes through a stress-plate modulator made of CaF_2. This modulator acts as a variable quarter-wave plate and alternately converts the beam to right and left-circularly polarized light at a frequency of 50 Khz. Next the light passes through the sample cell and is focussed on the photomultiplier which is equipped with a CaF_2 window. The detected light intensity is constant for an optically inactive sample but is modulated for an optically active sample since the absorption coefficient is now different for right and left-circularly polarized light. It can be shown that $\Delta\epsilon$ is proportional to the ratio V_{AC}/V_{DC} where V_{AC} is the 50 Khz AC-signal produced by the modulation which appears superimposed on the DC-signal designated by V_{DC}. The sensitivity

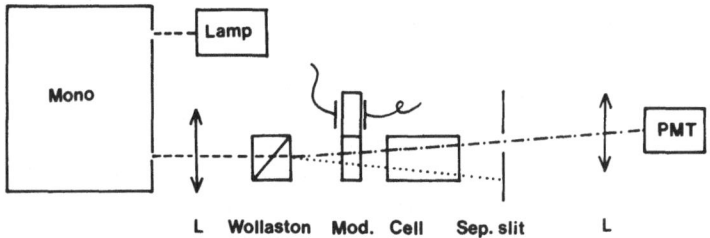

Figure 1

Schematic diagram of the optical system of the vacuum ultraviolet CD instrument.

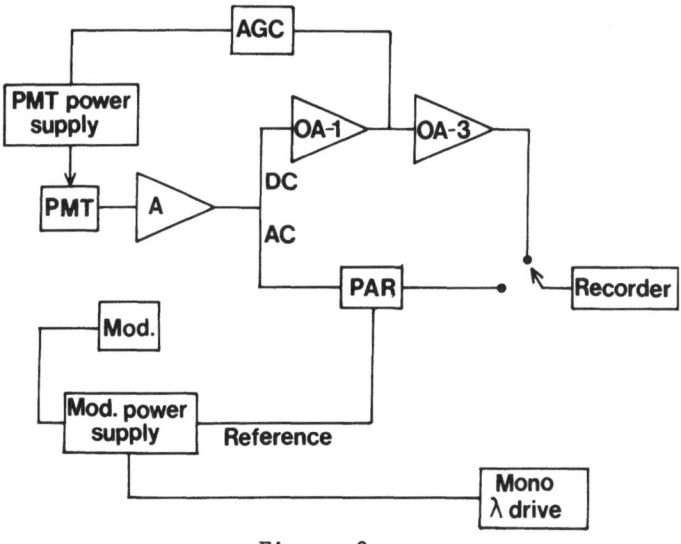

Figure 2

Block diagram of the electronics of the CD instrument.

of the instrument is determined by the capability to measure a
small AC-signal superimposed on a large DC-signal. The desired
sensitivity is of the order of $V_{AC}/V_{DC} \sim 10^{-5}$. To make such a
measurement possible, at least 10^{10} photons are required, and
therefore the source intensity is the major limiting factor.

In fig. 2, a schematic of the electronics system is presented.
The signal from the photomultiplier (PMT) is fed into a preamplifier
(A) which serves as current to voltage converter. Then the AC-
channel is fed to a PAR lock-in amplifier whose output is recorded.
The DC-channel is averaged by an operational amplifier (OA-1) which
has a long time constant. The resulting signal is kept at 1 volt
by an automatic gain control (AGC) feedback system which adjusts
the photomultiplier power supply. As a result the PAR output is
directly V_{AC}/V_{DC}. The reference signal to the PAR is provided by
the power supply of the stress-plate modulator (Mod.). The output
of this power supply is adjusted to keep the modulator at 1/4
wave retardation by a potentiometer geared to the wavelength drive
of the monochromator. It is seen that the electronics are very
routine and unsophisticated.

The sensitivity of the instrument is given in terms of the
limit on $V_{AC}/V_{DC} \sim 3 \times 10^{-5}$ in the wavelength region 300 - 150 nm.
In a number of instances it has been possible to make measurements
out to 140 nm at a lower sensitivity.

The instrument has been modified to permit MCD measurements
to be made by using a superconducting magnet producing a field of
40 - 50 kgauss. For this purpose it was necessary to increase
the distances between photomultiplier and sample and between sample
and polarizing-modulating optics. Also substantial magnetic shielding was added both around the magnet (and sample) as well as for
the photomultiplier.

III. THE NATURAL CD SPECTRA OF BENZENE CHROMOPHORES
Allen and Schnepp (1973a) investigated the CD spectrum of a
conformationally rigid benzene derivative 1-Methylindan (fig.3).
All measurements were made in the gas phase, Fig. 4 shows the
absorption and CD spectra in the B_{2u} region (280 - 240 nm). It is
seen that the CD curve follows closely the absorption in this region.
The difference in relative heights of the first band is an artifact
due to the wider slit for the CD record. Fig. 5 shows the spectra
in the B_{1u} region (230 - 200 nm). Here the contours are quite
different. In fact, the vibrational structure of the absorption
is not reproduced in the CD and moreover, the lowest frequency
peak of the CD occurs at a wavelength of low absorption. This
peak is followed by a deep minimum. The CD spectrum was interpreted
in terms of an additional state (in addition to the benzene B_{1u}

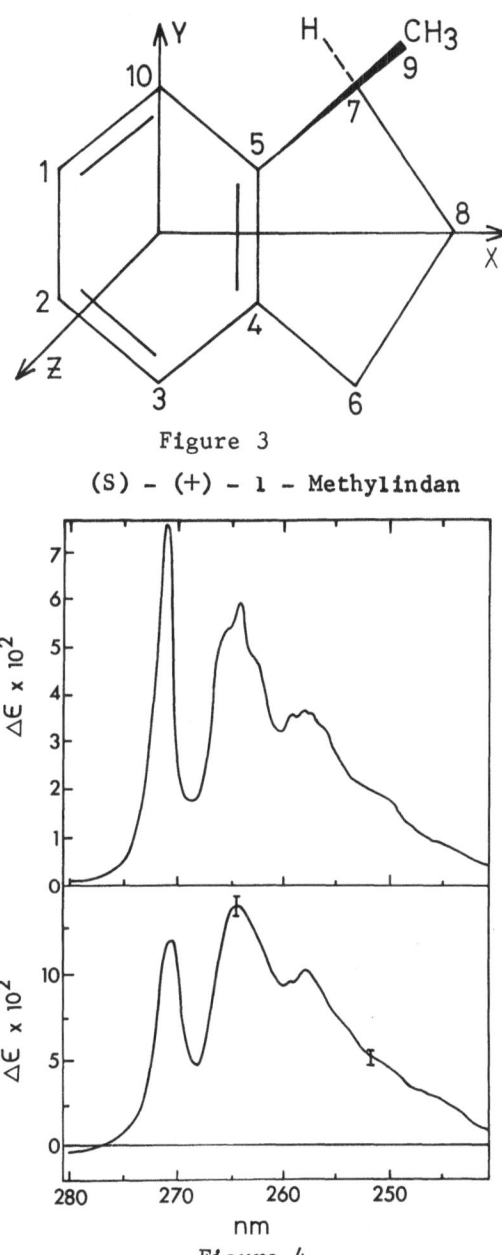

Figure 3

(S) – (+) – 1 – Methylindan

Figure 4

Absorption and CD spectra of 1-methylindan in the gas phase in the region 280–240 nm. Spectral slit width for the absorption is 0.8 nm and for the CD spectrum 1.6 nm.

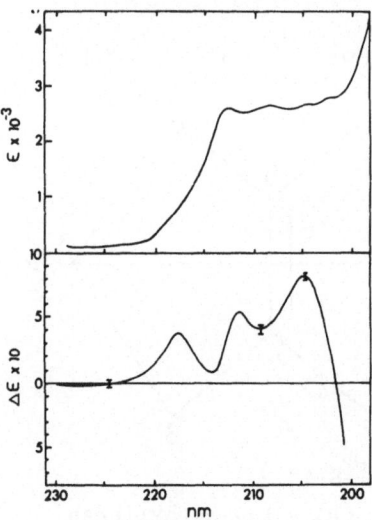

Figure 5

Absorption and CD spectra of 1 – methylindan in the gas phase in the region 230-200 nm. Spectral slit width for the absorption is 0.8 nm and for the CD spectrum 1.6 nm.

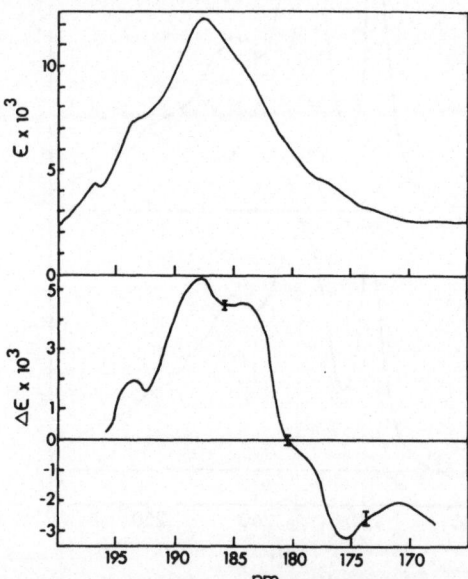

Figure 6

Absorption and CD spectra of 1 – methlindan in the gas phase in the region 195-170 nm (absorption covers a wider range). Spectral slit width is 1.6 nm.

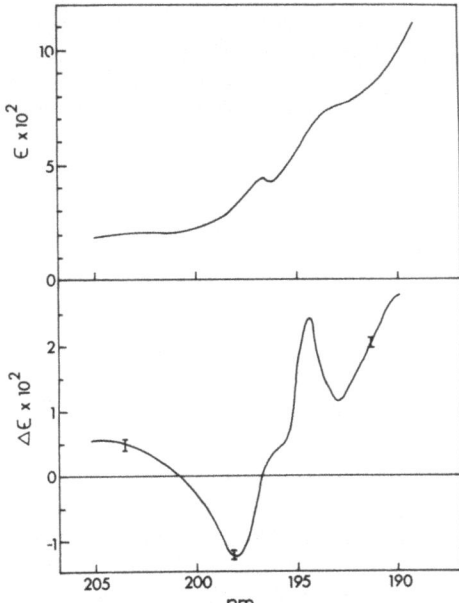

Figure 7

Absorption and CD spectra of gas phase 1 – methylindan in the region 205-190 nm.   Spectral slit width is 1.6 nm.

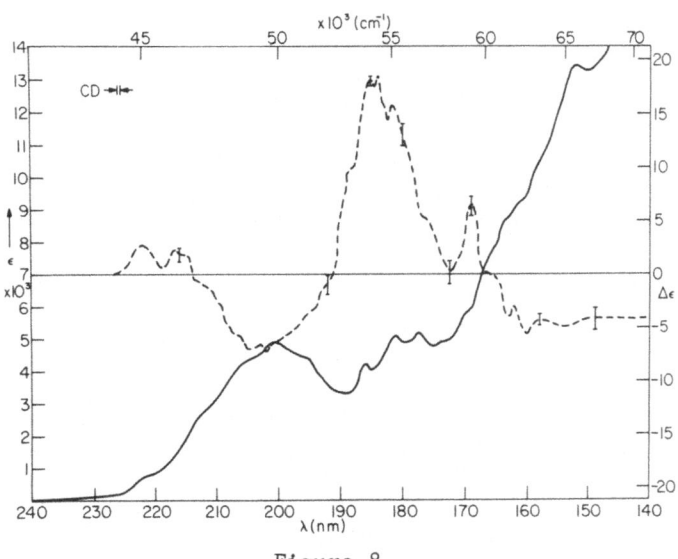

Figure 8

Absorption and CD spectra of gas phase α-pinene in the region 240-150 nm.

state) of type $\sigma - \pi^*$ or $\pi - \sigma^*$. In fact the observed CD spectrum can be reproduced by the superposition of a positive and a negative band close together on top of the B_{1u} contour. Such an additional feature provides evidence for a degenerate state in the symmetric chromophore which would be of symmetry E_{1g} for a magnetic dipole allowed state. The latter requirement is based on the large $\Delta\varepsilon/\varepsilon$ ratio observed here.

Fig. 6 shows the E_{1u} region of the spectrum of 1-methylindan (195-170 nm). It is seen that the removal of the orbital degeneracy by the asymmetric substitution does not manifest itself in the CD spectrum and we therefore conclude that the two components have either the same sign of CD or that they are very unequal in rotatory strength. This is not in agreement with theoretical prediction (Caldwell, 1962). In fig. 7 we present the absorption and CD spectra in the region 205 - 190 nm. The CD spectrum contains unambiguous evidence for at least one additional state. In fact, the absorption contains also weak shoulders corresponding to the CD peaks. This result was interpreted as evidence for the degenerate E_{2g} state of benzene. The $\Delta\varepsilon/\varepsilon$ ratio is similar to that of the other π-electron transitions. A low-lying E_{2g} state has been predicted consistently by theoretical calculations and is the result of configurational interactions between three higher energy excited configurations (Parr, Craig and Ross, 1950: Visscher and Falicov, 1970).

Calculations were carried out within the framework of the formalism given by Tinoco (1962) in the dipole-dipole approximation of separated chromophore interaction where the perturbing group is the C - C of the methyl substituent.

A feature similar to the above was observed in the CD spectrum of Secondary Butyl-benzene (Allen and Schnepp, 1973b). This result is taken as strong corroborative evidence for the additional electronic state interpreted as E_{2g}.

IV. THE NATURAL CD SPECTRA OF ETHYLENE CHROMOPHORES

Mason and Schnepp (1973) investigated the CD spectra of α-pinene, β-pinene and of trans-cyclooctene. In fig. 8, the CD and absorption spectra of α-pinene are presented. We note that the absorption spectrum contains a clear, intense, broad band peaking about 200 nm which is assigned as N - V $(\pi-\pi^*)$. This band is followed by a few weak features and to shorter wavelengths than 170 nm the absorption contour rises steeply. The CD spectrum contains appreciably more information. A negative peak is observed to coincide with the N - V band. At lower frequency than this band there appear two weak positive peaks which were assigned as the $\pi \rightarrow CH^*$ ($1b_{3u} - 4a_g$, $A_g - B_{3u}$) Rydberg transition (Robin et al., 1968). To shorter wavelength than the N - V band, the CD spectrum contains an intense and well resolved positive band which must be correlated with the absorption features in this region. The g-factor is found to be 5×10^{-3} whereas its value is 1.5×10^{-3} for the N -V band. This transition was assigned as $\pi \rightarrow CH^*$ (σ^*) ($1b_{3u} - 2b_{2u}$, $A_g - B_{1g}$). This state was predicted to be in the correct spectral region for ethylene by Buenker, Peyerimhoff and Kammer (1971) who found that it is composed of n=3 basis functions. Also Yaris et al. (1968) predicted this state to lie close to the N - V excited state in trans-cyclooctene. Buenker et al. (1971) also placed another state of suitable symmetry (magnitude dipole allowed) in this spectral region, namely $\pi - CC^*$ (σ^*) ($1b_{3u} - 3b_{1u}$, $A_g - B_{2g}$).

A weak but narrow positive CD band was observed at 169 nm which was assigned as $\pi - 4s$ ($1b_{3u} - 5a_g$, $A_g - B_{3u}$). The negative CD system of moderate intensity in the region 165 - 150 nm corresponds to very intense absorption. The assignment $1b_{3g} - 1b_{2g}$, $A_g - B_{1g}$ was proposed for this transition. Such a transition is better characterized for trans-cyclo octene.

The spectra of β-pinene corroborate the assignments proposed for α-pinene. Also here the N - V transition has CD spectrum opposite in sign to the $\pi-CH^*(\sigma^*)$ transition but both signs are reversed relative to α-pinene (both samples studied were the (-) conformers). In this case the high frequency region is negative and more intense than in α-pinene. Recently S - 3 - Methyl Pentene was investigated (Gross and Schnepp, 1973). Again the $\pi-CH^*$ transition appears well resolved from the N - V band and here the respective g-factors have values 2×10^{-3} and 2×10^{-4} respectively emphasizing the magnetic dipole character of the $\pi-CH^*$ transition. In this compound, however, the two transitions have the same sign CD spectrum.

The CD spectrum of trans-cyclo octene (Mason and Schnepp, 1973) is very intense in the N - V region reflecting the twisted nature of the ethylene chromophore in this molecule. Twisted ethylene is an intrinsically asymmetric chromophore. It is useful to note that both theoretical groups who have carried out calculations for the rotatory strengths for this molecule (Yaris et al., 1968; Robin et al., 1968) predicted a large negative band for the transition $CH(\sigma) - \pi^*$ ($1b_{3g} - 1b_{2g}$, $A_{1g} - B_{1g}$). Such a CD band was in fact found peaking at 156.5 nm and was so assigned.

V. MCD SPECTRA OF BENZENE AND TOLUENE

The MCD spectra of benzene and of toluene were measured in the vapor phase (Mason, Allen and Schnepp, 1973). In the N - V, E_{1u} region, the MCD spectrum of benzene is unexpectedly complex, presumably because of intense contributions due to Rydberg transitions in this region. This spectrum has not been successfully analysed to-date. It is however, known from studies of the

Rydberg transitions of benzene in the spectral region 170 - 140 nm
that these have very intense MCD features.

The MCD spectrum of toluene is, on the other hand, relatively
simple in the E_{1u} region (195 - 170 nm) and represents a somewhat
asymmetric A-term with some vibrational structure. The magnetic
moment of the state was determined to be -0.09 Bohr magnetons. A
calculation using LCAO theory predicts a value of -0.14 Bohr
magnetons.

VI. THE NATURAL CD SPECTRUM OF A SATURATED HYDROCARBON

Allen and Schnepp (1973c) have measured the CD spectrum of
(-) (3S:5S) - 2, 2, 3, 5 - tetramethyl heptane in the region
165 - 140 nm. The absorption rises monotonically in this region
and does not contain any discernible discrete features. The CD
spectrum has an intense negative peak ($\Delta\varepsilon_{max}$ = -21) at 148.5 nm.
This band is assigned as the $\sigma-\sigma^*$ transition.

REFERENCES:

Allen, S., and O. Schnepp,(1973a). J. Chem. Phys., in press.
Allen, S., and O. Schnepp, (1973b). Unpublished.
Allen, S., and O. Schnepp, (1973c). Unpublished.
Buenker, R. J., S. D. Peyerimhoff and W. E. Kammer, (1971). J.
    Chem. Phys. 55, 814.
Caldwell, D. J. (1962). Thesis, Princeton University. D. J.
    Caldwell and H. Eyring, Ann. Rev. Phys. Chem. 15, 281 (1964).
Gross, K., and O. Schnepp, (1973). Unpublished.
Johnson, W. C., Jr. (1971).
Mason, M. G., S. D. Allen and O. Schnepp (1973). Unpublished.
Mason, M. G. and O. Schnepp, (1973). J. Chem. Phys., in press.
Parr, R. G., D. P. Craig and I. G. Ross, (1950). J. Chem. Phys.
    18, 1561.
Robin, M. B., H. Basch, A. N. Kuebler, B. E. Kaplan and J. Meinwald,
    (1968). J. Chem. Phys. 48, 5037.
Schnepp, O.,S. Allen and E. F. Pearson, (1970). Rev. Sci. Instr.
    41, 1136.
Stephens, P. J. (1970). J. Chem. Phys. 52, 3489.
Stephens, P. J., R. L. Mowery and P. N. Schatz, (1971). J. Chem.
    Phys. 55, 224.
Tinoco, I., Jr. (1962). Adv. Chem. Phys. 4, 113.
Visscher, P. B. and L. M. Falicov, (1970). J. Chem. Phys. 52, 4717.
Yaris, M., A. Moscowitz and R. S. Berry (1968). J. Chem. Phys.
    49, 3150.

$(H_3C)_2C=C(CH_3)_2$, $(H_3C)_2BN(CH_3)_2$, $(H_3C)_2BF$ and $(H_3C)_3B$

FURTHER EVIDENCE FOR A $\pi \rightarrow \sigma^*$ ASSIGNMENT OF THE OLEFIN UV "MYSTERY" BAND [1,2]

W. FUß and H. BOCK

Chemistry Department, University Frankfurt/Main, Germany

SUMMARY: UV photoelectron spectra and far uv spectra of $R_2C=CR_2$, R_2BNR_2, R_2BF, and R_3B (R = CH_3) are assigned. Substracting uv excitation energies from corresponding ground state ionisation potentials lead to excited state ionisation potentials, the comparison of which lends further credit to the $\pi \rightarrow \sigma^*$ assignment of the olefin uv "mystery band". The excited state ionisation potentials can be satisfactorily calculated using the Rydberg series formula.

Helium (I) photoelectron spectroscopy[3] opens a new way to understand the optical excitations of σ electrons which are usually observed in the far uv spectral region. Particularly suited for investigations of this kind are molecules with single low lying unoccupied molecular orbitals, such as e.g. threefold coordinated boron compounds.

Another transition in which σ orbitals are possibly involved and which has been extensively discussed in the literature[4-6] is a weak band, which precedes the $\pi \rightarrow \pi^*$ band of alkylated olefins, and is superimposed on it in the far uv spectrum of ethylene. This so called "mystery band" was assigned by Berry[4] to a $\sigma \rightarrow \pi^*$ excitation in analogy to the n $\rightarrow \pi^*$ transitions of carbonyl compounds. On the other hand Robin et al.[5,6] collected convincing arguments in favor of a $\pi \rightarrow \sigma^*$ as-

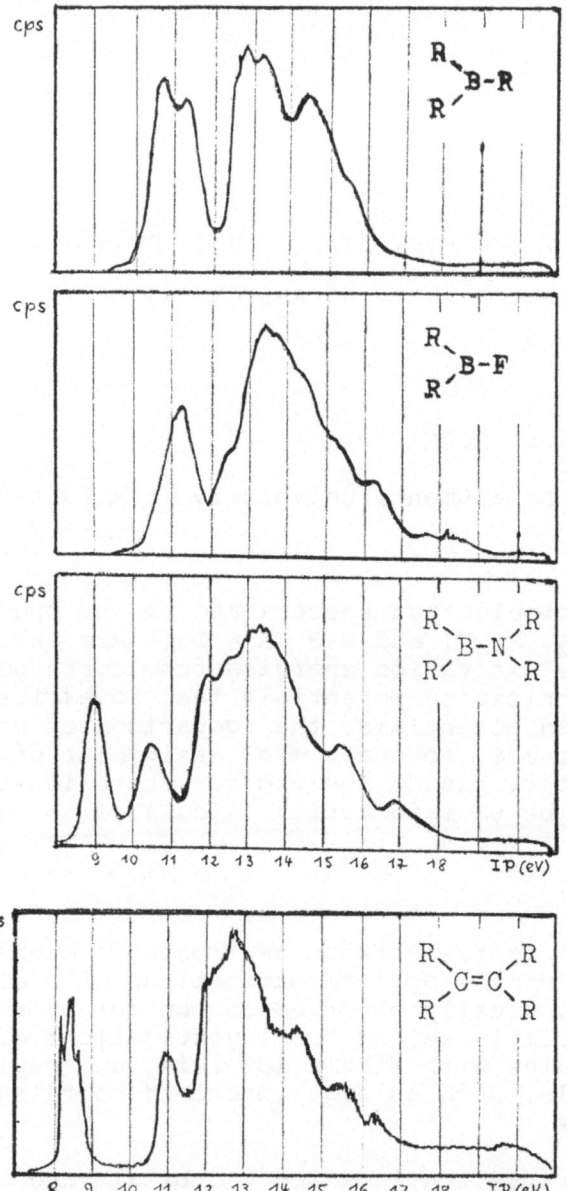

Figure 1: He(I) pe spectra of trimethylborane, dimethyl
Fluoro borane, dimethylamino dimethyl borane
and tetramethyl ethylene (R=CH$_3$)

signment, which presently seems to be prevailing opinion. But there still remain some inconsistancies; thus Ramsey[7] in assigning to $\sigma \rightarrow \pi*$ the first transition in the trimethylborane uv spectrum, supports the Berry interpretation, and Walsh in his review[8] still considers the question not settled. Therefore it may be of interest, that a comparison of the pe and far uv spectra of R_3B, R_2BF, R_2BNR_2, and $R_2C=CR_2$ favors the $\pi \rightarrow \sigma*$ character of the olefin mystery band, and not the Berry $\sigma \rightarrow \pi*$ assignment.

HELIUM(I) PE SPECTRA

In trimethylborane the boron/carbon bonds are the weakest; therefore it may be expected that they are also most easily ionized. The first pe band (figure 1) showing a 0.6 eV Jahn/Teller split is assigned[10] via Koopmans theorem[11] to the doubly degenerate BC bonding orbital 3e' (C_{3h}). This is supported by SCF MO calculations[2,12], according to which there follow the CH bonding orbitals le", 3a', 2e' and the BC bonding orbital 2a'.

The pe spectrum of dimethyl fluoro borane (figure 1) also fits in this assignment: On substituting a methyl group by fluorine, the 3e' orbital splits into b_2 and a_1 orbitals (C_{2h}), of which only the latter should be considerably lowered by the fluorine inductive effect[11] (figure 2). Thus the nearly unshifted first pe band and the low energy shoulder of the "CH mountains" are readily interpreted as b_2 and a_1 respectively.

The same arguments hold for dimethylamino dimethyl borane, the pe-spectrum of which (figure 1) has been assigned before[13] as follows $\pi_{BN}(3b_1)$, $\sigma_{CB}(6b_2)$, $\sigma_{BN}(7a_1)$.... The decrease of the second ionisation potential relative to the 3e' band of trimethyl borane results from a considerable antibonding NC admixture (figure 2).

Tetramethylethylene is isoelectronic with dimethylamino dimethyl borane and thus their pe spectra display a 1:1 correspondence (figures 1 and 2). The larger gap between π and σ orbitals in the olefin is easily understood; in the amino borane the π orbital is lowered due to the predominating N coefficient and the σ orbitals pushed upward due to the large BC bond contributions.

The correlation of the highest occupied σ orbitals in the four compounds (figure 1) gathered from the pe

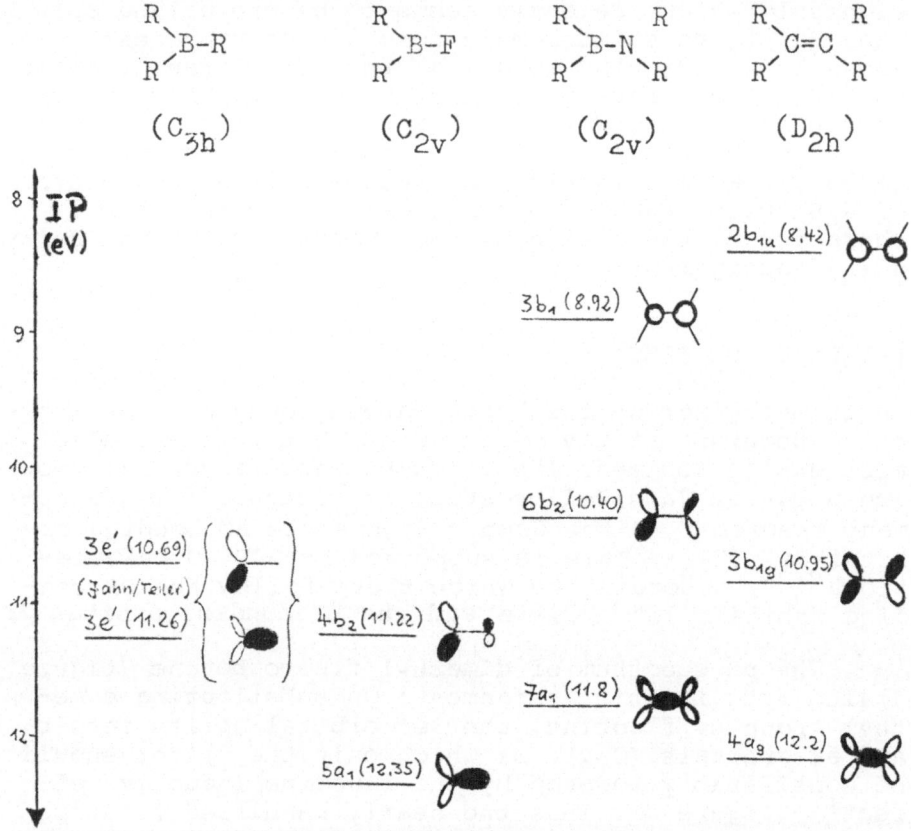

Figure 2:   Energy level scheme with vertical ionisation
            potentials (eV) and diagrams for the highest
            occupied orbitals of the four compounds

spectra can be used advantageously for interpretation
of their far uv spectra.

FAR UV SPECTRA

The far uv spectra are presented in figure 3 and
table 1.

Trimethyl borane shows in front of the alkyl ab-
sorption region ($\tilde{\nu} > 70000$ cm^{-1}) two distinguishable
bands. According to MO calculations[2,7,12] the lowest
unoccupied orbital is, as expected, the empty p orbital of a" symmetry, henceforth called π^*, followed
after a probably exaggerated orbital energy gap of more

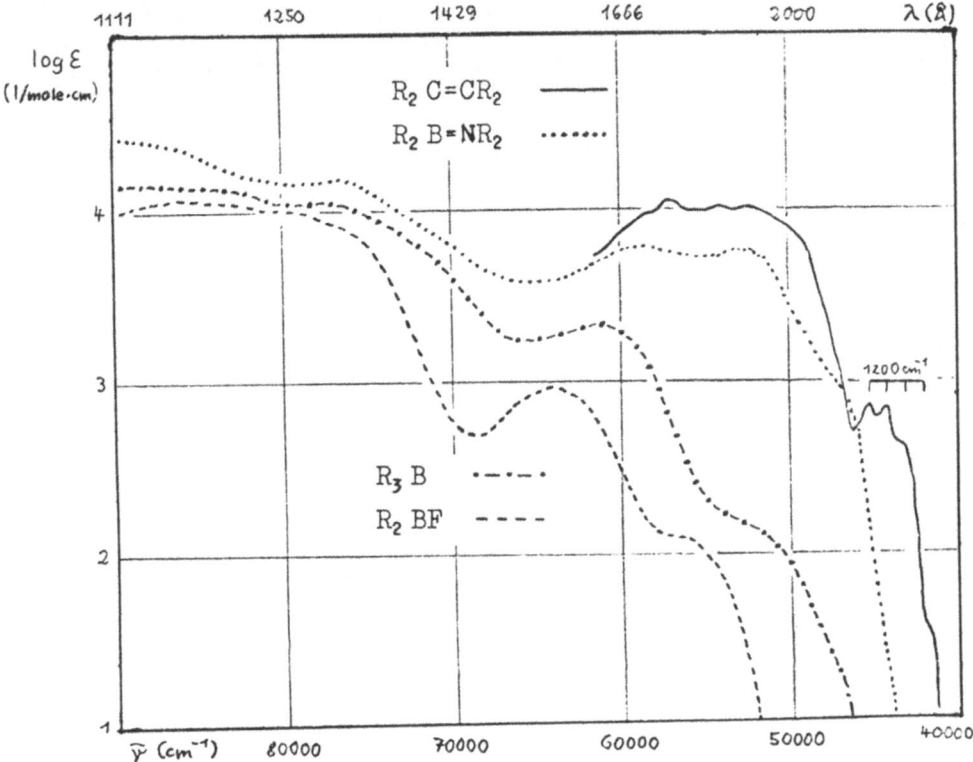

Figure 3:   Far uv spectra of R_3B, R_2BF, R_2BNR_2, and
$R_2C=CR_2$ (R=CH$_3$)

Table 1 :   UV band maximum wave numbers $\tilde{\nu}$ (cm^{-1}) and
molar extinctions ε (1/mole · cm)

R_3B	$\tilde{\nu}$	52 300	s	61 700		72 000	s	77 600
	ε	140		2 100		6 600		11 600
R_2BF	$\tilde{\nu}$	56 200		63 900				
	ε	120		950				
R_2BNR_2	$\tilde{\nu}$	47 000	s	52 900		59 200		76 300
	ε	1 100		6 000		6 100		14 000
$R_2C=CR_2$	$\tilde{\nu}$	44 300		54 400				
	ε	1 200		10 000				

than 3 eV, by several totally symmetric $\sigma^*(a')$ orbitals.
An SCF calculation including configuration interaction[12]
predicts the $\sigma \rightarrow \pi^*$ transition at 43200 cm^{-1} and then
above 70000 cm^{-1} $\pi \rightarrow \sigma^*(a'' \rightarrow a')$, $\sigma \rightarrow \pi^*(a' \rightarrow a'')$ and
$\pi \rightarrow \pi^*$ (e'' $\rightarrow$ a''), their large distance ($\sim$ 27000 cm^{-1})
probably being correct. Therefore the first uv band is
certainly due to the symmetry forbidden (e' $\rightarrow$ a'')
$\sigma \rightarrow \pi^*$ excitation; while the second band found experimentally in the middle of the calculated gap must then
be assigned to one or some symmetry allowed (e' $\rightarrow$ a')
$\sigma \rightarrow \sigma^*$ transitions (figure 4). Indeed, a recent calculation by a modified CNDO method including CI[14] sup-
ports this view and puts an additional forbidden $\sigma \rightarrow \pi^*$
(= $\sigma_{CH} \rightarrow \pi^*$?) transition below the strong $\sigma \rightarrow \sigma^*$. The
predicted excitation energies ($\sigma \rightarrow \pi^*$ = 35500 cm^{-1},
$\sigma \rightarrow \pi^*$ = 44400 cm^{-1}, both forbidden; $\sigma \rightarrow \sigma^*$ = 46000 cm,
$\sigma \rightarrow \sigma^*$ = 60500 cm^{-1}, both allowed), to be sure, are all
too low by more than 16000 cm^{-1}.

In dimethyl fluoro borane the π^* (b$_2$) orbital
should not be much affected by the fluorine substituent,
because for π type orbitals the inductive and mesomeric
effects of fluorine oppose each other (cf.[11] "perfluoro
effect"). The $\sigma^*(a_1)$ orbital is possibly lowered to a
greater extent, but being more diffuse (see concluding
remarks), not as much as the highest occupied σ orbital. Confirmingly, the shift of the first uv band relative to that of trimethyl borane is about the same as
the energy difference between the first pe bands of the
two compounds. The second uv band shifts by about the
same amount; implying, that σ^* is not much affected by
F substitution. Furthermore the intensity ratio of the
first two bands is reasonable for a symmetry forbidden
(b$_2 \rightarrow$ b$_1$) and a symmetry allowed (b$_2 \rightarrow$ a$_1$) transition.
The transition $\sigma \rightarrow \pi^*$(a$_1 \rightarrow$ b$_1$) starting from the second
highest orbital of dimethyl fluoro borane (figure 2),
has to be expected at energies higher by about 1 eV,
around 64000 cm^{-1}. Although symmetry allowed, it will
be weaker than the observed band of this energy, because
it has to be correlated with the low intensity (e' $\rightarrow$ a')
band of R$_3$B.

The far uv spectrum of dimethylamino dimethyl borane shows a long wavelength shoulder reminiscent of
the $\sigma \rightarrow \pi^*$ transitions of R$_3$B and R$_2$BF. There follow
two intense maxima, one of which is expected to be of
$\sigma \rightarrow \sigma^*$(b$_2 \rightarrow$ a$_1$) type, as in R$_2$BF, and the remaining, of
$\pi \rightarrow \pi^*$(b$_1 \rightarrow$ b$_1$) type. But according to the pe spectrum,
the highest occupied σ(b$_2$) level is more than 1.4 eV
below the occupied π(b$_1$) level. Therefore it is not

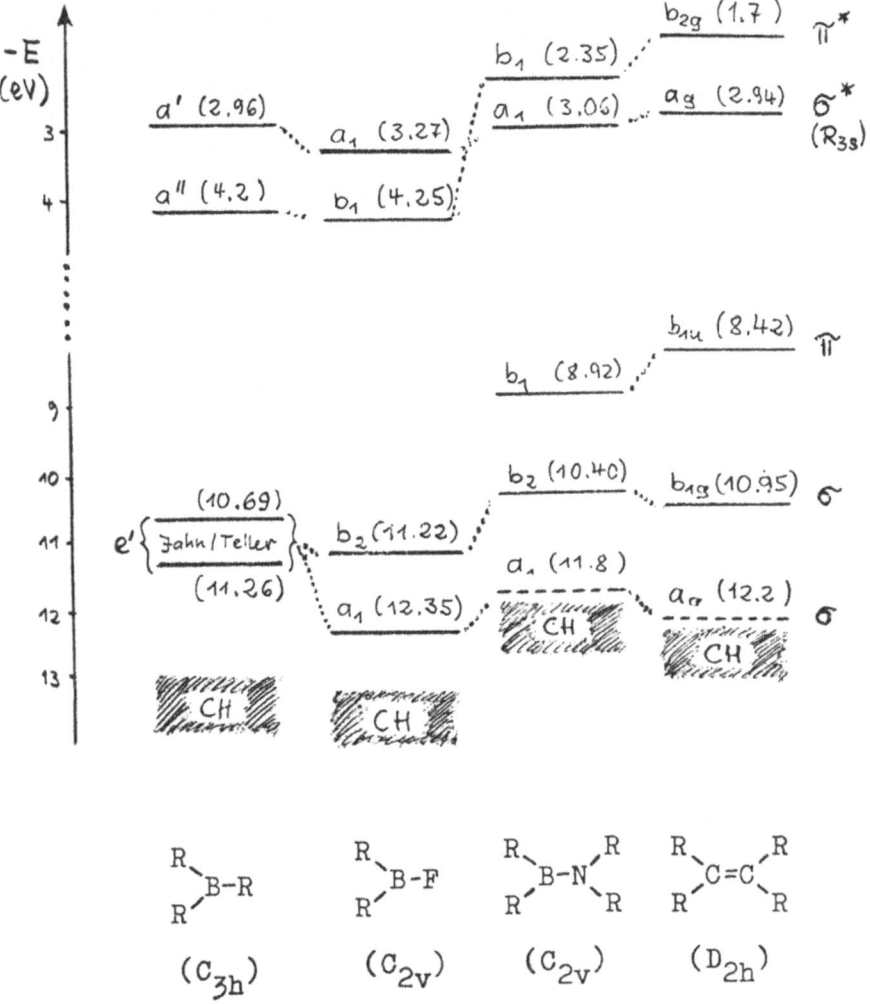

Figure 4: Ionisation potentials (eV) of the title compounds and of their low lying excited states (see text)

very probable that the σ → π* transition precedes the
π → π* transition in the uv spectrum, and that the long
wavelength shoulder originates from the σ → π* excitation. Instead, it will then originate from an excitation of a π electron to some vacant orbital different
from π*. So, if it is tentatively assigned to a π → σ*
transition, the first σ → σ* transition should follow
at a distance of about 1.4 eV, the energy difference
of the occupied π(b_1) and σ(b_2) orbitals. The band position thereby estimated coincides nearly with the second strong band, at about 59000 cm^{-1}, which may therefore be - not very cogently in the moment, but see concluding remarks - identified with the first σ → σ*
transition. This assignment can be further supported by
a comparison with R_3B and R_2BF: If in all three compounds the σ → σ* transitions have some similarity -
as suggested by CNDO calculations - the substitution
effect may be extrapolated from R_3B via R_2BF to R_2BNR_2.
Because fluorine leaves the σ* orbital almost unaffec-
ted, the more will do the NR_2 group. So, starting with
the pe energies of the σ orbitals, a shift of the σ → σ*
transition to 58000 - 60000 cm^{-1} is estimated. There
only the maximum already assigned to σ → σ* is found.

     Judging from the intensities in the uv spectrum of
R_2BNR_2, both the π → σ* and the σ → σ* transitions are
symmetry allowed. This is only possible if σ* is of symmetry type a_1 (the unoccupied CH orbital of symmetry
type a_2 can be eliminated because of poor overlap with
the occupied π, resulting in low intensity) and thus
π → σ* is of (b_1 → a_1) and σ → σ* of (b_2 → a_1) symmetry.
As in trimethyl borane and dimethyl fluoro borane the
lowest σ* orbital seems thus to be totally symmetric;
according to CNDO/2-calculations it is both BN and BC
antibonding with a rather large boron 2s coefficient
(cf. (1)). With the above assignment, even the intensity ratio of the π → σ* (b_1 → a_1) and the σ → σ*
(b_2 → a_1) bands may be understood:

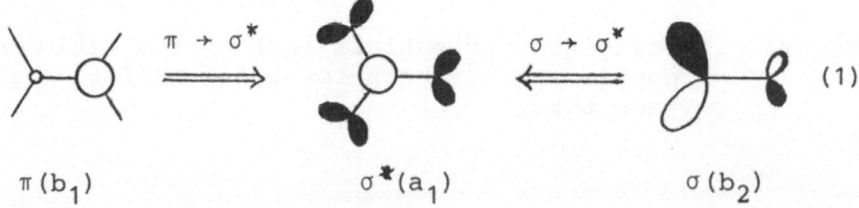

$\pi(b_1)$ $\sigma^*(a_1)$ $\sigma(b_2)$ (1)

Whereas σ and σ* orbitals are mainly localized near the
B atom, the π orbital has only a small B 2p coefficient.

So the transition moment integral $\langle \psi | X | \sigma^* \rangle$ is greater for $| \psi \rangle = | \sigma \rangle$ than for $| \psi \rangle = | \pi \rangle$.

There seems to be no doubt, that in the tetramethyl ethylene uv spectrum the broad intense band around 55000 cm^{-1} contains mainly the $\pi \to \pi^*$ transition, nearly coinciding in energy with the analogous transition of R_2NBR_2 discussed before. This means that π and π^* levels in tetramethyl ethylene are both raised by approximately 0.5 eV. On the other hand the occupied $\sigma(b_{1g})$ level is lowered by 0.5 eV in the olefin (figures 2 and 4). Thus the $\sigma \to \pi^*$ band should be found shifted to shorter wavelengths by about $|\Delta \pi^*| + |\Delta \sigma| \approx 1$ eV $\approx$ 8000 cm^{-1} and cannot be the "mystery band" preceding the $\pi \to \pi^*$ transition: this lower intensity band of tetramethyl ethylene at 44300 cm^{-1} (figure 3) is namely situated at longer wavelengths than any transition of the isoelectronic dimethylamino dimethyl borane. It must therefore be assigned to a different type of excitation. If an excitation of σ electrons then has to be excluded, a π electron must participate in the transition to an orbital lower than π^* and the olefin mystery band has to be assigned $\pi \to \sigma^*$.

CONCLUDING REMARKS

The "orbital energy diagram" in figure 4 was constructed using the pe ionisation potentials via Koopmans theorem[11] for the occupied levels. Adding to these the uv excitation energies, the first ionisation potentials of excited states were obtained, which in an one electron model correspond to energies of unoccupied levels. In doing so Koopmans' theorem is then also applied to previously unoccupied orbitals - keeping in mind, that for open shell systems there are restrictions[11,15]: It is known, that such a procedure for e.g. $n \to \pi^*$ and $\pi \to \pi^*$ transitions of the same molecule will not yield the same π^* level energy because of different two-electron integrals[16] for the individual configurations and eventual subsequent configuration interaction[16]. But in comparing similar transitions in related compounds as in figure 4 it may be assumed, that electron interactions are also comparable.

The occupied levels of the individual compounds have already been discussed in the pe section. The following remarks may be added concerning the unoccupied levels: Because the inductive effect of fluorine and the mesomeric effect of its low lying π type electron pairs oppose each other, π^* levels of R_3B and R_2BF are as expected nearly equal in energy. On the other hand the π interaction of the R_2N group with the BR_2 part dominates the inductive effect of nitrogen, so that in the BN compound the π^* level is raised appreciably. That it still lies lower than the π^* level in $R_2C = CR_2$, is again a consequence of the higher effective nuclear charge of nitrogen. In both compounds the $\pi \rightarrow \pi^*$ transition energies are nearly the same. In terms of Hückel theory, the difference $\alpha_N - \alpha_B$ compensates the lower β_{BN} ($< \beta_{CC}$) in the orbital energy difference $\Delta\varepsilon_J = \sqrt{(\alpha_B-\alpha_N)^2 + 4\beta^2}$, whereas the parallel shift of the π and π^* levels means that $(\alpha_B + \alpha_N)/2$ is lower than α_C by ≈ 0.5 eV.

The σ^* orbitals in figure 4 are strikingly constant; the explanation of which is not immediately evident. Mulliken[17] has pointed out that many unoccupied - particularly σ^* - orbitals may be looked upon as Rydberg orbitals of the united atom. Several authors[18],[23] propose to regard some low lying excited states even of saturated hydrocarbons as Rydberg states, although converging Rydberg series can not be observed in their spectra. Spectroscopists[19] widely use Rydberg's series formula for molecular transitions, which assigns to an unoccupied level the energy $-R/(n - \delta)^2$ relatively to the ionisation limit. Here $R = 109737$ cm^{-1} = 13.60 eV is the Rydberg constant, n is an integer normally $\geqslant 3$, and $\delta \geqslant 0$ is the socalled quantum defect. δ is a measure of the core penetration of the orbital and so depends mainly on its being of s, p or d type:

(2)

Rydberg orbital	limits of $\delta^{1)}$	n = 3	$\delta \left(\bigcirc \right)$	$\delta \left(\begin{matrix} R_2N \\ R_2N \end{matrix} B-X \right)$
ns	$0.9 \leqslant \delta \leqslant 1.2$	3s	0.8	0.7 - 0.8
np	$0.3 \leqslant \delta \leqslant 0.5$	$3p_\sigma$	0.46	0.4 - 0.55
		$3p_\pi$	0.16	0.20
nd	$0 \leqslant \delta \leqslant 0.1$	3d	0.10	0.0

The quantum defect, and so the energy of the Rydberg term, is not very sensitive to modifications of the molecule, which is a consequence of the diffuseness of

the Rydberg orbitals. As a rule electronegative consti-
tuents increase δ and so lower the orbital. For more
detailed studies on the shape of Rydberg orbitals and
their changes in series of chemically related compounds
we like to refer to Duncan[21], Robin[22] and Sandorfy[23]
et al.

If now, the lowest σ^* orbitals of the three boron
compounds and the olefin are regarded as 3s Rydberg orbitals, their energies require $\delta = 0.86$ for R_3B,
$\delta = 0.96$ for R_2BF, $\delta = 0.89$ for R_2BNR_2, and $\delta = 0.85$
for $R_2C = CR_2$. So all four δ values are found in the
ns region, and moreover, the 3s orbitals are totally
symmetric, as are the lowest σ^* orbitals. Now it is
also obvious, why in R_2BNR_2 the $\pi \to \sigma^*$ and the $\sigma \to \sigma^*$
transitions showed the same energetic distance as the
π and the σ ionisations: As a Rydberg level, σ^* is rather insensitive to the exact nature of the core, and
especially of the orbital from which the excited electron stems.

One may also try to identify the empty p orbital
of many boron compounds with a 2p Rydberg orbital.
Doing so, a quantum defect $\delta = 0.19$ results for R_3B and
$\delta = 0.20$, for R_2BF, which again agree with the figures
of table (2). But here it is obvious, that while its
energy and symmetry is correctly given by the Rydberg
formula, it cannot be a true Rydberg orbital. If it
were, it could not be as susceptible to interaction
with valence orbitals[17,22] as e.g. with the n_π orbital
of an amino substituent in R_2NBR_2 (figure 4).

Finally another consideration, not self-conclusive
but rounding off the above arguments may be worth some
attention. The first uv bands of all four compounds are
continuous; the only vibrational progression appears in
the spectrum of $R_2C=CR_2$, arises according to its spacings of 1200 cm^{-1} from a weakened CC double bond and
is also quite diffuse. So, probably the first excitations are all dissociative and lead to fission of one
B-C(H_3) or C-C(H_3) bond, respectively. The long wavelength uv absorption limits correspond to lower bounds
of dissociation energies, the equality signs holding
if the potential curves have no minimum[19]. The spectroscopic values are higher than the thermodynamic figures (3). Evidently, the dissociation leads to an excited boron or olefinic radical; the methyl radical having
only shorter wavelength excitations[19]. For $R_2B^\bullet$ produced
from R_3B there results a transition energy of about
2 eV, which is reasonable for a $\sigma \to \pi^*$ transition of

Dissociation Energy (eV)	R_3B	R_2BF	R_2BNR_2	$R_2C=CR_2$
Spectroscopic	$\geqslant 5.6$	$\geqslant 6.4$	$\geqslant 5.4$	$\geqslant 5.1$
Thermodynamic (estimated)		$\leqslant 3.8.$		$\leqslant 3.8$

$$(3)$$

the single electron (cf. 0.62 eV for BH_2 [19]). In $R\overset{\bullet}{B}F$ it is shifted to shorter wavelengths due to the lowering of the singly occupied σ- and the constancy of the π^* level - in total analogy to R_2BF. In both compounds the $\sigma \to \pi^*$ transition of the radical must be correlated with the $\sigma \to \pi^*$ of the parent compound for symmetry reasons [19]

  If for both $R_2NB\overset{\bullet}{R}$ and $R_2C=C\overset{\bullet}{R}$ the excitations were of $\sigma \to \pi^*$ type too, it would require more energy in the trimethyl vinyl case, because, as in the parent compounds, the σ orbital should be lowered and the π^* orbital should be raised relative to the boron radical. Actually it is found that the vinyl radical $R_2C=C\overset{\bullet}{R}$ has the smaller excitation energy. This may be understood by the assumption, that it is of bonding $\pi \to$ nonbonding σ type. For compared to $R_2C=C\overset{\bullet}{R}$, in $R_2NB\overset{\bullet}{R}$ the π orbital is lowered by its prevailing N contribution much more than the singly occupied nonbonding σ orbital, the N contribution to which is certainly not as significant. Because the $\pi \to \sigma$ transitions of the radicals (final state $\pi^1\sigma^2$) must be correlated with the $\pi \to \sigma^*$ transitions (final state $\pi^1\sigma^2(\sigma^*)^1$) of the parent compounds, this consideration again confirms the $\pi \to \sigma^*$ origin of the first singlet excitation of olefins, giving rise to the "mystery band".

REFERENCES

1) Part XXIV of Photoelectron Spectra and Molecular Properties; Part XXIII: H.BOCK, B.SOLOUKI, R.STEUDEL, and P.ROSMUS, Angew.Chem. 1973, 85; internat. Edit. 1973, 11, in print.
2) Part of the Thesis W.FUß, University of Frankfurt/ Main, 1971.
3) D.W.TURNER, C.BAKER, A.D.BAKER, and C.R.BRUNDLE, "Molecular Photoelectron Spectroscopy", Wiley-Inter-

science, London 1970.

4) R.S.BERRY, J.Chem.Phys. 1963, 38, 1934.

5) M.B.ROBIN, R.R.HART, and N.A.KUEBLER, J.Chem.Phys. 1966, 44, 1803; this reference also contains a survey of earlier assignments.

6) M.B.ROBIN, H.BASCH, and N.A.KUEBLER, J.Chem.Phys. 1968, 48, 5037; cf. R.HOFFMANN and R.B.DAVIDSON, J.Amer.Chem.Soc. 1971, 93, 5699.

7) B.G.RAMSEY, J.Phys.Chem. 1966, 70, 4097.

8) A.D.WALSH, Chem.Soc.(London), Ann.Rep.Progr.Chem. 1967, 63, 44.

9) H.C.BROWN, J.Amer.Chem.Soc. 1945, 67, 374; cf. P. LOVE, J.Chem.Phys. 1963, 39, 3044 and E.L.MUETTER-TIES (ed.), "The Chemistry of Boron and its Compounds", Wiley, New York 1967.

10) A.K.HOLLIDAY, W.READE, R.A.W.HOHNSTONE, and A.F.NE-VILLE, Chem.Comm. 1971, 51.

11) C.R.BRUNDLE and M.B.ROBIN, "Photoelectron Spectroscopy" in "Determination of Organic Structures by Physical Methods", 1971, 3, 1, edit. by F.C. NACHOD and J.J.ZUCKERMANN, Academic Press, New York.

12) D.R.ARMSTRONG and P.G.PERKINS, Theor.Chim.Acta 1968, 9, 412.

13) H.BOCK and W.FUß, Chem.Ber. 1971, 104, 1687.

14) G.KUEHNLENZ and H.H.JAFFÉ, J.Chem.Phys. 1973, 58, 2239.

15) W.RICHARDS, Internat.J.Mass Spectry.Ion Phys. 1969, 2, 419.

16) Cf. e.g. J.N.MURRELL, "The Theory of the Electronic Spectra of Organic Molecules", Methuen, London 1963.

17) R.S.MULLIKEN, J.Chem.Phys. 1935, 3, 517.

18) H.BASCH, M.B.ROBIN, N.A.KUEBLER, C.BAKER, and D.W. TURNER, J.Chem.Phys. 1969, 51, 52.

19) G.HERZBERG, "Molecular Spectra and Molecular Structure", Vol.3, van Nostrand, New York 1950.

20) E.LINDHOLM and B.-Ö.JONSSON, Chem.Phys.Letters 1967, 1, 501.

21) A.B.F.DUNCAN, "Rydberg Series in Atoms and Molecules", Academic Press, New York 1971.

22) M.B.ROBIN, Internat.J.Quant.Chem., in print.

23) D.R.SALAHUB and C.SANDORFY, Theor.Chim.Acta 1971,20, 227 and Chem.Phys.Lett. 1971, 8, 71.

24) T.L.COTTRELL, "The Strengths of Chemical Bonds", Butterworths, London 1958.

We gratefully acknowledge support by the Stiftung Volkswagenwerk and the Deutsche Forschungsgemeinschaft.

PHOTOELECTRON SPECTRA OF MODERATE SIZED MOLECULES

D.C. Frost, F.G. Herring and C.A. McDowell

Department of Chemistry
University of British Columbia
Vancouver 8, British Columbia

Introduction

Photoelectron spectroscopy (PES) provides information on the relative ordering of cationic states obtained by the removal of a single electron from successively higher binding energy orbitals. In the case of a closed shell parent molecule, therefore, these states correspond to excited states formed by promoting an electron from an inner doubly occupied orbital to the singly occupied orbital of the cation.

The relative ordering of these states will depend upon electronic effects variously described as electronegativity, conjugation, through bond and space orbital interactions, d-orbital participation etc. Generally speaking the chemist is interested in estimating the magnitude of relative dominance of one or other of these electronic interactions in the parent molecule. So that such information can be extracted from a PE spectrum it is necessary that the effect of the aforementioned electronic interactions on the relative ordering of the cationic states be translated into their possible importance in the parent molecule.

The attainment of this bridge between cationic states and the parent revolves around the concept of an orbital structure of the electronic configuration of the parent molecule, and that this orbital structure is largely unattained in the cation. The orbital structure may be conceived of in terms of valence bond configurations, equivalent orbitals or molecular orbitals (MO's); the last being the most popular. The idea that the orbitals do not rearrange on ionization is embodied in Koopmans' Theorem[1] when an SCF model is employed. The theorem, really an approximation, simply shows that the energy required to remove a single electron from an MO is equal in magnitude to the orbital energy.

The concept of an orbital energy arises from the fact that the Lagrangian multiplier matrix in the SCF method for closed shell molecules can be diagonalized and that these diagonal elements are made up of two parts, the kinetic energy and electron-nuclear attraction and the electronic Coulomb and exchange inter-actions (see for example reference 2). The orbital energies obtained from an SCF calculation correspond to vertical (no geometry change on ionization) ionization potentials (IP). The variation in the vertical IP's can then be explained in terms of the electronic interactions mentioned before, but, of course, a decision often has to be made as to which is the most important of these effects.

The very nature of such concepts as electronegativity, con-jugation etc. will render any interpretation of a PE spectrum based on them qualitative. The interpretation of PES in fact provides a severe test of such simple notions and their applica-bility to the description of changes in orbital configurations. The assignement of PE spectral bands is often aided by MO calcu-lations of varying degrees of sophistication, ranging from ex-tended Huckel[3], through zero differential overlap (ZDO) i.e. CNDO and INDO[4], to ab initio Slater type orbital (STO)[5] and Gaussian type orbital (GTO)[6] calculations and beyond[7]. In the majority of cases Koopmans' Theorem is employed to obtain an estimate of the vertical IP.

The alternative procedure to obtain a computational estimate of the vertical IP is the so-called ΔSCF method. The ΔSCF method entails a SCF calculation of both the ground state of the parent and the ground state of the cation and the appropriate excited states. The advantage of this method is that whatever electronic reorganization takes place on ionization will be accounted for within the scope of the basis set chosen for the calculation. However the difference in correlation energy between the parent and the cation will be neglected. In fact, it is the near can-cellation of electronic reorganizational effects and the differ-ence in correlation energy that leads to the success of Koopmans' Theorem. The main advantage of the ΔSCF method lies in the interpretation of the PE spectra of free radicals. The reason for this is the absence of a Koopmans' Theorem for open shell systems, although 'frozen orbital' estimates of vertical IP's can be made[8,9].

The object of this contribution is to outline the studies in PES made at the University of British Columbia in the light of the foregoing brief remarks. The research interests at U.B.C. range from the PES of small transient species[10] (not discussed here) to studies on the orbital interactions in large organic molecules. The work undertaken can be conveniently divided into the following six groups (a) the interpretation of the PE spectra of structurally similar molecules[11-14], (b) a comparison of the bonding of first and second (and subsequent) row p-block

elements[15,16], (c) the nature of the chemical bonding of a particular element[17-19], (d) the investigation of orbital interactions i.e. through space of bond[20-22], (e) the correlation of
PES with UV spectra and other techniques[23-25], (f) theoretical
studies[26]. Obviously parts (a), (b) and (c) overlap to a considerable degree, however, one or other of them usually provides
the motivation for a PES study. We turn now to a brief discussion of these topics.

(a) The interpretation of the PE spectra of structurally similar
molecules.

The section heading, of course, encompases all branches of
studies in PES; however, for the purposes of the present discussion it is taken to mean a study of those molecules which
contain a common grouping. Thus, for example, the dichloro-
compounds possessing C_{2v} symmetry, i.e. OCl_2, $H_2C=CCl_2$, $OCCl_2$
etc., form our first attempt[11] at the assignment of the cationic
states due to removal of those electrons which could be thought
of as chlorine lone-pairs. The complete interpretation of the
spectra of such molecules was confused by erroneous reportage of
spectra[27] and has recently been reinvestigated with the addition
of two new molecules CF_2Cl_2 and SCl_2[28]. Other studies that have
been carried out concern the bromo-ethylenes[12] and acetaldehyde
and acetyl halides[13].

In order to exemplify this type of study we shall briefly
discuss the results of a study on the phosphoryl (OPX_3) and thiophosphoryl molecules (SPX_3)[14], where X=Cl or Br. The results
obtained are summarized in the orbital correlation diagram shown
in Fig. 1. The interpretations contained in this diagram are in
complete accord with the work of Cox et.al.[29].

The molecules concerned have C_{3v} symmetry so that the
cationic states will correspond to $^2A_1$, $^2A_2$ and ^2E. The possibility of Jahn-Teller and spin-orbit splitting is present for the
states ^2E symmetry. In general the first few bands in the PE
spectra of these molecules will be due to ionizations from the
lone pairs on the halogens, the P-O or P-S bonding region and
the P-X bonding regions. The relative positions of these bonds
depends on the relative energies of the constituent atomic
orbitals.

Consider the molecular pair $OPCl_3$ and $SPCl_3$. The first
two bands in the PE spectrum of $OPCl$, are readily interpreted
as being due to ionizations from the e and a_2 lone pair orbitals.
In the absence of substantial d-orbital participation the energy
of the a_2 orbital will be determined largely by inductive
effects, so that the destabilization of this orbital on going
from $OPCl_3$ to $SPCl_3$ is expected on simple electronegativity
grounds. The e chlorine lone pair orbital on the other hand can
interact with orbitals on the phosphorous and either oxygen or
sulphur. In fact the characteristics of this e orbital are such
that there is essentially a node at the phosphorus atom. Thus

in the case of the $OPCl_3$ molecule, because of the energetic disparity between the oxygen and chlorine orbitals the e orbital is almost exclusively localized in the chlorine. On replacement of the oxygen atom by a sulphur the characteristics of the orbital change to that of a mixture of sulphur and chlorine orbitals because of the energetic similarity of the constituent atomic orbitals. The presence of this added interaction leads to a much greater destabilization of the first ^2E states on Cl_3PO^+ and Cl_3PS^+ relative to the destabilization of the $^2A_2$ states. The interpretation of the remainder of the spectra is not quite so simple and has been based on CNDO/2 calculations together with the observation of spin-orbit splitting, and the results contained in Fig. 1 indicate that some orbital rearrangements take place on replacing oxygen by sulphur in the chloro compounds but not the bromoderivatives. The replacement of the chlorine atoms by bromine atoms in phosphoryl chloride and thiophosphoyyl chloride lead to the expected reduction in binding energies. The overall conclusion that can be reached on the basis of these studies is that the main bonding characteristics (such as d_π-P_π back bonding) are essentially the same in all four molecules.
(b) A comparison of the bonding of first and second row p-block elements.

The comparison of the bonding exhibited by the light and heavy representative elements using PES has been carried out by a number of researchers, e.g. Cradock et.al.[30] Price et.al.[31] and ourselves[15,16]. Generally the comparison has centred on the Groups IVA to VIIA. It is perhaps appropriate to summarize the findings group by group.

A comparison of the Group IVA compounds EX_4 where $X=Cl$, Br, CH_3 etc. has shown that on replacing carbon by silicon that certain of the cationic states are stabilized by $(p \rightarrow d)_\pi$ bonding. A clear example of this is the stabilization of the $^2A_2$ state in H_2SiCl_2 (12.53 eV) relative to that in H_2CCl_2 (12.22 eV). These conclusions were reached by Cradock et.al.[30] and Frost et.al.[16]. The latter workers performed empirical one-electron calculations based on the method of Dixon et.al[32] which supported the contention that Si d-orbital participation is important. The same results were obtained for the germyl compounds[30]. It is important to remember, of course, that these results pertain to the cations and not necessarily to the parent molecules i.e. perhaps electronic reorganization requires the added flexibility imparted by d-orbital participation. In fact a computational study should be made but the cost may be prohibitive.

Many studies have been made on the compounds of Group VA[15,31]. The PE spectra of NH_3, PH_3 and AsH_3 are quite well understood[15,31]. The interpretation of the PES of the corresponding tris methyl, silyl and germyl derivatives leads to further indications of p→d bonding in the silyl and germyl amines[25].

The compounds of Group VIA elements have also been widely

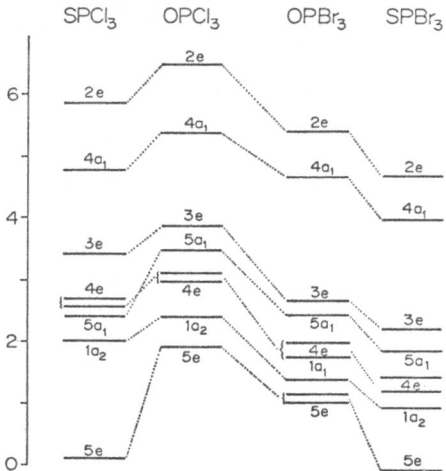

Figure 1

Orbital correlation diagram for the phosphoryl and thiophosphoryl halides.

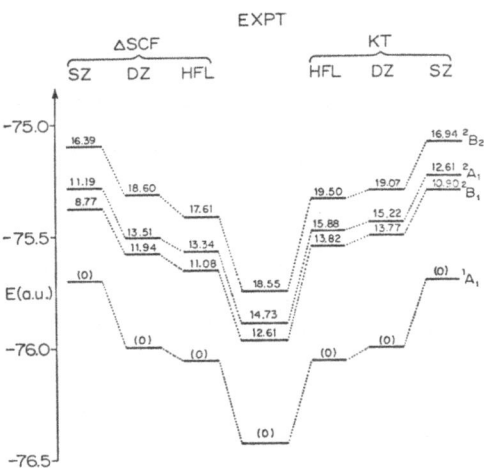

Figure 2

Experimental and computed ionization potentials of water.

studied[17,30]. The indications are that in divalent sulphur,
selenium and tellurium compounds d-orbital participation is
minimal, there appears to be one major exception to this (see
below).

Many PES studies on halogen compounds have been made. Of
particular interest are the interhalogen compounds and these
have been the subject of an extensive investigation in this
Laboratory[31] and by DeKock et.al.[35]. The interhalogens fall
outside the scope of this discourse as they, except for a few
notable exceptions, tend to consist of four atoms or less.
(c) Photoelectron studies of the bonding of particular elements.

The type of study indicated here is that in which extensive
investigations are made of the PES of compounds of a particular
element in varying chemical environments. Examples of this are
the studies of King et.al.[36] on boron compounds or our studies
on the compounds of sulphur[17-19]. The objective of such investigation is to ascertain the generality of such observations,
as that 'the bonding of divalent sulphur does not require
substantial sulphur d-orbital participation'[17,30]. Such sweep-
ing statements, of course, depend critically on the bonding
environment i.e. substantially different molecular conformations.
A good example is afforded by the interpretation of the PES
of ethylene sulphide and is borne out by the Rydberg series
observed high vacuum UV spectrum.[18(b)] The ground state of
ethylene sulphide is $^2B_1$ as is that in dimethyl sulphide (assuming C_{2v} geometry). The first ionization potential of ethylene
sulphide is 0.4 eV higher than that of dimethyl sulphide. On
the basis that formation of the carbon-carbon bond on going from
dimethyl sulphide to ethylene sulphide would tend to destabilize
the b_1 sulphur orbital due to increased orbital interactions
the 0.4 eV stabilization is unexpected. However, the CSC bond
angle in ethylene sulphide (48°) is much reduced relative to
dimethyl sulphide (84°) so that the effect of this change in
geometry must be considered. The spectroscopic evidence[18b] is
that the b_1 sulphur orbital is stabilized due to 3d orbital
participation.
(d) The investigation of "through space and through bond"
interactions.

The idea that functional groups can be thought of in terms
of localized groups of orbitals is well established. The fact
that two non-conjugated functional groups may interact with
each other either "through space" i.e. across a molecule or
"through bond" i.e. via the σ-framework has been discussed by
Hoffman[36]. Photoelectron spectroscopy provides an ideal tool
for the investigation of such effects as has been demonstrated
by the work of Heilbronner et.al.[38].

The principle of such studies is to obtain the PES of a
molecule containing the two interacting groups in question and
then compare the energy of the two states formed by ionization

from these groups in model compounds in which the specific
interaction is absent. Frost and Weiler[22] have performed such
studies in, for example, 2,5-dihydrofuran and 7-oxabicyclo
[2.21]-heptene-2. The results indicate that the interaction
between the π-double bond and the oxygen lone pair is greater
in the former molecule. The results of an interesting PES
study by Bunzli, Burak and Frost[38] on non-conjugated acyclic
dienes $H_2C = CH(CH_2)_n - CH = CH_2$ (n = S to 0) suggest that even
long chain dienes have a preferred conformation which enhances
the through space interaction.

The interpretation of the PES of such molecules revolves
around the unscrambling of which effect (i.e. overlap or σ-
interaction) dominates. In this way it is hoped that a better
knowledge of the physical factors that determine the relative
dominance of through space and through bond interactions.
(e) The correlation of PES with UV spectra and other techniques.

The PES of a molecule yields the energy separation between
the ground state and those excited states form by removing an
electron from a double occupied orbital and placing it in the
singly occupied orbital. Thus a direct correspondence between
PES and the UV spectrum of the cation occurs[24], except that there
are not the corresponding selection rules in PES and all states
are equally probable. Furthermore, a direct correspondence
occurs between PES and X-ray emission spectroscopy. Manne and
Klassen[27] have pointed out the usefulness of the results of
X-ray emission spectroscopy in the interpretation of PES.

The uv spectrum of the cation is not always available so
that often comparisons between pes and the uv spectrum of the
parent neutral are made. The procedure is difficult and relies
(see below) on either orbital energies or repulsion integrals
remaining constant over a series of molecules. The energy
difference between the ground state and an excited singlet state
formed by excitation of the i-th electron to the j-th virtual
level is given, in the frozen orbital approximation, by

$$\Delta E_{ij} = \epsilon_j - \epsilon_i - J_{ij} + 2K_{ij}$$

PES provides information about ϵ_i (via KT) so that if a series
of excitations to a common level are observed in the UV spectrum
and J_{ij} and K_{ij} remain constant then a correlation between PES
and UV spectroscopy can be made. Such studies have been carried
out by Hasselbach[39] and Bain and Frost[25]. The former workers
used INDO-calculations to provide estimates of J_{ij} and K_{ij} to
aid the correlation. Bain and Frost obtained an empirical
correlation between the PE and UV spectra of maleimicle, maleic
anhydride and 4-cyclopentene-1,3-dione which lead to a reassignment of the second band in the UV spectra of the last two
molecules. Such studies will no doubt be very useful in the
assignment of both forms of spectra.

(f)   Theoretical studies.
The main computational aid to interpretation that has been
applied in the foregoing studies is the approximate LCAO-SCF
scheme devised by Pople et.al., viz. CNDO/2 and INDO.   These
approximate schemes have been very useful in the interpretation
of PES using Koopmans' Theorem although caution must be exercised.
The main success of this theorem stems from the partial cancella-
tion of electronic reorganizational correlational effects.   The
alternative method of estimating IP's is to calculate, by separate SCF procedures, the total energy of both the cation and the
parent and to subtract the two, i.e. the so-called ΔSCF method.
The ΔSCF method, therefore, will take into account the reorganizational energy nn ionization but will neglect correlation
differences between the parent and cation.   Since the correlation
energy in the parent is expected to higher the ΔSCF method, as
opposed to that using Koopmans' Theorem, is expected to give
IP's too low by the difference in correlation energy (if the two
SCF calculations are at their respective Hartree-Fock Limit (HFL)).
The ΔSCF method at first sight appears to be the most
desirable but a reordering of states relative to Koopmans'
Theorem may occur if the reorganizational energy is very large
for a particular state.   Subsequently if corrections for correlation energy are applied the Koopmans' Theorem ordering may be
restored.   In order to investigate whether or not this is a
major concern, Chong et.al.[26] have performed ΔSCF calculations,
based initially on the INDO scheme.   The results are that, except
in two cases (H_2CO, F_2CO) for the molecules considered (H_2O, F_2O,
CH_2F_2, CH_2CF_2, H_2CO, F_2CO, etc.) reorganizational effects do not
appear to change the ordering of the first few states relative
to the Koopmans' Theorem order.   Subsequently, ab initio double-
zeta Slater type orbital (STO) ΔSCF calculations on, for example
H_2O and F_2O, have tended to confirm these findings, although
the study is far from complete.   The main objective of the ab
initio study is to investigate the effect of basis set variation,
i.e. the number and type of STO's on the estimation of IP's, as
very little work of this type has been carried out.
It is perhaps informative to consider the variation in the
estimate of the first three IP's of water using Koopmans' Theorem
and ΔSCF, as the basis set (STO) is changed from single-zeta (SZ)
through double-zeta (DZ) to approximately the Hartree-Fock Limit
(HFL).   The results are shown in Fig. 2.   The SZ and DZ computations were performed in this laboratory[41] while the near HFL
work is that of Dunning et.al.[40].
The vertical IPs of water as predicted by Koopmans' Theorem
at the HFL Limit are all greater than experiment which suggests
that reorganizational effects are greater than correlation effects.
The DZ calculation also exhibits this behaviour but this effect
is lost in the inadequacy of the SZ basis.   The ΔSCF estimates
of the IP's at the HFL are, as expected, all smaller than experiment reflecting the difference in correlation energy between

cation and parent. It is interesting that even the DZ basis cannot quite reproduce this for the third IP; again the SZ calculation is hopelessly inadequate. The results contained in Fig. 2, and other calculations, tend to suggest that DZ-STO calculations may be a reasonable compromise for the estimation of IPs in moderate sized molecules.

Acknowledgement

The authors wish to thank their many colleagues who have participated in this work, especially Dr. L. Weiler, Dr. D. Chadwick, Dr. J.C. Bunzli, Dr. A.B. Cornford, Dr. A. Katrib, Dr. R.A.N. McLean, Mr. S.T. Lee, Mr. A. Bain, and Mr. A.J. Burak. The financial support of the National Research Council of Canada is gratefully acknowledged.

References

1. T. Koopmans, Physica 1, 104 (1974).
2. See for example C.C.J. Roothan Rev. Mod. Phys. 23, 69 (1951).
3. R. Hoffmann, J. Chem. Phys. 39, 1397 (1963).
4. J.A. Pople and D.L. Beveridge 'Approximate Molecular Orbital Theory' (McGraw-Hill, 1970).
5. e.g. T.H. Dunning, R.M. Pitzer and S. Aung, J. Chem. Phys. 57, 5044 (1972).
6. e.g. C.R. Brundle, M.B. Robin and H. Basch, J. Chem. Phys. 53, 2196 (1970).
7. L.S. Cederbaum, G. Hohlneicher and S. Peyerimhoff, Chem. Phys. Lett., 11, 421 (1971).
8. O. Edquist, E. Lindholm, L.E. Selin, L. Asbrink, C.E. Kuyatt, S.R. Mielczanek and J.A. Simpson, Phys. Scripta, 1. 172, (1970).
9. A.B. Cornford, D.C. Frost, F.G. Herring and C.A. McDowell, J. Chem. Phys., 54, 1872 (1971).
10. A.B. Cornford, D.C. Frost, F.G. Herring and C.A. McDowell, Chemical Society, Faraday Division, Discussions 54, 1973.
11. D. Chadwick, A.B. Cornford, D.C. Frost, F.G. Herring, A. Katrib, C.A. McDowell and R.A.N. McLean, p.453, 'Electron Spectroscopy', ed. D.A. Shirley, North-Holland (1972)
12. D. Chadwick, D.C. Frost, A. Katrib, C.A. McDowell and R.A.N. McLean, Canad. J. Chem. 50, 2642 (1972).
13. D. Chadwick and A. Katrib (submitted).
14. J.C. Bunzli, D.C. Frost and C.A. McDowell (in press).
15. (a) G.R. Branton, D.C. Frost, F.G. Herring, C.A. McDowell and I.A. Stenhouse, Chem. Phys. Lett. 3, 581 (1969).
    (b) G.R. Branton, D.C. Frost, C.A. McDowell and I.A. Stenhouse, Chem. Phys. Lett., 5, 1 (1970).
16. D.C. Frost, F.G. Herring, A. Katrib, R.A.N. McLean, J.E. Drake and N.P.C. Westwood, Can. J. Chem. 49, 4033 (1971).
17. D.C. Frost, F.G. Herring, A. Katrib, C.A. McDowell and R.A.N. McLean, J. Phys. Chem., 76, 1030 (1972).

18. (a) D.C. Frost, F.G. Herring, A. Katrib and C.A. McDowell, Chem. Phys. Lett. (in press).
    (b) N. Basco and R. Morse, ibid.
19. D. Chadwick, D.C. Frost, F.G. Herring, A. Katrib, C.A. McDowell and R.A.N. McLean, Can. J. Chem. (in press).
20. D. Chadwick, D.C. Frost and L. Weiler, J. Am. Chem. Soc., 93, 4320 (1971).
21. D. Chadwick, D.C. Frost and L. Weiler, J. Amer. Chem. Soc., 93, 4962 (1971).
22. A.D. Bain, J.C. Bunzli, D.C. Frost, and L. Weiler, J. Amer. Chem. Soc., 95, 291 (1973).
23. A.D. Bain and D.C. Frost, Can. J. Chem. 51, 1245 (1973).
24. F.G. Herring and R.A.N. McLean, Inorg. Chem. 11, 1667 (1972).
25. A.D. Bain and D.C. Frost (to be published).
26. D.P. Chong, F.G. Herring and D. McWilliams (to be published).
27. M. Klasson and R. Manne, p.471, 'Electron Spectroscopy' ed. D.A. Shirley, North-Holland (1972).
28. J.C. Bunzli, D.C. Frost, F.G. Herring, S.T. Lee and C.A. McDowell (to be published).
29. P.A. Cox, S. Evans, A.F. Orchard, N.V. Richardson and R.J. Roberts, Chemical Society, Faraday Division Discussions, 54, (1973).
30. Cradock et.al. J.C.S. Faraday II, 68, 86, 281 (1972) and references contained therein.
31. A.W. Potts and W.C. Price, Proc. R. Soc. A326, 181 (1972).
32. R.N. Dixon, J.N. Murrell and B. Narayan, Mol. Phys., 20, 611 (1971).
33. A.B. Cornford, Ph.D. thesis, University of British Columbia, (1971).
34. R.L. DeKock, B.R. Higginson, D.R. Lloyd, A. Breeze, D.W. Cruickshank and D.R. Armstrong, Mol. Phys. 24, 1059 (1972).
35. G.H. King, S.S. Krishnamurthy, M.F. Lappert and J.B. Pedly, Chemical Society, Faraday Division Discussion 54, (1972).
36. R. Hoffman, Accts. Chem. Research, 4, 1 (1971).
37. e.g. M.J. Goldstein, S. Natowsky, E. Heilbronner and V. Hornung, Helv. Chim. Acta, 56, 294 (1973).
38. J.C. Bunzli, A.J. Burak and D.C. Frost, Tetrahedron Letters (in press).
39. E. Haselbach and A. Schmeltzer, Helv. Chim. Acta, 54, 1575 (1971) and 55, 1745 (1972).
40. T.H. Dunning, R.M. Pitzer and S. Aung, J. Chem. Phys. 57, 5044 (1972).
41. D.P. Chong, F.G. Herring and D. McWilliams (unpublished calculations using a programme kindly supplied by R.M. Stevens).

HAM - A SEMI-EMPIRICAL MO THEORY

L. Åsbrink, C. Fridh and E. Lindholm

Physics Department
The Royal Institute of Technology
S-100 44 Stockholm 70, Sweden

INTRODUCTION

The possibility to measure ionization potentials of molecules by use of molecular photoelectron spectroscopy has meant an important impetus to quantum chemistry. The first problem is then to develop methods for the interpretation of the photoelectron spectra and the second problem will be to develop more general theories which, while starting from the photoelectron spectra, make possible an understanding also of other properties of the chemical compounds such as energy, geometry, excitation, reactions and so on.

A year ago we presented a method, called SPINDO [1-5], for the first problem by parametrization of INDO. The SPINDO method appeared to be useful for interpretation of the photoelectron spectra of hydrocarbons but could not be extended to heteroatoms.

In the present paper a very preliminary description will be given of a new semi-empirical MO theory, called HAM, which in its present form seems to be useful for interpretation of photoelectron spectra of molecules, built-up from H, B, C, N, O, and F, but also seems to be useful for calculating energies and excitation processes.

THE HAM METHOD

The total wavefunction of a molecule will be described as a product of orthogonal molecular orbitals ψ_i, each containing the electronic charge q_i. These molecular orbitals are linear combinations of real atomic wavefunctions ϕ_μ centered on the different nuclei.

$$\psi_i = \sum_\mu c_{\mu i} \phi_\mu$$

Here $c_{\mu i}$ are real coefficients. The total charge in the molecular orbital ψ_i is then

$$q_i \int \psi_i^2 \, d\tau = q_i \sum_{\mu \nu} c_{\mu i} c_{\nu i} S_{\mu \nu}$$

where $S_{\mu \nu}$ denotes the overlap integral between ϕ_μ and ϕ_ν. As $q_i c_{\mu i} c_{\nu i} S_{\mu \nu}$ can be interpreted as the charge in the overlap region of ϕ_μ and ϕ_ν due to the charge in the molecular orbital ψ_i we find that the total charge in this region is given by

$$q_{\mu \nu} = \sum_i q_i c_{\mu i} c_{\nu i} S_{\mu \nu} = P_{\mu \nu} S_{\mu \nu}$$

with

$$P_{\mu \nu} = \sum_i q_i c_{\mu i} c_{\nu i}$$

giving

total number of electrons in molecule = $\sum_{\mu \nu} q_{\mu \nu}$.

We will now partition the total energy E of the molecule in the same way as the total number of electrons has been partitioned.

$$E = \sum_{\mu \nu} E_{\mu \nu}$$

In this expression $E_{\mu \nu}$ denotes the total energy of the charge $q_{\mu \nu}$ and is therefore proportional to $q_{\mu \nu}$. We treat the diagonal and off-diagonal energy elements separately.

The diagonal energy element $E_{\mu \mu}$ denotes approximately the total energy of an electronic charge $q_{\mu \mu}$ in ϕ_μ when it is attracted by the nuclear charge Z_A and repelled by the other electrons which also move around the nucleus A. A similar situation was described by Slater [6] who gave the total energy of one electron in an atom as

$$- (Z - s)^2/n^2$$

where n denotes the principal quantum number of the atomic orbital and s denotes the shielding from the other electrons in the atom. We will use a similar expression for $E_{\mu \mu}$ in the molecule and obtain with $q_{\mu \mu} = P_{\mu \mu}$

$$E_{\mu \mu} = - P_{\mu \mu} \cdot (Z_A - s_\mu)^2/n^2 = - P_{\mu \mu} \cdot \zeta_\mu^2$$

where ζ_μ means the orbital exponent.

For the off-diagonal energy elements $E_{\mu\nu}$ (with μ and ν on different atoms, A and B, respectively) we use the mean value of the Slater-type expressions and obtain

$$E_{\mu\nu} = - q_{\mu\nu} \cdot \frac{1}{2} (\zeta_\mu^2 + \zeta_\nu^2) \cdot f$$

where f is a factor to be chosen empirically. As it was found in SPINDO to be necessary to have different f for different cases of interaction (s-s, s-pσ, pσ-pσ, and pπ-pπ) we proceed in the following way.

We separate the charge element $q_{\mu\nu}$ according to the symmetry along the bond and obtain

$$q_{\mu\nu} = P_{\mu\nu} \cdot (S_{\mu\nu}^\sigma + S_{\mu\nu}^\pi)$$

where $S_{\mu\nu}^\sigma$ and $S_{\mu\nu}^\pi$ are the σ and π components, respectively, of the overlap integral $S_{\mu\nu}$. The energy element then becomes

$$E_{\mu\nu} = - P_{\mu\nu} \cdot \frac{1}{2} (\zeta_\mu^2 + \zeta_\nu^2) \cdot (S_{\mu\nu}^\sigma f_{\mu\nu}^\sigma + S_{\mu\nu}^\pi f_{\mu\nu}^\pi)$$

The expression for the total energy can be abbreviated

$$E = \sum_{\mu\nu} E_{\mu\nu} = - \sum_\mu T_\mu \cdot \zeta_\mu^2$$

with $T_\mu = P_{\mu\mu} + B_\mu$ and

$$B_\mu = \sum_\nu^{\neq A} \frac{1}{2} (P_{\mu\nu} + P_{\nu\mu}) (f_{\mu\nu}^\sigma S_{\mu\nu}^\sigma + f_{\mu\nu}^\pi S_{\mu\nu}^\pi)$$

In our study up till now a large number of attractions and repulsions were omitted. Most of these attractions and repulsions cancel each other so that only the electrostatic interaction between charged parts of the molecule remains. The final expression for the total energy is therefore

$$E = - \sum_\mu T_\mu \cdot \zeta_\mu^2 + \sum_{A>B} E_{AB}^c$$

We will now use the variational method to minimize E in the standard manner by varying the coefficients $c_{\mu i}$ and keeping the molecular orbitals orthonormal. Using the definition of $P_{\mu\nu}$ we conclude directly from the variational method that we have to solve the following set of linear equations (Roothaan's equations)

$$\sum_\nu (\partial E / \partial P_{\mu\nu} - \varepsilon_i S_{\mu\nu}) c_{\nu i} = 0$$

which means solving

$$(\bar{F} - \bar{S}\,\varepsilon_i)\,\bar{c}_i = 0$$

where $\bar{F}$ is a matrix of elements $\partial E/\partial P_{\mu\nu}$, $\bar{S}$ is the overlap matrix. ε_i is an eigenvalue and $\bar{c}_i$ is an eigenvector formed by the coefficients $c_{\mu i}$. Ordinary SCF procedure is used without neglect of overlap. The overlap matrix is calculated from Slater-type orbitals which depend upon the orbital exponents given above and is recalculated in each iteration.

SHIELDING (preliminary form)

In principle the shielding is the same as used by Slater [6]. The shielding experienced by an electron in ϕ_μ from a charge in ϕ_ν, both on atom A, is thus

$$q_{\nu\nu} \cdot \sigma_{\mu\nu} = P_{\nu\nu}\,\sigma_{\mu\nu}$$

where $\sigma_{\mu\nu}$ is a shielding constant to be found empirically. If ν equals μ the shielding is $(P_{\mu\mu} - 1)\,\sigma_{\mu\mu}$. The charge in the bonds between A and other atoms

$$q_{A\ bond} = \sum_\mu^A q_{\mu\ bond} = \sum_\mu^A \sum_\nu^{\neq A} \frac{1}{2}(q_{\mu\nu} + q_{\nu\mu})$$

(the other half is attributed to the centers B) gives the shielding $q_{A\ bond} \cdot \sigma_{\mu\ bond}$. The total shielding for the electrons in ϕ_μ is

$$s_\mu = (P_{\mu\mu} - 1)\,\sigma_{\mu\mu} + \sum_{\nu \neq \mu}^A P_{\nu\nu}\,\sigma_{\mu\nu} + q_{A\ bond}\,\sigma_{\mu\ bond}$$

ELECTROSTATIC INTERACTION (preliminary form)

A detailed study of the attractions and repulsions in the electrostatic interaction gives the same result as in e.g. CNDO (cf. [7] p. 68). With

$$P_A = \sum_\mu^A (P_{\mu\mu} + q_{\mu\ bond})$$

the interaction terms are

$$E_{AB}^c = (Z_A^{core}\,(1 + \delta_{AB}) - P_A)\,(Z_B^{core}\,(1 + \delta_{BA}) - P_B)\,\gamma_{AB}$$

where δ indicates the penetration

$$\delta_{AB} = K_1 \cdot \exp(-K_2 \zeta_B\,R_{AB})$$

and γ is calculated according to

$$\gamma_{AB} = 14.4/(R_{AB} + 0.5 \ (1/\zeta_A + 1/\dot{\zeta}_B)) \quad eV$$

Here ζ_A is the orbital exponent of the outermost orbital on A and K_1 and K_2 are empirical parameters.

RESULTS

Up till now only a very crude parametrization has been performed using a simple trial and error approach. Therefore the results presented here are very preliminary, but they show, however, that several properties are predicted with reasonable accuracy by our method.

The Figures 1-5 show the photoelectron spectra (PES) of benzene, pyridine, cyclohexane, difluorodiazine, and succinic anhydride together with the approximate IP's from the HAM calculations. In general, the distributions of the calculated and experimental IP's agree very well and our interpretations agree usually with other available interpretations.

In this preliminary parametrization we have assumed that the "differential ionization potential", $\partial E/\partial q_i = \varepsilon_i$ is the same throughout the ionization process, i.e. we have used Koopmans' theorem.

The valence excitation energies are approximated in the same way by taking the difference between the eigenvalues of the two orbitals involved. The results for pyrazine are shown in Figure 6 and for some other molecules in Table I.

For the difference in total energy between benzene and fulvene we get 41.7 kcal/mole. Ab initio calculations give 44 [8] and 40 [9]. Calculated barriers to internal rotation are given in Table II. Finally, the geometry and vibrational frequencies of methane were calculated using a harmonic oscillator approximation. The equilibrium geometry was tetrahedral and the result is shown in Table III.

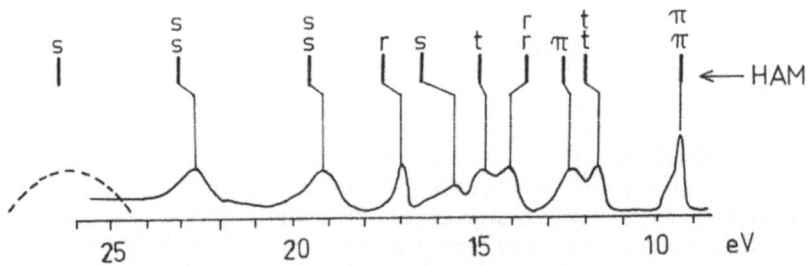

Fig. 1.   PES of benzene with 304 Å [10]   and Mg Kα[11]

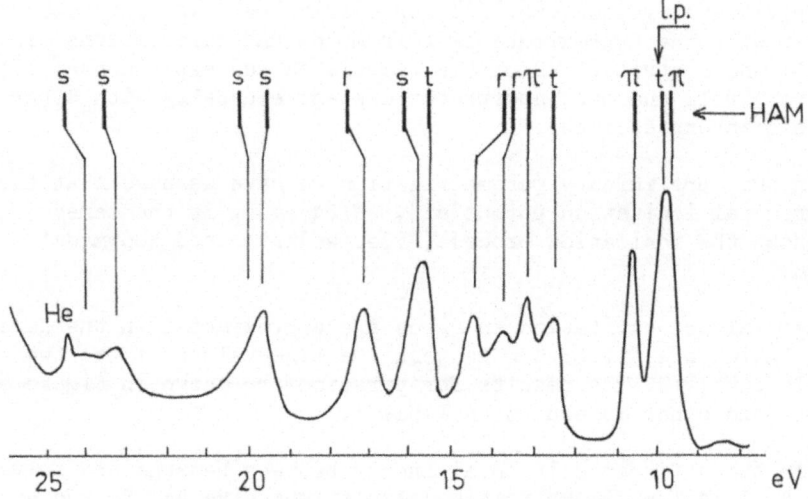

Fig. 2.   PES of pyridine with 304 Å.   Interpretation in agreement
          with a recent ab initio study [12].

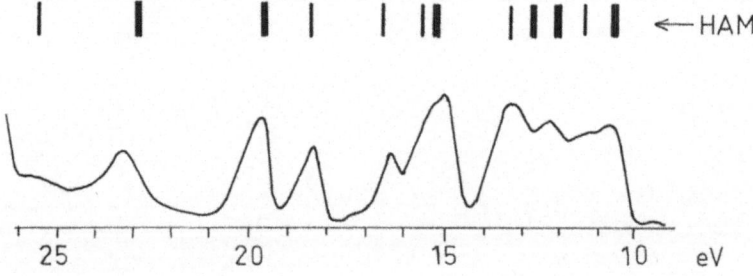

Fig. 3.   PES of cyclohexane.

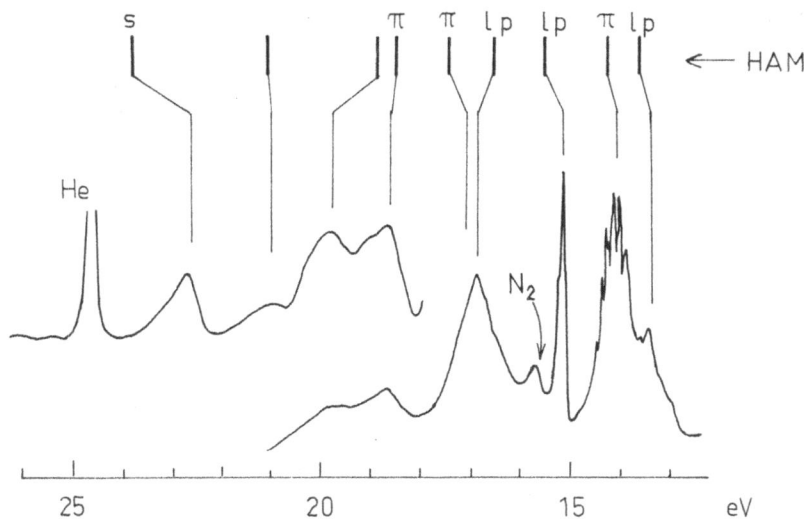

Fig. 4.   PES of difluorodiazine (N_2F_2)[13].

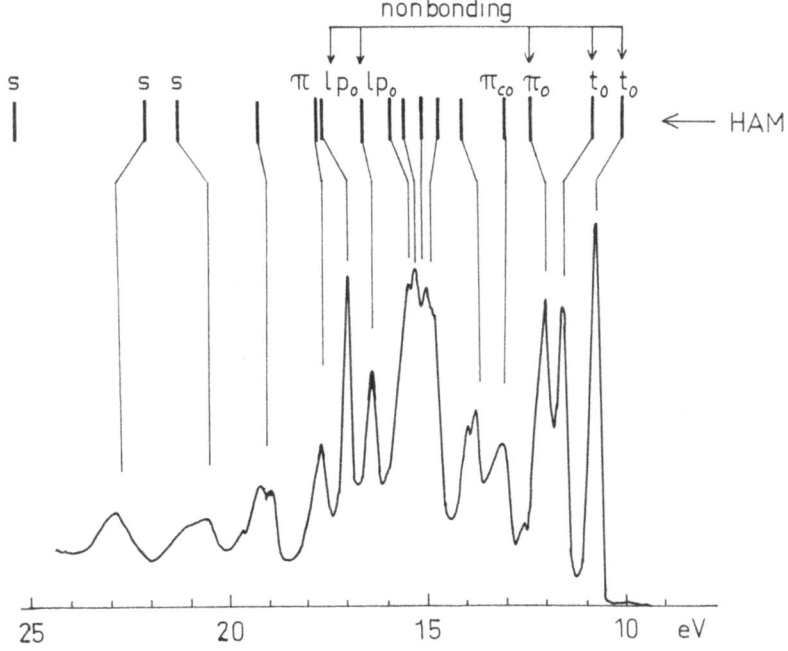

Fig. 5.   PES of succinic anhydride $(C_4O_3H_4)$.

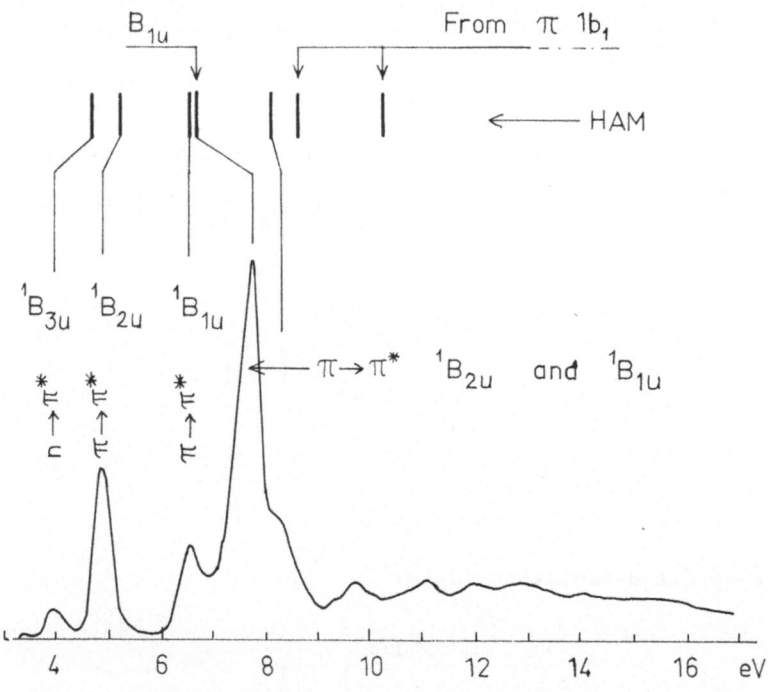

Fig. 6. Electron impact energy loss spectrum of pyrazine with
         200 eV electrons at 0° scattering angle [14]

TABLE I. Valence transitions (eV)

$\pi-\pi^*$	HAM		ab initio		Exp	
Benzene	6.8				6.9	
Butadiene	5.3				5.9	[15]
Ethylene	8.0				8.0	[16]
Fulvene	3.3				3.4	[17]
	4.7				5.1	[17]
Maleic anhydride	4.9				4.4	
Naphthalene	4.6	B_{1u}			4.5	[18,19]
	5.4	B_{2u}			5.9	[18,19]
	5.7	B_{2u}			6.1	[18,19]
	6.5	B_{1u}			7.6	[18,19]
$n-\pi^*$						
Acetone	5.6		4.4	[20]	4.4	[20]
Carbonyl fluoride	6.7		5.9	[20]	6.5	[20]
Formic acid	6.1		5.4	[20]	5.6	[20]
Tetrazine	3.4				2.3	[21]

TABLE II.   Barriers to internal rotation (kcal/mole)

	HAM	ab initio	Exp
Ethane	2.2	2.5-3.6 [22]	2.9 [22]
Ethylene	150*	139* [22]	65 [22]
Allene	61	92 [22]	
Methyl alcohol	1.34	1.12	1.07 [24]
		1.35	
		1.59	
Hydrogen peroxide cis	+3.1	+9 – +13	+7.6 [25]
trans	-0.5		+1.1 [25]

* Assuming no configuration interaction (cf. [23]).

TABLE III.   Methane

	HAM	Exp [26]
C-H distance (Å)	1.02	1.09
Vibrational frequencies (cm^{-1})		
ν_1	3060	2916
ν_2	2180	1533
ν_4	2000	1306

REFERENCES

1. C. Fridh, L. Åsbrink and E. Lindholm, *Chem. Phys. Letters 15*: 282, 1972.
2. L. Åsbrink, C. Fridh and E. Lindholm, *Chem. Phys. Letters 15*: 567, 1972.
3. C. Fridh, L. Åsbrink and E. Lindholm, *Chem. Phys. Letters 15*: 408, 1972.
4. L. Åsbrink, C. Fridh and E. Lindholm, *J. Am. Chem. Soc. 94*: 5501, 1972.

5. E. Lindholm, C. Fridh and L. Åsbrink, *Faraday Disc.* *54*:127,
    1972.

6. J.C. Slater, *Phys. Rev.* *36*:57, 1930.

7. J.A. Pople and D.L. Beveridge, "Approximate Molecular Orbital
    Theory", McGraw-Hill, New York, N.Y., 1970.

8. L. Praud, P. Millie and G. Berthier, *Theor. Chim. Acta 11*:169,
    1968.

9. S.D. Peyerimhoff, private communication (see R.E. Christoffer-
    sen, *J. Am. Chem. Soc. 93*:4104, 1971).

10. L. Åsbrink, O. Edqvist, E. Lindholm and L.E. Selin, *Chem. Phys.*
    Letters 5:192, 1970.

11. U. Gelius, C.J. Allan, G. Johansson, H. Siegbahn, D.A. Allison
    and K. Siegbahn, *Phys. Scripta 3*:237, 1971.

12. J. Almlöf, H. Johansen, B. Roos and U. Wahlgren, *J. Electron*
    Spectrosc. 2:51, 1973.

13. C.R. Brundle, M.B. Robin, N.A. Kuebler and H. Basch, *J. Am.*
    Chem. Soc. 94:1451, 1972.

14. C. Fridh, L. Åsbrink, B.Ö. Jonsson and E. Lindholm, *Int. J.*
    Mass Spectr. Ion Phys. 8:101, 1972.

15. O.A. Mosher, W.M. Flicker and A. Kuppermann, *Bull. Am. Phys.*
    Soc. 18:407, 1973.

16. J. Geiger and K. Wittmaack, *Z. Naturforsch. 20A*:628, 1965.

17. P.A. Straub, D. Meuche and E. Heilbronner, *Helv. Chim. Acta*
    49:517, 1966.

18. R.H. Huebner, S.R. Mielczarek and C.E. Kuyatt, *Chem. Phys.*
    Letters 16:464, 1972.

19. E.E. Koch, A. Otto and R. Radler, *Chem. Phys. Letters 16*:131,
    1972.

20. R. Ditchfield, J.E. Del Bene and J.A. Pople, *J. Am. Chem. Soc.*
    94:703, 1972.

21. C. Fridh, L. Åsbrink, B.Ö. Jonsson and E. Lindholm, *Int. J.*
    Mass Spectrom. Ion Phys. 9:485, 1972.

22. L. Radom and J.A. Pople, *J. Am. Chem. Soc. 92*:4786, 1970.

23. R.J. Buenker, S.D. Peyerimhoff and H.L. Hsu, *Chem. Phys.*
    Letters 11:65, 1971.

24. L. Radom and J.A. Pople, in "MTP International Review of
    Science, Physical Chemistry Series One, Volume 1, Theoretical
    Chemistry", ed. W. Byers Brown (Butterworths, London 1972),
    p. 71.

25. L. Radom, W.J. Hehre and J.A. Pople, *J. Am. Chem. Soc. 94*:2371,
    1972.

26. G. Herzberg, "Electronic Spectra of Polyatomic Molecules",
    Van Nostrand, Princeton, N.J., 1966.

AB INITIO CALCULATIONS FOR EXCITED STATES OF MOLECULES

S.D. Peyerimhoff and R.J. Buenker

Lehrstuhl für Theoretische Chemie
Universität Bonn, 53 Bonn, W. Germany

INTRODUCTION

While the calculation of molecules in their
ground state by ab initio methods has almost become a
routine operation, calculations of systems in their
elctronically excited states are still not very
numerous; they pose many more difficulties than do the
corresponding ground state treatments. In excited
state calculations, for example, it is required that
the excitation processes involving exclusively valence
orbitals are described equally as well as are the transitions into Rydberg MO's; similarly it is of great
importance that states of different multiplicity are
treated in an equivalent manner. Furthermore, since·
the structure of the molecule in its excited states
is in general different from its shape in the ground
state, it is also quite often necessary to consider
the entire excited state potential surface (or at least
a good portion thereof). The determination of correlation energy differences between the states is
necessary in order to obtain reliable values for the
energetical   differences between the various potential
surfaces, and there is also evidence that consideration
of the vibrational structure of a transition is also
quite often important for the prediction of the maximum in the absorption spectrum. All these points will
be treated in some detail in the following sections.

II. GEOMETRICAL ASPECTS

Although quite extensive calculations are necessary in order to obtain <u>quantitatively</u> reliable potential curves for excited states, it is nevertheless possible in many instances to obtain a good <u>qualitative</u> estimate of the general shapes of such potential surfaces once some relatively simple properties of the

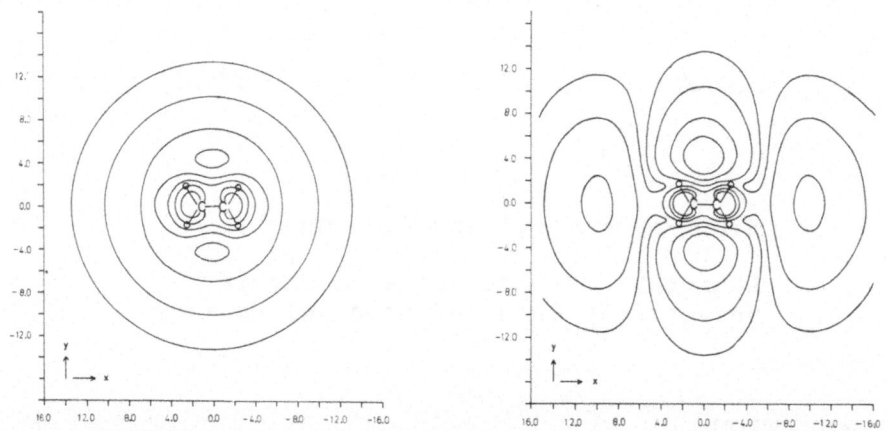

Fig.1. Charge density contours for the first members of the ns (left) and ndσ (right) Rydberg species in ethylene obtained from open-shell SCF calculations for the corresponding singlet states (Ref. 9).

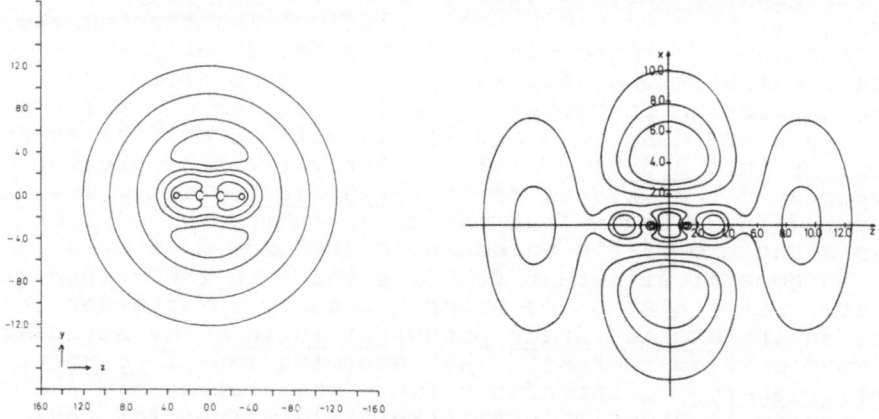

Fig.2. Charge density contours for the first members of the ns (left) and ndσ (right) Rydberg species in acetylene obtained from open-shell SCF calculations for the corresponding singlet states (Ref. 9).

orbitals involved in the excitation become known. By
way of example the calculated charge density contours
of the diffuse SCF orbitals of the first members of the
ns and the ndσ Rydberg series of ethylene are presented
in Fig. 1, which in turn can be compared with the
corresponding data for acetylene in Fig. 2. The united
atom character of the orbitals is quite obvious in
each case. The spatial extension is very large (especially compared to the nuclear framework) and these Rydberg
orbitals (as well as many others) are invariably found
to be quite similar regardless of whether they appear
in C_2H_2 or C_2H_4 (or other molecules of similar size
for that matter). In contrast the spatial extension
of the so-called valence orbitals (constructed from
valence AO functions in the LCAO framework) is much
smaller, as demonstrated in Figs. 3-6 for the π,n
and π^* orbitals of formaldehyde and thioformaldehyde
and also the π^* orbital in the $^3(\pi,\pi^*)$ state of
ethylene[1].

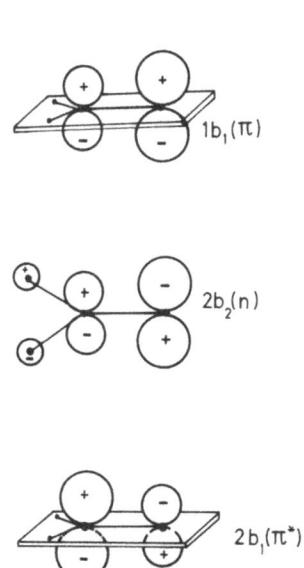

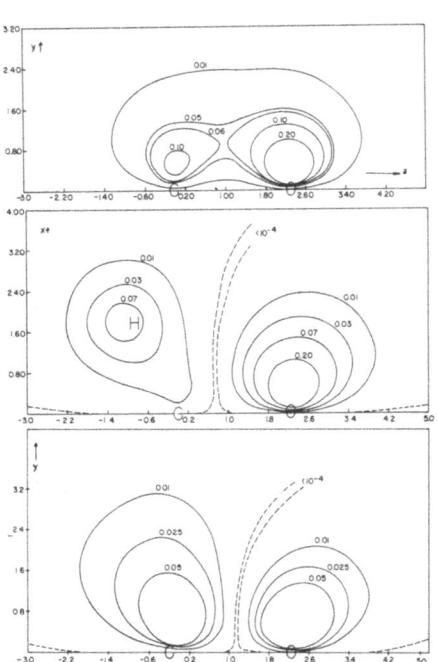

Fig.3. Schematic diagram
of the composition of
the π,n and π^* MO's in
an H_2AB system.

Fig.4. Calculated charge
density contours for the
$1b_1$ π, the $2b_2$ n and the
$2b_1$ π^* MO in formaldehyde
obtained from the ground
state SCF calculation
(Ref.[5]).

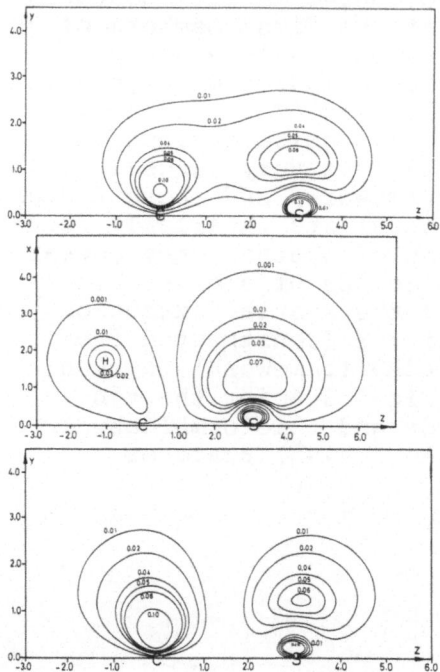

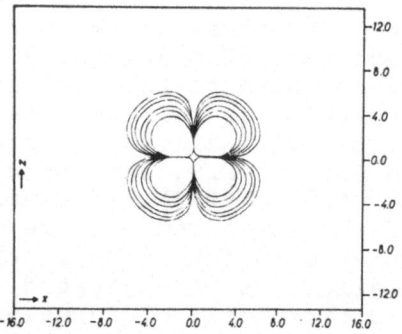

Fig.6. Charge density contours for the π^* orbital in ethylene obtained from the open-shell SCF calculation for the $^3(\pi,\pi^*)$ state (Ref.[17]).

Fig.5. Calculated charge density contours for the $2b_1$ π, the $3b_2$ n and the $3b_1$ π^* MO in H_2CS obtained from the ground state SCF calculation (Ref.[6]).

From the present observation it seems clear that any possible change in geometry occurring upon excitation into a Rydberg orbital should be very similar to that which results upon ionization; in both cases the change in geometry originates almost exclusively from the effect of depopulating a <u>single</u> valence orbital. Geometrical changes accompanying a valence-type transition by the same token can be traced essentially to the influence of <u>two</u> orbitals, one of which is depopulated and the other which becomes occupied during the process.

The effect upon the geometrical structure caused by addition or removal of electrons from an energy level (or orbital) in a molecule can in general be understood (or predicted) by means of MO theory; this

sort of argumentation was first introduced by Mulliken
[2] and later in considerable detail by Walsh [3].
A detailed description of these concepts, especially
as they relate to <u>ab initio</u> methods, has been given
elsewhere [4] and some of the main features of that
discussion will be summarized here and applied to a
few molecules of interest in spectroscopic studies.

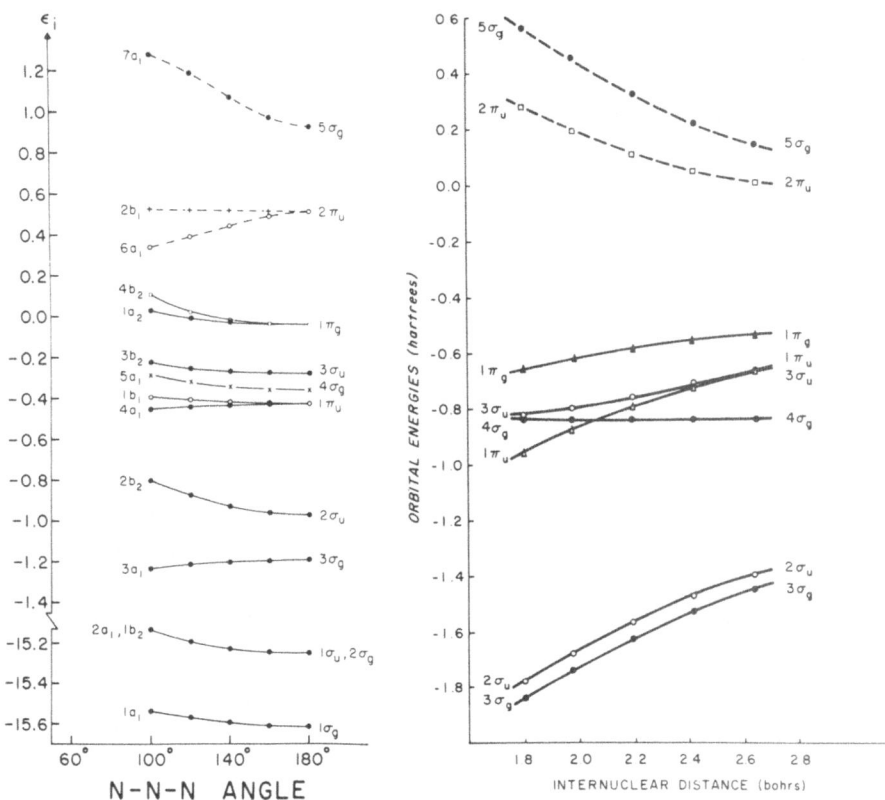

Fig.7. Typical angular (left) and AB stretching (right)
correlation diagram for a symmetric triatomic AB_2 molecule. (Plotted are the calculated canonical orbital
energies of N_3^- as a function of angle, and those of
CO_2 as a function of the CO distance; throughout this
work the energy values are given in hartree units unless specified otherwise).

A. Use of <u>ab initio</u> Walsh-type Diagrams

A typical calculated correlation diagram for a
symmetric triatomic molecule is given in Fig. 7a for
angular distortion and in Fig. 7b for the symmetrical

stretching motion; plotted in each case are the canonical SCF orbital energies. With the help of these two
diagrams it is possible to deduce the approximate
structure (relative bond lengths and angles) of a large
number of AB_2 (and with somewhat less certainty ABC
and H_nABC) molecules in ground and excited states without the need of further calculation. The behavior of
the orbital energies with the various geometrical
variables can also be understood in qualitative terms,
essentially from the nodal structure of the MO's, as
given schematically in Fig. 8: to a first approximation
increased in-phase overlap causes some stabilization
($2\pi_u$ - $6a_1$ with decreased angle and $1\pi_u$ with decreased
R as typical examples) while increased out-of-phase
overlap tends to make the orbitals less stable, as
illustrated quite emphatically by the change in the
$1\pi_g$ - $4b_2$ species on bending or by that in the $2\pi_u$
upon R stretch. More details can be found elsewhere[4].

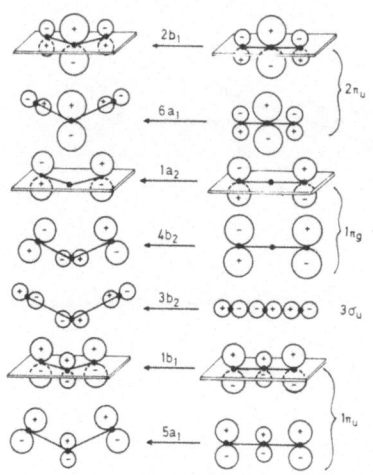

Fig.8. Schematic diagram indicating the constitution of
the valence orbitals for a
symmetric triatomic molecule
in linear and bent nuclear
conformations respectively.

     It is easily seen that Koopmans' theorem, which
relates the total energy of a given system A with that
of the ionized system A^+ and the energy ε_D of the
(differentiating) orbital from which ionization has
occured, in the form

(1) $E(A) = E(A^+) + \varepsilon_D$,

can be used to make the desired geometrical predictions.
From the derivative of this equation with respect to
some geometrical variable R (internuclear angle, bond
length or other quantity), namely

(2) · $\partial E(A)/\partial R = \partial E(A^+)/\partial R + \partial \epsilon_D/\partial R$

it follows that the slope of the total energy curve of the system A at the equilibrium structure of A^+ can be easily related to that of the differentiating orbital energy curve itself, namely by means of the following equation:

(2) $\partial E(A)/\partial R|_{R_0(A^+)} = 0 + \partial \epsilon_D/\partial R|_{R_0(A^+)}$.

This result naturally leads to the predictions:

(3a) $\partial E(A)/\partial R|_{R_0(A^+)} > 0$ if ψ_D is a bonding orbital,

or

(3a') $R_0(A) < R_0(A^+)$ if orbital ψ_D is bonding, and

(3b') $R_0(A) > R_0(A^+)$ if orbital ψ_D is antibonding.

In more pictorial terms these equations substantiate the intuitive rule used by Mulliken and Walsh, namely: the system A has a smaller bond length or bond angle (or other variable) than A^+ if the additional electron occupies a bonding orbital, i.e. one which becomes more stable for <u>smaller</u> geometrical variables, while of course the converse effect holds for the opposite orbital behavior. Thus it is clear that the following ordering in bond lengths must occur, for example:

$$R_0(CO_2, {}^1\Sigma_g^+, \ldots 1\pi_g^4) < R_0(CO_2^+, {}^2\Pi_g, \ldots 1\pi_g^3) < R_0(CO_2^+, {}^2\Pi_u, \ldots 1\pi_u^3);$$

similarly , if successive application of Koopmans' theorem is carried out, the following order in bond angles can be deduced:

$$\theta_0(CO_2, \ldots 1\pi_g^4) \geq \theta_0(CO_2^+, \ldots 1a_2^2 4b_2) \geq \theta_0(CO_2, \ldots 1a_2^2 4b_2 6a_1).$$

Indeed, if the Koopmans' theorem procedure is generalized somewhat more, as can be done, one can easily predict bond angle trends for different molecules entirely, as for example:

$$\theta_0(CO_2, .1\pi_g^4) > \theta_0(O_3, .1a_2^2 4b_2^2 6a_1^1) > \theta_0(F_2O, .1a_2^2 4b_2^2 6a_1^2 2b_1^1).$$

From a more quantitative point of view all that is really involved is the addition or subtraction of appropriate orbital energy curves (as determined by the corresponding MO occupation in the electronic configurations of interest) to the total energy curve of

some parent state. The resulting curve for the total
energy of a related state quite generally is very
similar in shape to that which ensues upon an actual
SCF calculation for the same species, as long as the
assumptions inherent in Koopmans' theorem itself remain
valid, that is, as long as the relative constitution
of the individual MO's does not differ greatly in the
systems for which geometrical comparisons are being
made. (For successive applications it is also necessary
that the shapes of corresponding orbital energy curves
do not change greatly from one system to another, but
if the first condition is valid the second is almost
always also satisfied to a sufficiently good approximation[4].)

Experience has shown that the desired behavior
fortunately occurs in the great majority of cases.
There are, however, well-known exceptions: the first
of these appears when the bonding nature can be characterized primarily as non-covalent, as in substances
containing fluorine atoms; the second exception, which
is more important in the present context, seems to
occur in states which arise from excitations (without
spin change) between orbitals occupying essentially
the same region of space, such as $\pi \rightarrow \pi^*$ excitations in
C_2H_4 and H_2CO, for which the π^* orbital in the excited
states is distinctly different from its counterpart in
the ground state or in the corresponding excited triplet
species. In both of the above cases the geometries (and/
or potential curves) associated with systems possessing
the same valence electronic configuration are going to
be rather dissimilar simply because the corresponding
MO's in each case are greatly different, thereby
vitiating the main use of Koopmans' theorem in such
connections.

B. Some Examples

The qualitative (or even semi-quantitative) procedure discussed above has been documented previously [4]
for many examples, particularly for the triatomic family
and related systems. In the present work it will be
exemplified for two more classes of systems, for which
detailed calculations of their actual potential curves
will be presented in the following sections. Figs. 9a,b
display a calculated angular and R-stretch correlation
diagram for H_2CO representing typical examples for H_2AB
systems. Since the curvature of the π^* orbital energy
curve in Fig. 9b is somewhat greater than that of the n
orbital, $n \rightarrow \pi^*$ excitations are expected to be accom-

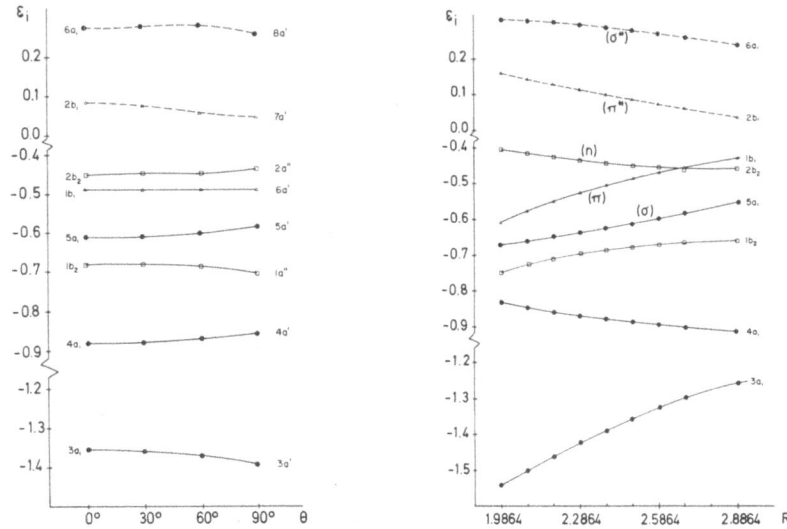

Fig.9. Angular (Fig. a, left) and CO stretching (Fig. b, right) correlation diagram for H_2CO obtained from a ground state calculation of H_2CO (Ref.[5]).

panied by an increase in AB bond lengths; this fact is well-known for H_2CO [5] and has also been demonstrated for H_2CS [6]. Excitations of $\sigma \rightarrow \pi^*$ and $^3(\pi \rightarrow \pi^*)$ type are characterized by a large increase in bond length accord- ing to the qualitative theory, while the analogous prediction with respect to the $^1(\pi, \pi^*)$ state requires somewhat more caution because Koopmans' theorem cannot be applied in the usual straigthforward way in this case, for reasons discussed above. More conventionally it is clear that Rydberg states originating from depopulating the n orbital should exhibit somewhat smaller bond lengths than the neutral molecule (since the n MO is slightly antibonding judging from Fig. 9b), close to that of the positive ion.

On the basis of the angular correlation diagram given in Fig. 9a, the singlet and triplet states resulting from n$\rightarrow \pi^*$ excitations should favor angular distortions much more than the ground state, and indeed it is well-known from experiment that the molecule possesses a slightly non-planar structure in these (n,π^*) states. According to the correlation diagram the effect should be somewhat smaller for the $^3(\pi, \pi^*)$ state and even less significant for the Rydberg species, in which the molecule is presumably still planar, but

with a smaller angular force constant than in the
ground state.

The second and final example in this section
dealing with qualitative geometrical details concerns
the molecular family of $H_2A_2H_2$ molecules, particularly
its most familiar member ethylene. A calculated angular
correlation diagram for this system (twisting angle
$\theta = 0°$ indicating the planar structure) is presented in
Fig. 10. Double occupation of the π orbital, as in
the C_2H_4 ground state, results in a planar conformation
whereas single occupation of this orbital, as in $C_2H_4^+$,
reduces the trend towards planarity so much that a
small angle of twist is actually preferred in the ion.
From previous considerations it is clear that all the
Rydberg species, which constitute most of the ethylene
low-lying states, should exhibit approximately the same
geometry as the positive ion. The first valence state
(π, π^*) should show a strong trend away from planarity
according to the qualitative theory, and it is of course
well-known that ethylene does prefer the $90°$ conformation of twisted hydrogens in this excited state.
Deviations from the qualitative rule might be expected
for the $^1(\pi, \pi^*)$ state, as mentioned before, and this
point will be discussed in detail in the next section
in which the results of quantitative calculations on
this subject are considered. One final comment on the
present subject concerns the fact that the geometry
of allene C_3H_4 in its ground and excited states is also
susceptible to interpretation on the basis of essentially

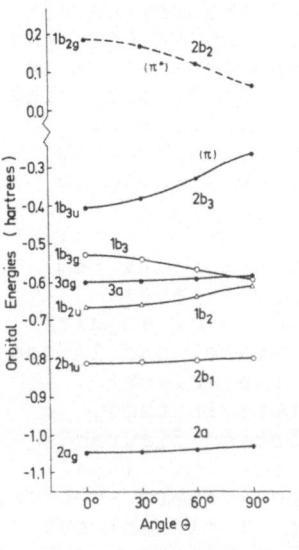

Fig. 10. Correlation diagram
for CH_2 twist in ethylene
obtained from the ground
state SCF calculation. (The
rotation angle is θ, with
$\theta = 0°$ corresponding to the
planar conformation of the
molecule, as in Ref.[7].

the same correlation diagram [7] (one need only note that in its ground state allene has effectively two more π electrons than does ethylene itself) as that shown in Fig. 10.

In conclusion of the present section it should thus be stressed that it is possible in numerous instances to predict in a reliable (albeit qualitative) way the nature of geometrical trends which occur upon excitation or ionization of a given system once a single correlation diagram for a representative mole- cule in this class (obtained by ab initio [8] or empirical techniques) is known to sufficient accuracy. This information can often be very helpful in inter- preting experimental results and thus this qualitative tool should not be neglected. While such information hardly can be said to represent a suitable alternative to that resulting from actual ab initio treatments it can serve as a highly useful and relatively simple indicator of the direction in which these more ela- borate calculations should proceed.

III. CALCULATION OF VERTICAL TRANSITION ENERGIES

In order to obtain a quantitatively reliable des- cription of excited states of both Rydberg and valence character by ab initio techniques it is necessary to use a considerably larger AO basis set than is neces- sary for ground state calculations; in particular diffuse or long-range functions representing the UA Rydberg orbitals have to be added. These functions can be located either at the nuclear centers, as is done in a more conventional way, or at the center of the molecule, which seems to be the more appropriate location from a more intuitive standpoint; a difference of only 0.001 hartree is obtained for the first ns Rydberg state in ethylene if the two methods are actually compared, however [9]. The orbital exponent of the Rydberg function is quite critical, as can be inferred from Fig. 11, in which the energy of the first singlet and triplet B_{1u} states (ns Rydberg state) in ethylene is shown as a function of the gaussion or- bital exponent α ; not at all surprisingly the corres- ponding change in ground state (^1A_g) energy is practi- cally negligible. From the results obtained so far a single function of proper symmetry seems to be sufficient for description of the energy of a Rydberg orbital (provided, of course, exponent optimization has been effected).

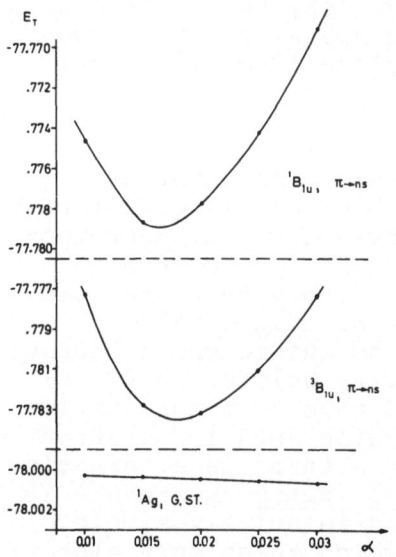

Fig.11. Total SCF energy
of three different states
in ethylene as a function
of the exponent α for the
3s Rydberg orbital (Ref.[9])

A number of excited states can also be found for
which the composition of the upper orbital involved in
the transition has changed considerably from its conventional make-up in the ground state (or even corresponding triplet). In these cases the orbital in question
is found to possess a considerably more expanded charge
distribution than is common in more conventional
valence-type states, but at the same time examination
of its AO composition makes it clear that classifying
it as a pure Rydberg species is not appropriate either.
In any event there is no question that such states cannot be treated satisfactorily unless use is made of a
flexible AO basis containing relatively diffuse functions.

Excitation energies cannot be obtained reliably
from a single-configuration (or SCF) treatment of the
excited states because of the different correlation
energy error inherent in the states of different
character (especially in comparison of closed and
open-shell species respectively); thus a successful
calculation of transition energies depends essentially
on the degree to which the difference in correlation
energies between states can be accounted for (if the
possibility of obtaining the total correlation energy
for all the states of a molecular system of reasonable
size is discarded). These inherent difficulties in the
SCF method can be overcome by means of a CI treatment
(of one sort or another). The definite advantage of

this method compared to calculations which treat
Rydberg states separately from valence-type species,
or which deal with triplet states in a different way
than for singlets, is clearly that such a procedure
allows for independent verification of the reliability
of the calculated results on a much broader basis.
Fortunately technical advances as well as theoretical
refinements have made such CI calculations quite
feasible at the present time, even on a relatively
large scale.

By way of example a summary of such results for the
molecule H_2CO is contained in Table I. The AO basis
set employed is of essentially double-zeta quality
augmented by the Rydberg-type functions discussed
above. The MO's of the parent configuration are
generally used in the CI expansion, i.e. the $^1A_2$ SCF
MO's for the $^1A_2$ calculation, the $^1B_2$ SCF MO's for the
CI treatment of $^1B_2$ etc.; details may be found elsewhere [5,10]. In this way the secular equation sizes
are kept relatively small (<200). Use of the proper
MO's is especially important if a relatively large
core of doubly occupied orbitals is maintained, as in
the present case (six MO's). Employing the MO's of the
ground state throughout [CI(GSMO)] leads to less realistic results unless the secular equation sizes are
increased to much greater values, as is done for
example in the calculation by Whitten and Hackmeyer [11]
for this molecule.

In general the agreement between the theoretical
values, calculated as the difference in the total
energy between two electronic states in the same (ground
state equilibrium) nuclear framework, and the observed
vertical transition energies is quite good; all theoretical values are too high by about 0.1 - 0.3 eV than
their experimental counterparts. In particular the first
and third Rydberg transition energies compare very well
with the corresponding experimental values. It seems
well established by now that the $\tilde{C}$ transition around
8 eV is a Rydberg transition and not the $^1(\pi,\pi^*)$ species
originally interpreted as being responsible for this
experimental band system. More recently experimental
support for the theoretical location of the $\sigma \longrightarrow \pi^*$
transition in the neighborhood of 9.0 eV has also been
reported [12]. The fact that the $^1(\pi,\pi^*)$ transition is
found at substantially higher energy is certainly a
surprising result from an historical standpoint,
especially in light of a number of semiempirical calculations which used such (π,π^*) states (also of other

State	H$_2$CO CI(PCMO)	Exptl.	H$_2$CS CI(PCMO)
$^1A_1$	0.0	$\tilde{X}$ 0.0	0.0
$^3A_2$ $\Big]$ $n \longrightarrow \pi^*$	3.41	$\tilde{a}$ 3.12–3.44	1.84
$^1A_2$	3.81	$\tilde{A}$ 3.50–5.39	2.17
$^3A_1$ $\pi \longrightarrow \pi^*$	5.56		3.28
$^3B_2$ $\Big]$ $n \longrightarrow 3S$	7.32	first Rydberg	5.72
$^1B_2$	7.38	$\tilde{B}$ 7.08–7.51	5.83
2 $^3A_1$ $\Big]$ $n \longrightarrow 3pn$	8.09	second Rydberg	6.58
2 $^1A_1$	8.11	$\tilde{C}$ 7.97	6.62
$^3B_1$ $\Big]$ $\sigma \longrightarrow \pi^*$	8.14		6.31
$^1B_1$	9.03		7.51
2 $^3A_2$ $\Big]$ $n \longrightarrow 3p\pi$	9.06		(7.80)
2 $^1A_2$	9.07		(7.88)
2 $^3B_2$ $\Big]$ $n \longrightarrow 3p\sigma$	8.29	third Rydberg	
2 $^1B_2$	8.39	$\tilde{D}$ 8.14	
3 $^1A_1$ $\pi \longrightarrow \pi^*$	11.31		7.92

Table I: Comparison of calculated vertical transition energies (in eV) of H$_2$CO with corresponding experimental data. The last column contains similar calculated results for H$_2$CS; for details see Refs. [10,6].

systems) to calibrate their parameterizations. There
seems to be little doubt in the meantime, however,
that the $^1(\pi,\pi^*)$ location is approximately at the
ionization limit [12]. One other interesting feature
of Table I is that the singlet-triplet splitting of
corresponding states is 1) very large when both orbitals
involved in the excitation have their charge density
maximum in the same plane, thus giving rise to a large
exchange integral, as is the case for the (π,π^*) states;
2) considerably smaller for the (σ,π^*) states for which
initial and final MO's possess charge distributions
which overlap far less greatly than in the first case;
and 3) very small for the Rydberg states for which the
pertinent exchange integral between the relatively com-
pact valence MO and the very diffuse Rydberg orbital
is not very large at all. Interestingly enough there is
a clear and easily understandable correlation between
the size of such triplet-singlet splittings and the
degree to which the singlet state in question assumes
a relatively more expanded charge distribution than
does its triplet counterpart.

The last column in Table I contains corresponding
calculated data for H_2CS. Basically the same pattern
is observed, only all transition energies are found to
be much smaller as result of the smaller stability of
the n and π orbitals; the effect is especially apparent
for the $^1(\pi,\pi^*)$ species, which should definitely be
found below the ionization limit in H_2CS, in distinct
contrast to the situation alluded to earlier for its oxy
analog. Details concerning the geometrical behavior of
these molecules in various low-energy states are presented in Table II. The agreement with experimental data
(whereever comparison is possible) is againquite satisfactory, and it is seen that the trends predicted by
the qualitative theory discussed in Sect. II are indeed
borne out in the results of the quantitative treatments.

Examination of the butadiene spectrum indicated
in Fig. 12 also shows quite good agreement between the
lowest experimental lines observed and those calculated
in the CI(PCMO) method; a comparison between Rydberg
states is somewhat difficult because of the lack of
experimental data, however. In this calculation it is
also quite obvious that readjustment of the core upon
excitation and use of diffuse functions in the AO basis
are quite important in the theoretical treatment,
especially for the description of the $\pi \rightarrow \pi^r$ singlet V_1
state. This conclusion rests on the observation that a
typical π-electron CI(GSMO) calculation [13] yields a

	State	Equilibrium θ_e (calc)	Angle θ_e (exptl)	Inversion barrier (eV) E (calc)	E (exptl)
H_2CO	$^1A_1 - {}^1A'$	$0.0°$	$0.0°$	0.0	0.0
	$^3A_2 - {}^3A''$	$32.7°$	$35°-36°$	0.090	0.096
	$^1A_2 - {}^1A''$	$31.9°$	$20°-31°$	0.073	0.044
	$^3A_1 - {}^3A'$	$25.9°$	—	—	—
H_2CS	$^1A_1 - {}^1A'$	$0.0°$	$0.0°$	0.0	0.0
	$^3A_2 - {}^3A''$	$15.9°$	—	0.014	—
	$^1A_2 - {}^1A''$	$15.2°$	—	0.011	—

Table II: Calculated equilibrium angles for the first excited states of H_2CO and H_2CS and corresponding inversion barriers compared with experimental data.

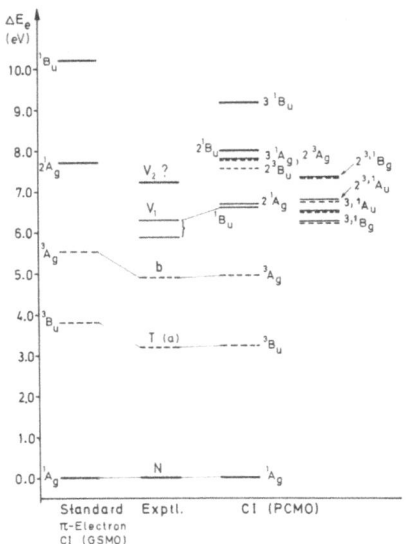

Fig.12. Energy levels of
the butadiene spectrum
obtained from two different
CI calculations and corresponding experimental data.

^1B_u state which lies much too high in the spectrum
and in fact corresponds more closely to the $3\ ^1B_u$
species in the extended calculation [14] than to the
$^1(\pi,\pi^*)$ species actually involved in the $N - V_1$
transition. Even in the CI(PCMO) treatment the description of this state is somewhat unsatisfactory and
it has been speculated [14] that the maximum in absorption might not directly correspond to the vertical
energy between the ^1A_g and ^1B_u potential curves as
calculated in the conventional way.

The discrepancy between the calculated finding and
the experimental location of the absorption maximum is
even more apparent in the study of the so-called V
state [$^1(\pi,\pi^*)$] of ethylene, this despite the fact that
all the other excited C_2H_4 states, including the Rydberg
and $^3(\pi,\pi^*)$ species, are seen to be represented quite
well by the CI treatment discussed so far (Table III).
The calculated energy difference between the ground
and V_u state potential curves is seen to considerably
overestimate the experimentally determined value of
7.66 eV. Speculation that the effect of the correlation
energy in the V state is so much different than in all
the other states does not seem to be justified, especially since much larger CI treatments do not lead to
noticeably different results (Table IV). Among treatments of the latter type are CI calculations which
allow excitations from all orbitals with the exception

State			E_e (calc.)	E (exptl.)
^1A_g	N		0.0	0.0
$^3B_{1u}$	T	(π, π^*)	4.19	4.4 - 4.7 or 4.22
$^3B_{3u}$	T_R	(π, ns)	6.96	—
$^1B_{3u}$	R	(π, ns)	7.10	7.11
$^1B_{1g}$	V_g	(π, py)	7.59	7.45
$^1B_{1u}$	V_u	(π, π^*)	8.27	7.66
$\left.\begin{array}{c}^1B_{1u}\\ ^1B_{2u}\end{array}\right\}$	R''	$\left\{\begin{array}{l}(\pi, d_{z^2-y^2})\\ (\pi, d_{zy})\end{array}\right.$	$8.66^{a)}$	8.62

a) value taken from Ref.[9]

Table III: Comparison of calculated vertical energies (eV) to the first excited states of ethylene with corresponding experimental data.

of the 1s core ($1a_g$ and $1b_{3u}$), and which include all configurations resulting from double excitations with respect to one or more main configurations which at the same time are capable of producing an energy lowering of at least 2×10^{-5} hartree upon inclusion in the CI treatment [15]. It is also noteworthy that the upper orbital in the V_u state is again more expanded in space than the corresponding orbital in the triplet [16,17,18] although there is a relatively small but definite trend to narrow this difference as larger and larger CI expansions are obtained. A considerably less diffuse $^1(\pi, \pi^*)$ species, which is much more similar in charge distribution to the $^3(\pi, \pi^*)$ state, is found in this work to lie about 1.6 eV higher. Furthermore, although polarization functions have not been used in this investigation, it is known that their inclusion in the SCF treatment has virtually no effect on the SCF value of the transition energy to the lowest $^1(\pi, \pi^*)$ species. In short such calculations tend strongly to refute the basic contention [19,20] that the charge distribution

Treatment	^1A_g	R	V_g	V_u	$^1(\pi,\pi^*)$ Valence
CI(PCMO) core a) = 5	-78.0661	-77.8048	-77.7872	-77.7622	-77.6912
	0.0	7.11	7.59	8.27	10.20
		Secular equation size up to 450			
B-D-S-G b) INO Core a) = 2				-77.8716 (1018)	
				V state 3.98 eV higher than T state, believed to lie about 4.22 eV above ground state (exp. 4.4 - 4.6 eV)	
2-Exc-CI core a) = 2 T e) = 2x10⁻⁵	-78.1776 (632, 2M) d)	-77.9125 (1502, 2M)	-77.8947 (933, 1M)	-77.8749 (1922, 3M)	-77.8159 (1922, 3M)
	0.0	7.21	7.708	8.24	9.84
2-Exc-CI core a) = 2 T→0 extrapolated c)	-78.182 (2105)	-77.917 (4043)	-77.900 (2282)	-77.880 (6043)	-77.821 (6043)
	0.0	7.21	7.67	8.22	9.82

a) core = m indicates that the first m orbitals were always kept doubly occupied in the CI treatment

b) Ref.[18], INO-treatment

c) T indicates the threshold below which the configuration were discarded (see text)

d) The symbol 2M indicates that double excited configurations with respect to two main configurations were included in the CI calculations.

e) The extrapolated values result essentially from summation of all individual energy lowerings for each of the unselected configurations in the small test CI; see also Ref.[15].

Table IV: Comparison of total energies obtained from several theoretical treatments for important states of ethylene (values are given in hartrees). The corresponding vertical excitation energies ΔE (in eV) are also given; the size of the secular equation solved ise is shown in parentheses.

of the $^{3,1}(\pi,\pi^*)$ species become very nearly identical
as better account of correlation and polarization
effects is taken. The only way known so far to achieve
a satisfactory transition energy between theses states
while allowing them to possess essentially the same
charge distributionremains the ad hoc adjustment of
parameters in semiempirical treatments, unless of
course one resorts to considerably more restricted CI
treatments which use basis sets which do not include
realistic diffuse functions in the first place [21].
Given these findings the obvious question arises as to
whether the known experimental facts cannot be explained
satisfactorily on the basis of the ab initio calculations
which are now at hand. If one adopts this approach it
seems clear from the outset that, as Mulliken has
asserted many years earlier, consideration of the geometrical characteristics of the states involved must
also be taken into account.

From qualitative arguments discussed in Sect. II it
is clear that the structure of ethylene in its excited
states is different from the planar form observed in the
ground state. Quantitative aspects are given in Fig. 13,
which contains the potential curves actually calculated
as a function of the twisting angle θ using the CI(PCMO)
method [22]. As expected, the Rydberg state geometry is
very similar to the structure of the ion; also the T
state energy minimum is found at 90°, as a result of the
exchange in the population of π and π^* MO's, with their
(Fig. 10).

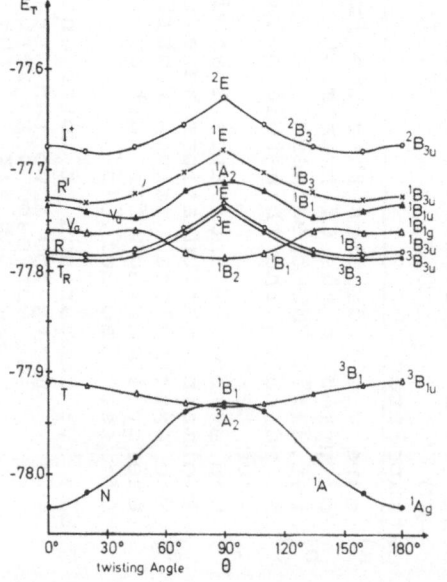

Fig.13. Potential curves for
several states of ethylene
as a function of the twisting angle θ obtained from
the CI(PCMO) treatment (R_{CC}
is optimized at each point,
see Ref.[22]).

Nevertheless a novel feature of the calculations [22] is apparent, when attention is directed toward the results for the $^1(\pi,\pi^*)$ species ($^1B_{1u}$), for which according to symmetry arguments an avoided crossing with a lower energy ($^1B_{1g}$) state is indicated as the CH_2 twisting motion proceeds. The reduced symmetry of the partially rotated ethylene structures thereby is seen to result in a situation in which the $^1(\pi,\pi^*)$ twisting potential curve is even more different from that of the corresponding triplet species than would have been expected just on the basis of the aforementioned distinctions in their respective charge distributions. While this result obviously depends quite critically on the actual energy difference between the $^1B_{1u}$ (denoted by V_u) and $^1B_{1g}$ (V_g) states, there is independent experimental evidence that the lower-lying of these two species does in fact exist in the neighborhood of 7.45 eV [23,24], that is, below the expected location of the $^1(\pi,\pi^*)$ species; furthermore very elaborate CI calculations again do not lead to any change in the aforementioned theoretical ordering for these two states (see Tables III and IV). Indeed a statement that the $^1B_{1g}$ state lies very much higher than the 7.6 eV result calculated is contrary to the seemingly secure hypothesis [25] that, as a pure Rydberg species, one should be able to easily ascertain the relative position of this state in the spectrum with respect to a host of other Rydberg states already discussed. Thus while some skepticism about such unusual findings, particularly a desire to explain them away as mere artifacts of overly restrictive calculations, is naturally expected, all our experience about the effects of extending the scope of the theoretical treatments suggests on the other hand that a significant change in the present calculated findings is unlikely.

Our conclusion is that both the V_g and V_u states must be involved in producing the known broadness of the N - V absorption bands in ethylene; the unusual character of the twisting potential curves, in particular that of the V_u state, suggests strongly that speculations regarding the non-vertical character of these transitions [22,25] (at 7.66 eV) is in fact correct. To investigate this whole question further it is clear that little is served by continuing to escalate the level of the CI treatments undertaken; instead a much more fruitful approach involves examination of the vibrational structure associated with these electronic transitions, a first attempt at which is reported in the following section.

IV. CONSIDERATION OF THE VIBRATIONAL STRUCTURE OF ELECTRONIC TRANSITIONS

In the Born-Oppenheimer approximation the total wavefunction is represented by a product of the electronic wavefunction ψ_e (obtained from the CI treatment) and a function ψ_v which depends only on the nuclear coordinates. In the sample case to be discussed, only two important vibrations, namely ethylene CC stretching and CH_2 twisting, have been included in the treatment of the nuclear motion, under the additional assumption that both of these modes are completely separable, i.e.

(4) $\qquad \Psi = \psi_e(\vec{r}_i; R, \theta) \psi_v(R) \psi_v(\theta)$.

The vibrational functions have been obtained by expansion techniques ($\sin 2n\theta$ or $\cos 2n\theta$ series for the angular motion and Hermite-type polynomials for CC stretch) calculated in the potential represented by the curves obtained by the CI method discussed above (Fig. 13).

The transition moment between ground state Ψ'' and excited state Ψ' is then given by

(5) $\vec{R}_{e'v'e''v''} = \iint \psi_{v'}^*(R) \psi_{v'}^*(\theta) \left[\vec{R}_{e'e''}(R,\theta) \right] \psi_{v''}(R) \psi_{v''}(\theta) dRd\theta$

whereby the expression in brackets is the electronic transition moment, involving integration only over electronic coordinates, namely:

(6) $\vec{R}_{e'e''}(R,\theta) = \int .. \int \psi_{e'}^*(\vec{r}_i; R,\theta) \{ \sum \vec{r}_i \} \psi_{e''}(\vec{r}_i; R,\theta) \, d\vec{r}_i$.

In the case of ethylene this quantity depends very much on the twisting angle θ (for example it is zero at angles $\theta = 0^\circ$ and $\theta = 90^\circ$ for the $V_g - A_g$ transition but noticeably different from zero for the other values) so that the usual Franck-Condon approximation, namely

(6) $\vec{R}_{e'v'e''v''} = \vec{R}_{e'e''} \int \psi_{v'}^*(R) \psi_{v''}(r) \, dR \int \psi_{v'}^*(\theta) \psi_{v''}(\theta) \, d\theta$

should be replaced by the more general formula

(7) $\vec{R}_{e'v'e''v''} = \int \psi_{v'}^*(R) \psi_{v''}(R) \, dR \int \psi_{v'}^*(\theta) \vec{R}_{e'e''}(\theta) \psi_{v''}(\theta) \, d\theta$;

the final calculations show, however, that eq. 6 can still be used for qualitative reasoning. Oscillator strengths are calculated using eq. 7 as

(8) $f_{e'v'e''v''} = 2/3\Delta E |R_{e'v'e''v''}|^2$.

Vibrational levels thus calculated for the first Rydberg state in ethylene [26] are presented in Fig. 14 (the scale accentuates the curvature much more than in Fig. 13); corresponding wavefunctions are compared in Fig. 15 with their counterparts for the electronic ground state. Qualitatively it is seen that the overlap of the $v_4'' = 0$ ground state curve is large with both the $v_4' = 0$ and $v_4' = 2$ function and smaller with the $v_4' = 4$ function. In other words, the greatest transition probability from the ground state is to the 0 and 2 vibrational levels respectively in this electronic transition; on the basis of these qualitative arguments it is also understandable that the energy difference between ground state curve at $\theta = 0°$ and Rydberg state curve at $\theta = 0°$ is a relatively good approximation to the true maximum in absorption from $v_4'' = 0$ to $v_4' = 0$ and $v_4' = 2$. Quantitative details are given in Table V, in which also excitations to excited stretching levels ($v_2' = 1,2$) are taken into account. The progression of doublets in the stretching frequency v_2', which is observed experimentally is quite apparent from these data.

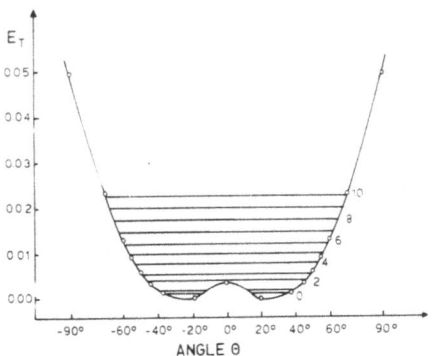

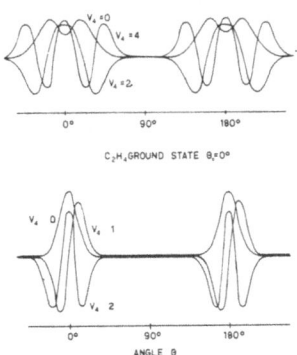

Fig.14. Calculated CH_2 twisting potential curve and vibrational levels for the lowest $\pi \longrightarrow$ ns Rydberg state in ethylene [26].

Fig.15. Comparison of important vibrational wavefunctions for the CH_2 twisting mode in the R ($^1B_3$) state with corresponding ones for the electronic ground state of ethylene.

$v_4{}'$	$v_2{}' = 0$	$v_2{}' = 1$	$v_2{}' = 2$
0	0.0192	0.0138	0.0060
2	0.0196	0.0141	0.0062
4	0.0091	0.0065	0.0029
6	0.0026	0.0019	0.0008
8	0.0008	0.0005	0.0002
10	0.0002	0.0002	0.0001
12	0.0001	0.0001	0.0000
14	0.0000	0.0000	0.0000
16	0.0000	0.0000	0.0000
$\sum\limits_{v_4{}'}$	0.0516	0.0371	0.0162

$$\sum\limits_{v_4{}'}\ \sum\limits_{v_2{}'}\quad f = 0.114$$

Table V: Tabulation of the oscillator strengths for the R - N transition of ethylene. All results are obtained relative to the lowest vibrational level of the ^1A_g ground state ($v_2'' = v_4'' = 0$)

The situation is somewhat different for the next two higher singlet states, referred to as V_u and V_g in our adopted notation (Fig. 16). The overlap of the first ground state vibrational function with the first eight vibrational twisting functions for the V_g state is extremely small. On the other hand while the overlaps between the $v_4'' = 0$ function and the corresponding vibrational functions of the V_u state is negligible for the level $v_4' = 0$, it is seen to be very large for $v_4' = 2$, an energy level which is considerably <u>below</u>

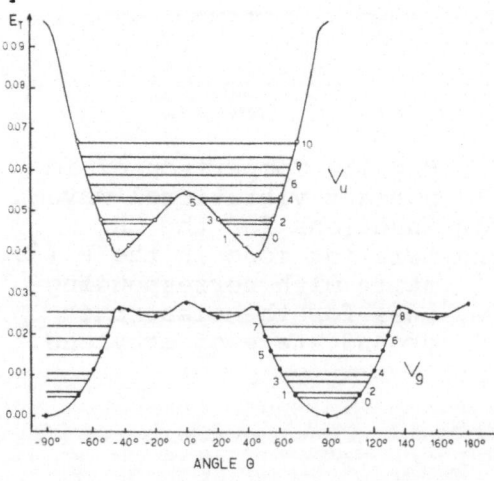

Fig.16. Calculated CH_2 twisting potential curves and vibrational levels for the V_u and V_g states in ethylene.

the energy of the vertical difference ΔE_e between ground and V_u states for $\theta = 0°$. Details of the calculations, in which again the oscillator strengths are evaluated by taking the dependence of the electronic transition moment on the twisting angle into account, are given in Table VI. In calculating the energy values it is assumed therein that the vertical energy between states is 8.25 eV, probably slightly too large; nevertheless it can be seen very clearly that the energy of the most probable transition, ΔE_{max}, is considerably lower than the vertical electronic energy difference ΔE_e, and falls much more in the neighborhood of the 7.66 eV absorption maximum found experimentally. The total oscillator strength for this transition (Table VI) is calculated to be 0.289 in very good agreement with the experimental value of 0.34, although a much greater discrepancy would not have been surprising.

Closer examination of these results leads to a very simple explanation for the calculated non-verticality of this absorption. The calculated V_u twisting energy surface is very similar to a double-well potential with maxima and minima separated by less than $40°$ (Fig. 16) due to the avoided crossing undergone by this species with the lower-energy V_g state. Mathematically the results are thus very nearly identical to what one finds in elementary studies of tunneling processes (Fig. 17). At the same time, despite the fact that the ΔE_{max} value for this transition is in much better agreement with that reported experimentally for the N - V system than what one obtains simply by using ΔE_e for this purpose, it still is clear from Table VI that the present treatment of the vibrational details of this band system does not succeed in reproducing the broad pattern of absorption which is well-known to characterize the experimental data. Again from a purely mathematical point of view it is easy to see that the reason for the calculated sharpness is the fact that the stretching potential curves assumed for the N and V_u states (taken as cuts through the surfaces of their respective optimum

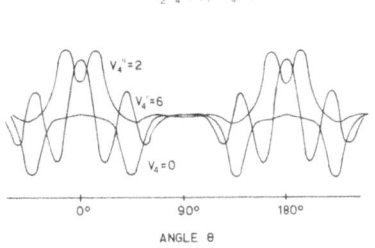

Fig.17. Calculated important vibrational wavefunctions for the CH_2 twisting mode in the $2\ ^1B_1\ (V_u)$ excited state of ethylene.

v_4'	f	$\Delta E(eV)$	f	$\Delta E(eV)$	f	$\Delta E(eV)$
	$v_2'=0$		$v_2'=1$		$v_2'=2$	
0	0.0005	7.813	0.0001	7.925	0.0000	8.036
2	0.1405	7.894	0.0239	8.005	0.0009	8.116
4	0.0109	8.049	0.0019	8.160	0.0001	8.271
6	0.0644	8.154	0.0109	8.265	0.0004	8.376
8	0.0157	8.275	0.0027	8.387	0.0001	8.497
10	0.0081	8.427	0.0014	8.539	0.0001	8.650
12	0.0035	8.580	0.0006	8.692	0.0000	8.803
14	0.0014	8.749	0.0002	8.861	0.0000	8.972
16	0.0006	8.924	0.0001	9.035	0.0000	9.146
18	0.0002	9.093	0.0000	9.205	0.0000	9.316
20	0.0001	9.249	0.0000	9.360	0.0000	9.471
$\sum_{v_4'}$	0.2459		0.0418		0.0016	
$\sum_{v_2'}$	$f = 0.289$					

Table VI: Tabulation of transition energies ΔE and oscillator strengths f for the 2 $^1B_{1u}$ (V) state of C_2H_4. All results are obtained relative to the lowest vibrational level of the (^1A_g) ground state $(v_2'' = v_4'' = 0)$. The corresponding calculated electronic vertical transition energy is 8.250 eV.

θ values) are very nearly parallel to one another. In
fact this assumption does not appear to be a good one
since the twisting and stretching motion is seen to be
rather strongly coupled in both the initial and final
electronic states in this instance. There remains then
the distinct possibility that solution of the coupled
vibrational problem, while not significantly altering
the location of ΔE_{max} itself, would produce a much more
diffuse absorption pattern than has been found under
the assumption of complete separability of these two
types of nuclear motion.

Regardless of the absorption pattern in the V_u
system it is obvious that the existence of a lower-lying
V_g state calls into sharp conflict any attempt to explain the appearance of the N - V absorption bands exclusively on the basis of a single excited electronic
state. There seems to be no question, for example, that
the O-O transition for this system involves the ethylene
(π, π^*) species with an antiplanar ($\theta = 90°$) arrangement of the methylene groups; calculations at the level
of those discussed in Table IV place the energy for this
transition at 6.4 eV, in reasonably good agreement with
McDiarmid's experimental estimate[27] of roughly 6.0 eV.
Thus the extremely weak vibrational transitions in the
neighborhood of the O-O species would have to be ascribed
to the N - V_g electronic system from the point of view
of the ab initio calculations.

In summary we consider this study of the influence
of the vibrational structure on the electronic spectrum
of ethylene as a first attempt towards the treatment
of such problems. Although there might be questions
about the reliability of the potential curves used for
this study, in particular with respect to the validity
of the Born-Oppenheimer approximation in the area of
the curve crossings (a question which is related to the
uncertainty of connecting the potential curves) there
seems to be nevertheless the clear indication that consideration of the vibrational structure might well be
of considerable importance in such a study. In particular the deviation of the most probable transition energy
ΔE_{max} from that of the vertical electronic energy difference ΔE_e usually given exclusive consideration in
such theoretical treatments can easily be in the order
of several tenths of an electron volt.

More generally it is our conclusion that the
advancement of our knowledge of the electronic spectra
of molecules can be greatly aided by consideration of

both the qualitative inferences from MO theory and the
quantitative results of <u>ab initio</u> methods. It is becoming increasingly evident that the role of <u>ab initio</u>
calculations is considerably more than one of simply
reaffirming existing interpretations of experimental
results; they seem to have arrived at a point where
they can give enough quantitative information to be
of significant aid in interpreting the electronic spectra
of molecules.

References and Footnotes

[1] Orbitals which are calculated from a different
Fock operator constructed in such a way that the
orbital energy difference is approximately equal
to the transition energy have very similar
appearance (E.R. Davidson, J. Chem. Phys. 57,
1999 (1972); S.T. Elbert, S.R. Langhoff and
E.R. Davidson, ibid. 57, 2005 (1972) and private
communication)

[2] R.S. Mulliken, for example: Rev. Mod. Phys. 14,
204 (1942)

[3] A.D. Walsh, J. Chem. Soc. 1953, 2260

[4] R.J. Buenker and S.D. Peyerimhoff: "Molecular
Geometry and the Mulliken-Walsh MO Model : An
Ab Initio Study", to appear in Chem. Rev. Jan.1974

[5] R.J. Buenker and S.D. Peyerimhoff, J. Chem. Phys.
53, 1368 (1970)

[6] P.J. Bruna, S.D. Peyerimhoff, R.J. Buenker and
P. Rosmus, "Non-Empirical SCF and CI Study of the
Ground and Excited States of Thioformaldehyde",
in press

[7] R.J. Buenker, J. Chem. Phys. 48, 1368 (1968)

[8] For best results one should abstract information
about any given orbital energy curve in such a
generalized diagram from calculations in which
the corresponding MO is actually occupied; use of
virtual orbital energies for this purpose can be
misleading if large basis sets are employed which
contain many long-range functions. In practice because of the aforementioned regularity in the
shapes of occupied canonical orbital energy curves,
regardless of the system in which they are actually
calculated, this stricture does not represent any
particular difficulty in arriving at a reliable
correlation diagram for use in the Mulliken-Walsh
model.

[9] U. Fischbach, Diplomarbeit, Mainz, Jan. 1973

[10] S.D. Peyerimhoff, R.J. Buenker, W.E. Kammer and
H. Hsu, Chem. Phys. Letters 8, 129 (1971)

[11] J.L. Whitten and M. Hackmeyer, J. Chem. Phys. 51, 5584 (1969)

[12] J.E. Mentall, E.P. Gentien, M. Krauss and D. Neumann, J. Chem. Phys. 55, 5471 (1971)

[13] R.J. Buenker and J.L. Whitten, J. Chem. Phys. 49, 5381 (1968)

[14] S. Shih, R.J. Buenker and S.D. Peyerimhoff, Chem. Phys. Letters 16, 244 (1972)

[15] More details may be found in: S.D. Peyerimhoff and R.J. Buenker, Chem. Phys. Letters 16, 235 (1972) and in "Configuration Selection in General CI Calculations", to be published

[16] T.H. Dunning Jr., W.J. Hunt and W.A. Goddard III, Chem. Phys. Letters 4, 147 (1969)

[17] R.J. Buenker, S.D. Peyerimhoff and W.E. Kammer, J. Chem. Phys. 55, 814 (1971)

[18] C.F. Bender, T.H. Dunning Jr., H.F. Schaefer III and W.A. Goddard, Chem. Phys. Letters 15, 171 (1972)

[19] H. Basch and V. McKoy, J. Chem. Phys. 53, 1628 (1970)

[20] R.S. Mulliken, Chem. Phys. Letters 14, 141 (1972)

[21] J.A. Ryan and J.L. Whitten, Chem. Phys. Letters 15, 119 (1972)

[22] R.J. Buenker, S.D. Peyerimhoff and H.L. Hsu, Chem. Phys. Letters 11, 65 (1971)

[23] A.J. Merer and R.S. Mulliken, Chem. Rev. 69, 639 (1969)

[24] K.J. Ross and E.N. Lassettre, J. Chem. Phys. 44, 4633 (1966)

[25] A similar result has been predicted even on the basis of a single potential minimum; see L. Salem, "The Molecular Orbital Theory of Conjugated Systems", W.A. Benjamin, Inc., 1966, p. 364

[26] S.D. Peyerimhoff and R.J. Buenker, Theoret. Chim. Acta (Berl.) 27, 243 (1972)

[27] R. McDiarmid, J. Chem. Phys. 55, 4669 (1971)

THRESHOLD ELECTRON-IMPACT SPECTROSCOPY

H.H. Brongersma

Philips Research Laboratories

Eindhoven - The Netherlands

INTRODUCTION

Molecules in their excited states play an important role in many kinds of processes such as encountered in photochemistry, in energy degradation in radiation chemistry and in astrophysics. Among the electronically excited states especially those states, for which the decay to the ground state by photon emission is forbidden, are of interest. This is due to the fact that these states will have longer lifetimes and therefore, more possibilities for other reactions. Triplet states form an important group of these "optically forbidden" states since the presence of two parallel electron spins and in consequence lifetimes of the order of $10^{-4}-1$ sec make these species important reaction partners. Unfortunately, their optical study is hampered by the same selection rule which impedes their decay. Even in the more favourable cases where the optical extinction coëfficient is large enough to detect the state, it is often difficult to be absolutely sure that the absorption process is due to a singlet triplet excitation and not to a normal absorption by some trace impurity.

Although more suitable techniques making use of phosphorescence, of enhanced photon absorption due to external perturbation (heavy atoms or paramagnetic molecules) or of flash photolysis can be used, the desired information is still rather scanty.

An alternative way of exciting molecules is the use of electron impact. A trivial, but important, difference between excitation by photons and by

electrons can be noted. Upon excitation of a
molecule all the energy of a photon is absorbed and
the photon disappears. An electron, however, may
transfer part of its energy to the molecule and will
not disappear. This provides a much greater flexibility
for excitation by electron impact. One and the same
excitation process (= energy loss of the electron)may
be studied for various incident energies and scattering angles of the electron. The selection rules for
excitation by electron impact depend heavily on the
experimental conditions. In general, the rules differ
markedly from those for photon absorption. This is
especially true when the energy of the incident
electrons just exceeds the energy necessary for the
excitation process (threshold energy). A most
important feature being that singlet(S)-triplet (T)
transitions are allowed. Therefore, when investigating
the energy level diagram of a molecule, threshold
electron-impact spectroscopy is an attractive tool:
it often complements the information obtained by
photon-absorption or high-energy electron-impact
studies.

INTERACTION OF ELECTRONS AND MOLECULES

The interaction of an electron with a molecule
may give rise to a number of processes. Among these
are elastic scattering (no energy transfer),
excitation or ionisation of the target molecule,
electron exchange, negative ion formation and
combinations of these processes. At present we will
restrict our interest to excitation processes.

The selection rules for the excitation processes
depend strongly on the experimental conditions which
are used. At incident energies above a few hundred
electron volts the optically allowed (dipole) transitions dominate the spectra. At lower incident energies
the quadrupole transitions become increasingly
important (peaking around 50 eV). Besides scattering
it is also possible that the incident electron is
captured into an empty orbital of the target molecule
and a molecular electron is ejected. Thus the multiplicity of the target molecule can change while the
electron spin angular momentum of the colliding
system is conserved.

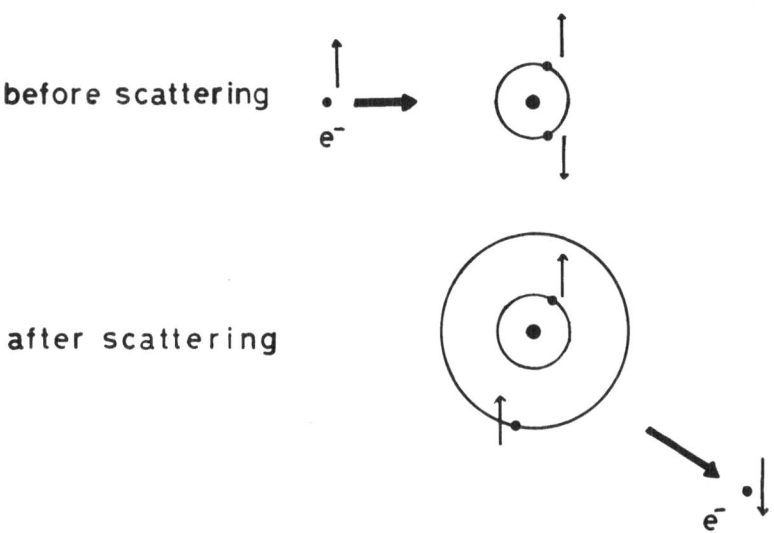

In fig. 1 it is pictured how such electron exchange
may give rise to the formation of a triplet state.

The possibility of electron exchange gives a
new dimension to electron scattering in comparison
with photon absorption. Electron exchange and thus
triplet excitation which depends on this process is
important when the interaction is very strong between
the incident electron and the molecule ("the incident
electron gets so close to the molecule that one
cannot distinguish it anymore from the molecular
electrons"). Such a strong interaction is likely to
occur for electrons which are scattered over large
angles and indeed the relative intensities for triplet with
respect to singlet excitations increase with increasing
scattering angle. In the case of 25eV electron
impact on acetylene, for example, Trajmar et.al.(1)
find an increase by a factor of 100 for the triplet
with respect to the singlet excitation when the
scattering angle increased from 10 to 80 degrees.The
absolute intensities decrease with scattering angle.

A long interaction time is obtained by using
low-energy electrons for the excitation process.
Therefore, singlet-triplet transitions will be most
important near threshold.

Unfortunately experiments very close to threshold and for large scattering angles are not easy to perform. Therefore, experimentalists restrict themselves for these studies to techniques where use is made of all scattered electrons rather than to differential techniques where only the electrons scattered into a specific direction are analysed.

In the next section the basic principles of the Trapped-Electron (TE), the Scavenger (Sc) and the Modulated Trapped-Electron (MTE) techniques which are used to study threshold excitation are discussed. In addition some attention is paid to the possibilities of ion scattering (I). Although the energies used in these experiments are well above threshold, a close resemblance with the threshold electron-impact spectra is often observed.

Many interesting results have been obtained by Hamill et al. for a great variety of molecules using another threshold technique (LEER). These results are not discussed here, since they have been obtained for films rather than for isolated molecules. This would complicate a comparison somewhat.

EXPERIMENTAL TECHNIQUES

The trapped-electron method, introduced by Schulz (2), is one of the oldest and most widely used of the techniques for studying electron scattering near threshold. The apparatus which is employed in these studies consists essentially of a gun for generating a beam of mono-energetic low-energy electrons and a collision chamber where the electrons may transfer part of their energy to a target molecule.

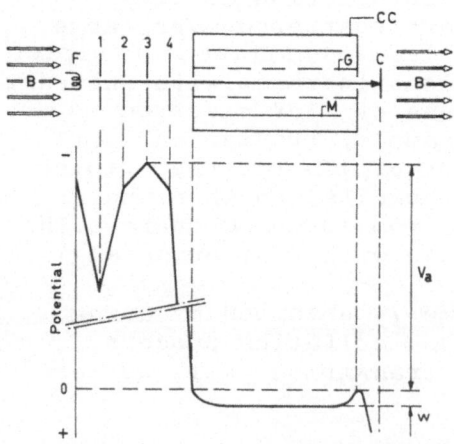

Trapping of electrons is a method to energy select the scattered electrons. It is performed by crossed magnetic and electric fields. The magnetic field is parallel to the axis of the tube preventing the elastically scattered electrons from reaching the collector M.Fig. 2 Diagram of electrodes and potentials.

A potential well is effected in the axial direction by operating the coaxial cylinder M at a positive potential with respect to the collision chamber. A fraction of this potential penetrates through the grid G into the center of the tube. This prevents electrons which have lost practically all of their initial energy in an inelastic collision from leaving the collision chamber along the axis of the tube. These trapped electrons will oscillate until their motion is displaced out of the center of the tube by subsequent collisions. This mechanism by which the electrons can reach the collector M requires many collisions (≈ 100).

The spectra presented in the next section are recordings of the trapped-electron current as a function of the accelerating voltage V_a. The energy of the incident electrons in the collision chamber corresponds to V_a plus the well depth. In general the well depth, which is only defined at the axis 1 of the tube, will be a function of 1. For simplicity we first assume, however, that the well has a constant value W. The trapped-electron current will show an onset at $V_a = \frac{E_{ex}}{e}$ - W, since there the energy of the primary electrons in the collision chamber is just equal to the excitation energy E_{ex} of the target molecule (e is the elementary charge). Only electrons having an energy less than eW are being trapped. Therefore, the excitation curve will decrease abruptly as soon as the electrons have an energy of eW after the inelastic collision. In this case: $eV_a + eW = E_{ex} + eW$, thus $eV_a = E_{ex}$. This implies that the shape of the well depth has no influence on the accelerating voltage at which the peak is reached.

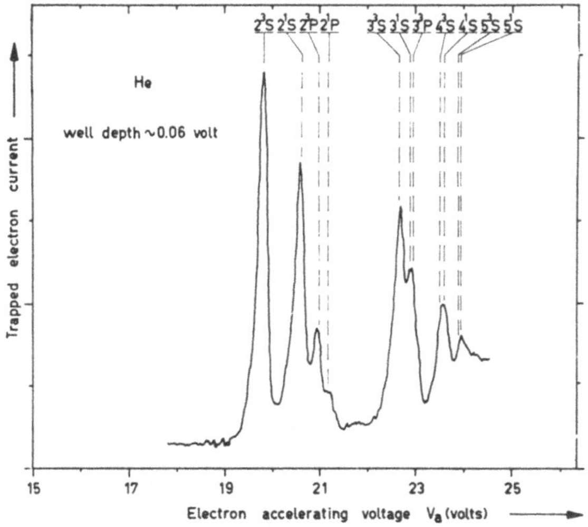

Fig. 3.TE excitation spectrum of helium. The character of the excited states is presented above the peaks. The $1^1S \rightarrow 2^3S$ transition has been used to calibrate the incident electron-energy scale.

The shape of the well does influence, of course, the shape of the peak.

Fig. 3 shows such an excitation obtained for helium (3). The spectrum clearly shows the great difference with photon absorption spectroscopy where only transitions to the n^1P states are allowed.

The energy calibration of the excitation spectra is of great importance for a correct interpretation of the measurements. In general the energy of the electrons reaching the collision chamber will differ from the energy eV_a + eW. This difference finds mainly its origin in contact potentials which are different at the various surfaces of the electrodes. In order to correct for these contact potentials the position of the $1^1S \rightarrow 2^3S$ transition of helium is generally used to calibrate the energy scale. It is essential that this calibration is performed for a <u>mixture</u> of the compound under investigation and helium, since contact potentials depend on the nature and pressure of the target gas. Helium is particularly suitable as a calibration gas, since it has no energy levels in the region which is of interest for molecules (1-16eV).

Since the trapped-electron method collects all electrons with energies up to eW, this technique is only suitable for studies close to threshold (within 1 - 2 eV depending on the energy resolution required).

The strong magnetic field which is necessary for the method limits the choice of the monoenergetic electron guns. Originally a Fox electron gun (4) was used but nowadays most researchers prefer the trochoidal monochromator (5).

The scavenger technique makes use of special molecules, instead of an artificial well, to energy select the scattered electrons. An example of such a molecule is sulfurhexafluoride which captures select-ively electrons with approximately zero kinetic energy with high efficiency:

$$e^- (0\ eV) + SF_6 \longrightarrow SF_6^-$$

Various types of mass spectrometers have been used to separate the ions from other ions and from the electrons. Since electron-capture processes are only efficient for electron energies below 1eV, therefore the scavenger technique is also limited to threshold studies.

The modulated trapped-electron method is a more recent development (6,7) in this field. Essentially, the height of the potential barrier at the end of the collision chamber (Fig. 2) is modulated with a small potential ΔV. This causes the contribution of electrons with energies between $e(W- \Delta V)$ and $e(W+ \Delta V)$ to vary accordingly. When using a phase sensitive detector, one studies effectively electrons which have an energy of eW after scattering. This extra energy selection makes it possible to study excitation functions up to several electronvolts above threshold (sometimes more than 10eV). Since the intensities of singlet-triplet transitions are expected to vary drastically near threshold, studies of the threshold behaviour provide an extra possibility for identification.

Another important application of this method is the study of molecules which capture electrons efficiently. These molecules cannot been studied with the trapped-electron method since the negative ions (which are also trapped in the well) may dominate the excitation spectrum. The modulated trapped-electron method removes this difficulty. In combination with a Fox electron source, this technique is called the D.R.P.D. (double retarding potential difference)method.

The ion scattering experiments are generally performed with monoenergetic ions of 100-10.000 eV which are scattered over a very small angle ($\sim 0°$). Accurate comparison of the energy of the ions before and after collision provides the excitation spectrum. The close resemblance of this type of spectroscopy with threshold spectroscopy is caused by the fact that ions (higher mass m) have a very low velocity v compared to electrons with the same energy (kinetic energy $eV = \frac{1}{2} mv^2 \Rightarrow v = \sqrt{\frac{2eV}{m}}$). Even at high energies, the period of interaction may be sufficiently long that electron exchange can occur with appreciable probability.

The choice of the type of ions provides an extra feature. Protons having no electrons, can only excite singlet-singlet transitions. Comparison of the H^+ and the He^+ impact energy-loss spectra can, therefore, be used for differentiating between $S \rightarrow S$ and $S \rightarrow T$ transitions (8). Heavy particle impact can also be used to study Franck-Condon forbidden transitions(9).

APPLICATIONS OF THRESHOLD METHODS

During the last decade the trapped-electron and the scavenger method have been used by various groups in the world. Unfortunately, in contrast to high-energy scattering, no good simple theory exists to predict the transition probabilities ("cross sections") near threshold. Therefore, the methods have mostly been applied to atoms and diatomic molecules in order to try and obtain phemenological information on the selection rules.

A striking feature of practically all spectra is the importance of $S \rightarrow T$ and other optically forbidden transitions. Often these transitions dominate the spectra. For simple diatomics the transitions could often be resolved into their vibrational fine structure. For the excitation to the $B^3 \Pi g$ state of N_2 and the $a^3 \Pi$ state of CO it could be shown that, when taking the experimental line shape into account, the relative intensities for the vibrational levels correspond within experimental error ($\sim 5\%$) with the theoretical Franck Condon factors (10). So even at these very low energies the collision times are short enough to guarantee vertical excitation.

Recently van Veen has modified our apparatus to improve the energy resolution. At present vibrational fine structure has already been observed for several larger molecules. Hall et al. (11) have succeeded in obtaining an even   better energy resolution with the TE method.

For a selection of the larger molecules which have been studied, the use of the threshold techniques is illustrated. In order to facilitate the comparison of the excitation spectra with the usual ultra-violet and vacuum ultra-violet absorption spectra, a wave-length scale is shown below the spectra. Together with the references the applied techniques are indicated.

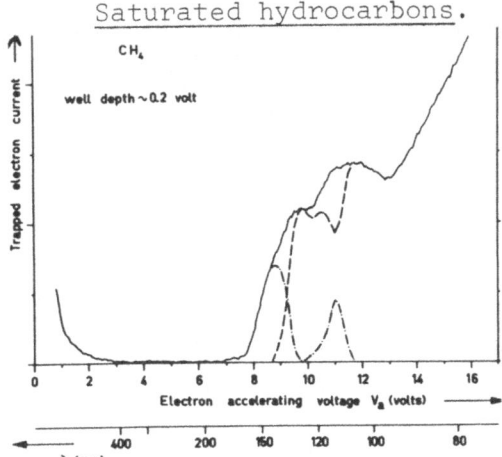

Saturated hydrocarbons.

Fig. 4. Methane, Tentatively
the optical excitation spectrum (--) has been normalized
to the TE spectrum ( ). It
is likely that the difference
(-,-) is a result of singlet-
triplet and (perhaps) other
optically forbidden transitions.

The excitation spectrum
of methane (12,TE) ob-
tained with the TE method
is presented in fig.4.
For comparison the vacuum
ultraviolet absorption
spectrum of Raymonda and
Simpson (13) is shown in
broken lines. A striking
difference is observed
between the threshold ex-
citation spectrum on the
one hand and the optical
and high-energy electron
impact measurements on
the other hand. The onset
in the figure is at approximately 7.5 eV while the
onset in the high-energy
electron-impact and op-
tical spectra is always
at 8.5 eV. This differ-
ence in onset is far too
large to be result of an
inaccurate energy cali-
bration. Moreover,

Bowman and Miller (14,TE) also using the TE method reported the same value (7.5 eV) although they don't seem
to have noted this divergency. In the cases of ethane,
cyclopropane similar displacements of the absorption
threshold to lower energies have been observed (12,TE).
It seems plausible to assume that $S \rightarrow T$ transitions are
involved.

For methane the relative intensities in the maxima
around 9.8 and 11.7 eV are the same for the optical(13),
the high-energy electron-impact data (15) and our threshold spectrum. This suggests that for all spectra the
same transitions are observed at these energies. Tentatively the optical excitation spectrum has been normalized to the TE spectrum. Subtraction shows two bands
peaking at 8.8 and 11.0 eV which are most likely due to
$S \rightarrow T$ and (perhaps) other optically forbidden transitions.
Katagiri and Sandorfy (16) using Pariser-Parr type approximations have calculated energy levels of alkanes.
They found that the energy difference between a $S \rightarrow T$ and
the related $S \rightarrow S$ transition can amount to 1 eV. For the
first and second electronic excitation of methane they
found $(E_S - E_T) = 0.95$ and 0.38 eV in good agreement with
our values of 1.0 and 0.7 eV respectively.

For more complicated alkanes a normalization of the
optical on the TE spectrum is not possible. The large
differences between the two types of spectra suggest,
however, that the contributions from S   T transitions
will be important.

Olefins

The excited states of ethylene and its derivatives
have received considerable interest in recent years.
Olefins which have been studied are ethylene (14,17,
TE; 18, Sc; 8,19,I), cis butene-2 (17,TE; 8, I),
trans butene-2 (20, TE; 8,I) and butene-1, monochoro-
ethylene, 1,1 dichloroethylene, tetrachloroethylene,
monofluoroethylene, methoxyethylene (8,I) and 2,3
dimethylbutene-2, isoetraline, hexamethylbicyclo [2,2,0]
hexa-2,5-diene, norbornadiene (20,TE) and 1,2
dimethylcyclohexene(21,TE)

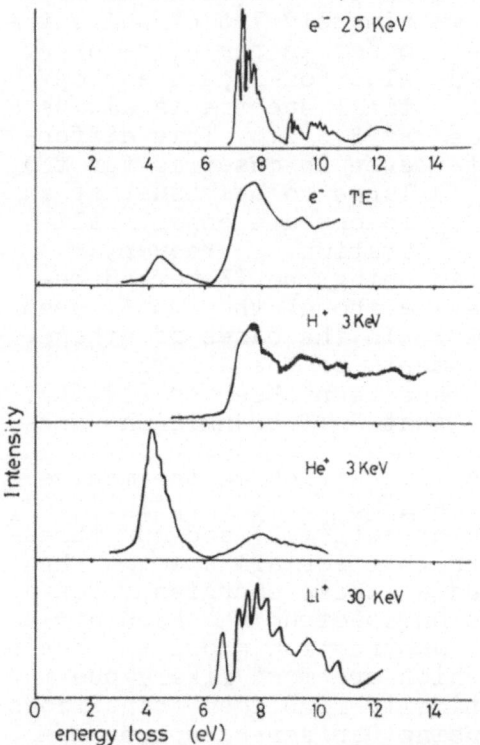

Fig.5.Ethylene.A comparison of the
scattering of 25 keV (a,43),thres-
hold electrons (b,17), 3keV H+ions
(d,8) and 30 keV Li+ ions(e,28)
illustrates how heavily the selection
rules depend on the energy and
species used for the excitation.

Fig. 5 shows for
ethylene how the
selection rules
vary from one
technique to the
other. This clearly
illustrates the
importance of
using several
techniques.

A special aim of
the olefin studies
was the so-called
"olefin mystery
band". This band
which shows up
weakly in light
absorption spectra
around 6 eV has
been assigned to
σ(C-H)$\rightarrow$ $\pi^{*}$and
$\pi \rightarrow \sigma$(C-H)*
transitions.
Using the TE
method the mystery
band was observed
at 6.1 eV for cis
butene-2 (17,TE)
and at 5.5 eV for
2,3-dimethyl
butene-2 (20,TE).

We failed, however, to detect a similar band in
molecules such as ethylene, transbutene-2 and 1,6
heptadiene (17,20,TE). Similar to optical spectroscopy
steric hindrance of the substituents may perturb the
symmetry of the molecule and thus enhance the intensity
of a symmetry forbidden transition. Another explanation
is that in of the molecules the mystery band remains
hidden under the strong $\pi \rightarrow \pi^*$ transition at 7 eV (see
fig. 5b for ethylene). The latter explanation is
probably the correct one. Hubin-Franskin and Collin
(18,Sc) using a deconvolution technique for their
TE spectrum have found an indication for a mystery
band in ethylene.Moreoever, Van Veen (22, TE) after
improving the energy resolution of our equipment has
clearly resolved a peak at 6.7eV.

Although more and more reliable theories become
available for ethylene (22-24) the situation is still
somewhat confusing. A low-lying Rydberg state has been
observed in light absorption at 7.1 eV in agreement
with theory. The corresponding S-T transition is
predicted by Merer and Mulliken(22) at 6.15 eV and
by Peyerimhoff (25) at 6.96 eV. On the basis of the
TE spectra it seems therefore, reasonable to attribute
the peak at 6.6-6.7 eV to the excitation of the lowest
triplet Rydberg state. This assignment seems to be
in contradiction to the results of Nikolai (26,I),
although he gives the same interpretation. Nikolai
studied the scattering of 30 KeV Li^+ ions. The energy
loss spectrum of ethylene shows a strong peak at
6.6 eV. This peak is not observed by other authors
using 400 eV and 3000 eV He^+ and H^+ ion scattering
(18, 19,I). When   comparing these spectra it is
important to realise that a groundstate Li^+ ion has
two electrons with antiparallel spins. Although electron
exchange with an ethylene molecule may occur, a change
in the spin state of the ion would require much
energy (at least 59 eV): due to the Pauli exclusion
principle one of the electrons should go to an excited
state. This implies that such an exchange process
cannot contribute to the energy loss spectrum of
ethylene shown in fig. 5.
In principle Li^+ impact could induce $S \rightarrow T$ transitions
if we assume that the selection rule for conservation
of spin angular momentum brakes down while only the
total angular momentum is conserved. However, this
process is generally unimportant. The Li^+-scattering
experiment itself also indicates the unimportance of
$S \rightarrow T$ transitions since the low-lying triplet state
at 4.4 eV is not observed (in contrast to He^+ and e^-

scattering experiments). Heavy ion impact is known to
violate the Franck-Condon rule. An alternative interpretation for the band would, therefore, be a non-
vertical excitation to a singlet level.

   Knoop studying another Olefin, 1,2 dimethylcyclo-
hexene, was able to detect besides the singlet and
triplet $\pi \rightarrow \pi^*$ transitions at 7.9 and 4.2 eV respectively,
two additional transitions at 5.4 and 6.2 eV.
In analogy ethylene these states can be assigned to a
triplet and a singlet Rydberg. The observed S-T splitting
of 0.8 eV is somewhat larger than for ethylene, which
can be understood since the Rydberg is more tightly
bound.
Ross and Lassetter (27) using 150 and 400 eV electron
scattering showed that the peak around 7-8 eV in ethylene consists of two transitions: one at 7.45 eV and
the dipole allowed $\pi \rightarrow \pi^*$ transition at 7.65 eV. Recently
van Veen 22, TE was also able to resolve the peak
using the TE method. It is found that the relative
intensities for the two transitions depend markedly on
the excess energy above threshold at which the spectrum
is taken. This may explain why Collin and Hubin-Franskin
only observe a peak at 7.45 eV.

Other unsaturated hydrocarbons

   Many unsaturated hydrocarbons have been studied
during the last decade: butadiene (17,TE; 8,I), benzene
(17,TE;18, 31, Sc), acetylene (14, 21, 29, TE; 21,MTE),
propyne (14, 21, TE; 21, MTE), 1,1-diphenylethylene
(17,TE), 1-butyne (14. TE) and 7-dehydrocholesterol,
1,3,5-transhexatriene, cyclooctatetraene, allene
(21,TE) naphtalene,(31,Sc). In all cases triplet states
could be identified.

   For butadiene an energy level at 3.2 eV is observed
by threshold electron and He^+ ion scattering but not
by photon absorption and H^+ ion scattering. This clearly
indicates the triplet character of this level (in
agreement with theory). The corresponding singlet level
around 5.9 eV is observed by all four methods. From
the peaks observed in the TE spectrum at 3.8 and 4.8
eV the former may correspond to a second triplet state
and the latter to the mystery band of butadiene (30),
although other assignments have been suggested.

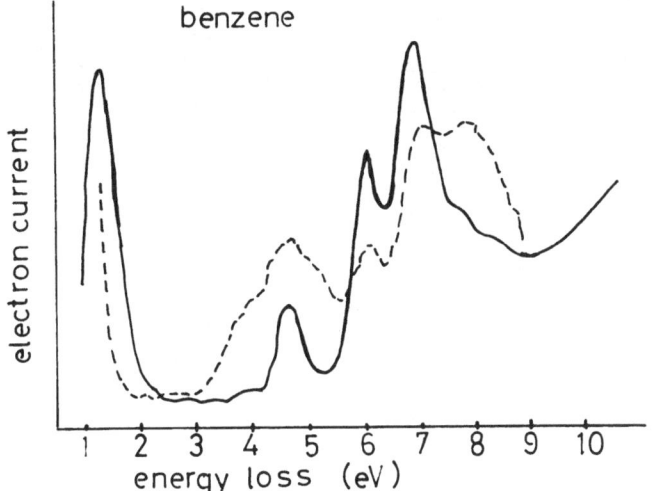

Fig.6.Excitation spectra of benzene
measured with the TE method(17) for a well
depth of 0.06eV (--) and with the MTE
method(21) for a well depth of 0.9 eV.

The TE and MTE spectra of benzene are presented in fig. 6. The transition to the $^1E_{1u}$ state at 6.9 eV,which is symmetry allowed for dipole transitions, is the most pronounced peak. The symmetry transition for-bidden at 6.0 eV is,however, of comparable intensity.

The spectrum (drawn line in fig. 6) agrees with the
scavenger spectrum obtained by Hubin-Franskin and
Colling (18, Sc) but differs markedly from the
scavenger spectrum by Compton et al. (30, Sc).
The discrepancy is not understood, Knoop (21, MTE)
was able to indentify three triplet states in benzene
with the MTE.(broken line in fig. 6). The benzene
molecule has also been studied extensively with non-
threshold techniques (32, 33)

Molecules containing a carbonyl group
    In light absorption spectra of aldehydes and ketones
a weal (symmetry forbidden) transition is found at
approximately 270- 330 mµ (3.8-4.6 eV) which is
ascribed to a n $\rightarrow \pi^*$ excitation of the carbonyl group.
In acetone this transition peaks at 4.51 eV. The TE
excitation spectrum (fig.7) of acetone (20,TE) shows
a maximum at 4.15 eV. It is not likely that such a
difference (0.35 eV) is a result of an inaccurate
energy calibration. This is also emphasized by the
good agreement found for the transitions at 6.3 and
7.45 eV with the optical values of 6.36 and 7.45 eV
respectively found by Lake and Harrison (31). Assuming
that the Franck-Condon principle still holds at these
low energies the discrepancy of 0.35eV can only be
explained by ascribing the band to a vertical triplet
$n \rightarrow \pi^*$ excitation. The orthogonality of the one

electron n-orbital and the one electron　-orbital
explains the very small S-T splitting. Experimentally
a 0.4 eV separation is found in the case of formaldehyde (35). Recent studies of acetone with the
scavenger technique (26, Sc) lead to the same
conclusion.

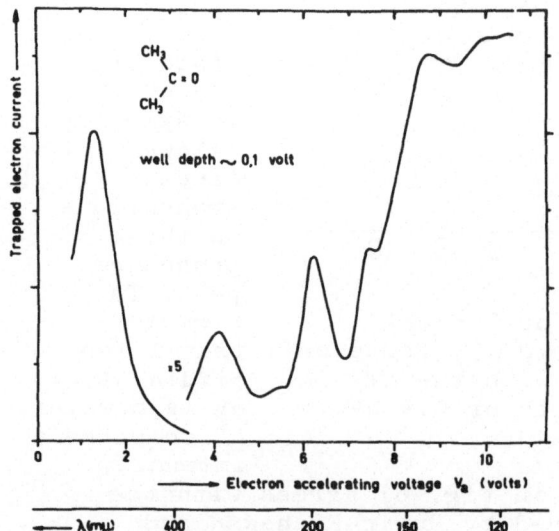

Fig.7. TE spectrum of acetone (20).
For the right hand side of the
spectrum the sensitivity of the
detection system has been increased
by a factor of 5.

Another triplet state, corresponding to the $\pi \to \pi^*$ excitation is observed at 5.5 eV. This is especially clear when taking the peak shapes of the other transitions into account. Van Veen(22,TE) determining TE spectra for formaldehyde could resolve the vibrational fine structure of the $\pi \to \pi^*$ transition. A detailed study of especially the Rydberg series has been made using higher-

energy electron scattering (37).
   Other molecules which have been studied with
threshold techniques are: acetaldehyde, urea (36,Sc)
and benzaldehyde, acetophenone (20, TE).

Molecules containing a nitrogen atom

   The spectrum of pyridine has been studied
extensively (20, 38, TE); 39, I) by the TE method
and by higher-energy electron and ion scattering.

Although the results obtained with the higher-energy electron and ion scattering are in good agreement with our TE spectrum (fig. 8). the results of Pisanias et al. also using the TE technique are rather different. Van Veen recently repeated our TE measurements and obtained the same spectrum. Van Veen(22,TE) found a similar discrepancy for pyridazine. The differences are not understood.

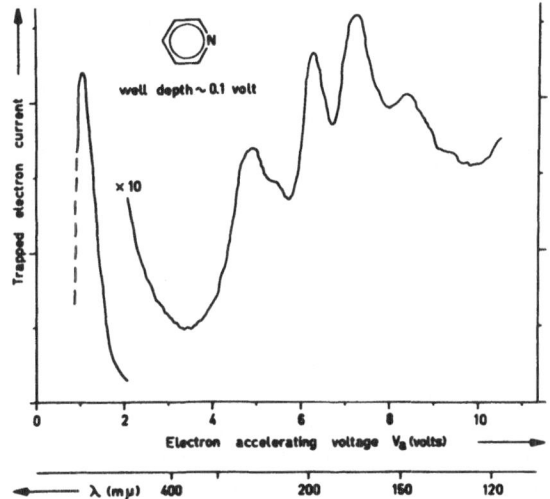

Fig.8. TE spectrum of pyridine (20) For the right hand side of the spectrum the sensitivity of the detection system has been increased by a factor of 10.

Other nitrogen molecules which have been studied are pyrimidine, sym-triazine (38, Sc), quinuclidine, triethylamine, ammonia (20, TE) and quinoline, isoquinoline (40,Sc) and urea (36, Sc).

Other molecules

A variety of other molecules has also been studied: borontrichloride (41,Sc), methanol (42,TE; 42,MTE), and borazine, dimethylether (21,TE),1-chloronaphtalene (43,Sc) and of course a great number of molecules containing less than 4 atoms. One of these smaller molecules which has received considerable attention is water. A low-lying triplet state at 4.5 eV has been detected by various authors. Recently a second triplet state was found at 7.2 eV (42,TE; 42, MTE). Similar states have been observed in methanol (42,TE; 42, MTE) and dimethylether (21,TE). Present theoretical calculations do not predict the lower state.

CONCLUSION

Much information on optically forbidden transitions of molecules has been obtained by the threshold techniques. Unfortunately no good simple theory exists at these low energies to predict the cross sections for the excitation processes. Although very much work remains to be done for the trapped-electron and scavenger techniques, angular and energy (e.g. MTE) dependent studies will prove to be of great importance for the assignment of the higher energy levels.

Recently (44) we have built a new instrument which allows the angular and energy dependent studies of both electron and ion scattering from gaseous and solid targets. On the one hand the solid introduces the complication of possible multiple collisions. On the other hand, however, the solid (or adsorbed) molecule) provides a new challenge: the study of a target in a well defined orientation. This would enable the determination of the directions of transition moments for optically forbidden transitions.

REFERENCES

1. S. Trajmar, J.K.Rice, P.S.P. Wei and A.Kuppermann, Chem. Phys. Lett. 1 (1968) 703
2. G.J. Schulz, Phys. Rev. 112 (1958) 150.
3. H.H. Brongersma, A.J.H.Boerboom and J.Kistemaker, Physica 44 (1969) 449.
4. R.E. Fox, W.M.Hickam, D.J.Grove and T.Kjeldaas, Rev. Sci. Instr. 26 (1955) 1101.
5. A. Stamatovic and G.J. Schulz, Rev. Sci. Instr. 41 (1970) 423.
6. F.W.E. Knoop, H.H.Brongersma and A.J.H.Boerboom, Chem. Phys. Lett. 5 (1970) 450.
7. H.H. Brongersma, F.W.E. Knoop and C. Bakx, Chem. Phys. Lett. 13 (1972) 16.
8. J.H. Moore, J. Phys. Chem. 76 (1972) 1130.
9. J.H. Moore and J.P. Doering, Phys. Rev. 182 (1969) 176.
10. H.H.Brongersma and L.J. Oosterhoff, Chem.Phys. Lett. 1 (1967) 169.
11. R.I. Hall, J. Mazeau, J. Reinhardt and C.Schermann, J. Phys. B3 (1970) 991.
12. H.H. Brongersma and L.J. Oosterhoff, Chem. Phys. Lett. 3 (1969) 437.
13. J.W. Raymonda and W.T. Simpson, J. Chem. Phys. 47 (1967) 430.
14. C.R. Bowman and W.D. Miller, J. Chem. Phys. 42 (1965) 681.
15. E.N. Lassettre and S.A. Francis, J. Chem. Phys. 40 (1964) 1208.
16. S. Katagari and C. Sandorfi, Theor. Chim. Acta (Berlin) 4 (1966) 203.
17. H.H. Brongersma, J.A. v.d. Hart and L.J.Oosterhoff, in: Fast reactions and primary kinetics, Nobel Symposium 5, ed. S Claesson (Interscience, New York, 1967) 211.
18. M.J. Hubin-Franskin and J.E. Collin, Int.J. Mass. Spectrom. and Ion Phys. 5 (1970) 163.
19. J.H. Moore and J.P. Doering, J.Chem. Phys. 52 (1970) 1692.
20. H.H. Brongersma, Thesis Leiden (1968).
21. F.W.E. Knoop, Thesis Leiden (1972).
22. E.H. v. Veen, private communication.
23. A.J.Merer and R.S. Mulliken, Chem. Rev. 69 (1969)639.
24. J. Fischer-Hjalmars and J. Kowalewski, Theoret·Chim. Acta 27 (1972) 197.
25. S.D. Peyerimhoff and R.J. Buenker, Theoret. Chim. Acta (Berlin) 27 (1972) 243.
26. S.D. Peyerimhoff, NATO-Advanced Study Institute on Chemical Spectroscopy (1973).

27. R. Nikolai, Z. Phys. $\underline{228}$ (1968) 16.
28. K.J. Ross and E.N. Lassettre, J. Chem. Phys. $\underline{44}$ (1966) 4633.
29. D.F. Dance and I.C. Walker, Chem. Phys. Lett. $\underline{18}$ (1973) 601.
30. M.B. Robin, R.R. Hart and N.A. Kuebler, J. Chem. Phys. $\underline{44}$ (1966) 1803.
31. R.N. Compton, R.M. Huebner, P.W. Reinhardt and L.G. Christiphorou, J. Chem. Phys. $\underline{48}$ (1968) 901.
32. J.P. Doering, J. Chem. Phys. $\underline{51}$ (1969) 2866.
33. E.N. Lassettre, A. Skerbele, M.A. Dillon and K.J. Ross, J. Chem. Phys. $\underline{48}$ (1968) 5066.
34. J.S. Lake and A.J. Harrison, J. Chem. Phys. $\underline{30}$ (1959) 361.
35. J.R. Henderson and M. Muramoto, J. Chem. Phys. $\underline{43}$ (1965) 1215.
36. W.T. Naff, R.N. Compton and C.D. Cooper, J. Chem. Phys. 57 (1972) 1303
37. M.J. Weiss, C.E. Kuyatt and S. Mieczarek, J. Chem. Phys. $\underline{54}$ (1971) 4147.
38. M.N. Pisanias, L.G. Christophorou, J.G. Carter and D.L. Corkle, J. Chem. Phys. $\underline{58}$ (1973) 2110.
39. J.P. Doering and J.H. Moore, J. Chem. Phys. $\underline{56}$ (1972) 2176.
40. M.N. Pisanias, L.G. Christophorou and J.G. Carter Chem. Phys. Lett. $\underline{13}$ (1972) 433.
41. J.A. Stockdale, D.R. Nelson, F.J. Davis and R.N. Compton, J. Chem. Phys. $\underline{56}$ (1972) 3336.
42. F.W.E. Knoop, H.H. Brongersma and L.J. Oosterhoff, Chem. Phys. Lett. $\underline{13}$ (1972) 20.
43. R.N. Compton and R.H. Huebner, in Advances in Radiation Chemistry, vol. 2, ed. M. Burton and J.C. Magee (J. Wiley and Sons, 1970) 281.
44. H.H. Brongersma and P.M. Mul, Surf. Sci. $\underline{35}$ (1972) 393.
45. J. Geiger and K. Wittmaack, Z. Naturf. $\underline{20a}$ (1965) 628.

SPECTRA OF FREE RADICALS AND MOLECULAR IONS PRODUCED BY VACUUM
ULTRAVIOLET PHOTOLYSIS IN LOW-TEMPERATURE MATRICES

Dolphus E. Milligan and Marilyn E. Jacox

National Bureau of Standards
Washington, DC 20234, USA

GENERAL PRINCIPLES

The matrix isolation technique has proved to be an extremely
valuable tool for the direct spectroscopic observation of photochemical reaction intermediates [1]. In a typical experiment, a
gas mixture of a suitable reactive molecule precursor with a large
excess of an inert diluent is frozen onto a cryostat window maintained at a temperature sufficiently low to afford complete inhibition of molecular diffusion, and the sample is subjected to
vacuum ultraviolet radiation to produce the species of interest.
The inert diluent, or matrix material, is usually argon or another
of the rare gases or nitrogen. These substances yield rigid deposits, transparent over a very wide infrared and ultraviolet
spectral range, at temperatures between 4° and 20°K, conveniently
attained using liquid helium or liquid hydrogen as coolants in
conventional cryostats. Suitable sources of vacuum-ultraviolet
radiation have included a microwave-excited discharge through
hydrogen, which provides an intense source of 1216-Å radiation,
and xenon, krypton, and argon resonance lamps. In order to prevent filtering of vacuum-ultraviolet radiation from the inner
portions of the deposit by trace impurities and by photolysis
products, it is very advantageous to subject the sample to vacuum
ultraviolet radiation concurrently with the sample deposition.

The spectrum of a molecule isolated in a matrix is considerably simpler than that characteristic of the gas phase. Because
of the very low sample temperature, absorptions arising from
excited vibrational and electronic energy levels ("hot bands") do
not appear. Furthermore, except for a few simple hydride molecules of formula AH_n isolated in rare-gas matrices, the rotational

structure also is completely quenched. Infrared absorptions are generally sharp, often with half widths of only a few cm^{-1}, and in rare-gas matrices they almost always appear within 5 or 10 cm^{-1} of the gas-phase band center. Molecules which are potentially hydrogen-bonding or which have bonds with appreciable ionic character show larger matrix shifts when isolated in solid nitrogen. Although the quenching of rotational structure results in the loss of the valuable information provided by the gas-phase band contour and by the detailed rotational band structure, the sharpening of the bands often permits the resolution of peaks contributed by the various isotopic species present in the sample. By analyzing the infrared data for these isotopic molecules, it has frequently been possible to obtain not only a positive identification of the reaction intermediate but also semiquantitative information on its structure. Most non-Rydberg electronic transitions are red-shifted by no more than a few hundred cm^{-1}. The upper-state vibrational spacings, like their ground-state counterparts, are closely similar to those characteristic of the gas-phase molecule.

The "cage effect" places important restrictions on processes which can lead to the stabilization of neutral or charged reaction intermediates in a matrix environment. Because molecular diffusion through the matrix is completely inhibited, unless there is a significant activation energy for the process, the reverse recombination of diatomic or polyatomic molecular fragments will occur. On the other hand, photoprocesses involving atom production can lead to the stabilization of an appreciable yield of the radical of interest even if the reverse recombination occurs readily, since atoms can undergo at least limited diffusion from the site of their photoproduction. The mobility of electrons through a rare-gas lattice has also been found to be sufficiently great to permit the observation of cations produced on photoionization of the parent species and the monitoring of the destruction of anions by photodetachment. A supplementary atom or electron source is frequently added to the sample, leading to the stabilization of still other reaction intermediates upon reaction of a photoproduced atom or electron with a molecule trapped in the matrix. Occasionally, the matrix itself may react with a photoproduced reaction intermediate, providing the sole exception to the exclusion of molecule-molecule reactions as a suitable source of free radicals for observation in a matrix environment. The utilization of these principles has led to the stabilization of over 70 free radicals and of approximately 25 molecular ions in concentrations suitable for direct spectroscopic study in our laboratory. Since 1965, when a hydrogen-discharge lamp was developed which photolyzed NH_3 yielding a sufficient concentration of NH_2 for direct observation of its infrared spectrum in argon and nitrogen matrices [2], the majority of our studies have been focused on the _in situ_ photoproduction of reaction intermediates in a matrix environment utilizing vacuum ultraviolet radiation. The following discussion

will present specific examples of the successful coupling of
vacuum-ultraviolet photolysis with matrix isolation observations,
with emphasis on both the potential and the limitations of the
technique for use by photochemists.

STABILIZATION AND SPECTRA OF FREE RADICALS

On vacuum-ultraviolet photolysis of HCN isolated in an argon
matrix, very prominent absorptions corresponding to the violet
system of CN appeared, and a weak infrared absorption due to
ground-state CN was also identified [3]. A sufficient concentration of HNC, first identified by the observation of five different
isotopic species on prolonged mercury-arc photolysis of matrix-
isolated CH_3N_3, also resulted for detection of all three of its
vibrational fundamentals. The vacuum-ultraviolet photolysis of
FCN, ClCN, BrCN, or C_2N_2 in an argon matrix also led to the appearance of infrared absorptions due to the corresponding isocya-
nide. As the mass of the halogen atom was increased, successively
less CN absorption was observed, and no CN absorption appeared on
vacuum-ultraviolet photolysis of matrix-isolated cyanogen. These
observations illustrate two important points. First, the diffusion of atoms through a rare-gas matrix is strongly dependent on
their mass, and diffusion of even small diatomic molecules does
not occur. Second, the appearance of a reactive molecule such as
HNC does not imply that it is necessarily a photochemical intermediate, since its stabilization may result from cage inhibition
of diffusion and from rapid collisional deactivation by the matrix.
However, the observation of such a species does imply that it possesses a stable ground state and often provides valuable information regarding its vibrational and electronic energy levels.

Gas-phase studies have established that CH_2 is the major
primary product of the vacuum-ultraviolet photolysis of CH_4,
although H-atom detachment to produce CH_3 also plays a significant
role. On photolysis of CH_4 in an argon or a nitrogen matrix using
1216-Å radiation, a prominent product absorption appeared near
610 cm^{-1} [4]. Studies of isotopically enriched samples of methane
demonstrated the presence of one carbon atom and three hydrogen
atoms in the molecule which contributes this absorption. More
quantitative analysis of the frequency pattern of the isotopic
absorptions indicated that the absorption is contributed by the
out-of-plane deformation of a planar CH_3 structure. Although the
high-resolution analysis of the 2160-Å band system of CH_3 had indicated that the molecule must be planar or nearly planar in its
ground state, the observation of a "negative anharmonicity" for
the infrared absorption is consistent only with a planar structure.
The 1503-Å Rydberg band of CH_3 has also been observed in argon-
matrix studies. Despite the formation of a significant concentration of CH_3, no CH_2 absorptions have been identified in the

matrix studies. The formation of CH_2 is, however, inferred by the appearance of CH_2N_2 absorptions in photolyzed samples of methane in a nitrogen matrix. There is reason to believe that the reverse recombination of CH_2 and H_2 in the matrix may account for the failure to detect CH_4; in studies of the reaction of C atoms with H_2 in an argon matrix, absorptions due to CH_4 and to C_2H_2 were readily detected. Thus, the methane photolysis studies provide a further example of the danger of using matrix isolation observations to infer which primary process predominates in gas-phase photolysis studies.

In gas-phase studies of the photolysis of methyl chloride, the C-Cl bond ruptures most easily. However, in matrix studies of this photolysis system [5] only a weak absorption due to CH_3 appeared, whereas absorptions due to H_2CCl, $HCCl$, and CCl were prominent. In fact, in samples containing a relatively low concentration of CH_3Cl the CCl absorption predominated. Not only does the relatively heavy Cl atom experience substantial inhibition of its diffusion through the matrix, but secondary photolysis processes also play an important role. The prolonged exposure of the sample to photolyzing radiation typical of matrix isolation experiments strongly favors the occurrence of secondary photolysis.

In contrast to the results already discussed, in matrix isolation studies of the vacuum-ultraviolet photolysis of methanol it has been possible to obtain information regarding the structure of the primary photolysis product [6]. Gas-phase studies of the vacuum-ultraviolet photolysis of methanol have yielded a product distribution which is invariant over an extremely wide wavelength range, but they have not definitively established whether the primary photolysis product is CH_3O or CH_2OH. In matrix-isolation experiments of this photolysis system, prominent infrared absorptions of CO and HCO and weaker absorptions of H_2CO appeared. In addition, several peaks were present which disappeared together when the sample was subjected to radiation in the 2500-2800-Å spectral range. The yield of the molecule which contributes these absorptions was maximized when a xenon resonance lamp was used as the source of vacuum-ultraviolet radiation. The most prominent absorption appeared at 1183 cm^{-1}, and its shift in carbon-13 substitution experiments was appropriate for its assignment to a carbon-oxygen stretching absorption. Since this peak appeared at a frequency somewhat higher than that of the carbon-oxygen stretching fundamental of methanol, it was inferred that the carbon-oxygen bond may possess partial double bond character. Another prominent absorption appeared close to the OH stretching fundamental of methanol, requiring its assignment to an OH stretching fundamental. Still another absorption, near 480 cm^{-1}, behaved appropriately in isotopic substitution experiments for assignment to a torsional vibration. Taken together, these results dictate a CH_2OH structure. When CH_3OD was photolyzed, the resulting spectrum was consistent

with the formation of CH_2OD, with no evidence for the presence of
other deuterium-substituted species which might be formed by the
reaction of D atoms with H_2CO. Very little HDCO was present.
Similarly, very little HDCO resulted on photolysis of CD_3OH in a
matrix, and the infrared spectrum of the new product was consistent
with a CD_2OH structure. A more detailed analysis of the infrared
data has confirmed that the C-O bond of CH_2OH is somewhat stronger
than that of methanol and that the torsional barrier is somewhat
higher, in accord with the conclusions of earlier studies of the
esr spectrum of CH_2OH.

All of the data in the isotopic substitution experiments are
consistent with the photolytic rupture of a single C-H bond; there
is no evidence for isotopic "scrambling" due to secondary reactions
in the matrix, nor can the results be explained by rearrangement
of an initially formed CH_3O radical. It is inferred that CH_2OH is
the primary photolysis product. There remains the problem of explaining the appearance of a significant concentration of H_2 on
photolysis of CD_3OH in the gas phase, previously interpreted as
inferring the formation of CD_3O as the primary photolysis product.
However, this observation is consistent with the matrix data if
CD_2OH possesses a continuum absorption in the vacuum ultraviolet
associated with its photodecomposition into CD_2O + H.

The reaction between OH and CO provides the principal path by
which CO_2 is formed in combustion. A very small activation energy
has frequently been postulated for this reaction. However, very
recent data suggest that the earlier gas-phase studies do not
necessarily require a nonzero activation energy. Since we had
found that FCO is stabilized by the reaction of F atoms with CO at
cryogenic temperatures, it was of interest to study the reaction
of OH with a CO matrix, using the vacuum-ultraviolet photolysis of
H_2O as a source of OH radicals [7]. As expected, prominent absorptions due to HCO and CO_2 appeared, and some H_2CO was stabilized.
However, several other prominent absorptions were attributable to
an intermediate in the reaction of OH with CO. There are three
possible structures for a molecule of formula HCO_2--one in which
the H atom is bonded to the carbon and the two oxygen atoms are
equivalent and two stereoisomers of the HOCO chain structure.
Since two absorptions appeared in the OH stretching region and two
other absorptions in the C=O stretching region, it was inferred
that both the cis- and the trans- stereoisomers of HOCO contribute
to the product spectrum. Detailed isotopic substitution studies
have confirmed this assignment.

STABILIZATION AND SPECTRA OF MOLECULAR IONS

For several years, absorptions which appeared near 5200 and
4725 Å on trapping the products of a discharge through methane or

of the vaporization of graphite in an argon matrix were attributed
to the Swan bands of triplet C_2. However, subsequent to the initial assignment of these matrix bands, it was found that C_2 possesses a singlet state which lies some 610 cm^{-1} below the lowest
triplet state. Furthermore, in the matrix observations the upper-
state band separation was some 200 cm^{-1} greater and the lower-
state band separation about 150 cm^{-1} greater than the values for
the Swan bands of gaseous C_2. At the time of the initial assignment of the bands to the Swan system of C_2, such vibrational
spacing anomalies were not considered surprising, since relatively
few molecules had been studied in a matrix environment. As further
matrix data accumulated, such large deviations from the gas-phase
vibrational spacings acquired no precedent.

In an attempt to clarify the assignment of these bands, we
undertook a series of experiments on the vacuum-ultraviolet
photolysis of acetylene [8]. In the infrared spectral studies,
one infrared absorption due to HC_2 was positively identified; all
eight isotopic species containing carbon-12, carbon-13, hydrogen,
and deuterium were observed. In the visible and ultraviolet
spectral regions, the Mulliken and Phillips bands of singlet C_2
were prominent, excluding a crossover in the energies of the lowest singlet and triplet levels of C_2 upon isolation in a rare-gas
matrix. The 5200- and 4725-Å bands also appeared, and isotopic
studies required that they be assigned to a species of formula C_2.
Upon exposure of the sample to radiation of wavelength between
2000 and 3000 Å these bands disappeared.

Shortly after these results were reported, Herzberg and
Lagerqvist [9] observed a new band system with origin near 5400 Å
in studies of a flash discharge through gaseous methane. Isotopic
studies demonstrated that the species which contributed these bands
possessed two carbon atoms and no hydrogen. The vibrational band
spacings agreed within experimental error with those observed for
the 5200-Å band system in the matrix. Since there remain no un-
assigned low-lying electronic states of C_2 which could account for
this transition, they suggested the assignment of the band system
to C_2^-, which supplementary mass spectrometric observations demonstrated to be present in the flash discharge. They noted previous
experimental evidence that C_2 possesses an exceptionally high
electron affinity.

The report of the flash-discharge experiments prompted further
matrix studies designed to test the possible assignment of the
5200-Å band system to C_2^- [10]. In these experiments, an atomic
beam of cesium was codeposited with the $Ar:C_2H_2$ sample, and the
sample was subsequently exposed to 1216-Å radiation to produce C_2
and photoelectrons. The contrast with the experiments performed
in the absence of cesium was dramatic; the ratio of the peak optical density of the 5200-Å band to that of the 2382-Å Mulliken band

of singlet C_2 was enhanced by a factor of more than twenty by the addition of a small concentration of cesium to the sample. Furthermore, upon subsequent irradiation of the sample with the full light of a medium-pressure mercury arc, the intensity of the 5200-Å band remained unchanged. These observations are fully consistent with the assignment of the 5200-Å band system to C_2^-. The cesium atoms provide a photolytic source of electrons which may be captured by C_2, resulting in the enhancement of the 5200-Å band system at the expense of the C_2 absorptions. On mercury-arc irradiation of the sample in the absence of cesium atoms, photodetachment occurs, and the 5200-Å band system disappears. However, in the presence of cesium atoms, mercury-arc irradiation results in a steady state between photodetachment and photoelectron capture. The reassignment of the 5200-Å band system to C_2^- has recently received still further support from several other matrix and gas-phase studies. Nevertheless, several questions remain to be answered. Most important of these is the nature of the positively charged species in the experiments in which cesium is not present.

In studies of the vacuum-ultraviolet photolysis of NO_2 in an argon matrix using 1236- or 1216-Å radiation [11], a new infrared absorption appeared at 1244 cm^{-1}, with a satellite absorption at 1247 cm^{-1}. Since isotopic studies dictated the assignment of this absorption to a species of formula NO_2, with two equivalent O atoms, and since the peak did not fall near an absorption of the neutral molecule, this observation was, at first, puzzling. With the demonstration that small molecular ions can be stabilized in an argon-matrix environment, it was noted that the photolyzing radiation used in these experiments considerably exceeded the ionization potential of NO_2. The known high electron affinity of NO_2 and the often-observed strong absorption near 1275 cm^{-1} of the nitrite anion in an ionic crystal environment suggested that the 1244-cm^{-1} peak of the argon-matrix experiments might be contributed by NO_2^-. To check on this possibility, other experiments were performed involving the electron bombardment of an Ar:NO_2 deposit and the codeposition of Ar:NO_2 samples with atomic beams of the various alkali metals [12]. In each of these systems, a prominent absorption appeared at 1244 cm^{-1}, with a 1247-cm^{-1} satellite. Indeed, it was unnecessary to irradiate the samples containing an alkali metal to produce the 1244-cm^{-1} absorption, consistent with the occurrence of strong charge-transfer interaction between the alkali metal and NO_2 in an argon matrix. Although NO_2^- has often been studied in ionic crystals and in polar solvents, its identification in the inert, nonionic argon matrix environment is of considerable interest, since its behavior under these conditions should more closely approximate that of NO_2^- in the gas phase.

The vacuum-ultraviolet photolysis of $HCCl_3$ in an argon matrix has provided a source of several free radicals and molecular ions. The gas-phase vacuum-ultraviolet spectral studies of $HCCl_3$ by Zobel and Duncan [13] have demonstrated that this molecule possesses three distinct absorption regions. A relatively weak band appears near 1750 Å, a more prominent absorption maximum lies at 1330 Å, and a still more prominent absorption maximum appears at 1170 Å. Associated with the 1330-Å band are two higher frequency satellite absorptions which they tentatively attributed to the excitation of a highly anharmonic CH stretching vibration. Beyond 1110 Å the absorption is continuous. On 1216-Å photolysis of $HCCl_3$ in an argon matrix, an extremely prominent infrared absorption of CCl_3 appeared at 900 cm^{-1} [14]. Another prominent absorption at 1037 cm^{-1}, showed a chlorine-isotopic splitting appropriate for its assignment to a CCl_3^+ molecule of D_{3h} symmetry [15]. This absorption was unshifted in studies of the 1216-Å photolysis of $DCCl_3$ in an argon matrix, and its behavior in studies of the photolysis of samples enriched in $H^{13}CCl_3$ was consistent with its assignment to the cation. The mechanism by which the cation is formed is clear. In accord with the observations of Zobel and Duncan, the primary photolysis process at 1216 Å involves hydrogen-atom detachment. Since the ionization potential of CCl_3 is considerably exceeded by radiation of this wavelength, the photoionization of CCl_3 can and does occur.

On the other hand, when $HCCl_3$ samples isolated in an argon matrix were subjected to 1067-Å argon resonance radiation, very little CCl_3 appeared, but two previously identified absorptions of $HCCl_2$ were recognized. Two other absorptions behaved appropriately in isotopic substitution experiments for assignment to $HCCl_2^+$. Since the ionization potential of $HCCl_2$ is also considerably exceeded, again the mechanism for formation of the cation is straightforward. Although Rydberg transitions are sometimes greatly perturbed for molecules isolated in rare-gas matrices, these observations require that such series possess significant absorption intensities for CCl_3 and for $HCCl_2$ in an argon matrix, since the probability for the occurrence of two successive photolytic processes during deposition of the sample is extremely small.

Still other infrared absorptions were common to both the 1216- and the 1067-Å photolysis studies. These peaks were also formed on interaction of photoelectrons with $HCCl_3$ isolated in an argon matrix, suggesting that they are contributed by a negatively charged species. Analysis of the infrared data requires that they be contributed by $HCCl_2^-$ rather than by $HCCl_3^-$. In studies of electron capture by $HCCl_3$ in the gas phase, Cl^- has been detected, inferring that $HCCl_2$ is also formed. However, in studies of the interaction of photoelectrons with $HCCl_3$ in a matrix only weak $HCCl_2$ absorptions appeared. It is suggested that processes forming $HCCl_2 + Cl^-$ and $HCCl_2^- + Cl$ occur at similar energies and that polarization of $HCCl_2$ by Cl^- favors cage recombination of these products.

Further studies have been conducted on the vacuum-ultraviolet photolysis of $HCCl_2F$ and of $HCClF_2$ isolated in an argon matrix [16]. In each of these systems the corresponding trihalomethyl radical has been identified. Two different anions have been identified in each of these systems, and both anions retain the hydrogen atom of the parent molecule. However, no absorptions attributable to halomethyl cations have been observed in these experiments.

In a number of our photolysis experiments having in common only the production of H or D atoms in an argon matrix by 1216-Å radiation, a peak appeared at 905 (H) or at 644 (D) cm^{-1}. In all of these experiments either oxygen or a halogen was also present [17]. Bondybey and Pimentel [18] have also observed the 905- and 644-cm^{-1} peaks in the spectra of the trapped products of a microwave discharge through $Ar:H_2$ and $Ar:D_2$ mixtures, respectively. In studies of a $^{36}Ar:D_2$ sample, they obtained a small argon-isotopic shift. However, this shift was much too small for $^{36}ArD^+$, demonstrating that any molecular product must contain two or more Ar atoms. Bondybey and Pimentel noted that on prolonged exposure of the sample to 900-cm^{-1} radiation the two peaks disappeared, but the rate of disappearance of the 644-cm^{-1} peak was much slower than that of the 905-cm^{-1} peak and was dependent on whether the species responsible for the 905-cm^{-1} peak was present in the sample. They explained this observation by postulating that the absorption results from H atoms trapped in very special sites in the argon lattice. However, such an explanation poses a problem in accounting for the sharpness and relatively great intensity of the two peaks, which would be much more adequately understood by assigning the absorptions to molecular vibrations in which there is a large change in the dipole moment.

In our studies of the behavior of these two peaks, we have noted that they disappear not only on exposure of the sample to 900-cm^{-1} radiation but also on exposure to 2537-Å radiation. Since neither H atoms nor argon absorb in this spectral region, this observation requires the reassignment of the infrared peaks to a molecular vibration. Further consideration of the chemistry of argon-hydrogen mixtures in gas-phase discharge processes suggests that the peaks may be contributed by ArH_n^+ and ArD_n^+, respectively, with n = 2, 4, or 6. The reaction

$$H\ (2^2S) + Ar \rightarrow ArH^+ + e \tag{1}$$

has been proposed by Chupka and Russell [19]. Its occurrence would be consistent with the appearance of the peaks <u>only</u> in the 1216-Å photolysis experiments, since such radiation would be necessary to provide a source of H (2^2S) for reaction with Ar. In an argon lattice, the further reaction

$$ArH^+ + (n-1)Ar \rightarrow HAr_n^+ \tag{2}$$

would be expected to occur. Although the species HAr_2^+ is prominent in the mass-spectrometric analysis of the products of a discharge through argon-hydrogen mixtures, the occurrence of still further clustering in an argon lattice cannot be excluded.

In summary, great care is necessary in inferring the mechanism of photolysis through matrix-isolation observations, since cage inhibition of important gas-phase processes is commonly observed and since conditions are virtually ideal for the occurrence of secondary photolysis processes. Nevertheless, it is sometimes possible to shed light on photolytic mechanisms through careful analysis of matrix-isolation observations, and the technique has provided an extremely powerful tool for the determination of vibrational and electronic energy levels of both neutral and charged reaction intermediates.

REFERENCES

[1] D. E. Milligan and M. E. Jacox, Molecular Spectroscopy: Modern Research, K. N. Rao and C. W. Mathews, Eds. (Academic Press Inc., New York, 1972) pp. 259-286.

[2] D. E. Milligan and M. E. Jacox, J. Chem. Phys. 43, 4487 (1965).

[3] D. E. Milligan and M. E. Jacox, J. Chem. Phys. 47, 278 (1967).

[4] D. E. Milligan and M. E. Jacox, J. Chem. Phys. 47, 5146 (1967).

[5] M. E. Jacox and D. E. Milligan, J. Chem. Phys. 53, 2688 (1970).

[6] M. E. Jacox and D. E. Milligan, J. Mol. Spectrosc. 47, 148 (1973).

[7] D. E. Milligan and M. E. Jacox, J. Chem. Phys. 54, 927 (1971).

[8] D. E. Milligan, M. E. Jacox, and L. Abouaf-Marguin, J. Chem. Phys. 46, 4562 (1967).

[9] G. Herzberg and A. Lagerqvist, Can. J. Phys. 46, 2363 (1968).

[10] D. E. Milligan and M. E. Jacox, J. Chem. Phys. 51, 1952 (1969).

[11] D. E. Milligan, M. E. Jacox, and W. A. Guillory, J. Chem. Phys. 52, 3864 (1970).

[12] D. E. Milligan and M. E. Jacox, J. Chem. Phys. 55, 3404 (1971).

[13] C. R. Zobel and A. B. F. Duncan, J. Amer. Chem. Soc. 77, 2611 (1955).

[14] E. E. Rogers, S. Abramowitz, M. E. Jacox, and D. E. Milligan, J. Chem. Phys. 52, 2198 (1970).

[15] M. E. Jacox and D. E. Milligan, J. Chem. Phys. 54, 3935 (1971).

[16] M. E. Jacox, D. E. Milligan, J. H. McAuley, and C. E. Smith, J. Mol. Spectrosc. 45, 377 (1973).

[17] D. E. Milligan and M. E. Jacox, J. Mol. Spectrosc. 46, 460 (1973).

[18] V. E. Bondybey and G. C. Pimentel, J. Chem. Phys. 56, 3832 (1972).

[19] W. A. Chupka and M. E. Russell, J. Chem. Phys. 49, 5426 (1968).

RYDBERG STATES OF DIATOMIC AND POLYATOMIC MOLECULES

V. McKoy and T. Betts

Contribution No. 4757 from the Arthur Amos Noyes
Laboratory of Chemical Physics, California Institute
of Technology, Pasadena, California, 91109, U.S.A.

INTRODUCTION

Rydberg states clearly play an important role in the
study of the electronic structure of atoms and molecules in the
vacuum ultraviolet. Moreover many molecules, including
diatomic and simpler polyatomic types of particular importance
in astrophysics, aeronomy and photochemistry--such as H_2, O_2,
NO, N_2, CO, CO_2, H_2O, C_6H_6--have their most important and
strongest electronic bands in the vacuum ultraviolet.[1] Interest
in this spectral region has been stimulated both by the improved
resolution of optical absorption studies and by the development
of techniques such as electron impact spectroscopy. From
electron impact spectroscopy one can now obtain accurate
spectra of molecules well into the vacuum ultraviolet quite
readily. In the analysis of such electron impact spectra one
must know the location and nature of several Rydberg series.
From several recent experiments it has also become obvious
that the Rydberg series in many molecules can perturb the
intravalence region of the spectrum and lead to some unusual
features in the term values and intensity distribution.[2,3] We
will return to this point shortly. Rydberg series also lead to
several important features in photoionization cross sections.
With the increased resolution and activity in the measurement
of photoionization cross sections we can expect a renewed
interest in this role of Rydberg series. Finally a knowledge of
the quantum defects in a Rydberg series can provide useful
information on the phase shifts for electron-molecule scattering.

Our interest in developing a simple method for obtaining
Rydberg term values stems from our broader program in which

we have been studying excitation energies, intensities, and Born cross sections for electron scattering by several molecules. It became obvious that Rydberg series would occur in or near regions of strong intravalence absorptions, e.g., the valence-like $b'\,^1\Sigma_u^+$ state and the Rydberg-like $c'\,^1\Sigma_u^+$ state of N_2. Interactions between these states lead to complex perturbations in the resulting spectrum, e.g., irregular intensity distributions and large quantum defects. This Rydberg-valence mixing will also be reflected in the behavior of the generalized oscillator strength as a function of momentum transfer. We therefore decided to develop a simple method with which we could predict the location of Rydberg series in diatomic and polyatomic molecules with modest computing effort.

In the next section we will give a brief discussion of the pseudopotential method and the model approximations to these pseudopotentials that we have used in our molecular calculations. We will then discuss the results of these calculations for several molecules including N_2, O_2, CO_2, N_2O, H_2CO, and C_6H_6 and the assignment and interpretation of several of the Rydberg series in these molecules. We will also compare these results with those of nonempirical calculations on some molecules in order to confirm some important trends.

THEORY

In a Rydberg transition, an electron is excited to an orbital large in size relative to a singly-charged core. Moreover, the energy levels of these states can be expressed as

$$E_n = \text{I.P.} - \frac{R}{(n-\delta)^2}, \tag{1}$$

where I.P. is the ionization potential, R is the Rydberg constant, and δ is the quantum defect that may vary slightly with n. From Eq.(1) we know that Rydberg states are essentially hydrogenic in character, which implies that the molecular core produces the potential of a monopole. The quantum defect δ measures the departure from this simple hydrogenic model and correlates with how the nodes and loops of the Rydberg orbital are displaced relative to analogous hydrogenic nodes and loops. With these properties it is obvious how one should apply a pseudopotential formalism to study Rydberg states. In an independent-particle approximation, the Rydberg orbital, ψ_R, must be orthogonal to the core solutions of the one-electron Hamiltonian, H, i.e.,

$$H\psi_R = \epsilon_R \psi_R \tag{2a}$$

$$H\psi_c = \epsilon_c \psi_c \tag{2b}$$

$$\langle \psi_R, \psi_c \rangle = 0. \tag{2c}$$

We wish to solve the eigenvalue equation, Eq. (2a) subject to the orthogonality condition, Eq. (2c). The orthogonality of the valence eigenfunction, ψ_v, to the core eigenfunctions, ψ_c, requires that ψ_v must have several oscillations in the core region. Phillips and Kleinman[4] have shown that one can transform the eigenvalue equation

$$H\psi_v \equiv (T + V)\psi_v = E_v\psi_v \tag{3a}$$

with

$$\langle \psi_v, \psi_c \rangle = 0 \tag{3b}$$

and

$$H\psi_c = E_c\psi_c \tag{3c}$$

to the form

$$(H + V_s)\phi_v = E_v\phi_v \tag{4}$$

where ϕ_v is the pseudo-wavefunction and is not subject to the orthogonality constraints, Eq. (3b) but the eigenvalue, E_v, is unchanged. The pseudopotential V_s of ref. 4 is a repulsive nonlocal potential defined by

$$V_s\phi_v = \sum_{core} (E_v - E_c) \langle \phi_c, \phi_v \rangle \phi_c \tag{5}$$

Eq. (5) defines the Phillips-Kleinman pseudopotential. An excellent discussion of these equations and their generalization and of the use of pseudopotentials in atomic and molecular problems has been given by Weeks et al.[5] In the pseudo-potential method one replaces orthogonality constraints in the eigenvalue equation by adding a nonlocal pseudopotential V_s to H without changing the eigenvalue, E_v.[5]

Eq. (4) with the V_s of Eq. (5) is probably no simpler to solve at a practical level than is Eq. (2a) with its constraints Eq. (3c). However, as stressed by Hazi and Rice,[6] we can construct much simpler model potentials which reproduce the main characteristics of the formal pseudopotential. Model potentials are flexible and usually contain some adjustable para-meters. The resulting equations are much simpler than those of Eqs. (2). Clearly the model potential must go as $-1/r$ at large r, must have the symmetry of the molecular framework and should account for penetration and exclusion effects.

For atomic Rydberg states one could use the model potential suggested by Abarenkov and Heine[7]

$$V_M = A \qquad r < r_0$$

$$= \delta Z/r \qquad r \geqslant r_0 \tag{6}$$

where δZ is the effective charge of the atom. This potential is constant inside a sphere of radius r_0 and Coulombic outside of the sphere. The model potential is an approximate representation of the effective potential $V + V_S$. For a reasonable choice of r_0 we can use V_M of Eq. (6) to solve for the atomic term values as a function of the parameter A. Comparison of the predicted term values with the experimental spectrum gives a unique choice for A. The only absolute requirement on a pseudo- or model potential is that it reproduces the eigenvalues of the real potential. V_M of Eq. (6) is not a correct representation of the true $V + V_S$ at all points. Outside the core it is probably a good representation of the effective potential but it certainly does not represent any meaningful potential inside the core. The model wavefunction should be a good approximation to the Rydberg orbital only outside the core but its behavior inside the core is probably not physically significant. [5]

For molecular applications we want a model potential with no additional adjustable parameters. To apply this model potential, Eq. (6) to molecular Rydberg states we assume that the molecular model potential is a simple superposition of the atomic potentials. All parameters of the atomic V_M are fixed by calibration to atomic term values for each atom separately. Hence

$$V_{mol} = \sum_{atoms} V_{atom} . \tag{7}$$

We have found that including a term in Eq. (7) to represent the hydrogen atom contribution has only a small effect on the eigenvalue spectrum. We also use Gaussian basis functions on the atomic and other centers to provide speed and flexibility in the numerical solution of the one-electron molecular problem. Although one has to use many such functions due to their incorrect asymptotic behavior, we feel that their other properties, especially the ease of computation with high n values and their general diffuseness outweigh other considerations. We use only S, P_x, P_y, and P_z basis functions, but by placing these functions at other than atomic centers we can calculate molecular Rydberg states of various symmetries.

ATOMIC CALIBRATION

The solution of the one-electron Schrödinger equation for an atomic Rydberg state with V_M of Eq. (6) is simple. Writing

$$\psi = \frac{1}{r} R_{n\ell}(r) Y_{\ell m}(\theta, \phi) \qquad (8)$$

we have

$$\frac{d^2 R_{n\ell}}{dr^2} + \left\{ b - \frac{\ell(\ell+1)}{r^2} \right\} R_{n\ell} = 0 \qquad r < r_0 \qquad (9)$$

and

$$\frac{d^2 R_{n\ell}}{dr^2} + \left\{ a + \frac{2}{r} - \frac{\ell(\ell+1)}{r^2} \right\} R_{n\ell} = 0 \qquad r \geqslant r_0 \qquad (10)$$

where δZ has been set equal to unity and

$$b = 2(E_v - A)$$

$$a = 2E_v. \qquad (11)$$

The solutions of Eq. (9) and (10 are well known. We determine the eigenvalues E_v by requiring the logarithmic derivatives of these solutions to be equal at the boundary $r = r_0$. In calibrating the atomic model potential we choose a series of A's and compare the eigenvalue spectrum with the experimental values. Table I gives the parameters A and r_0 of the model potential for the carbon, nitrogen, and oxygen cores in our calculations.

RESULTS

We now discuss the application of this model potential to the calculation of various Rydberg series of the diatomic molecules N_2 and O_2 and the polyatomic molecules CO_2, N_2O, H_2CO, and C_6H_6. We have studied several other molecules including carbon monoxide, nitric oxide, acetylene, allene, and butadiene. With the results of these model calculations we have been able to assign and interpret most of the Rydberg series in the molecules we have studied. Recall that all adjustable parameters are fixed by calibration to atomic data and that the molecular model potential has no additional adjustable parameters. Details of the Gaussian basis sets are discussed in refs. 8 and 9.

The ground state of $N_2(X\,{}^1\Sigma_g^+)$ has the electron configuration $(1\sigma_g)^2(1\sigma_u)^2(2\sigma_g)^2(2\sigma_u)^2(1\pi_u)^4(3\sigma_g)^2$. We consider the $ns\sigma_g$, $np\sigma_u$, $np\pi_u$, and $nd\sigma_g$ Rydberg series. The experimentally observed Rydberg states correspond to an electron in one of these orbitals and one of the following three cores: $[\,(core)(2\sigma_u)^2(1\pi_u)^4$

TABLE I.   Parameters for atomic model potentials.

Atom	A^a	$r_0{}^a$
C	0.375	2.5
N	0.115	2.5
O	0.045	2.5

[a] In atomic units.

$3\sigma_g$] $X\,{}^2\Sigma_g{}^+$, [(core)$(2\sigma_u)^2(1\pi_u)^3(3\sigma_g)^2$] $A\,{}^2\Pi_u$, and [(core) $2\sigma_u(1\pi_u)^4$ $(3\sigma_g)^2$] $B\,{}^2\Sigma_u{}^+$. In our simple model we assume these three cores to be identical and therefore specified by the same atomic model potentials. Table II gives the results for N_2. In our model these results are the term values measured in electron volts below any of the three ionization potentials referred to above. The results are encouraging. We can readily identify three Rydberg series in agreement with available experimental data. These are the ns series with a quantum defect $\delta = 1.1$, the $np\sigma_u$ series with $\delta = 1.71$ ($n = 4, 5\ldots$) and the $np\pi_u$ series with $\delta = 0.73$ ($n = 3, 4\ldots$). With this model potential and Gaussian basis functions we can obtain the first four members of several series with a few minutes of computer time. The $3s\sigma_g$ is the only member of the ns series well identified experimentally. The term value of 3.65 listed as the experimental value for the $3s\sigma_g$ is actually an average of term values for states with this Rydberg orbital in several series leading to different states of the ion. Within the limitations of our model this is probably the simplest procedure and we use it to estimate the term values for many series in other molecules. Further details are given in references 8 and 9. For the $np\sigma$ series the $4p\sigma_u$ to $7p\sigma_u$ members have been identified.[10, 11] The calculated term values for this series agree well with these results. The calculated term value for the lowest member of the $nd\sigma_g$ series seems large. This is probably due to the lack of nd functions in our basis. This root of our σ_g spectrum may be spurious. We will see that the $4d\sigma_g$ term value is closer to the value one expects for the first member of the $nd\sigma$ series.

In the fourth column we list the term values obtained from direct Hartree-Fock calculations.[12] The agreement between these term values and those of model potential calculations is reasonable. The term values for the higher members of the Rydberg series can be obtained more economically by our procedure but, on the other hand, the term values for the early members of the series, in particular the first, where Rydberg-valence interactions can be important, are predicted more reliably by a nonempirical approach.

Next we consider the Rydberg spectrum of O_2^+. The ground state of $O_2({}^3\Sigma_g{}^-)$ has the electron configuration $(1\sigma_g)^2$ $(1\sigma_u)^2(2\sigma_g)^2(2\sigma_u)^2(3\sigma_g)^2(1\pi_u)^4(1\pi_g)^2$. We calculate term values for states with Rydberg orbitals of $ns\sigma_g$, $np\sigma_u$, $np\pi_u$, $nd\sigma_g$, and $nd\pi_g$ symmetry. The observed Rydberg series correspond to an electron in one of these orbitals and one of the following ions: [(core)$(2\sigma_u)^2(3\sigma_g)^2(1\pi_u)^4(1\pi_g)$] $X\,{}^2\Pi_g$, [(core)$(2\sigma_u)^2(3\sigma_g)^2(1\pi_u)^3$

TABLE II. Rydberg levels for N_2.[a]

Rydberg orbital symmetry	Calculated term values[b]	Observed term values	Calculated term values[g]
$3s\sigma_g$	3.80	3.65[d]	3.69[h]
4s	1.61		1.69
5s	0.88		
6s	0.53		
$4p\sigma_u$	2.60	2.65[e]	2.48[i]
5p	1.23	1.26	
6p	0.72	0.74	
7p	0.45	0.48	
$3p\pi_u$	2.63	2.52[f]	2.58[j]
4p	1.26		1.18[j]
5p	0.77		
6p	0.47		
$4d\sigma_g$	2.70[c]		
5d	1.35		1.39[k]
6d	0.79		
7d	0.40		

Footnotes to Table II.

---

[a] Internuclear distance of 2.11 a.u.

[b] All term values in electron volts.

[c] Probably a spurious root. The term value of 1.35 eV for the $5d\sigma_g$ series seems to correspond to the first true member of this series. See text.

[d] Average of the $3s\sigma_g$ term value in several series. See Ref. 8 for details.

[e] Reference 11.

[f] Average of term values for the $z\,^1\Delta_g$ and $x\,^1\Sigma_g^-$ states.

[g] Hartree-Fock calculations of Ref. 12.

[h] For the $ns\sigma_g$ series converging to the A $^2\Pi_u$ state of N_2^+.

[i] $np\sigma_u$ series converging to the X $^2\Sigma_g^+$ state of N_2^+.

[j] $np\pi_u$ series converging to the X $^2\Sigma_g^+$ state of N_2^+.

[k] Converging to the A $^2\Pi_u$ state of N_2^+. See text.

$(1\pi_g)^2]$ a $^4\Pi_u$ and A $^2\Pi_u$, $[(core)(2\sigma_u)^2(3\sigma_g)(1\pi_u)^4(1\pi_g)^2]$ b $^4\Sigma_g^-$ and B $^2\Sigma_g^-$, and $[(core)(2\sigma_u)(3\sigma_g)^2(1\pi_u)^4(1\pi_g)^2]$ c $^4\Sigma_u^-$. The ionization potentials of these states are 12.08 eV $(^2\Pi_g)$, 16.11 and 16.82 eV $(^4\Pi_u, {}^2\Pi_u)$, 18.17 and 20.31 eV $(^4\Sigma_g^-, {}^2\Sigma_g^-)$, and 24.56 eV $(^4\Sigma_u^-)$. Again the simple model potentials we use do not distinguish between these various cores. In Table III we therefore list the term values only by the symmetry of the Rydberg orbital.

From the results of Table III we can identify four Rydberg series in agreement with available experimental data and the Hartree-Fock calculations of Leclercq.[13] These are the $ns\sigma_g$ series with a quantum defect $\delta = 1.15$, $np\sigma_u$ (n = 4, 5, ...) with $\delta = 1.73$, $np\pi_u$ (n = 3, 4, ...) with $\delta = 0.74$, and the $nd\pi_g$ with $\delta = 1.04$. Unlike N_2, the $ns\sigma_g$ series is well identified from n = 3 to 6. The n = 4, 5, and 6 numbers agree well with the ns series converging to the c $^4\Sigma_u^-$ state of O_2^+ reported by Codling and Madlen.[14] Moreover, they assign an observed state with a term value of 3.70 eV to an $nd\pi$ series. Our results for the ns and $nd\pi$ series clearly show that the state with a term value of 3.70 eV belongs to an ns series. Again the agreement with the results of available Hartree-Fock calculations for the low members of the series is good. The $3d\sigma_g$ member is also probably a spurious root and we have assigned the $4d\sigma_g$ member as the first of the $nd\sigma_g$ series.

We now look at the results of applying this same model potential to study the Rydberg series of the small polyatomic molecules CO_2, N_2O, and H_2CO. Here we are looking at molecules which are large enough to make extensive nonempirical calculations of Rydberg series quite difficult but in which several Rydberg series have been extensively studied. Moreover the vacuum ultraviolet regions of the spectra of CO_2 and H_2CO are important in understanding the roles of these molecules in planetary atmospheres and interstellar clouds respectively.

In CO_2 Rydberg series have been observed converging to four different ions at 13.8, 17.3, 18.1, and 19.4 eV respecrively. Four Rydberg series, two strong and two weak, converge to the first ionization limit. The strong series have quantum defects of 0.65 while the weak series have quantum defects of 0.97 and 0.57. Two more series, known as Henning's sharp and diffuse series, with quantum defects of 0.01 and 0.34 respectively converge to the third ion. Finally series with quantum defects of 0.71, 0.56, and 0.05 converge to the ion at 19.4 eV.[15] From the results in Table IV we can assign most of the observed series. Henning's sharp series[16] with term values of 1.60, 0.80, and 0.54 must be assigned as the $ns\sigma_g$ series while we can assign Henning's diffuse series as $nd\pi_g$. The calculated term values of 1.91 and 0.94 eV's agree well with the observed values of 1.90 and 1.0 eV's, respectively. We have also assigned the series with members at 10.98, 12.47, and 13.03 eV (term values of 2.81, 1.32, and 0.76 eV respectively) as an $np\pi_u$ series and the states at 11.39, 12.64, and 13.11 eV (term values of 2.40, 1.15, and 0.68 eV's) as a $np\sigma_u$ series. Both of these series are superimposed on a strong continuum-like absorption[17] which is probably predominantly due to intravalence $1\pi_g \rightarrow \pi_u$ excitation. The $^1\Sigma_u^+$ states, i.e., the $np\pi_u$ series, can be expected to be more strongly perturbed by this underlying continuum, if its a $^1\Sigma_u^+$ state, leading to a lowering of the hypothetical unperturbed $np\pi_u$ levels and hence larger term values. Also recent calculations which allow for the perturbation of the $^1\Sigma_u^+$ ($1\pi_g \rightarrow 3p\pi_u$) state by a valence-like $^1\Sigma_u^+$ state put the $3p\pi_u$ level below the $3p\sigma_u$.[18]

For nitrous oxide Rydberg series are observed converging to ions at 12.9, 16.4, and 20.1 eV.[15] The first Rydberg state appears as a single sharp peak at 1292 Å near the maximum of a broad $\tilde{X} \rightarrow \tilde{C}$ absorption. Two Rydberg series with δ's of 0.60 and 0.68 converge to the lowest state of N_2O^+. Four series with quantum defects of 0.31, 0.06, 0.68, and 0.58 respectively, converge to the ion at 20.1 eV.[19] There are about eight observed series in N_2O.[19] From our model calculations we can consistently assign most of these series. Table V gives these assignments. Experimental term values for the $ns\sigma$ series leading to the $^2\Pi$ and $^2\Sigma$ ions at 12.90 and 16.40 eV respectively are shown in Table V. The difference between these term values for the $n = 3$ member of the $^2\Pi$ and $^2\Sigma$ series, i.e., 3.40 and 4.03 eV's may indicate a perturbation of the $^2\Sigma(3s\sigma)$ state by another state. From recent electron impact data[20] we can confirm Duncan's series as an $ns\sigma$ progression.

Four Rydberg series with quantum defects of 1.04, 0.70, 0.70, and 0.4 converge to the lowest ion of formaldehyde at 10.9 eV.[15] Table VI shows the term values for these series. The series with a quantum defect of 1.04 is clearly an ns series while the two series with quantum defects of 0.70 are npa_1 and npb_2 series. The observed term value for the $n = 3$ member of the npb_2 series is larger than the calculated term value, i.e., 2.91 eV compared to 2.59 eV. This is probably due to a perturbation of the $^1A_1(2b_2 npb_2)$ Rydberg state by a valence $^1A_1(\pi \rightarrow \pi^*)$ state.[2] We have assigned the series with a δ of 0.40 as an nd series.

Finally we have obtained the term values for several Rydberg series in benzene. Table VII compares these calculated term values with the observed series.[21] The model

TABLE III. Rydberg levels for O_2.[a]

Rydberg orbital symmetry	Calculated term values[b]	Observed term values	Calculated term values
$3s\sigma_g$	3.97	3.71[d]	3.60[g]
4s	1.67	1.66[e]	1.62
5s	0.90	0.93	
6s	0.53	0.59	
$4p\sigma_u$	2.64[c]	2.54[f]	2.61[h]
5p	1.25	1.25	
6p	0.73	0.74	
7p	0.45	0.49	
$4d\pi_g$	1.55	1.49[e]	1.48[i]
5d	1.02	0.83	
6d	0.55	0.53	
$3d\sigma_g$	2.82		
4d	1.34		1.41[i]
5d	0.72		

Footnotes for Table III.

---

[a] Internuclear distance of 2.28 a.u.

[b] All term values in electron volts.

[c] The results for the $np\pi_u$ and $np\sigma_u$ series are almost identical, but the $np\pi$ series begins at $n=3$ with $\delta = 0.74$. Experimentally the two series have not been definitely distinguished since in most cases $np\sigma_u$ and $np\pi_u$ both lead to allowed transitions.

[d] Average of term values for series leading to the $^4\Sigma_u^-$, $^4\Pi_u$, and $^2\Pi_u$ states of O_2^+. See Ref. 8.

[e] Reference 14.

[f] See Reference 8.

[g] $[(O_2^+, A\ ^2\Pi_u)3s\sigma_g$ and $4s\sigma_g]^3\Pi_u$ state. Hartree-Fock calculations of Ref. 13.

[h] $[(O_2^+, b\ ^4\Sigma_g^-)3p\sigma_u]^3\Sigma_u^-$ state of Ref. 13.

[i] See Ref. 13.

TABLE IV. Rydberg series in CO_2.[a]

Rydberg orbital	Calculated term values[b]	Observed term values
$3s\sigma_g$	3.39^c	
$4s$	1.50	1.56^d
$5s$	0.84	0.83
$6s$	0.52	0.54
$3p\sigma_u$	2.47	2.40^e
$4p$	1.19	1.15
$5p$	0.70	0.68
$6p$	0.45	0.45
$3p\pi_u$	2.37	2.81^f
$4p$	1.11	1.32
$3d\pi$	1.91	1.90^g
$4d$	0.94	1.0

[a] Ground state configuration:

$$1\sigma_u^2 1\sigma_g^2 2\sigma_g^2 3\sigma_g^2 2\sigma_u^2 4\sigma_g^2 3\sigma_u^2 1\pi_u^4 1\pi_g^4 - {}^1\Sigma_g^+ .$$

[b] In electron volts.

[c] $\delta = 1.0$.

[d] Averages of the term values for the $1\pi_g^3\ ns\sigma_g$, $1\pi_u^3\ ns\sigma_g$, and $3\sigma_u\ ns\sigma_g$ series

[e] Term values for the $1\pi_g^3\ np\sigma_u$ series. See Ref. 16.

[f] Term values for the $1\pi_g^3\ np\pi_u$ series. Reference 16.

[g] $\delta = 0.32$. Term values are those of Henning's diffuse series.

TABLE V. Rydberg series in N_2O.[a]

Rydberg orbital	Calculated term values	Experimental term values	
$3s\sigma$	3.55^b	3.40^c	4.03^d
4s	1.58	1.52	1.56
5s	0.87	0.85	0.86
6s	0.53		0.52
$3p\sigma$	2.46^e		2.40^f
4p	1.19		1.16
5p	0.70		1.71
6p	0.45		0.49
$3p\pi$	2.54^g		2.56
4p	1.24		1.24
5p	0.75		0.73
6p	0.47		0.49
$3d\pi$	1.58^i		1.80^j
4d	1.06		0.98
5d	0.65		0.61

[a] Ground state configuration: $1\sigma^2\,2\sigma^2\,3\sigma^2\,4\sigma^2\,5\sigma^2\,6\sigma^2\,1\pi^4\,7\sigma^2\,2\pi^4 - {}^1\Sigma_g^+$.

[b] $\delta = 1.04$.

[c] From the electron impact spectrum of Ref. 20.

[d] Series III of Ref. 19.

[e] $\delta = 0.65$.

[f] Series II and IX of Ref. 19.

[g] $\delta = 0.68$.

[h] Series VIII of Ref. 19.

[i] $\delta = 0.24$.

[j] Average of series IV and VI of Ref. 19.

TABLE VI. Rydberg series in Formaldehyde.[a]

Rydberg orbital	Calculated term values		Observed term values
$3sa_1$	3.67^b	3.60^c	3.80^d
4s	1.57		1.63
5s	0.86		0.87
6s	0.52		0.55
$3pa_1$	2.64^e	2.76	2.74^f
4p	1.32	1.33	1.25
5p	0.77		0.74
6p	0.48		0.48
$3pb_2$	2.59^g	2.73	2.91^h
4p	1.25	1.30	1.29
5p	0.77		0.74
6p	0.47		0.48
3d	2.25^i		2.01
4d	1.17		1.04
5d	0.71		0.63

[a] Ground state electron configuration: $1a_1^2 2a_1^2 3a_1^2 4a_1^2 1b_2^2 5a_1^2 1b_1^2 2b_2^2 - {}^1A_1$.

[b] $\delta = 1.08$.

[c] Results in this column are from Ref. 3. See also Ref. 2.

[d] $\tilde{B}\ {}^1B_2$ state.

[e] $\delta = 0.73$.

[f] $\tilde{D}\ {}^1B_2$ state.

[g] $\delta = 0.71$.

[h] This state is probably perturbed.

[i] $\delta = 0.57$.

TABLE VII. Rydberg series in benzene.[a]

Rydberg orbital	Calculated term values	Experimental term values
$3sa_{1g}$	2.52[b]	
4s	1.21	
5s	0.71	
6s	0.47	
$3pa_{2u}$	2.16	2.32[c]
4p	1.09	1.10
5p	0.66	0.66
6p	0.44	0.44
$3pe_{1u}$	2.03	1.84[d]
4p	1.02	0.88
5p	0.62	0.56
6p	0.42	0.39
$3de_{1g}$	1.40	1.63[e]
4d	0.87	0.87
5d	0.60	0.56
$3de_{2g}$	1.38	1.26[f]
4d	0.86	0.86
5d	0.58	0.55

[a] A C-C distance of 2.64 a.u.

[b] $\delta = 0.68$.

[c] nR series of Ref. 21.

[d] nR′ series.

[e] nR″ series of Ref. 21. Our assignment.

[f] nR‴ series of Ref. 21. Our assignment.

potential of the positive ion core is again approximated by the sum of model potentials of the six carbon atoms. With the molecule in the x-y plane the electron configuration of the ground state is $\{$ core$\}$ $2e_{2g}^4\ 3a_{1g}^2\ 2b_{1u}^2\ 1b_{2u}^2\ 3e_{1u}^4\ 3e_{2g}^4\ 1a_{2u}^2\ 1e_{1g}^4$ with core denoting $1a_{1g}^2\ 1e_{1u}^4\ 1e_{2g}^4\ 1b_{1u}^2\ 2a_{1g}^2\ 2e_{1u}^4$. Wilkinson[21] has observed four series converging to the ion at 9.25 eV. These series all correspond to excitation out of the $1e_{1g}$ orbital. The transitions of the nR and nR' series are strong and are clearly dipole allowed. The upper states must be either $^1E_{1u}$ (xy polarized) or $^1A_{2u}$ (z polarized) corresponding to transitions $1e_{1g} \rightleftharpoons npa_{2u}$ and $1e_{1g} \rightleftharpoons npe_{1u}$, respectively. It has usually been assumed that the upper states of the nR series are $^1A_{2u}$ and the nR series assigned as $1e_{1g} \rightarrow npe_{1u}$. The nR' series must then be $1e_{1g} \rightarrow npa_{2u}$ ($^1E_{1u}$ states). Our calculated term values suggest the reverse assignment, namely that the nR series is $1e_{1g} \rightarrow npa_{2u}$ ($^1E_{1u}$) and the nR' series $1e_{1g} \rightarrow npe_{1u}$ ($^1A_{2u}$). The calculated quantum defects are 0.49 and 0.35 for the npa_{2u} and npe_{1u} series respectively compared to Wilkinson's[21] observed δ's of 0.46 and 0.16. Note that this assignment puts a $^1E_{1u}$ Rydberg state on top of the N $\rightarrow$ V absorption. The V state is definitely $^1E_{1u}$ ($1e_{1g} \rightarrow 1e_{2u}$). The nR'' and nR''' series are weak and probably correspond to forbidden transitions. The transitions in these series can be $1e_{1g} \rightarrow nd$. We can tentatively assign the R'' and R''' series as $1e_{1g} \rightarrow nde_{1g}$ and nde_{2g} respectively. These series have δ's of 0.04 and 0.02.

CONCLUSIONS

The main purpose of this work has been to develop and use simple model potentials to calculate the Rydberg series of diatomic and polyatomic molecules. We represented the molecular model potential by a sum of atomic model potentials which are calibrated to atomic Rydberg term values and hence the molecular potential contains no additional parameters. The results are encouraging. We have been able to assign many of the Rydberg series in several diatomic and polyatomic molecules including N_2, O_2, CO_2, H_2CO, and C_6H_6. The s, p, and d series emerge very distinctly. These results can be improved by explicit including nd basis functions in the calculations and by closer calibration of the molecular potential.

REFERENCES

[1] P. G. Wilkinson, J. Mol. Spectry. $\underset{\sim}{6}$, 1 (1961).

[2] J. E. Mentall, E. P. Gentien, M. Krauss, and D. Neumann, J. Chem. Phys. $\underset{\sim}{55}$, 5471 (1971).

[3] D. L. Yeager and V. McKoy, "The Equations of Motion Method: Excitation Energies and Intensities in Formaldehyde," J. Chem. Phys., April (1974).

[4] J. C. Phillips and L. Kleinmann, Phys. Rev. $\underset{\sim}{116}$, 287 (1969).

[5] J. D. Weeks, A. Hazi, and S. A. Rice, Advan. Chem. Phys. $\underset{\sim}{16}$, 283 (1969).

[6] A. U. Hazi and S. A. Rice, J. Chem. Phys. $\underset{\sim}{48}$, 495 (1968).

[7] I. V. Abarenkov and V. Heine, Phil. Mag. $\underset{\sim}{11}$, 529 (1969).

[8] T. Betts and V. McKoy, J. Chem. Phys. $\underset{\sim}{54}$, 113 (1971).

[9] T. Betts and V. McKoy, "Rydberg States of Polyatomic Molecules Using Model Potentials," J. Chem. Phys., April (1974).

[10] R. S. Mulliken, The Threshold of Space (Pergamon, New York, 1957) p. 169.

[11] M. Ogawa and Y. Tanaka, Can. J. Phys. $\underset{\sim}{40}$, 1593 (1962).

[12] H. LeFebvre-Brion and C. M. Moser, J. Chem. Phys. $\underset{\sim}{43}$, 1394 (1965).

[13] J. Leclercq, Ann. Astrophys. $\underset{\sim}{30}$, 93 (1967).

[14] K. Codling and R. Madlen, J. Chem. Phys. $\underset{\sim}{42}$, 3935 (1965).

[15] G. Herzberg, Electronic Spectra of Polyatomic Molecules, Van Nostrand, Princeton, New Jersey 1966.

[16] Y. Tanaka, A. S. Jursa, and F. LeBlanc, J. Chem. Phys. $\underset{\sim}{32}$, 1199 (1960).

[17] M. Krauss, S. R. Mielczarek, D. Neumann, and C. E. Kuyatt, J. Geophys. Research, $\underset{\sim}{76}$, 3733 (1971).

[18] C. W. McCurdy, Jr., and V. McKoy, "Differential Cross

Sections for the Scattering of Electrons by CO_2," J. Chem. Phys. (to be submitted).

[19] G. R. Cook, A. S. Jursa, and F. J. LeBlanc, J. Chem. Phys. 32, 1205 (1960).

[20] E. N. Lassettre, A. Skerbele, M. A. Dillon, and K. J. Ross, J. Chem. Phys. 48, 5066 (1968).

[21] P. G. Wilkinson, Can. J. Phys. 34, 596 (1956).

THEORY OF INTRAVALENCY AND RYDBERG TRANSITIONS IN MOLECULES

Oktay Sinanoğlu

Institut für Organische Chemie,
Technische Universität, München, W. Germany
Sterling Chemistry Laboratory, Yale University, New
Haven, Connecticut 06520, U.S.A.
Boğaziçi Evrenkenti (Üniversitesi), Istanbul, Türkiye

CONTENTS

I.   Questions
II.   Theory of Excited States, Non-Closed Shell States
        A. Semi-internal Orbitals
        B. The "Charge Wave Function" ψ_c of a Non-Closed Shell State
        C. Energy of a Non-Closed Shell State
III.   Relation to the Intravalency and Rydberg States of Spectra
IV.   Classification of Transitions into Intravalency ($\bar{V}$), Pre-Rydberg (pR), and Rydberg (R) Types
V.   Spectral Form of the Charge Wave Function of an Intravalency State
VI.   The Charge Wave Function of a Pre-Rydberg State
VII.   Electric Dipole Oscillator Strengths of Intravalency ($\bar{V}$) and Pre-Rydberg (pR) Transitions
VIII.   Inner Ionization Potentials; Excited Core Rydberg Series Limits
IX.   Theory of Excited States of Molecules – Many Center Aspects
        A. Intravalency and Pre-Rydberg States of Molecules with RHF - MO(SCF) / NCMET
        B. The Intravalency AO-Pool
        C. Molecular Semi-internal Orbitals from the $\bar{V}$-AO-Pool; Problems of the United Atom
        D. The Spectral Charge Wave Function of a Molecular Intravalency State ($\bar{V}$) and of a Pre-Rydberg State (pR)
X.   What is a Rydberg Orbital?

I.   QUESTIONS

A number of questions come up in vacuum <u>uv</u> molecular
spectroscopy:

1 - Among the first few electronic transitions of a molecule
which are "Rydberg", which are "intravalency"? Can these always
be clearly defined?

2 - If a Rydberg orbital is to be defined with respect to a
united, or semi-united atom, what is its relation to the $n \geq 3$
atomic orbitals (AO's) of each center (for KL-shell atoms, i.e.
{H, B, C, N, O, F}) and to some intravalency MO's (like C-H $\sigma*$ in
C_2H_4, O-H $\sigma*$ in H_2O, etc.)?

3 - Do intravalency and Rydberg transitions, Rydberg and
other Rydberg transitions in molecules mix? How much? Can we
deal with such mixings without involving more and more configurations and just large computations? For how many members of a
Rydberg series would the mixings be important?

4 - Can we predict inner ionization potentials (inner I.P.'s),
for <u>uv</u> photoelectron spectroscopy and for ESCA, from molecular
orbital (MO) theory (accurate non-empirical MO SCF, i.e. RHF
("Restricted Hartree-Fock") or approximate and/or semi-empirical
ones like "limited basis set LCAO MO SCF"[1], CNDO, EHMO, MINDO,
etc.).

5 - With accurate MO's (i.e. RHF, MO SCF) could we calculate
accurate allowed and forbidden optical transition probabilities
(E1 ≡ electric dipole; E2 ≡ electric quadrupole; M1 ≡ magnetic
dipole; SOAE1 ≡ spin-orbit allowed electric dipole; CD ≡ circular
dichroism)?

6 - With MO theory (accurate or semi-empirical) can we
calculate electronic potential energy (P.E.) curves and surfaces?
What about C.I. ("configuration-interaction")?

These and related questions have a <u>many-electron</u> aspect and
a <u>many-center</u> (many nuclei) aspect. The many-electron aspect
exists in atoms as well.

MO's provide a language for molecular spectroscopy. They
are also a starting point in discussing the exact wave function
(w.f.; ψ) and energy (E) of a many-electron system. Along with
MO's we have electron correlation. Electron correlation (CORR)
is essential in low lying spectra (cf. table I).

Problems above require a theory of many-electron non-closed
shell states. All stable molecule ground and excited states, even

TABLE I

Orbital Theory (MO, SCF, ...) vs. Experiment – Importance of
Electron Correlation on Spectroscopic Properties

Property	Orbital theory[a] (Hartree-Fock, SCF MO)	Experiment[a] or NCMET[a] (Hartree-Fock + Correlation)
Oscillator strength f_{E1} of phosphorus ion P^+ $(3p^2\ ^3P \rightarrow 3p^3\ ^3P^\circ)$	0.292 (RHF)	0.040 (exp.) 0.041 (NCMET)
Oscillator strength f_{E1} of nitrogen ion, N^+ $(2p^2\ ^3P \rightarrow 2p^3\ ^3D^\circ)$	0.236 (RHF)	0.101 (exp.) 0.102 (NCMET)
Term Splitting Ratio $(^1S - ^1D)/(^1D - ^3P)$ for O^{+4} ion $1s^2 2p^2$ configuration	1.47 (RHF)	3.15 (exp.) 3.66 (NCMET)
Dissociation Energy of F_2 (D_e)	-1.37 eV. (RHF)	$+1.68$ eV. (exp.)
Electron Affinity of Oxygen atom	-0.54 eV. (RHF)	$+1.465$ eV. (exp.)

a. See references in text to NCMET and comparison with RHF
    and experiment therein, with references to diverse RHF and
    experimental sources, as well as other examples.

$^1\Sigma_g^+$, ^1A_g ones are non-closed shell states (see IX B). This last
fact is responsible for most problems encountered with MO theory
(used by itself, without CORR). Non-closed shells have large
correlation effects specific to each state.

Theory of excited states with the exact many-electron ψ and
E has been developed recently [2] and applied to many atomic
properties (and some molecular [3]) with good experimental tests
[2,4]. We can now use it for problems in molecular spectroscopy
with confidence.

We outline below the theory of excited states (also ground,
non-closed shell states). We see it leads naturally to a treatment of Rydberg and intravalency transitions and their mixings.

II.   THEORY OF EXCITED STATES, NON-CLOSED SHELL STATES

The exact wave function of a state is [5]

$$\psi = \phi_{RHF} + \chi_{CORR}; \quad \text{with } \langle\phi_{RHF} / \chi_{CORR}\rangle = 0. \tag{1}$$

The RHF [6] means a different Hartree-Fock solution for each term
of each configuration (with orbitals "restricted" to transform as
representations of the symmetry group $O(3) \; \otimes \; SU(2)_{spin}$ in
atoms; point group $G_{pt} \; \otimes \; SU(2)_\sigma$ in molecules).

The Non-Closed Shell Many-Electron Theory of atoms and
molecules (NCMET) by this writer shows [2] there are three
physically distinct correlation effects in such states

$$\chi_{CORR} = \chi_{INT} + \chi_F + \chi_U \tag{2}$$

INT $\equiv$ "internal correlations" $(ij \to i'j')$

  F $\equiv$ "semi-internal" correlations $(ij \to k\hat{f}_{ij;k}(x_2))$ plus orbital
     polarizations $(i \to \hat{f}_p^i(x_1))$

  U $\equiv$ "all-external correlations" $(ij \to \hat{u}_{ij}(x_1,x_2))$.

These are completely determined once the "Hartree-Fock sea", the occupied <u>and</u> unoccupied RHF set of MO's i, j, k, ... (or AO's) are specified.

We distinguish between <u>intravalency states</u> $(\overline{V})$ [8] and <u>inter-states</u>:

For the chemically interesting 1st and 2nd row atom set {H, B, C, N, O, F} = "CHEMSET", intravalency states are all terms of configurations [7] $1s^k 2s^m 2p^n$ (k + m + n = N, number of electrons). Then the intravalency RHF - (RHF-AO set) is $\{k\}_{\overline{V}}$

$$\{k\} = \{k\}_{\overline{V}} = \{1s\alpha, 1s\beta, 2s\alpha, \ldots 2p_z\alpha\} \qquad (3)$$

Number of spin orbitals in $\{k\}_{\overline{V}}$:   Dim $\{k\}_{\overline{V}}$ = M (=10 for KL)

"Inter-states" for the CHEMSET are atomic terms of configurations KL <u>and</u> M, N, etc. shells, for example the $(1s^p 2s^q 2p^r n\ell;$ n > 2) atomic Rydberg states, and others [9]. We shall return to these later. Now we continue with $\overline{V}$-states.

A.   Semi-internal Orbitals

Then, for a given KL-shell $\overline{V}$-state:

χ_{INT} contains all the (no. $\leq \binom{M}{N}$) Slater determinants (dets) made of the $\{k\}_{\overline{V}}$ set. [i j → i' j'; {i,j,i',j'} $\in$ {k} $_{\overline{V}}$].

χ_F contains all dets made of the new "semi-internal orbitals" $\hat{f}_{ij;k}(\chi_1), \hat{f}_p^i(\chi_1)$ of NCMET [2], s.t. electrons go virtually from RHF-occupied orbitals i, j into unoccupied RHF AO's k <u>and</u> a new $\hat{f}(\chi_1)$. [{i,j,k} $\in$ {k}$_{\overline{V}}$].

The $\hat{f}$ if expanded in symmetry orbitals truncate at a certain symmetry type; e.g. in KL-atoms:

$$\hat{f}(\bar{r}) \quad = \quad f_s(r)Y_{00}(\bar{\Omega}) + f_p(r)Y_{1m}(\bar{\Omega}) + f_d(r)Y_{2m'}(\bar{\Omega})$$

$$+f_f(r)Y_{3m'} \tag{4}$$

with semi-internal radial orbitals for s, p, d, f types only.

Now when calculated with our fully automated algebraic and computational CDC-6600 program system [2] (for atoms Be through Sc, Ti, ...Fe) [system "ATOM"], the $\hat{f}$'s turn out to good approximation to be mainly (for KL-shells):

$$f_s \quad \sim \quad 3s - like$$
$$f_p \quad \sim \quad 3p - like$$
$$f_d \quad \sim \quad 3d - like$$
$$f_f \quad \sim \quad 4f - like \tag{5}$$

i.e. n = 3 or 4, but with new fully optimized Slater exponents in each. Adding higher n > 3 or 4 (resp. for s, p, d or f) do not affect the accuracy. So for KL = $\bar{V}$, we have $\{f\}_{\bar{V}} = \{f_s, f_p, f_d, f_f\}$ as in equation (5).

B. The "Charge Wave Function, ψ_c" of a Non-Closed Shell State

It will be clear now that χ_F will contain also only a finite number of determinants, those made up for N-electrons of the set [10] $\{k\}_{\bar{V}} \bigcup \{f\}_{\bar{V}}$.

We define the charge wave function [2,3], ψ_c :

$$\boxed{\psi_c \equiv \phi_{RHF} + \chi_{INT} + \chi_F} \tag{6}$$

This is the only part of ψ that affects the charge distribution significantly [2] (also transition charge density in El, etc.) as all specifically non-closed shell type correlation effects--which involve some virtual transitions within the same RHF $\{k\}_{\overline{V}}$ set-- are in ψ_c. It has moreover only a finite number of determinants in it! All of ψ_c is calculated automatically by our program system which also finds by itself all the determinants that come in.

Examples of our system "ATOM" direct output ψ_c's are shown for some atomic $\overline{V}$-states in figures 1-4.

All "charge-like properties", i.e. one electron operators Q_1 expectation values and transition matrix elements can be calculated well with ψ_c alone [2,3].

Now we have a new and physically more meaningful subdivision of the exact ψ :

$$\boxed{\psi = \psi_c + \chi_U} \qquad (\text{with } \langle\psi_c \mid \chi_U\rangle = 0) \qquad (7)$$

ψ_c = the charge wave function containing all specifically non-closed shell type correlation and the RHF-effects. Gives $\langle\psi \mid Q_1 \mid \psi\rangle \cong \langle\psi_c \mid Q_1 \mid \psi_c\rangle$

χ_U = the remaining "closed-shell like" correlations with main effect on energy only, not on one e^- matrix elements. $\langle\psi_c \mid Q_1 \mid \psi_c'\rangle$.

We now state a key result:

Theorem 1: The ψ_c is the proper generalization to non-closed shell states of the orbital (RHF; MO; etc.) wave function ϕ_{RHF} of the closed shells (where $\phi_{RHF} = \Delta_1$ a single determinant of all

NITROGEN II   1S2   2S   2P3   TRIPLET S   09/27/73

								COEFFICIENT	CONTRIBUTIONS (A.U.)
DET HF=A	(1SA	1SB	.2SA	1SB	2P+B	2P0A	2P-A)	.28867513	7.13815595E-08
DET HF=A	(1SA	1SB	2SA	2SA	2P+A	2P0B	2P-A)	.28867513	-1.38660455E-02
DET HF=A	(1SA	1SB	2SA	2SA	2P+A	2P0A	2P-B)	.28867513	-6.93302277E-03
DET HF=A	(1SA	1SB	2SB	2SA	2P+A	2P0A	2P-A)	-.86602541	1.39044322E-07
DET 2=A	(1SA	1SB	2SB	2P-A	2P0A	2P0A	3P+A)	-.00052475	7.92803735E-31
DET 3=A	(1SA	1SB	2SA	2P-A	2P0A	2P0A	3D-2A)	-.08065934	-6.93302277E-03
DET 4=A	(1SA	1SB	2SA	2P-A	2P0B	2P0A	3D-1A)	-.05703477	-1.38660455E-02
DET 5=A	(1SA	1SB	2SA	2P-A	2P0A	2P+A	3SB)	-.00508921	-1.15825801E-04
DET 6=A	(1SA	1SB	2SA	2P-A	2P0A	2P+A	3D0B)	-.00000000	-4.46795498E-04
DET 7=A	(1SA	1SB	2SA	2P-A	2P0A	2P+B	3D-1A)	-.05703477	-1.15825801E-04
DET 8=A	(1SA	1SB	2SA	2P0A	2P+B	2P+A	3D-2A)	-.08065934	-1.15825801E-04
DET 9=A	(1SA	1SB	2SA	2SA	2P0B	2P+A	3P+A)	-.00666122	0.
DET 10=A	(1SA	1SB	2SA	2SA	2P-A	2P+A	3P0B)	-.01279770	-4.46795498E-04
DET 11=A	(1SA	1SB	2SA	2SA	2P0A	2P+A	3P+A)	-.00666122	-1.54501265E-08
DET 12=A	(1SA	1SB	2SA	2SA	2P0A	2P+A	3P0A)	-.00000000	-2.31100759E-03
DET 13=A	(1SA	1SB	2SA	2SA	2P0A	2P+A	3P-B)	.01279770	-4.46795498E-04
DET 14=A	(1SA	1SB	2SA	2P-A	2P0A	2P+B	3SA)	-.00169640	-1.15825801E-04
DET 15=A	(1SA	1SB	2SA	2P-A	2P0A	2P+A	3D0A)	-.03292904	-1.15825801E-04
DET 16=A	(1SA	1SB	2SA	2SA	2P-A	2P0A	3P+B)	.01279770	-6.93302277E-08
DET 17=A	(1SA	1SB	2SA	2SA	2P0A	2P+B	3P0A)	-.00666122	7.13815594E-08
DET 18=A	(1SA	1SB	2SB	2SA	2P0A	2P-B	3P-A)	-.00666122	-6.93302277E-03
DET 19=A	(1SA	1SB	2SB	2P-B	2P0A	2P+A	3P0A)	-.00052475	1.54501265E-08
DET 20=A	(1SA	1SB	2SA	2P-A	2P0B	2P+A	3D-1A)	.05703477	-1.15825801E-04
DET 21=A	(1SA	1SB	2P0A	2SA	2P0B	2P+A	3P0A)	-.00666122	-6.93302277E-03
DET 22=A	(1SA	1SB	2SA	2P-A	2P0B	2P+A	3SA)	-.00169640	-1.15825801E-04
DET 23=A	(1SA	1SB	2SA	2SA	2P0B	2P+A	3D0A)	-.06585808	1.54501265E-08
DET 24=A	(1SA	1SB	2SA	2SA	2P0B	2P+A	3P-A)	-.00666122	-9.24403036E-03
DET 25=A	(1SA	1SB	2SA	2P0A	2P0B	2P+A	3P-A)	-.00052475	-1.15825801E-04
DET 26=A	(1SA	1SB	2SB	2SA	2P0A	2P+A	3SA)	-.00169640	7.13815594E-08
DET 27=A	(1SA	1SB	2P-B	2P0A	2P0A	2P+A	3D0A)	-.03292904	-1.54501265E-08
DET 28=A	(1SA	1SB	2P-B	2P0A	2P0A	2P+A	3D0A)		-2.31100759E-03

EHF(A.U.) =	-5.31517746E+01	
ETOTAL(A.U.) =	-5.32231398E+01	
DIFFERENCE(A.U.) =	-7.13651694E-02	

RENORMALIZATION FACTOR =   .98374387

EXPONENTS		CONTRIBUTIONS
3S =	2.30000	1.8539470E-07
3P =	3.33800	-2.0351272E-3
3D =	2.28300	-6.9330228E-02
4F =	2.30000	0.

Figure 1 – Example of a NCMET charge wave function as given directly by our fully automated program system "ATOM". The dets that come out are all the theoretically predicted ones (NCMET; cf. references in text). They are found algebraically by the program itself. (All energies are in a.u.)

NITROGEN I   1S2 2S2 2P3   QUARTET S   09/25/73

DET	HF=A (	1SA	1SB	2SA	2SB	2P-A	2P0A	2P+A)	COEFFICIENT	CONTRIBUTION (A.U.)
	HF=A (	1SA	1SB	2SA	2SB	2P-A	2P0A	2P+A)	1.00000000	
DET 2=A	1SA	1SB	2SA	2P-B	2P0A	2P+A	3P+A		.00313510	-3.35029928E-05
DET 3=A	1SA	1SB	2SA	2P0B	2P0A	2P+A	3P0A		.00313510	-3.35029907E-05
DET 4=A	1SA	1SB	2SA	2P0A	2P+B	2P+A	3P-A		.00313510	-3.35029883E-05
DET 5=A	1SA	1SB	2SA	2SA	2P0A	2P+A	3SA		.01992055	-5.17337789E-04
DET 6=A	1SA	1SB	2SA	2SA	2P0A	2P+A	3D0A		.02665997	-1.52329217E-03
DET 7=A	1SA	1SB	2SA	2P0B	2P0A	2P+A	3D-1A		.04617643	-4.56987650E-03
DET 8=A	1SA	1SB	2SA	2SA	2P0A	2P+R	3D-2A		.06530333	-9.13975302E-03
DET 9=A	1SA	1SB	2SA	2P-A	2P+A	2P+A	3D+1A		.04617643	-4.56987651E-03
DET 10=A	1SA	1SB	2SA	2SA	2P0B	2P+A	3SA		.01992055	-5.17337738E-04
DET 11=A	1SA	1SB	2SA	2SA	2P0B	2P+A	3D0A		.05331994	-6.09316869E-03
DET 12=A	1SA	1SB	2SA	2P-A	2P+A	2P+A	3D-1A		.04617643	-4.56987551E-03
DET 13=A	1SA	1SB	2SA	2SA	2P-A	2P0A	3D-2A		.06530333	-9.13975303E-03
DET 14=A	1SA	1SB	2SA	2P-A	2P0A	2P+B	3D+1A		.04617643	-4.56987551E-03
DET 15=A	1SA	1SB	2SA	2SA	2P0A	2P+B	3SA		.01992055	-5.17337718E-04
DET 16=A	1SA	1SB	2SA	2SB	2P0A	2P+R	3D0A		.02665997	-1.52329217E-03
DET 17=A	1SA	1SB	2SA	2SB	2P0A	2P+A	3SA		.02594183	-1.00911824E-03
DET 18=A	1SA	1SB	2SA	2SA	2P0A	2P+A	3D0A		.00000000	-2.95655393E-27
DET 19=A	1SA	1SB	2SA	2SA	2P0A	2P+A	3SB		.03381982	-1.31934076E-03
DET 20=A	1SA	1SB	2SA	2SA	2P0A	2P+A	3D0B		.00000000	-1.89369549E-27
DET 21=A	1SA	1SB	2SA	2SB	2P0A	2P-A	3P-A		-.00067196	3.47732700E-08
DET 22=A	1SA	1SB	2SA	2SB	2P-A	2P+A	3P0A		-.00067186	3.47732678E-08
DET 23=A	1SA	1SB	2SA	2SB	2P-A	2P0A	3P+A		-.00067186	3.47732668E-08

EHF(A.U.) =   -5.44000968E+01
ETOTAL(A.U.) =   -5.44505864E+01
DIFFERENCE(A.U.) = -4.96796418E-02

EXPONENTS		CONTRIBUTIONS
3S = 1.45900		-3.88047227E-03
3P = 2.05000		-1.00404655E-04
3D = 2.15600		-4.56987657E-02

RENORMALIZATION FACTOR = .98803776

Figure 2 – Example of a NCMET charge wave function as given directly by our fully automated program system "ATOM". The dets that come out are all the theoretically predicted ones (NCMET; cf. references in text). They are found algebraically by the program itself. (All energies are in a.u.)

OXYGEN III   1S2 2S2 2P2   SINGLET D   09/27/73

										COEFFICIENT	CONTRIBUTIONS (A.U.)
DET HF=A	(1SA	1SB	2SA	2SB	2P+A	2P+B	2P+A	2P+B	)	1.00000000	
DET 2=A	(1SA	1SB	2P-A	2SB	2P+A	2P+B	3P+B		)	.00774284	-4.83692644E-04
DET 3=A	(1SA	1SB	2P-B	2SB	2P+A	2P+B	3P+A		)	-.00774284	-4.83692644E-04
DET 4=A	(1SA	1SB	2P0A	2SB	2P+B	2P+B	3P0B		)	-.00744751	-4.65243194E-04
DET 5=A	(1SA	1SB	2P0B	2SB	2P+A	2P+B	3P0A		)	-.00744751	-4.65243194E-04
DET 6=A	(1SA	1SB	2P-A	2SB	2P-B	2P+B	3D+2A		)	.02575686	-2.74755473E-03
DET 7=A	(1SA	1SB	2P-B	2SB	2P-A	2P+B	3D+1A		)	-.01454908	-1.09742281E-03
DET 8=A	(1SA	1SB	2P-A	2SB	2P-A	2P+A	3D+2B		)	-.04987944	-9.74034976E-03
DET 9=A	(1SA	1SB	2P-B	2SB	2P-A	2P+A	3D+1B		)	.02412258	-2.13738395E-03
DET 10=A	(1SA	1SB	2P0A	2SB	2P+A	2P+A	3D+2A		)	-.03527009	-4.87017488E-03
DET 11=A	(1SA	1SB	2P0B	2SB	2P+B	2P+A	3D+1A		)	-.02072101	-1.29823920E-03
DET 12=A	(1SA	1SB	2P0A	2SB	2P-B	2P+A	3D+2B		)	-.02412258	-2.13738395E-03
DET 13=A	(1SA	1SB	2P-B	2SA	2P-B	2P+A	3D+1B		)	-.04987944	-4.87017488E-03
DET 14=A	(1SA	1SB	2SA	2P-B	2P+B	3D+2A			)	-.02072101	-1.29823920E-03
DET 15=A	(1SA	1SB	2SA	2P-B	2P+B	3D+1A			)	.03527009	-4.87017488E-03
DET 16=A	(1SA	1SB	2SA	2P-A	2P+A	4F+3A			)	.02575686	-2.74755473E-03
DET 17=A	(1SA	1SB	2SA	2P-A	2P+A	4F+2A			)	-.01454908	-1.09742281E-03
DET 18=A	(1SA	1SB	2SB	2P-A	2P+B	3D+2B			)	.01925950	-1.76688709E-03
DET 19=A	(1SA	1SB	2SB	2P-A	2P+A	4F+3B			)	-.01925950	-1.76688709E-03
DET 20=A	(1SA	1SB	2SB	2P0A	2P+A	4F+2B			)	-.01111948	-5.88962365E-04
DET 21=A	(1SA	1SB	2SA	2P0B	2P0B	4F+2A			)	.01111948	-5.88962365E-04
DET 22=A	(1SA	1SB	2SB	2P+A	2P+B	3SA			)	.00271752	-5.31392923E-06
DET 23=A	(1SA	1SB	2SB	2P+A	2P+B	3D0A			)	-.00356336	2.62836187E-05
DET 24=A	(1SA	1SB	2SA	2P+A	2P+A	3SB			)	-.00271752	-5.31392923E-06
DET 25=A	(1SA	1SB	2SA	2P+A	2P+B	3D0B			)	.00356336	2.62836187E-05
DET 26=A	(1SA	1SB	2SB	2P+B	2P+B	3P+A			)	-.00112809	-1.95483429E-06
DET 27=A	(1SA	1SB	2SB	2P+A	2P+A	4F+A			)	-.00497278	-1.17792473E-04
DET 28=A	(1SA	1SB	2SA	2P+A	2P+B	3P+B			)	.00112809	-1.95483429E-06
DET 29=A	(1SA	1SB	2SA	2P0A	2P+A	4F+B			)	.00497278	-1.17792473E-04
DET 30=A	(1SA	1SB	2P0A	2SB	2P+A	3P+B			)	-.00029533	0.
DET 31=A	(1SA	1SB	2P0A	2SB	2P+B	3P+A			)	-.00029533	0.
DET 32=A	(1SA	1SB	2SB	2P0A	2P0B	3D+2A			)	-.00518135	0.
DET 33=A	(1SA	1SB	2SA	2P0A	2P0A	3D+2B			)	.00518135	0.
DET 34=A	(1SA	1SB	2P0A	2P0A	2P+A	2P+B			)	-.13500676	-2.45846676E-02

EHF(A.U.) =   -7.29972034E+01
ETOTAL(A.U.) =   -7.30723774E+01
DIFFERENCE(A.U.) =   -7.51740441E-02

RENORMALIZATION FACTOR =   .98493280

EXPONENTS		CONTRIBUTIONS
3S =	3.02800	-1.06278548E-05
3P =	3.91710	-1.90178135-03
3D =	3.02800	-4.37296838-02
4F =	3.60770	-4.94728390-03

Figure 3 – Example of a NCMET charge wave function as given directly by our fully automated program system "ATOM". The dets that come out are all the theoretically predicted ones (NCMET; cf. references in text). They are found algebraically by the program itself. (All energies are in a.u.)

OXYGEN III   1S2 2S 2P3   SINGLET P   09/27/73

	COEFFICIENT	CONTRIBUTIONS (A.U.)
DET HF=A (1SA 1SB 2SA 2P-B 2P+A 2P+B)	.50000000	
DET HF=A (1SA 1SB 2SB 2P-A 2P+A 2P+B)	-.50000000	
DET HF=A (1SA 1SB 2SB 2P0A 2P0B 2P+A)	-.50000000	
DET HF=A (1SA 1SB 2SA 2P0A 2P0B 2P+B)	.50000000	
DET 2=A (1SA 1SB 2SB 2P+A 2P+B 3P-A)	-.01511138	-2.07038977E-04
DET 3=A (1SA 1SB 2P-A 2P+A 2P+B 3SB)	.00849473	-2.61991058E-06
DET 4=A (1SA 1SB 2P-A 2P+A 2P+B 3D0B)	.01527261	-7.76463033E-06
DET 5=A (1SA 1SB 2P0A 2P+A 2P+B 3D-1B)	.00126696	-7.75939737E-06
DET 6=A (1SA 1SB 2P0B 2P+A 2P+B 3D-1A)	-.00126696	-7.75939736E-06
DET 7=A (1SA 1SB 2P-A 2P-B 2P+B 3D+2A)	-.04099361	-5.81515272E-03
DET 8=A (1SA 1SB 2P0A 2P-B 2P+B 3D+1A)	-.00126696	-7.75939740E-06
DET 9=A (1SA 1SB 2SB 2P-B 2P+A 3P+A)	.00768364	-6.99996493E-06
DET 10=A (1SA 1SB 2P-A 2P-B 2P+A 3D+2B)	-.04099361	-5.81515271E-03
DET 11=A (1SA 1SB 2P-B 2P0A 2P+A 3D+1B)	.02898686	-2.90757636E-03
DET 12=A (1SA 1SB 2P-B 2P0B 2P+A 3D+1A)	-.02771990	-2.61072333E-03
DET 13=A (1SA 1SB 2SB 2P+D 3SA)	.10345547	-3.59500290E-03
DET 14=A (1SA 1SB 2SA 2SB 2P+D 3D0A)	-.01814122	-6.79015185E-04
DET 15=A (1SA 1SB 2SA 2P-A 2P+D 3P+B)	.00268364	-6.90996423E-06
DET 16=A (1SA 1SB 2SA 2P-A 2P+D 4P+B)	-.00317159	-3.52566585E-05
DET 17=A (1SA 1SB 2SA 2P0A 2P+D 3P0B)	-.00252905	-1.07252799E-05
DET 18=A (1SA 1SB 2SA 2P0A 2P+D 4P0B)	.00188439	-5.28849878E-05
DET 19=A (1SA 1SB 2SA 2P0B 2P+D 3P0A)	-.01005328	-5.24700586E-05
DET 20=A (1SA 1SB 2SA 2SB 2P+D 4P0A)	.00000000	1.10667289E-27
DET 21=A (1SA 1SB 2SA 2SB 2P+A 3SB)	-.10346547	-3.59500277E-03
DET 22=A (1SA 1SB 2SA 2SB 2P+A 3D0B)	.01814122	-6.79015187E-04
DET 23=A (1SA 1SB 2SA 2P0B 2P+A 3P0B)	.01258233	-1.19029228E-04
DET 24=A (1SA 1SB 2SA 2P0B 2P+A 4P0B)	-.00388439	-5.28849878E-05
DET 25=A (1SA 1SB 2SA 2SB 2P-D 3D+2A)	-.04443672	-4.07409112E-03
DET 26=A (1SA 1SB 2SA 2P-A 2P-B 4P+3B)	-.01224353	-5.28849878E-04
DET 27=A (1SA 1SB 2SA 2P-B 2P0A 4P+2B)	-.00709190	-1.76283293E-04
DET 28=A (1SA 1SB 2SA 2P-B 2P0B 4P+2A)	.00000000	-5.18418100E-28
DET 29=A (1SA 1SB 2P-B 2P+A 2P+B 3SA)	-.00849473	-2.61991087E-06
DET 30=A (1SA 1SB 2P-B 2P+A 2P+B 3D0A)	-.01527261	-7.76463033E-04
DET 31=A (1SA 1SB 2SA 2P-D 2P0B 3P-B)	.01511138	-2.07038959E-04
DET 28=A (1SA 1SB 2SA 2P-D 2P0B 4P+2A)	.00000000	-5.18918180E-28
DET 29=A (1SA 1SB 2P-B 2P+A 2P+B 3SA)	-.00849473	-2.61991087E-06
DET 30=A (1SA 1SB 2P-B 2P+A 2P+B 3D0A)	-.01527261	-7.76463033E-04
DET 31=A (1SA 1SB 2SA 2P+A 2P+B 3P-B)	.01511138	-2.07038959E-04
DET 32=A (1SA 1SB 2SA 2P-B 2P+B 4F-B)	-.00317159	-3.52566585E-05
DET 33=A (1SA 1SB 2SA 2P-B 2P+B 3P+A)	.01005328	-5.24700579E-05
DET 34=A (1SA 1SB 2SA 2P-B 2P+B 4F+A)	.00000000	3.98814129E-28
DET 35=A (1SA 1SB 2SA 2P-B 2P+A 3P+B)	.01273692	-9.27312277E-05
DET 36=A (1SA 1SB 2SA 2P-B 2P+A 4F+B)	-.00317159	-3.52566585E-05
DET 37=A (1SA 1SB 2P-A 2P0B 2P+B 3D+1B)	-.02771990	-2.61072333E-03
DET 38=A (1SA 1SB 2P-A 2P0B 2P+A 3D+1A)	.02898686	-2.90757636E-03
DET 39=A (1SA 1SB 2P-A 2P0B 2P+A 3D+1A)	-.00126696	-7.75939742E-06
DET 40=A (1SA 1SB 2SB 2P0A 2P+A 3P0A)	.01258233	-1.19029228E-04
DET 41=A (1SA 1SB 2SB 2P0A 2P+B 4F0A)	-.00388439	-5.28849878E-05
DET 42=A (1SA 1SB 2SB 2P-B 2P+A 4P+A)	-.00317159	-3.52566585E-05
DET 43=A (1SA 1SB 2SB 2P0A 2P+A 3P0B)	-.01005328	-5.24700568E-05
DET 44=A (1SA 1SB 2SB 2P0A 2P+A 4F0B)	.00000000	-2.30818787E-27
DET 45=A (1SA 1SB 2SB 2P0B 2P+A 3P0A)	-.00252905	-1.07252799E-05
DET 46=A (1SA 1SB 2SB 2P0B 2P+A 4P0A)	.00188439	-5.28849877E-05
DET 47=A (1SA 1SB 2SA 2SB 2P-A 3D+2B)	.04443672	-4.07409112E-03
DET 48=A (1SA 1SB 2SB 2P-A 2P-B 4P+3A)	.01224353	-5.28849878E-04
DET 49=A (1SA 1SB 2SB 2P-A 2P0A 4P+2A)	.00000000	9.92140769E-28
DET 50=A (1SA 1SB 2SB 2P-A 2P0B 4P+2A)	-.00709190	-1.76283292E-04
DET 51=A (1SA 1SB 2SB 2P-A 2P+A 4F-A)	-.00317159	-3.52566585E-05
DET 52=A (1SA 1SB 2SB 2P-A 2P+B 3P+A)	-.01273692	-9.92721292E-05
DET 53=A (1SA 1SB 2SB 2P-A 2P+B 3P+A)	.00317159	-3.52566585E-05
DET 54=A (1SA 1SB 2SB 2P-A 2P+A 3P+B)	.01005328	-5.24700570E-05
DET 55=A (1SA 1SB 2SB 2P-A 2P+A 4F+B)	.00000000	2.52446149E-28
DET 56=A (1SA 1SB 2SA 2P0A 2P0B 3P+B)	-.00015459	2.57545424E-07
DET 57=A (1SA 1SB 2P-A 2P0A 2P0B 3D+2B)	-.03920186	-5.22144666E-03
DET 58=A (1SA 1SB 2P-B 2P0A 2P0B 3D+2A)	-.03920186	-5.22144666E-03
DET 59=A (1SA 1SB 2P0A 2P0B 2P+B 3SA)	-.00849473	-2.61491041E-06
DET 60=A (1SA 1SB 2P0A 2P0B 2P+B 3D0A)	-.01746705	-1.07331606E-03
DET 61=A (1SA 1SB 2SA 2SB 2P0B 3D+1A)	.03142151	-2.03704556E-03
DET 62=A (1SA 1SB 2SA 2SB 2P0A 3D+1B)	.03142151	-2.03704556E-03
DET 62=A (1SA 1SB 2SA 2SB 2P0A 3D+1B)	.03142151	-2.03704568E-03
DET 63=A (1SA 1SB 2SB 2P-B 2P0A 4P+2A)	.00709190	-1.76283293E-04
DET 64=A (1SA 1SB 2P0A 2P0B 2P+A 3SB)	.00849473	-2.61491079E-06
DET 65=A (1SA 1SB 2P0A 2P0B 2P+A 3D0B)	.01746705	-1.07331606E-03
DET 66=A (1SA 1SB 2SB 2P0A 2P0B 3P+A)	.00015459	2.57545424E-07
DET 67=A (1SA 1SB 2SB 2P0A 2P0B 4P+A)	.00634319	-1.41026634E-04
DET 68=A (1SA 1SB 2SA 2P-A 2P0B 4P+2B)	.00709190	-1.76283293E-04
DET 69=A (1SA 1SB 2SA 2P0A 2P0B 4P+B)	-.00634319	-1.41026634E-04

	EXPONENTS	CONTRIBUTIONS
EHF(A.U.) = -7.21050463E+01	1S = 1.60000	-7.2004853E-03
ETOTAL(A.U.) = -7.21662308E+01	3P = 1.44500	-1.0953163E-03
DIFFERENCE(A.U.) = -6.11844652E-02	3D = 2.79000	-5.0420698E-02
	4P = 1.41400	-2.4679661E-03

RENORMALIZATION FACTOR = .97999226

Figure 4 - Example of a NCMET charge wave function as given directly by our fully automated program system "ATOM". The dets that come out are all the theoretically predicted ones (NCMET; cf. references in text). They are found algebraically by the program itself. (All energies are in a.u.)

doubly occupied SCF MO's). All $\overline{V}$-orbital effects with their
large charge distribution affecting role, are contained in ψ_c.

Proof: Take the closed shell limit of ψ , equations (6) and
(7), i.e. fill up the RHF-sea till N=M in equation (3). Then as
there is no room left in $\{k\}_{\overline{V}}$ for virtual transitions, in the
closed shell limit:

$$\chi_{INT} \rightarrow 0$$
$$\chi_F \rightarrow 0$$
$$\psi_c \rightarrow \phi_{RHF} \text{ (single det; } ^1\Sigma_g^+, \, ^1A_g, \text{ etc.)}$$
$$\chi_U \rightarrow \chi_U$$
$$\psi \rightarrow \phi_{RHF} + \chi_U = \psi_{closed \, shell} \qquad (8)$$

Remark 1: Hence χ_U are "closed shell-like" correlations.
The theory of $\psi_{closed \, shell}$, i.e. of χ_U was worked out in detail
earlier by Sinanoğlu ("MET" $\equiv$ Many-Electron Theory of Atoms and
Molecules [11]) and has been applied by many workers to atoms and
molecules since then (E. Davidson, R. Bender, R.K. Nesbet,
Kutzellnigg, F. Harris and others[12]).

Remark 2: The non-closed shell Hartree-Fock method's
$\phi_{RHF} = \sum_K c_K \Delta_K$ containing in general several dets and the main
orbital "reorganization" effects is part of the new ψ_c, but it is
not the full generalization of the orbital theory's ϕ_{RHF} of closed
shells. Up to now MO theory has dealt with ϕ_{RHF} on excited states,
but has omitted the important effects which are in ψ_c (for experimental consequences in atomic physics, see [2]).

C. Energy of a Non-Closed Shell State

From equation (7), the exact (non-relativistic) energy of the state is:

$$E = E_c + E_U \qquad (9)$$

with $E_c = E_{RHF} + E_{INT} + E_F$ coming out of the finite ψ_c calculations as in figures 1-4. We have shown in MET/NCMET that [2,11]

$$E_U = \sum_{i>j}^{M} \rho_{ij} \, \varepsilon_{ij}^{U} + \sum_{(i>j) \neq (k>\ell)}^{M} \rho_{ij;k\ell} \, \varepsilon_{ij;k\ell}^{U} \qquad (10)$$

ε_{ij}^{U} = all-external <u>pair correlation energy</u> [11] of spin-orbital pair i,j $\in \{k\}_{\overline{V}}$, and

$\varepsilon_{ij}^{U} = \sqrt{2} \, \langle ij | 1/r_{12} | \hat{u}_{ij} \rangle$

$\varepsilon_{ij;k\ell}^{U}$ = all-external "<u>cross pair</u>" energy of NCMET [2]

i,j,k,$\ell \in \{k\}_{\overline{V}}$ $\varepsilon_{ij;k\ell}^{U} = \sqrt{2} \, \langle ij | 1/r_{12} | \hat{u}_{k\ell} \rangle$

(i,j)$\neq$(k,ℓ)

ρ_{ij} resp. $\rho_{ij;k\ell}$ = pair resp. cross pair occupation probabilities [2] in the RHF wave function ϕ_{RHF} of the state. These are group theoretic numbers determined by the coefficients c_K in $\phi_{RHF} = \sum_K c_K \Delta_K$ of RHF dets.

Equation (11) is in terms of the "reducible pairs" and "cross pairs" of MET/NCMET [2], i.e. in terms of spinorbital products ij.

For closed shells, equation (10) becomes:

$$\rho_{ij} \rightarrow 1 \quad ; \quad \sum_{(i,j) \neq (k,\ell)}^{M=N} \rho_{ij;k\ell} \, \varepsilon_{ij;k\ell} \rightarrow 0$$

(i.e. cross pairs cancel out)   and

$$E_U \rightarrow \sum_{i>j}^{M=N} \varepsilon_{ij}^{U} = \sum_{i>j}^{N} \varepsilon_{ij} \qquad (11)$$

yielding the now familiar MET result of Sinanoğlu [11].

In the non-closed shell NCMET general case, we can also reduce the i $\otimes$ j products under the (point group or 0(3)) $\otimes$

$SU(2)_\sigma$ of the system, to get the "irreducible pairs" $\varepsilon_I^{irr,U}$ introduced in NCMET [2].

$$E_U = \sum_I{}' \rho_I^{irr} \; \varepsilon_I^{irr,U} + \text{(one or very few cross} \atop \text{irr. pairs [2])} \qquad (12)$$

Now, for a collection of $\overline{V}$-states of the CHEMSET, there is one set of eleven pair correlation values in atoms. These being nearly transferable from state to state in the $\overline{V}$, they are evaluated once [13] and tabulated in [2]. The ρ_{ij}, etc. come from group theory on MO or AO SCF, also easily figured out and tabulated.

Thus, for any $\overline{V}$-state use table of ε_I^{irr}'s and ρ^{irr}'s [2] to get its E_U. We'll need the E_U added to E_c to get inner I.P.'s, E.A.'s, $\Delta E_{excitation}$, singlet-triplet splittings, etc.

In molecules, E_U is obtained semi-empirically from atomic ε_{ij}'s by the MET/EPCE method [14].

III.   RELATION TO THE INTRAVALENCY AND RYDBERG STATES OF SPECTRA

The effects in the exact wave function (INT, F, and U beyond RHF) were derived mathematically [2] from the Schrödinger equation for properties of stationary states. It turns out now however, we can recast them in the spectroscopic language and this will tell us immediately about low lying transitions and their mixings.

Consider again the low lying states of the CHEMSET atoms as we are still on the many-electron aspects. We shall go into additional, strictly molecular, many-center aspects later.

The low lying KL-CHEMSET E1-transitions are of the types:

i) $\overline{V} \rightarrow \overline{V}'$: $1s^2 2s^2 2p^m \rightarrow 1s^2 2s 2p^{m+1}$

$(\Delta S=0,\ \Delta L=0,\ \pm 1,\ J=0 \nrightarrow J=0)$ (13a)

or

ii) $\overline{V} \rightarrow F$: $1s^2 2s^2 2p^m \rightarrow 1s^2 2s^2 2p^{m-1}$ ns

$1s^2 2s^2 2p^m \rightarrow 1s^2 2s^2 2p^{m-1}$ nd

$1s^2 2s^2 2p^m \rightarrow 1s^2 2s 2p^m$ np

$(\Delta S=0,\ \Delta L=0,\ \pm 1,\ J=0 \nrightarrow J=0)$ (13b)

For positive ions (13a) comes first, in neutrals (13b) comes before (13a) in general. In figure 5 the ΔE's of the first three singlet transitions from the ground are plotted vs. atomic number Z. Note the crossing of the curves towards the neutrals.

Now in ψ_c of a given state as given by NCMET, we observe and state as a theorem that:

Theorem 2: a) The χ_{INT} part of the charge wave function ψ_c of any intravalency state $\in \{\overline{V}\}$ contains all the intravalency states $\in \{\overline{V}\}$ of the same symmetry.

b) The χ_F part of the ψ_c of a given intravalency state $\in \{\overline{V}\}$ of certain symmetry contains all the Rydberg states and series of the same symmetry.

Proof: All parts of ψ in equation (6) are orthogonal to each other, $\langle \phi_{RHF} | \chi_{INT} \rangle = 0$; $\langle \chi_{INT} | \chi_F \rangle = 0$, etc. Each part transforms the same under the symmetry group G of the Hamiltonian, $[G,H] = 0$; i.e. each part $(\phi_{RHF},\ \chi_{INT},\ \chi_F,\ \chi_U)$ have the same symmetry, the same overall state quantum numbers (e.g. L, S, M_L, M_S). These are basis properties of NCMET [2].

Then: a) $\chi_{INT} \in \{\overline{V}\}$ (a vector in the $\overline{V}$-space). So are all $\overline{V}$-states of same symmetry. χ_{INT} can be reexpressed in terms of the $\overline{V}$-state terms and configurations made of the $\overline{V}$-RHF AO set.

b) Similarly the dets in χ_F, with $\hat{f}_{ij;k}(\vec{x}_1)$ and $\hat{f}_p^{(i)}(\vec{x}_1)$ expanded in $R_{n\ell}(r)\ Y_{\ell m}(\vec{\Omega})$ (with $n \geq 3$) are of the same types as those of equation (13b) but of same symmetry.   Q.E.D.

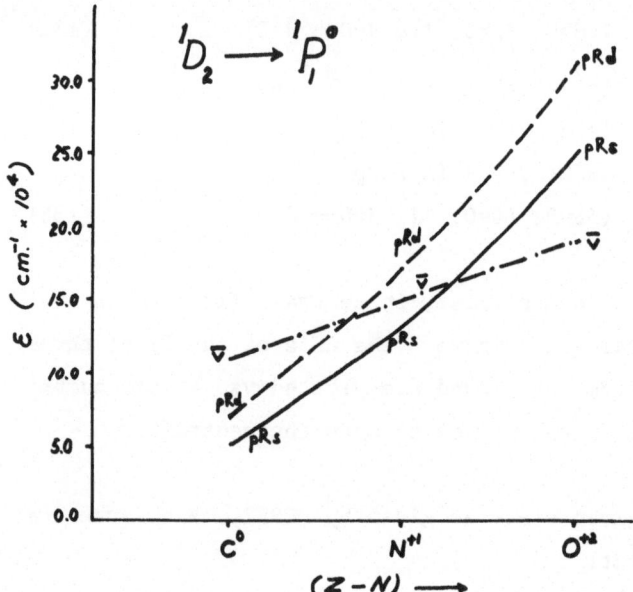

Figure 5 - First three singlet transitions from the ground
singlet ^1D, of type $^1D_2 \rightarrow {}^1P_1^{\circ}$ vs. atomic net charge in the
Carbon I - isosequence.

$$\varepsilon_{\overline{V}} \equiv E(^1P_1^{\circ}; \ 2s2p^3) - E(^1D_2; \ 2s^2 2p^2)$$
$$\varepsilon_{pRs} \equiv E(^1P_1^{\circ}; \ 2s^2 2p3s) - E(^1D_2; \ 2s^2 2p^2)$$
$$\varepsilon_{pRs} \equiv E(^1P_1^{\circ}; \ 2s^2 2p3d) - E(^1D_2; \ 2s^2 2p^2)$$

In the neutral, the preRydbergs (pRs and pRd) are the first
electric dipole excitations of the symmetry. The intravalency
transition ($\overline{V} \rightarrow \overline{V}'$) comes next. However, already at net charge
about +0.75, the $\Delta\overline{V}$ and ΔpRs cross over. In O^{+2}, intravalency
transition is first. Atoms in neutral molecules are in the
cross-over region (charge withdrawal into overlap region).
Strong intravalency $\leftrightarrow$ preRydberg mixings and crossovers
(depending on calculational accuracies too) are therefore
expected.

Corollary 1: By looking at the ψ_c of an intravalency state $\overline{V}$, we can see which intravalency states (in the χ_{INT} part) and which Rydberg states (in the χ_F) of same symmetry would mix strongly with each other due to Coulomb repulsions [15]. If ψ_c of $\overline{V}$ shows a strong virtual Rydberg component R, then the actual spectroscopic state R of the same symmetry shows a strong virtual component $\overline{V}$. This follows from the H-sum rule.

Corollary 2: Although the semi-internal correlations $(ij \rightarrow k\hat{f}_{ij;k})$ resp. orbital polarizations $(i \rightarrow \hat{f}_p^{(i)})$ in the χ_F represent mixings of $\overline{V}$ with entire "excited core Rydberg" resp. usual Rydberg series, NCMET ψ_c calculations such as those in figures 1-4, on several hundred CHEMSET atomic $\overline{V}$-states, have shown the significant mixings to be with the KL'3s, KL'3p, KL'3d, and KL'4f Rydberg states only. Thus mixing with higher Rydberg (n > 3 for s, p, d; n > 4 for f, and any nℓ with ℓ > 3) states of KL-shells $\overline{V}$ states (and vice versa) can be neglected.

In tables II and III we show for two different types of intra-valency states, the $\ddot{V}$ and pR (low Rydberg) configurations that occur in their respective NCMET charge wavefunctions.

The above theorem and its corollaries lead us to a practical reclassification of low lying spectral transitions looking ahead at the molecular problem as well.

IV. CLASSIFICATION OF TRANSITIONS INTO INTRAVALENCY, PRE-RYDBERG, AND RYDBERG TYPES

Finding from NCMET that large Coulomb mixings should occur only among intravalency states $\{\overline{V}\}$ and 3s, 3p, 3d, 4f Rydbergs, we reclassify the spectral states into:

TABLE II

The intravalency ($\overline{V}$) and preRydberg (pR) configurations that occur
in the NCMET charge wave function ψ_c of Carbon ($1s^2 2s2p^3$; ^3S).
The coefficients of these and their energy contributions to the
ψ_c and E_c are easily obtained from the ψ_c direct output from our
computer program system "ATOM", which gives the ψ_c as a linear
combination of all the corresponding determinants (as in
Figures 1-4).

Type	Configuration and Term	No. of terms with same LS but different parentage in subconfigurations
ϕ_{RHF}	$1s^2 2s2p^3$; ^3S	1
χ_F^s	$1s^2 2s2p^2 3p$; ^3S	2
(semi-int. +	$1s^2 2s2p^2 4f$; ^3S	0
orbi. polariz.)	$1s^2 2p^3 3s$; ^3S	1
	$1s^2 2p^3 3d$; ^3S	1
	$1s^2 2p3s$; ^3S	0
	$1s^2 2p3d$; ^3S	0

(Note: due to symmetry there is no "internal correlation" for
        $2s2p^3$; ^3S)

TABLE III

The intravalency ($\overline{V}$) and preRydberg (pR) configurations that occur in the NCMET charge wave function ψ_c of Nitrogen ($1s^2 2s^2 2p^3$; ^2P). The coefficients of these and their energy contributions to the ψ_c and E_c are easily obtained from the ψ_c direct output from our computer program system "ATOM", which gives the ψ_c as a linear combination of all the corresponding determinants (as in Figures 1-4).

Type	Configuration and Term	No. of terms with same LS but different parentage in subconfigurations
ϕ_{RHF}	$1s^2 2s^2 2p^3$; ^2P	1
χ_{INT}	$1s^2 2p^5$; ^2P	1
χ_F^s	$1s^2 2s^2 2p^2 3p$; ^2P	3
(semi-int. +	$1s^2 2s^2 2p^2 4f$; ^2P	1
orbi. polariz.)	$1s^2 2s 2p^3 3s$; ^2P	2
	$1s^2 2s 2p^3 3d$; ^2P	4
	$1s^2 2p^4 3p$; ^2P	3
	$(1s^2 2p^4 4f$; $^2P)*$	$(1)*$

* No direct interaction with ϕ_{RHF}, only indirect.

$$\boxed{\begin{aligned}
\{\overline{V}\} &\equiv \text{ intravalency states} \\
\{pR\} &\equiv \text{ PreRydberg states} \\
\{R\} &\equiv \text{ Rydberg states}
\end{aligned}}$$

(14)

For KL-shell atoms:

$\{\overline{V}\} \quad \supset \quad 1s^k 2s^m 2p^n$ - terms   (15)

$\{pR\} \quad \supset \quad 1s^p 2s^q 2p^r 3s,\ 3p,\ 3d,\ [4f]$ (16)

$\{R\} \quad \supset \quad 1s^r 2s^s 2p^t ns,\ np,\ nd,\ [n'f]$ (17)

$\qquad\qquad\qquad (n > 3); \quad (n' > 4)$

[the [nf] are not accessible by E1 strongly from $\{\overline{V}\}$].

Note that $\{pR\}$ contains excited core ([1s]-hole, [2s]-hole, etc.) Rydbergs as well as "outer electron" originated Rydbergs.

More generally given the definition of $\overline{V}$ states (e.g. KLM-shells in Aℓ, Si, S, P, Cℓ, Sc, Fe, etc.), the pR are found from the corresponding $\hat{f}_{ij;k}$ and $\hat{f}_P^{(i)}$ of the ψ_c's. (The allowed Coulomb mixings are approximately given by $\langle ij | 1/r_{12} | k\hat{f}\rangle$)[2].

As a corollary to this classification we can also express ψ_c and χ_U in the :

V.   SPECTRAL FORM OF THE CHARGE WAVE FUNCTION OF AN INTRAVALENCY STATE

$$\psi = \psi_c^s + \chi_U^s ; \quad \langle \psi_c^s | \chi_U^s \rangle = 0 \tag{18}$$

with

$$\psi_c^s \equiv \phi_{RHF} + \chi_{INT} + \chi_F^s \tag{19}$$

$$\phi_{RHF} + \chi_{INT} \text{ (or } \phi_{GRHF}; \text{ see footnote [5])} \equiv \psi_{\{\overline{V}\}} \tag{20}$$

$$\chi_F^s \equiv \chi_{\{pR\}} = \{\text{dets of 1s,2s,2p,3s,3d,4f only}\} \tag{21}$$

$$\chi_U^s \equiv \chi_U + \chi_F^{\{R\}}; \quad \chi_F^R = \{\text{dets with Rydbergs of } n \neq$$
$$(3s,3p,3d,4f)\} \tag{22}$$

Thus we have:

$$\psi = \psi_{\{\overline{V}\}} + \chi_{\{pR\}} + \chi_F^{\{R\}} + \chi_U \tag{23}$$

But since $\chi_F^{\{R\}} \approx 0$, the charge wave function ψ_c is

$$\psi_c \cong \psi_{\{\overline{V}\}} + \chi_{\{pR\}} \tag{24}$$

as figures 1-4 and many more similar cases show. We state the result:

<u>The charge wave function of an intravalency state contains only the Coulomb mixings of all intravalency states and of all pre-Rydberg states of the same symmetry.</u>

We have shown this clearly for the $\overline{V}$-state. We need to look at an **actual** pR-state itself further however to show that its ψ_c too contains mainly these mixing, not any new ones even though its RHF-sea is now larger (KLM instead of KL).

VI. THE CHARGE WAVE FUNCTION OF A PRE-RYDBERG STATE

Consider pre-Rydberg states of say the CHEMSET, i.e. of the type of equation (16). With states of KL-core and an electron say in $(3\ell) \in$ M-shell, the NCMET RHF-sea now would be that of KLM, i.e. $\overline{V}_{KLM}$;

$$\{k\}_{KLM} = \{ \ \{k\}_{KL} \ \bigcup \ \{3s\alpha, 3s\beta, 3p_{-1}\alpha, \dots 3d_{+2}\beta\} \ \} \tag{25}$$

and the ψ_c again is clearly defined. However as there is only one electron in the M-shell, while in general the KL-core is unfilled, the ψ_c will have different important effects in it as compared to say the ground terms of actual KLM atoms $\{KL \ 3s^n 3p^m 3d^k\}$ (Mg, Al, Si,...and iron series ions [16]) with their KL-cores full.

In the actual KLM atoms' intravalency states $\{\overline{V}_M\}$ with no K or L-holes,

$$\psi_c^{KLM} \approx \psi_c^M \tag{26}[16]$$

with the KL $\rightarrow$ M type internal effects belonging approximately in the

$$\psi_c^{KL \rightarrow M} \subset \chi_U^{KL\text{-core-like ion}} \tag{27}$$

so that

$$\psi^{KLM} \cong \psi_c^M + \chi_U^M + \chi_U^{KL\text{-core}} \tag{28}$$

The pre-Rydbergs $\{pR\}$ of the CHEMSET on the other hand, have L $\leftrightarrow$ M intershell ψ_c effects which are crucial, the L-shell being non-closed too [17].

The types of ψ_c effects for a pR-state (KL to M of CHEMSET) are displayed in figure 6. This exemplifies the ψ_c effects of _interstates_ which have two non-closed consecutive shells, inner hole states of photoelectron spectroscopy and ESCA being also of this type.

Figure 6 shows in the ψ_c (pR),

a) KL-Core Effects:

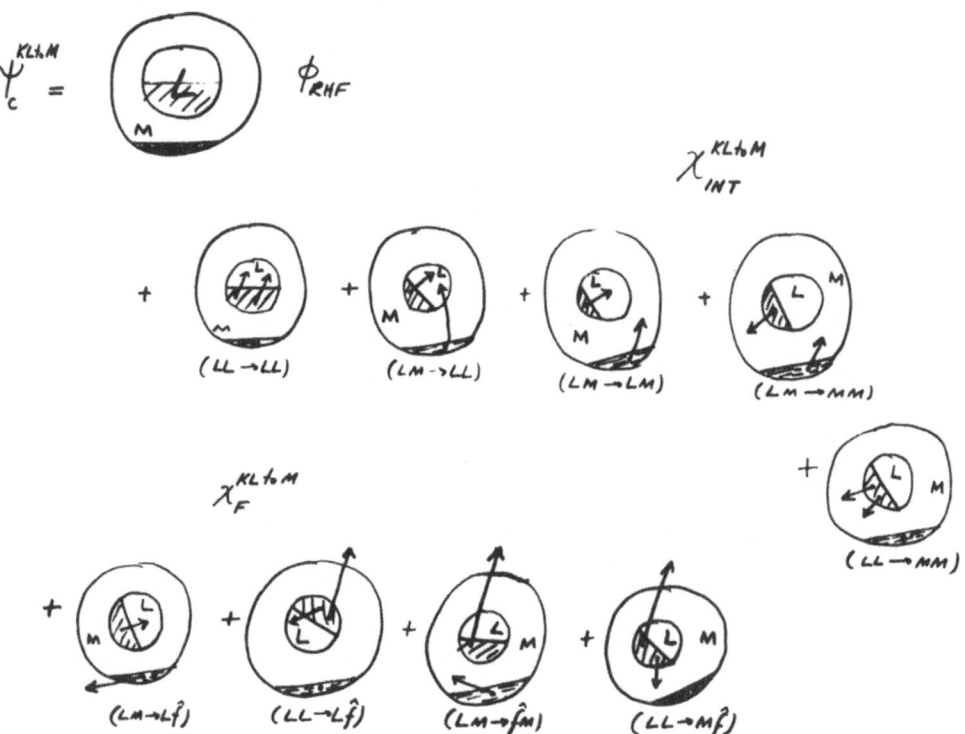

Figure 6 - The charge wave function of a KL to M   pR-state.
(e.g. $1s^2 2s2p^3 3d$) $\psi_c^{KL\ to\ M} = \phi_{RHF} + \chi_{INT}^{KL\ to\ M} + \chi_F^{KL\ to\ M}$

(LL → LL):   core-internal correlation (as in $1s^2 2s 2p^3$
              by itself).

(LL → MM):   core-all-external correlation for the KL-sea
              itself. Little affected by the pre–Rydberg
              electron (3d here).

(LL → M$\hat{f}$): core-all-external correlation for the KL-sea
              itself. Little affected by the pre–Rydberg
              electron (3d here).

(LL → L$\hat{f}$): A smaller effect in the ψ_c^s of core itself.

b) Pre-Rydberg – Intravalency Mixing
   (LM → LL) :   pR $\leftrightarrow \overline{V}_L$ (A crucial effect; converse of the
                 $\overline{V}_L \leftrightarrow$ pR in the ψ_c of $\overline{V}$-states).

c) Pre–Rydberg – Excited Core Pre-Rydberg Mixing
   (LM → LM):   pR $\leftrightarrow$ (L')(pR)'   (An important effect)

d) Core Polarization of the Closed Shell–like Core part
   (LM → MM) :   Can be treated as if an all–external effect
                 of the KL-core.

   (LM → $\hat{f}$M): Can be treated as if an all–external effect
                 of the KL-core.

e) Non–Closed Shell Core – Polarization
   (LM → L$\hat{f}$): Sizable effect, but similar to the Rydberg
                 states of the same series as this pR.
                                                                  (29)

We see that the main mixing effects, and specific to the
pR-state, are again those found in the $\overline{V}$-states earlier, i.e.
effects (b) and (c). Other effects are of the core or of core-
polarization [11a] types which are sizable in magnitude, but quite
similar in the corresponding Rydberg states (of n > 3) and expected
little to change the quantum defect δ of the series.

For spectroscopic purposes we define the significant part of the ψ_c of a pR, $\psi_c^S(pR)$, <u>to contain all the mixings which are found in the ψ_c^S of the intravalency states $\{\overline{V}\}$ (any one) of the same symmetry</u>.

$$\psi_c^S(pR) \equiv \phi_{RHF}(pR) + \chi_c^S \; [\{\overline{V};pR'\}; \text{ same dets as in}$$
$$\psi_c^S(\overline{V})]' \qquad (30)$$

then

$$\psi(pR) = \psi_c^S(pR) + \chi_U^S(pR); \; \psi_c^S(pR) \cong \psi_c(pR) \qquad (31)$$

<u>Remark 1</u>: All $\{\overline{V}\}$ and $\{pR\}$ states are now treated on the same footing. They all mix strongly. Thus the $\{pR\}$ deviate from the R – Rydberg series. The regular Rydberg states $\{R\}$ (i.e. KL $n\ell$ with $n\ell \neq 3s,3p,3d,4f$) are not expected to mix much and should follow the $1/(n-\delta_\ell)^2$ series.

<u>Remark 2</u>: To calculate the $\{\overline{V}\} \cup \{pR\}$ states, take the $\{k\}_{\overline{V}}$ $\cup \; \{\hat{f}\}_{\overline{V}}$ (e.g. for CHEMSET, $\{1s\alpha,1s\beta,2s\alpha,\ldots2p_z\beta,3s\alpha,\ldots4f_{+3}\beta\}$) as the orbital set. Calculate the $\psi_c^S(\overline{V})$ and $\psi_c^S(pR)$ of the same symmetry from the same set of dets that will arise (e.g. given by the auto-program system "ATOM" for one of the states).

<u>Remark 3</u>: Although in the above discussion we concentrated on the KL and M electrons, the odd member of the $\{f\}_{\overline{V}}$ -set 4f is also included on the same footing (but not 4s,4p,4d, physically because their role is better played and eliminated by 3s,3p,3d semi-internals and/or RHF-orbitals).

VII. ELECTRIC DIPOLE OSCILLATOR STRENGTHS OF INTRAVALENCY $(\overline{V})$ AND
      PRE-RYDBERG (pR) TRANSITIONS

The strong charge wave function NCMET mixings of $\overline{V}$ and pR-

states affect their energy splittings, charge-like properties,
expectation values (hyperfine constants, electric moments, form
factors) and their E1-transition probabilities A_{E1}.

In the CHEMSET atoms, the multiplet oscillator strengths
f_{E1} ($\gamma LS \rightarrow \gamma L'S$) of the intravalency transitions $\overline{V} \rightarrow \overline{V}'$
($1s^2 2s^2 2p^m \rightarrow 1s^2 2s 2p^{m+1}$) come out too low by factors like 2-3 with
the RHF method only, compared to recent experiments and to NCMET
$\psi_c(\overline{V}) \rightarrow \psi_c(\overline{V}')$ full calculations [4,2]. For third row atoms the
RHF values are too low by factors like 7-30 for weak type
($f_{E1} < 0.1$) $\overline{V} \rightarrow \overline{V}'$ (KLM → KLM') transitions [4]. Table IV
shows a sample of the $\overline{V} \rightarrow \overline{V}'$ transitions for the CHEMSET. A
comprehensive table for nearly all of the CHEMSET $\overline{V} \rightarrow \overline{V}'$ along
isoelectronic sequences can be found in [4].

The large decreases in the $\overline{V} \rightarrow \overline{V}'$ oscillator strengths occur
partly due to $\overline{V} \leftrightarrow \overline{V}$ mixings in each ψ_c, and partly due to the
$\overline{V} \leftrightarrow pR$ mixings in each state. Corresponding increases in
oscillator strength should take place therefore in the $\overline{V} \rightarrow pR'$.

The fundamental intermingling of $\overline{V}$ and pR states for each
symmetry (here LS) is thus such that one may view these states as
members of an NCMET ψ_c charge supermultiplet. Then within

$$\{ \{\overline{V}\} \cup \{pR\} \}_{LS} \xrightarrow{E1} \{ \{\overline{V}'\} \cup \{\overline{pR}'\} \}_{L'S} \qquad (32)$$

an LS-NCMET supermultiplet making transitions to another (L'S)
NCMET supermultiplet, there will be strong intensity borrowings
between the supercomponents (actual spectroscopic γLS-states;
$\gamma \equiv$ configuration).

An example of a CHEMSET supertransition (E1) would be
Table V:

TABLE IV

Intravalency transitions ($\overline{V} \to \overline{V}'$) oscillator strengths ($f_{E1}$) as given by i) orbital theory (RHF of each state), ii) NCMET (this work), and by iii) new experiments (beam-foil spectroscopy, phase-shift method, etc.).

Species	Transitions	λ(Å)	f (RHF)	f (Z) (note g)	f (NBS)	f (MET)	f (EXP)	f (EXP)
CII	$1s^2 2s^2 2p\ ^2P \to 1s^2 2s 2p^2\ ^2D$	1335	0.263	0.204	0.17[a1] (0.121)[a2]	0.125	0.114	0.114 (±.011)[b]
NII	$1s^2 2s^2 2p^2\ ^3P \to 1s^2 2s 2p^3\ ^3D$	1085	0.236	0.192	0.17[c]	0.100	0.109 / 0.101	0.109 (±.011)[b] / 0.101 (±.006)[d]
NII	$1s^2 2s^2 2p^2\ ^3P \to 1s^2 2s 2p^3\ ^3P$	916	0.170	0.213	0.22[c]	0.137	0.131	0.131 (±.007)[d]
NII	$1s^2 2s^2 2p^2\ ^3P \to 1s^2 2s 2p^3\ ^3S$	645	0.334	0.244	0.23[c]	0.218	0.189	0.189 (±.016)[d]
NIII	$1s^2 2s^2 2p\ ^2P \to 1s^2 2s 2p^2\ ^2D$	991	0.213	0.167	0.18[e]	0.114	0.103	0.103 (±.010)[d]
NIII	$1s^2 2s^2 2p\ ^2P \to 1s^2 2s 2p^2\ ^2P$	686	0.577	0.415	0.45[e]	0.399	0.416	0.416 (±.075)[d]
OIII	$1s^2 2s^2 2p^2\ ^3P \to 1s^2 2s 2p^3\ ^3D$	834	0.200	0.162	0.15[c]	0.100	0.102	0.102 (±.002)[f]
OIII	$1s^2 2s^2 2p^2\ ^1D \to 1s^2 2s 2p^3\ ^1D$	600	0.534	–	0.37[c]	0.297	–	–
OIV	$1s^2 2s^2 2p\ ^2P \to 1s^2 2s 2p^2\ ^2D$	789	0.179	0.141	0.15[e]	0.106	0.091	0.091 (±.002)[f]

a1. A.W. Weiss, (Reported in NBS Tables). a2. A.W. Weiss, Phys. Rev. 162, 71 (1967).

b. G.M. Lawrence and B.D. Savage, Phys. Rev. 141, 67 (1966).

c. A.B. Bolotin, I.B. Levinson and L.I. Levin, Soviet Physics – JETP 2, 391 (1956).

d. L. Heroux, Phys. Rev. 153, 156 (1967).

e. A.B. Bolotin and A.P. Yutsis, JETP 24, 537 (1953).

f. W.S. Bickel, Phys. Rev. 162, 7 (1967).

g. M. Cohen and A. Dalgarno, Proc. Roy. Soc. A280, 258 (1964).

TABLE V

The optical E1-transitions $\{^1D_2\} \rightarrow \{^1P_1^\circ\}$ in C°, N^{+1}, O^{+2}, ... as viewed as "supertransitions" between two NCMET charge super multiplets with strong mixings, intensity borrowings among members of a supermultiplet.

#	Initial Supermultiplet $\{^1D_2\}$	Observed terms in atom	#	Final state Supermultiplet $\{^1P_1^\circ\}$	Observed terms in atom
①	$1s^2 2s^2 2p^2$ $\,^1D_2$	C°, N^+, O^{++}	①	$1s^2 2s 2p^3$ $\,^1P_1^\circ$	$(C^\circ), N^+, O^{++}$
②	$1s^2 2p^4$ $\,^1D_2$	O^{++}	②	$1s^2 2s^2 2p 3s$ $\,^1P_1^\circ$	C°, N^+, O^{++}
③	$1s^2 2s^2 2p 3p$ $\,^1D_2$	C°, N^+, O^{++}	③	$1s^2 2s^2 2p 3d$ $\,^1P_1^\circ$	C°, N^+, O^{++}
④	$1s^2 2s^2 2p 4f$ $\,^1D_2$	N^+	④	$1s^2 2s 2p^2 3p$ $\,^1P_1^\circ$	O^{++}
⑤	$1s^2 2s 2p^2 3s$ $\,^1D_2$	–	⑤	$1s^2 2s 2p^2 4f$ $\,^1P_1^\circ$	–
⑥	$1s^2 2s 2p^2 3d$ $\,^1D_2$	–	⑥	$1s^2 2p^3 3s$ $\,^1P_1^\circ$	–
⑦	$1s^2 2p^3 3p$ $\,^1D_2$	–	⑦	$1s^2 2p^3 3d$ $\,^1P_1^\circ$	–

VIII. INNER IONIZATION POTENTIALS; EXCITED CORE RYDBERG SERIES LIMITS

In photoelectron spectra one observes $\overline{V}(N) \to \overline{V}(N-1)+e^-$, the same kind of final state as in photoionization after a R-series $\{\overline{V}(N-1)n\ell\} \to \overline{V}(N-1) +e^-$. Inner I.P.'s correspond to hole-states within the $\{\overline{V}\}$.

In CHEMSET atoms, the hole-states are intravalency states $\{1s^2 2s2p^n\}$, K[L], in vacuum uv, or [K]L = $\{1s2s^n 2p^m\}$, interstates [17] in ESCA.

For third row atoms KL[M] are intravalency $\{\overline{V}_M\}$ states, the K[L]M are interstates [17].

The inner I.P.'s for intravalency hole states need to be calculated with the $\psi_c \cong \psi_c^S$, the $\overline{V} \leftrightarrow pR$ mixings altering the RHF results appreciably. As with all energy quantities however, with I.P.'s too all-external correlation energies E_U also must be added (cf. II C).

$$-\text{I.P.} = \Delta E_c + \Delta E_U = \Delta E_c^S + \Delta E_U^S \text{ with } \Delta E = E(X^+) - E(X) \quad (33)$$

The kinds of intershell NCMET correlations that enter the I.P.'s for hole states of interstate type were shown in figure 7. (In addition, with more than one e^- in the M-shell, one would now have (MM $\to$ LL), (MM $\to$ LM), (MM $\to$ MM), (MM $\to$ L$\hat{f}$), and (MM $\to$ M$\hat{f}$) types).

Both types of inner I.P.'s are calculated in a clear way with the ψ_c (and E_U) methods given above, the crucial fact being the important contribution of $\overline{V} \leftrightarrow \overline{V}$ and $\overline{V} \leftrightarrow pR$ mixings which make correlation energies of hole states highly anomalous (magni-

tude of $|E_{CORR}|$ increasing when number of electrons decreased by
one! [18]). An unexpected anomaly discovered via NCMET the same
year (1963) as uv-photoelectron spectroscopy started.

Table VI shows some inner I.P.'s as given by Koopman's Rule
(ground state ϕ_{RHF}, $(I.P.)_i \approx -\varepsilon_i$), by ΔE_{RHF} of initial atom and
final ion, by ΔE_c of NCMET, and by $\Delta E_c + \Delta E_U = -I.P.$, the full
NCMET and exact.

IX. THEORY OF EXCITED STATES OF MOLECULES - MANY CENTER ASPECTS

The results and effects on many-electron aspects above apply
the same way to a molecule in the RHF - MO SCF picture. There
are a set of intravalency RHF-MO's $\{k\}_{\overline{V}}$, just one electron
spin-orbitals $k(\underset{\sim}{x}_i)$ regardless of how they may relate to AO's.
With these strictly molecular MO's (e.g. functions in elliptic
coordinates for diatomics), non-closed shell many-electron theory
would be entirely parallel to the atomic structure NCMET as in
the examples above. In this form one sees only the many-electron
aspects of the problem. Let us look at NCMET of molecules briefly
this way, then go into LCAO MO forms so we can discuss united
atom type molecular Rydberg orbitals and other questions raised
in section I.

A. Intravalency and Pre-Rydberg States of Molecules with RHF -
    MO (SCF)/NCMET

Take the CHEMSET molecules (each atom's intravalency state:
$\{1s_H, (1s,2s,2p)_{B,C,N,O,F}\}$ (one may omit for simplicity the closed
K-shells except for ESCA problems).

There is a finite accurate RHF-MO set $\{k\}_{\overline{V}}$ for the

TABLE VI

Inner Ionization Potentials (in electron volts) as given by Koopman's Rule (KT; ground state RHF; $(I.P.)_i \approx -\varepsilon_i$), by ΔE_{RHF} (Hartree Fock of initial and final states separately), by ΔE_c (Difference of energies of charge wave functions only), and by full NCMET containing all the correlation effects (needed in energetics) and experimental ($\Delta E \equiv E(X^+) - E(X)$; I.P. = $-\Delta E$)

1. Boron 2s-I.P.	(BI $1s^2 2s^2 2p$ $^2P°$ → BII $1s^2 2s2p$ $^{1,3}P°$)	
As given by	$^2P → {}^3P$	$^2P → {}^1P$
KT	13.46	13.46
Δ(RHF)	11.13	16.77
$\Delta(\psi_c)$	12.46	17.28
Δ(full NCMET; $E_c + E_U$)	12.92[a]	17.31[a]
Δ(exp; E_i)	12.93[b]	

2. Carbon 2s-I.P.	(CI $1s^2 2s^2 2p^2$ ^3P → CII $1s^2 2s2p^2$ $^{2,4}P$)			
As given by	$^3P → {}^4P$	$^3P → {}^2P$	($^3P → {}^2D$)	($^3P → {}^2S$)
KT	19.20	19.20	19.20	19.20
Δ(RHF)	14.36	24.70	19.72	22.40
$\Delta(\psi_c)$	15.73	24.70	19.92	23.60
Δ(full NCMET; $E_c + E_U$)	16.43[a]	24.87[a]	20.45[a]	23.11[a]
Δ(exp; E_i)	16.60[b]			

3. Nitrogen 2s-I.P.	(NI $1s^2 2s^2 2p^3$ ^4S → NII $1s^2 2s2p^3$ $^{3,5}S$)	
As given by	$^4S → {}^5S$	$^4S → {}^3S$
KT	25.72	25.72
Δ(RHF)	17.78	33.99
$\Delta(\psi_c)$	19.14	33.44
Δ(full NCMET; $E_c + E_U$)	20.14[a]	33.63[a]
Δ(exp; E_i)	20.40[b]	

a. Full NCMET calculation. ΔE_c from charge wave function calculation $+\Delta E_U$ from transferable all-external pair correlation values of O. Sinanoğlu and W. Luken (to be published)
b. Term values from C. Moore's tables "Atomic Energy Levels" (U.S. NBS, Circular 467)

intravalency states $\{\overline{V}\}$ of each molecule. The $\{k\}_{\overline{V}}$ are labeled by point groups G_{pt}. We will define the accurate RHF-MO sea such that the number (occupied or unoccupied) of spin-orbitals M is

$$Dim\{k\}_{\overline{V}} = M = 2n_H + 10n_X \tag{34}$$

n_H = number of hydrogens in the molecule.
n_X = number of other CHEMSET atoms ($X \in \{B,C,N,O,F\}$) (This does not necessarily imply the LCAO MO's).

Accurate RHF-MO's have been obtained (analogous to atomic structure RHF [6]) for diatomics and a few polyatomics like C_2H_6. All other MO calculations ("ab initio" or not) $[\widetilde{MO}]$ are approximate with often $\Delta E(\widetilde{MO}-RHF)$ error larger than ΔE_{CORR}.

Take diatomics with the RHF-MO sea for $\overline{V}$-states:

$$\{k\}_{\overline{V}} = \{1\sigma_g\alpha, 1\sigma_g\beta, \ldots 3\sigma_u\beta, \ldots 1\pi_g^+\beta\}$$
$$(M = 2 \text{ for } H_2; \quad M = 20 \text{ for } X_2) \tag{35}$$

Then the entire exact wave function ψ of an intravalency state is fully specified as derived by NCMET [2,3].

$$\psi(\overline{V}) = \psi_c + \chi_U \tag{36a}$$

$$\psi_c = \phi_{RHF}(\overline{V}) + \chi_{INT} + \chi_F \tag{36b}$$

Again $\phi_{RHF} + \chi_{INT}$ contains all N-electron dets made of the $\{k\}_{\overline{V}}$ intravalency MO set (yielding the overall $G_{pt} \otimes SU(2)_\sigma$ symmetry of the $\overline{V}$-state at hand).

The semi-internal MO's, $\hat{f}_{ij;k}(x_1)$ and $\hat{f}_p(x_1)$ are again fully determined once the $\{k\}_{\overline{V}}$ is known. For diatomics of CHEMSET they are only of the symmetries [19]:

$$\{\hat{f}\}_{\overline{V}} = \{f_\sigma, f_\pi, f_\delta, f_\varphi\} \tag{37}$$

each f_λ being $f_{\lambda g}$ and $f_{\lambda u}$.

Recall that the semi-internals are orthogonal to all of the RHF-MO sea:

$$\langle \hat{f} | k \rangle = 0, \quad \text{for } k \in \{k\}_{\overline{V}} \tag{38}$$

Now the $\psi_c(\overline{V})$ is given by a finite number of all the dets made of the charge wave function set $\{ \{k\}_{\overline{V}} \cup \{\hat{f}\}_{\overline{V}} \}$.

Molecular excited states are then again classified into $\{\overline{V}\}$-states, $\{pR\}$-states, and $\{R\}$-states as in equation (14).

Definitions: All states with nominal configurations made of the RHF-MO $\{k\}_{\overline{V}}$-set are intravalency states.
 All states with nominal configurations approximatable (cf. IX C) within the $\{ \{k\}_{\overline{V}}^{N-1} \otimes \{\hat{f}\}_{\overline{V}} \}$ are pre-Rydberg states.
 All other singly excited states are true Rydberg (R) states.

To go further we need to get the $\{\hat{f}\}$ and relate them as well as the RHF-MO's to conventional MO theory, and to united atom and separated atoms.

B.   The Intravalency AO Pool

The RHF-MO-based NCMET is fine for accurate calculation of physical properties, lifetimes, hyperfine structure, etc. in just the same way as in the new atomic structure theory [2]. A more chemical, and spectroscopic theory is obtained however by going back to atomic orbitals (AO's).

The RHF-MO sea is now replaced by:

Definition: The intravalency AO-pool is the set $\{k^A\}_{\overline{V}}$ of all up through valence shell AO's, occupied or unoccupied, of all atoms of a molecule.

For the CHEMSET $\{\overline{V}\}$-states:

$$\{k^A\}_{\overline{V}} = \{1s_H, \{1s,2s,2p\}_X\} \tag{39a}$$

Remark: This constitutes a linearly independent (cf. Appendix), though non-orthonormal (non-O.N.) basis set for the $\overline{V}$-space. As shown in the Appendix, in all molecular $\overline{V}$-states of interest the $\{k^A\}_{\overline{V}}$-set remains linearly independent as all overlaps $S_{ab} = \langle a|b \rangle$ between CHEMSET KL-shell AO's are not close to unity.

$$\langle k^A|k^{A'} \rangle = S_{k^A k^{A'}} \neq 1 \tag{39b}$$

Well-known corollary of this and basis of LCAO-MO theory is that:

Theorem 3: The Dim $\{k^A\}_{\overline{V}}$ = M, equation (34), number of $\overline{V}$-AO's lead to M LCAO MO's at the usual $(\vec{R} \approx \vec{R}_{equil})$ molecular geometries.

$$\text{Dim}\{k^A\}_{\overline{V}} = \text{Dim}\{k\}_{\overline{V}} = M \tag{40}$$

Proof: To get M LCAO MO's $\{k\}_{\overline{V}}$ from the M $\{k^A\}_{\overline{V}}$ AO's (with a linear transformation), the determinant of the overlap matrix $|\underset{\sim}{S}|$ must be $\neq 0$ (cf. Appendix). At usual geometries (not

approaching bond lengths → 0, no united atoms) in the $\overline{V}$-AO set
equation (39b) holds; and $|\underset{\sim}{A}| \neq 0$.

Corollary 1: From the $\{k^A\}_{\overline{V}}$ -set we can construct many types
of M - as many spinorbitals for the $\overline{V}$; e.g. orthogonalized AO's
(OAO's), localized MO's (LO's), etc. All are related by linear
transformations which exist since $|\underset{\sim}{A}| \neq 0$.

Corollary 2: For NCMET of molecular states we can start
either with $\{k^A\}_{\overline{V}}$, the $\overline{V}$-AO pool, or the $\{k\}_{\overline{V}}$, the $\overline{V}$-LCAO MO
pool.

In conventional MO theory, point groups G_{pt}, play a key role.
They provide the MO-designations. They define what is a "closed
shell", what is a "non-closed shell" molecular species or state.
The lowest doubly occupied state belonging to the totally
symmetric representation ($^1\Sigma_g^+$, ^1A_g, etc.) of the G_{pt} is said to
be "closed shell" system. This is an inadequate concept inconsistent with the chemical reactivities and spectroscopic properties of molecular "closed shell" ground states as compared to the
atomic "closed shell" systems (He, Ne). The ambiguity between
intravalency and low Rydberg states in MO theory is also retrac-
able to this.

Point groups G_{pt} not being fundamental to the chemical and
spectroscopic behaviour of molecules (as evident from distorted
or substituted or isotopic molecules' [7] behaviour with no G_{pt},
to similar molecules with a G_{pt}) but only convenient, we prefer
to base the concept of "closed" or "non-closed-shellness" of a
molecule rigorously, on the intravalency AO-pool, stating as a:

Theorem 4: All stable molecules even doubly occupied ground
states (and even totally symmetric ones with respect to a point

group) are non-closed shell systems.

Proof: The most stringent cases are the CHEMSET molecules
(whose valency shells do not expand into higher valencies which
higher atoms like P, S, Sb, . . .and most metal atoms do). The
intravalency AO-pool (or LCAO MO-pool) is clearly defined (KL
shells).

In any CHEMSET neutral atom X, the number of its electrons
$N_X < M_X$ its $Dim\{k\}_{\overline{V}_X}$.

In any CHEMSET molecule,

i) $Dim \{k^A\}_{\overline{V}} = M = M_H n_H + M_X n_X = 2n_H + 10n_X$

ii) $n_H, n_X \geq 0$,   iii) $N_H < M_H; N_X < M_X$,   iv) $N = n_H N_H + n_X N_X$

thus, $N < M$.   Q.E.D.

Regardless of any point group symmetry, stable molecules
always have fewer electrons than $\overline{V}$-AO's (intravalency spin-orbitals). The same is true in the LCAO MO (or RHF MO) description
because of theorem 3.

Corollary 1: The only closed shell molecules are He_2, Ne_2
and multiply charged negative ions $(F_2^{--}, (NH_3)^{-6}$, etc.) which are
not stable.

Corollary 2: Stabilities of molecules result from their AO-
pool non-closed-shellness.

Corollary 3: All stable molecule ground (and of course
excited) states need be treated by non-closed shell theory (NCMET).
Considering the totally point group symmetric ground states as
closed shells would miss the very important non-closed shell
correlation effects (which are in NCMET). [This makes it no

longer surprising that F_2 ($^1\Sigma_g^+$) comes out unstable (Table I) in the ("closed shell")-MO theory, but stable with NCMET.]

Corollary 4: In all stable molecules there will be unoccupied valence shell $\overline{V}$-set MO's leading to reactivity, and to a richness of intravalency states $\{\overline{V}\}$, maximum possible number being the binomial coefficient $\binom{M}{N}$, reduced considerably however below this number for each symmetry.

Corollary 5: "Closed shell" in MO theory meant totally symmetric (lowest) representation of a point group $G_{pt} \otimes SU(2)_\sigma$. We see here this is a necessary but not sufficient condition. All true AO-pool (NCMET) closed shell states are also totally symmetric under any molecular point group, but not vice versa.

In third row atoms (Si, S, P, Cℓ, Ar, Mn^{++}, Fe^{++}, etc.) and heavier, with their $\overline{V}$-AO-pool being of KLM-shells or more, we see how multiple valency arises as compared to second row CHEMSET; also that Ar, Kr, Xe, are not closed shell systems while He and Ne are.

C. Molecular Semi-Internal Orbitals from the $\overline{V}$-AO-Pool; Problems of the United Atom.

The exact ψ of a $\overline{V}$-state, equation (36), can also be rewritten now as

$$\psi = \psi_c^A (\overline{V}) + \chi_U^A \tag{40}$$

with

$$\psi_c^A = \phi_{LCAO\ MO} + \chi_{INT}^A + \chi_F^A \tag{41}$$

Instead of rigorous RHF-MO's in ψ_c we have LCAO MO's in it.

These MO's correspond to what is called "limited basis set LCAO MO SCF" [1a] if we use one STO per AO of the atom (we could also have one atomic RHF AO for each, with little difference to content).

Recall that χ_F had the semi-internal correlations $(ij \rightarrow k\hat{f}_{ij;k})$ as well as $(i \rightarrow \hat{f}_p^i)$ orbital polarization terms in it. It will turn out that with LCAO MO's instead of actual RHF ones, some $\hat{f}_p^i$ terms ordinarily small with RHF, now grow. They correct for

$$i_{LCAO\ MO} \rightarrow i_{RHF\ MO} \tag{42a}$$

by

$$\frac{\left(i_{LCAO\ MO} + \hat{f}^{i,A}\right)}{\sqrt{1 + |\hat{f}^{i,A}|^2}} = i_{RHF\ MO} \tag{42b}$$

For CHEMSET KL-atoms, the molecular f's will turn out to be made of (2s,2p) -like for H and 3s,3p,3d and 4f-like for the others, as in the NCMET atomic structure theory. But this is just the additional basis STO's many workers have found needed to get RHF MO's by experience on the computer. Thus we make the

Observation: While "Limited Basis Set LCAO MO SCF" [1a] are the LCAO MO's from the $\overline{V}$-AO-pool, the NCMET shows that in "Extended Basis Set LCAO MO SCF" [1a] $\cong$ RHF MO, the $\{1s_H, (2s,2p)_X \cup (2s,2p)_H, (3s,3p,3d,4f)_X\}$ AO's will occur to give close to actual Hartree-Fock MO's giving a justification to the appearance of these AO's in computer studies in getting RHF MO's.

The AO-pool based (in the $\{k^A\}_{\overline{V}}$ or $\{k\}_{\overline{V},LCAO\ MO}$ basis forms) semi-internal orbitals $\hat{f}_{ij;k}$ can be found either from triplets of AO's, $k_A \otimes k_B \otimes k_C \rightarrow \hat{f}^A$ or from triplets of MO's $k_1 \otimes k_2 \otimes k_3 \rightarrow \hat{f}(LCAO\ MO)$, there resulting the same number of them either way. In either form,

$$\langle \hat{f} | k^A \rangle \; = \; 0 \qquad \text{for all } k^A \in \text{AO-pool} \tag{43}$$

From the extensive NCMET atomic structure experience we expect the $\hat{f}^A$ molecular basis to consist approximately of the first atomic virtual excited orbitals:

$$\{\hat{f}^A\} \; \cong \; \{\vec{f}^A\} \; \equiv \; \{(2s,2p)_H \; \bigcup \; (3s,3p,3d,4f)_X\} \tag{44}$$

(one STO each with fully optimized exponent).

The $\hat{f}$ in MO forms are linearly related to these,

$$\hat{f}(\text{LCAO MO}) \; = \; \sum_i c_i^A \; \vec{f}_i^A \tag{45}$$

but with one major difficulty, a difficulty intimately related to those of Rydberg orbitals in molecules.

Theorem 3, the conservation of the number of AO's upon linear transformation into LCAO MO's holds for the intravalency orbitals $\{k\}_{\overline{V}}$. In conventional MO theory this is assumed to hold for higher MO's as well, and/or for small internuclear distances down to the united atom (UA).

It does not seem to have been noted earlier however that theorem 3 breaks down whenever some overlap integrals $S_{AB} \to 1$, i.e. for higher (n > 2 for CHEMSET) $n\ell m$ AO's as in "Rydberg" states (at $\vec{R} \approx \vec{R}_{eq}$ already), and/or near UA or semi-united atoms (SUA). We referred to this during the Val Morin Summer School as the "$\underset{\sim}{\Lambda}$ - castastrophe" and state it now as

Theorem 5: With LCAO MO's, whenever, some AO-AO' overlaps $S_{AB} \to 1$ and/or the overlap matrix $\underset{\sim}{\Lambda}$ has $|\underset{\sim}{\Lambda}| = 0$, the number of linearly independent MO's M_Λ, that can be constructed from the initial M as many AO's is less than M.

$$\text{If} \quad |\underset{\sim}{A}| = 0, \qquad M_\Delta < M. \tag{46}$$

Proof follows simply from the Appendix.

Example: (United Atom)   Consider already H_2.

$$1\sigma_g = c_g (1s_a + 1s_b)$$
$$1\sigma_u = c_u (1s_a - 1s_b) \tag{47}$$

$$S_{ab} = \langle 1s_a | 1s_b \rangle \; ; \quad \underset{\sim}{A} = \begin{pmatrix} 1 & S_{ab} \\ S_{ab} & 1 \end{pmatrix}$$

$$|\underset{\sim}{A}| = 1 - S_{ab}^2 \tag{48}$$

At internuclear distance $R \neq 0$, $S_{ab} < 1$; hence two AO's give two independent MO's.

But at $\quad R \to 0, \quad S_{ab} \to 1, \quad |\underset{\sim}{A}| \to 0$

and $\tag{49}$

$$1\sigma_g \to 1s_a = 1s_b$$

the LCAO MO becoming a one center AO very similar to each of the separate atom AO's, while

$$1\sigma_u \to 0 \tag{50}$$

Two AO's yield only one MO at $R \to 0$. As AO's become linearly dependent, some redundant basis vectors must be eliminated.

One overlooks this difficulty in LCAO MO theory, by looking at the normalized $1\sigma_u$,

$$1\sigma_u = \frac{1}{\sqrt{2(1-S_{ab})}} \ (1s_a - 1s_b) \to \frac{0}{0} \ \text{as} \ R \to 0. \tag{51}$$

One then says, the limit analytically taken in R will resolve the indeterminate form $\frac{0}{0}$ to yield a UA $2p_u$ orbital.

There are several objections to this procedure: (i) Taking the limit is not an operation within the linear vector space on which the AO $\to$ MO, LCAO theory is based. The $2p_u$ cannot be expressed by any coefficients at $R \sim 0$ in terms of the original AO's,

$$2p_u^{(UA)} \neq (c_1 1s_a - c_2 1s_b)_{R \sim 0} \tag{52}$$

One of the two AO's has actually disappeared due to linear dependence. Needing still two independent orbitals, one has simply invoked from outside of the 2-dim. linear vector space, a new one, $(2p_u)$. (ii) There is difficulty even with the analytic limit, eq.(51), which is not $2p_u$. (iii) By normalizing $(1s_a - 1s_b)$ near $R \to 0$, we took a very "small" residual orbital (compared to a $1s_a$) and blew it up to full size (to make again $<k|k> = 1$). Thus the new orbital may be a physically unrealistic one.

Of course with atoms other than hydrogen, UA becomes ambiguous to use in any case due to existence of $(1s_a^2)$, $(1s_b^2)$,... cores. One invokes various stages of semi-united atoms $(C_2H_4 \to O_2 \to Si,$ etc.) with resulting ambiguities of which to use to represent an actual $\overline{V}$-LCAO MO (like the (C-H) $\sigma*$'s).

But the $\mathcal{A}$ - catastrophe will occur in the out-of-the $\overline{V}$-set, higher MO's, i.e. semi-internal MO's, equation (45), which are also pre-Rydberg MO's and certainly for Rydberg MO's, if constructed out of AO's on each atomic center. [The overlaps S_{ab} are already large ($\to 1$) at $R \approx R_{eq}$, in CHEMSET molecules even with 3ℓ and $4f$].

Note that in actual computer calculations, the $\mathcal{A}$-catastrophe

will cause difficulties. The "ab initio" calculations are carried
out (LCAO MO or C.I.) by matrix diagonalizations which means
we stay within the M-dimensional linear vector space. If two
or more vectors (AO's, MO's) become linearly nearly dependent
($|\underset{\sim}{\Delta}| \cong 0$), the calculation will break down with very large or
very small numbers.

Resolution of the $\underset{\sim}{\Delta}$ - catastrophe: To construct a proper
semi-internal orbital set $\{\overline{f}\}$ (and Rydberg orbitals) (1) start
with the semi-internals $\{\overline{f}^A\}$ = $\{(2s,2p)_H, (3s,3p,3d,4f)_X\}$ of
each center. (2) Look at the overlaps S_{ab} (and $|\underset{\sim}{\Delta}|$). If an
$S_{ab} \approx 1$ (e.g. $<3p\sigma_{C_1}|3p\sigma_{C_2}>$ in C_2H_4), then keep only the rein-
forcing (bonding) combination and throw out the anti-bonding
combination, e.g.

$$3s_a + 3s_b \rightarrow \overline{f}_{3s\sigma_g} \tag{53a}$$
$$3s_a - 3s_b \rightarrow 0 \tag{53b}$$

The (+)-combination will also be an orbital very similar to that
centered in the middle of the molecule, $3s_{\overline{C}}$ (such as the super
AO Rydbergs visualized by the spectroscopists), since

$$3s_a \approx 3s_b \approx 3s_{\overline{C}} \qquad \text{(when } S_{ab} \approx 1 \text{)} \tag{53c}$$

(3) If additional $\overline{f}$ are needed (by symmetry, etc.), construct
new, independent ones by invoking the next higher atomic AO's
on each center ($n > 3,4$) and again applying equations (53).

The higher Rydberg MO's are constructed the same way from
higher R-AO's of each atom.

We now have a well-behaved molecular semi-internal set $\{\overline{f}\}$
along with the $\overline{V}$-set. This set we call the $\{\overline{f}^s_\Delta\}$.

D. The Spectral Charge wave function of a Molecular Intravalency
    State $(\overline{V})$ and of a pre-Rydberg State (pR)

The many-center difficulties having been taken into account,
the NCMET ψ_c wave functions, energies, and definitions of the
intravalency $(\overline{V})$, pre-Rydberg (pR) and Rydberg (R) states now
proceed as in the NCMET atomic structure theory (III - VI).
We have

$$\psi_{exact}^{(\overline{V})} = \psi_c^s\,(\overline{V}) + \chi_U^s \tag{54}$$

$$\psi_c^s\,(\overline{V}) = \phi_{LCAO\ MO}(\overline{V}) + \chi_{INT}^A(\{\overline{V}\}) + \chi_F^s(\{pR\})$$

the complete finite CI of all N-e$^-$ dets from the $\{\ \{k\}_{\overline{V}} \bigcup \{f_\Delta^s\}\ \}$
set. The ψ_c^s is equivalent to the ψ_c of Section IX A(χ_F^s making up
the LCAO MO $\rightarrow$ RHF MO difference as discussed). The ψ_c^s also gives
correct dissociation of the molecule (P.E. curves)[3].

The same set gives the pR states

$$\psi_c^s(pR) = \phi_{LCAO\ MO}(pR) + \chi_c^A(\{\overline{V}\} \bigcup \{pR'\}) \tag{55}$$

All $\overline{V}$ and pR states of same symmetry again mix strongly.

True Rydbergs (R) are constructed by the Δ-procedure from
AO's of each atom with $n\ell \neq$ (3s,3p,3d,4f, $\{k^A\}_{\overline{V}}$,and others
required in the pR by Δ). For these

$$\psi(R) = \psi_c(R) + \chi_U(R) \tag{56a}$$

$$\psi_c(R) \cong [\phi_{RHF\ MO}(R) \cong \phi_{LCAO\ MO}(R) \cong \phi_{R\ MO}] + \chi_c(core) \tag{56b}$$

with R MO meaning "Rydberg orbital" a super atom AO and core
LCAO MO's in $\overline{V}$.

$\chi_U(R)$ contains the free ion core like, all external correlations (the non-closed shell core ones being in χ_c(core)) and the all-external type (closed shell like core aspect) core-polarization.

X.   WHAT IS A RYDBERG ORBITAL?

In conclusion, (cf. I),

i) in a CHEMSET molecule (and similarly for the larger $\overline{V}$-set heavier atom cases), all MO's and states constructed out of only the intravalency AO-pool are intravalency states ($\overline{V}$)and σ, $\sigma*$, π, $\pi*$, n-type $\overline{V}$ MO's. In C_2H_4, there are 28 $\overline{V}$-AO's (including Carbon 1s) and 28 $\overline{V}$-MO's ($\sigma,\sigma*,\pi,\pi*$ types).

ii) The MO's constructed out of $(2s,2p)_H$ and $(3s,3p,3d,4f)_c$ (in C_2H_4)(allowing for the Δ-eliminations), $\{\overline{f}\}$ are pre-Rydbergs and yield pre-Rydberg states which mix strongly with the same symmetry $\overline{V}$ ones. These may not fall onto $1/(n-\delta)^2$ series.

iii) All higher MO's (constructed as in (ii) but from higher AO's) are Rydbergs which mix hardly any with the $\{\overline{V}\}$ and the $\{pR\}$, hence should follow regular $1/(n-\delta)^2$ series.

iv) We have not had to invoke any UA or SUA's; all MO's are unambiguously $\overline{V}$, pR or R. Ambiguity of whether a Rydberg orbital is centered on the molecule as a whole or on a given atom also disappears [20] (equation (53c)).

We thank the lecturers and Professor C. Sandorfy and the Organizing Committee of the Val Morin NATO Summer School for their most stimulating lectures. The author thanks also Professor I. Ugi, Dr. Hans Gasteiger and Dr. P. Gillespie for their hospitality and stimulation at the Institut Organisch-Chemisches Laboratorium where part of this work was done. Support of the work by a U.S. National Science Foundation grant is acknowledged.

REFERENCES

[1a]   For various non-empirical MO methods ("limited basis set", "extended basis set", etc) see O. Sinanoğlu and D.F. Tuan, Ann. Rev. Phys. Chem. 15, 251 (1964).

[1b]   For semi-and non-empirical MO methods see e.g. O. Sinanoğlu and K.B. Wiberg and other authors, Sigma MO Theory, (Yale Press, New Haven and London, 1970).

[2]   O. Sinanoğlu, "Atomic Structure, Transition Probabilities and Theory of Electron Correlation in Ground and Excited States" in Atomic Physics, I, V.W. Hughes,B. Bederson, V.W. Cohen, F.M.J. Pichanick, eds., (Plenum, New York, 1969) p. 131, and references to our earlier work therein.

[3]   O. Sinanoğlu, "Prediction of Molecular Excited State Properties, Potential Energy Curves and the Non-Closed Shell Many-Electron Theory" J. Mol. Struct. (1973) [Special proceedings issue of IUPAC Conference on Molecular Spectroscopy, Wrɵclaw, Poland, September 1972].

[4]   O. Sinanoğlu, "Beam Foil Spectroscopy and New Atomic Structure Theory with a Survey of Results Since 1970" Nuclear Instrs. and Methods 110, 193 (1973) [Special proceedings issue for the Third International Conference on Beam-Foil Spectroscopy, October, 1972].

[5]   This is adequate in atoms and stable molecules. For molecules near dissociation we need $\psi = \phi_{GRHF} + \chi'_{CORR}$ where

$$\phi_{GRHF} \cong \frac{\phi_{RHF} + \chi_{INT}}{\sqrt{1 + |<\chi_{INT}|\chi_{INT}>|^2}} \quad .$$

[6]   C.C.J. Roothaan, Rev. Mod. Phys. 32, 179 (1960).

[7]   We include $1s^k$ for ESCA.

[8]   We denote intravalency states by $\overline{V}$. In Mulliken notation in molecules $\overline{V} \supset \{N, V, T, \ldots\}$.

[9]   We shall not discuss inter-states of doubly excited {D**}
        type in this paper, e.g. {1s^{m}2s^{n}2p8 nℓ n'ℓ; {n,n'} > 2},
        although such D** are now being observed in <u>vacuum uv</u>
        atomic spectra and in beam-foil spectra (cf. e.g. reference
        4 for review) and their classification with new group
        theory is interesting.

[10]   Notation: { } ≡ set; $\subset$ ≡ contained in; $\in$ ≡ belongs to;
        $\cup$ ≡ union of sets.

[11a]   O. Sinanoğlu, J. Chem. Phys. <u>33</u>, 1212 (1960).

[11b]   O. Sinanoğlu, J. Chem. Phys. <u>36</u>, 706 (1962).

[11c]   cf. also R.G. Parr, <u>Quantum Theory of Molecular Electronic</u>
        <u>Structure</u>, (Benjamin, New York, 1963).

[12]   see e.g. K. Freed, Ann. Rev. P. Chem. <u>22</u>, 313 (1971). Also,
        O. Sinanoğlu and K.A. Brueckner, <u>Three Approaches to</u>
        <u>Electron Correlation in Atoms</u>, (Yale Press, New Haven and
        London, 1970).

[13]   NCMET/MET gives pair equations [11,2,12] to individually
        evaluate each ε_{ij} by itself, non-empirically or semi-
        empirically from atomic energy level data. They have
        been worked out both ways by many [12] with similar
        results.

[14]   O. Sinanoğlu and H.Ö. Pamuk, JACS <u>95</u>, 5435 (1973).

[15]   More precisely, these mixings, correlation effects are
        "switched on" by the "<u>fluctuation potentials</u>"[2,11] m_{ij}
        between Hartree-Fock electrons i and j. <u>Fluctuation</u>
        <u>potential</u> introduced in MET [11] is the difference between
        $1/r_{ij}$ and virtual Hartree-Fock orbital potentials.

[16]   Charge wave functions and optical transition probabilities
        of these atoms we have calculated [4]. On the latter there
        is some LM intershell ψ_c effects which are included for
        high accuracy. For a cruder approximation, these are
        omitted in equations (26) - (28).

[17]   This is why we distinguish between "intravalency states"
        (here $\bar{V}_M$) and "interstates" like the pR or some L-hole
        states of third row (KLM)-atoms. Interstates have two
        non-closed shells (here L and M).

[18]   O. Sinanoğlu, "Anomalous Effects of Electron Correlation
        in the Energies of Excited States" Comm. At. Molec. Phys.
        <u>1/#4</u>, 116 (1971).

[19]   Strictly speaking there are different $\hat{f}_\lambda$'s for each $\hat{f}_{ij;k}$, i.e. each (k_i, k_j, k_k) – three RHF_MO combination. However in $\overline{V}$-states usually only one f_λ of each kind arise (by symmetry etc.) for the L-shells (there are new ones for K-electrons virtually excited cases important only in ESCA and in hyperfine (hfs) structure). One can also assume all f_λ 's from the same shell to be nearly the same with little sacrifice in content of theory or accuracy.

[20]   These Rydberg orbitals seem also consistent with the definition of C. Sandorfy (cf. Salahub and Sandorfy, Theo. Chim. Acta 20, 227 (1971).), but not with that of Mulliken (cf. Merer and Mulliken, Chem. Rev. 69, 639 (1969)).

APPENDIX

LINEAR DEPENDENCE OF ATOMIC ORBITALS IN LCAO MO AND IN RYDBERG
ORBITALS

The vectors $\{1a>, 1b>, 1c>, . . .\}$ are linearly independent if
and only if there exist no coefficients $\{c_i\}$ other than all zeros,
which satisfy

$$c_1|a> + \ c_2|b> + \ c_3|c> + \ ... \ = 0 \tag{A1}$$

Given a set of vectors how do we determine practically, if
there are some non-zero c_i or not   To get a practical criterion,
form the scalar products of equation (A1) first with $<a|$, then
with $<b|$, etc. obtaining

$$
\begin{aligned}
c_1 \ S_{aa} + c_2 \ S_{ab} + c_3 \ S_{ac} + ... \ &= 0 \\
c_1 \ S_{ba} + c_2 \ S_{bb} + c_3 \ S_{bc} + ... \ &= 0 \\
c_1 \ S_{ca} + \ . \ . \ . \quad\quad\quad\quad &= 0 \\
. \ \ . \ \ . \ \ . \ \ .
\end{aligned}
\tag{A2}
$$

i.e. the matrix form of (A1) as:

$$\mathcal{A} \ \mathcal{c} \ = \ 0 \tag{A3}$$

where
$$
\mathcal{A} \ = \ \begin{pmatrix} S_{aa} & S_{ab} & \cdots \\ S_{ba} & S_{bb} & \cdots \\ \vdots & & \end{pmatrix}
\qquad \mathcal{c} \ = \ \begin{pmatrix} c_1 \\ c_2 \\ \vdots \end{pmatrix}
$$

is the overlap matrix and $\mathcal{c}$ the unknown $\{c_i\}$.

For the homogeneous equation, equation (A3) to have a non-
trivial solution (i.e. $\mathcal{c} \neq 0$), the condition is

$$|\mathcal{A}| \ = \ 0 \tag{A4}$$

Thus is the determinant of the overlap matrix is zero, the set of
vectors is linearly dependent. Some of the vectors of the set will
then be redundant in the construction of a basis set.   Q.E.D.

ON THE ASSIGNMENT OF MOLECULAR RYDBERG SERIES[1]

Petr Hochmann, Hung-tai Wang, W. Sidney Felps, Stephen Foster and Sean P. McGlynn.
The Coates Chemical Laboratories, The Louisiana State University, Baton Rouge, Louisiana 70803.

INTRODUCTION

The strategy behind the identification of Rydberg series, and the reliability of such identifications, is a function of both the complexity and the resolution of the Rydberg spectral region (i.e., that region of the spectrum, usually in the vacuum ultraviolet, extending from photo-ionization limit to lower energies). Two extreme cases illustrate this view.

Consider a spectrum which exhibits resolved rovibronic structure over major parts of the R-region. As a result, the spectrum may be analysed to an extent which permits the assignment of rotational and vibrational progressions, as well as the identification of the "pure" excited electronic states with regard to both orbital configuration and spin multiplicity ($\Lambda\Omega$ coupling assumed). An R-spectrum of this sort permits Rydberg series identifications without any recourse to Rydberg term values, $T_n = R/(n + \delta)^2$. In turn, the term value equation may then be used to determine the parameters of individual R-series (i.e., ionization limit, I, quantum defect, etc.,) and to validate the prior assignment of series members.

Spectra suited to the above approach are generated only by very simple molecules.[2] The R-region of larger molecules is usually very dense, exhibiting much overlap, not only of individual vibronic bands but even of entire rovibronic band systems. For such molecules, identification techniques analogous to those for highly-resolved systems are not feasible: The spectra are too complex. As a result, the identification of R-transitions relies heavily on observed band shapes and intensities[3] and on correlations with related, simple systems. Consequently, R-series parameters are not usually determinable with much accuracy. Instead, the process is usually reversed:

I is obtained from PES data, δ is abstracted from studies of related molecules, and the assignment of an R-member is validated or
not depending on how it fits the term-value formula.

Our interest lies in spectra intermediate to these two limiting
cases: Spectra with unresolved rovibronic[4] fine-structure but with
resolved or partially-resolved vibronic structure over the entire
R-region. Such spectra are provided by most diatomics and by many
polyatomics. Since these spectra often contain one or more regions
of sharp, well-defined bands and since this sharpness and definition
is often preserved in related molecules, it follows that such
molecules provide an ideal means of studying (R-chromophore)-
(R-chromophore) interactions, (R-chromophore)-substituent interactions, etc. Consequently, their analysis is of some interest.

The approach adopted here is based on the fact that these
moderately resolved spectra often yield several hundred bands which,
with minimal difficulty, may be measured to an accuracy
$20 > \Delta\bar{\nu} > 10 cm^{-1}$ over the entire R-region. Consequently, the identification of R-series by "trial and error" selections from the ob-
served experimental set and by comparison of such selections with
the term-value requirements of a Rydberg formula is possible, at
least in principle. The inevitable ambiguities in a series obtained by such means may then be resolved by cross-correlation with
spectra of related molecules.

The practice required by such an approach implies computeri-
zation. Consequently, prior to provision of any details on this
aspect of things, it is well to emphasize the basic assumptions involved.

THE RYDBERG EQUATION
The frequency of the nth member of the αth R-series which
converges toward the ionization limit (I) is supposedly given by

$$\nu_{\alpha,n} = I - R/\{n + \delta_{\alpha}(n)\}^2 \qquad \ldots\ldots 1$$

where R is the Rydberg constant and $\delta_{\alpha}(n)$ is a quantum defect. The
observed frequencies, $\nu_{\alpha,n}$ may be compared with $\nu_{\alpha,n}$ of Eq. 1 in a
"point-by-point" fashion using quantities such as $\nu_{\alpha,n} - \nu_{\alpha,n}$, or
in a "whole-series" fashion using, for example, the quadratic deviation

$$D = \sum_{n} W_n \{\nu_{\alpha,n} - \nu_{\alpha,n}\}^2 \qquad \ldots\ldots 2$$

where the W_n are weighting factors.[5] It is this latter tactic
which is followed here. If the quantities $\nu_{\alpha,n}$ and the n-seriali-
zation within the series be known for a sufficient number of R-
series members, D may be written as a function $D = D\{I, \delta_{\alpha}(n)\}$ and
the quantities I and $\delta_{\alpha}(n)$ obtained by a minimization of D: That is,

by least-squares fitting procedures. In so doing, however, certain restrictions must be imposed on the $\delta_\alpha(n)$.

The simplest assumption, and the one normally made, is that the quantum defect is a constant for a given progression: That is, $\delta_\alpha(n) = \overline{\delta}_\alpha$. The degree to which the least-squares fitting reproduces the observed frequencies may then be studied using $\nu_{\alpha,n} - \overline{\nu}_{\alpha,n}$, the magnitude of D_{min}, or by comparing $\overline{\delta}_\alpha$ with $\underline{\delta}_\alpha(n)$ where

$$\underline{\delta}_\alpha(n) \equiv \{R/(I - \underline{\nu}_{\alpha,n})\}^{\frac{1}{2}} - n \qquad \ldots\ldots 3$$

with I being the least-square value of the ionization limit. If $\underline{\delta}_\alpha(n)$ exhibits a sufficiently small and adequately random variation about $\overline{\delta}_\alpha$, the original assumption (i.e., $\delta_\alpha(n) = \overline{\delta}_\alpha$) may be considered to be acceptable. Usually, however, the variations are neither small nor random. For example, for the s, p and f R-series of Neon through Radon, $\delta_\alpha(n)$ shows small but significantly systematic deviations from δ_α and these deviations are usually largest at small n. Similar observations also exist for molecular R-series. In such cases, the variations of $\delta_\alpha(n)$ are adequately reproducible using a power series expansion in 1/n

$$\delta_\alpha(n) = \sum_{k=0}^{K} \frac{\delta_{\alpha,k}}{n^k} \qquad \ldots\ldots 4$$

with $K \leq 3$. The parameters $\delta_{\alpha,k}$, as estimated by minimization of D, usually satisfy relations

$$\delta_{\alpha,0} \gg \delta_{\alpha,1}; \quad \delta_{\alpha,2}; \quad \ldots$$

$$\delta_{\alpha,0} \simeq \overline{\delta}_\alpha$$

In other instances (the d R-series of the noble gases, for example), the variations of the $\underline{\delta}_\alpha(n)$ are often larger than the differences $\underline{\delta}_\alpha(n) - \underline{\delta}_\beta(n)$ between different series and cannot be satisfactorily reproduced by Eq. 4 with any reasonably small value of K.

The observed behavior of the $\underline{\delta}_\alpha(n)$ probably results from a combination of factors. The highly excited R-states of an atom are presumably describable by an effective Hamiltonian of the sort discussed by Sommerfeld.[7] As n decreases, various perturbations[2] which make diagonal and off-diagonal correction to the original Hamiltonian become important. The diagonal corrections are held responsible for the smooth variation of $\underline{\delta}_\alpha(n)$ with n which sets in at small n whereas the off-diagonal terms, which destroy the "purity" of individual R-series members, cause the random variations of $\underline{\delta}_\alpha(n)$ about the smooth values.

In molecules, another effect is present, one presumably productive of smooth variations of $\underline{\delta}_\alpha(n)$ with n; this effect is associable with the "purely-electronic" or atomic nature of Eq. 1.

Molecular Rydberg transitions contain contributions from both rotational and vibrational motions and it is difficult to envision
the pertinence of Eq. 1 to such transitions. However, as indicated
by Eq. 3, the quantum defect is determined by the term value
$T_{\alpha,n} = I - \nu_{\alpha,n}$ or, equivalently, by the energy difference of a
molecular cationic and a molecular rovibronic R-level. At large n,
where the Rydberg orbital is largely non-bonding, it is expected
that the equilibrium geometry and vibrational frequencies of the
R-state and the cation are roughly identical. Consequently, the
rovibrational contributions to $T_{\alpha,n}$ largely cancel in taking the
difference $I - \nu_{\alpha,n}$ and the resultant values of $T_{\alpha,n}$ are surely
close to "pure-electronic." Indeed, providing that, the frequencies
$\nu_{\alpha,n}$ contain the same vibrational excitation v ", v ' for all of n,
the optimized ionization potential will be an $I_{v'',v'}$ value and
the quantum defects $\delta_\alpha(n)$ will be almost virgin of rovibronic effects.
If, however, the coupled vibrational frequency is a function of n,
this may result in different Franck-Condon band shapes for different
R-series members, particularly at low n, and both effects (variation
of band shape and of vibrational frequency) may result in a smooth
variation of $\delta_{\alpha,n}$ with n at low n.

    In view of the above, it follows that the most felicitous
choice of an R-series is one which involves excitation of a ground
state electron which is nearly non-bonding and almost atomic-like.
For such a series, the intensity will be concentrated in the 0", 0'
vibronic band and the v " and v ' vibrational frequencies will be almost identical. The best approximation to such a system we could
find is that available in iodine-containing molecules such as HI,
CH_3I, etc.

SERIES IDENTIFICATION
    The procedure used can be summarized as follows:
A.   Based on symmetry considerations, the photoelectron spectrum,
     and the nature of the cationic states, the number and the ap-
     proximate low-energy termini of the R-series can be estimated.
     In this connection, a previous analysis of a related molecule
     is exceedingly useful.
B.   The following quantities are now calculated for all observed
     frequencies, ν, and a trial value of I

$$n^*(\nu,I) = \{R/(I-\nu)\}^{\frac{1}{2}} \qquad \ldots\ldots 5$$

$$n = \text{Int}\{n^*(\nu,I)\} \qquad \ldots\ldots 5'$$

$$\Delta(\nu,I) = n^*(\nu,I) - n \qquad \ldots\ldots 5''$$

Providing that the frequency ν corresponds to the nth member of the
α^{th} R-series and that the trial value I is close to the series
convergence limit, $\Delta(\nu,I)$ equals $\delta_\alpha(n)$ of Eq. 3. Thus, from plots
of $\Delta(\nu,I)$ against n it is possible to make a trial selection of

series members: The points belonging to a given series should
group along a relatively flat curve. This procedure, which we
term "screening", is illustrated in Figure 1. The following

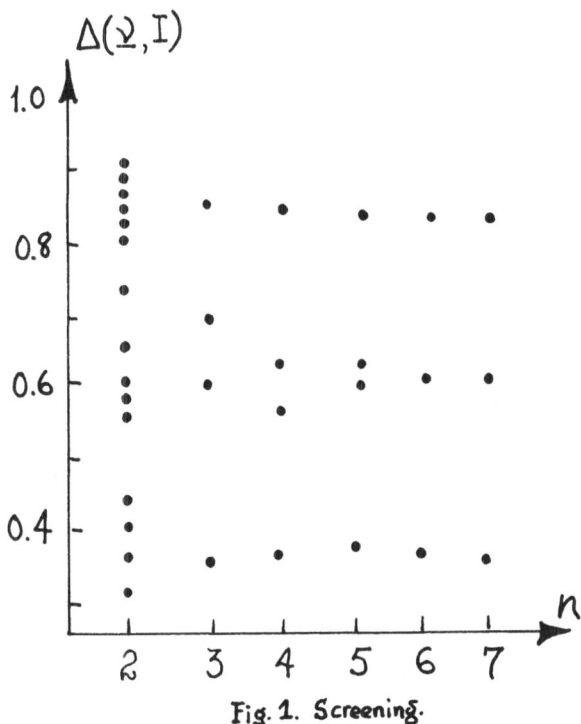

Fig. 1. Screening.

difficulties may be encountered in the screening process:
(i) The $\Delta(\underline{\nu}, I)$ versus n plot may contain many points representing
members of series with ionization limits different from I; some
points may not refer to R-series members at all; and points representing vibrationally-excited R-series members may provide further
complexity.
(ii) The shape of the $\delta_\alpha(n)$ versus n and the average value $\bar{\delta}_\alpha$ are
not a priori known.
(iii) The values of $\delta_\alpha(n)$ at large n are very sensitive to errors
in frequency, $|\Delta\underline{\nu}|$, and in the ionization limit, $|\Delta I|$. In fact,

$$|\Delta\underline{\delta}| \simeq \frac{(|\Delta\underline{\nu}| + |\Delta I|)\, n^3}{2R} \qquad\qquad \dots\dots 6$$

and, via Eq's 5, 5' and 5", $\Delta(\underline{\nu}, I)$ is limited to the range
$0 \leq \Delta(\underline{\nu}, I) \leq 1$. Since PES ionization potentials are usually in
error by more than $100\,cm^{-1}$, it follows that $|\Delta\delta| > 0.5$ for
$n \geq 10$, even if $|\Delta\underline{\nu}| = 0$. Thus, PES values for I provide only a

very primitive, initial selection of R-series members. The over-
crowding of the screening plots can be reduced by vibrational
analyses (which are usually possible for R-transitions in the low-
energy R-region) and by deductions based on the spectra of isotopic
molecules (CH_3I and CD_3I, for example). Since the variations of
$\delta_\alpha(n)$ are usually large only for low n, difficulty (ii) may be removed by considering that region of n values where $\delta_\alpha(n) \simeq \delta_{\alpha,0}$
and where difficulty (iii) has not yet grown inhibitory. In fact,
since errors in I produce deviations which are cubic in n, the
direction and even the approximate magnitude of the ΔI correction
can be estimated from the screening plots.

Finally, after several trial series have been selected, the
values of I and the corresponding $\delta_{\alpha,0}$ can be evaluated by minimization of the expression

$$D' = \sum_\alpha \sum_n W_{\alpha,n} \{\nu_{\alpha,n} - \frac{R}{(n+\delta_{\alpha,0})^2}\}^2 \qquad \dots\dots 7$$

where the weighting factors $W_{\alpha,n}$ depress the influence of the
frequencies corresponding to low or very large n as well as the less-
reliable selections.

C. The trial series are now compared with the measured spectra.
    Various criteria such as an assumed similarity of band shape,
    a regularity of band intensity and even of band environment,
    all for members of a given series, may be used at this point.
    Regularities relating the same members of different series
    (for example, the E_1 - E_3 separation of $\sim 5000 cm^{-1}$ in the
    iodides) may also be inspected. If discrepancies are noted,
    these may be investigated further or removed by repeating
    step B.
D. As a final check, the trial series may now be compared with
    those of isochromophoric systems which have been analysed by
    the same or different means (or investigators). Thus, the
    R-series of CH_3I may be compared with those of CD_3I, HI or Xe.
    This stage has proven to be very efficient in removing am-
    biguities which remain even after steps B and C.

CONCLUSION
    The results of an analysis of the CH_3I spectrum are given
in Table 1 for series terminating on the E_3 ionization limit.
Eleven series are identified. Since an equal number terminating
on the E_1 limit is also known, it follows that we imply an increase
of series from the previously known   six[8,9] to a new total   of
twenty two.

    The result of a comparison of similar series among isochromo-
phoric R-chromophores is shown in Figure 2. This comparison is
impressive and justifies our coinage of a new terminology.

    A comparison of previous ionization limits[8,10] with the

TABLE 1.

$\delta_{\alpha,}(n)$ FOR SERIES MEMBERS OF CH_3I TERMINATING ON THE $E3_{3/2}$ IONIZATION LIMIT[a]

n	A	B	C	D	E	F	G	H	I	J	K
1	1.01	0.96									
2	1.00	0.97	0.87	0.81	0.71	0.67	0.51	0.47	0.52	0.50	0.29
3	1.05	0.96	0.88	0.81	0.74	0.66	0.47	0.51	0.54	0.50	0.28
4	1.07	0.94	0.89	0.81	0.72	0.66	0.49	0.51	0.53	0.48	0.22
5	1.05	0.96	0.86	0.81	0.72	0.68	0.54	0.51	0.49	0.45	0.24
6	1.10	0.93	0.86	0.80		0.71	0.54	0.50	0.49	0.40	0.27
7	1.08	0.93	0.83	0.81		0.69	0.60	0.51		0.43	0.28
8	1.05	0.94		0.83				0.54			0.28
9	1.05			0.81				0.55		0.42	0.23
10		0.92		0.82				0.56		0.44	0.24
11	1.03			0.80	0.72						
12	1.02			0.79	0.74						
13	1.07	0.93	0.86		0.72						
14	1.06				0.71						
15		0.95									0.24
16						0.77					0.28
17						0.73					

a $I = 76960 \, \text{cm}^{-1}$

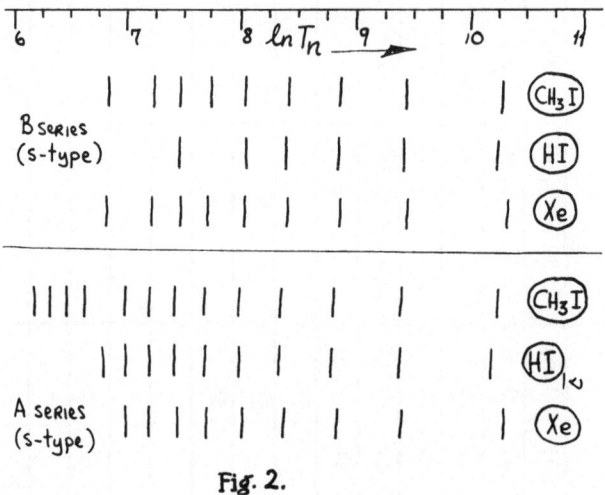

Fig. 2.

optimized values obtained here is given in Table 2. It is clear

TABLE 2
IONIZATION POTENTIALS
OF CH_3I and HI

CH_3I	HI	Source
$E_{3/2} = 76930\text{cm}^{-1}$	$\pi_{3/2} = 83560\text{cm}^{-1}$	Price (Ref. 8)
$E_{3/2} = 76960\text{cm}^{-1}$	$\pi_{3/2} = 83460\text{cm}^{-1}$	Present Work[a]
$E_{1/2} = 81990\text{cm}^{-1}$	$\pi_{1/2} = 89130\text{cm}^{-1}$	Price (Ref. 10)
$E_{1/2} = 82050\text{cm}^{-1}$	$\pi_{1/2} = 89130\text{cm}^{-1}$	Present Work

[a] The accuracy is $\pm 20\text{cm}^{-1}$.

that the precision is well beyond PES capabilities and that it is
excellent even for molecules such as CH_3I which exhibit only
moderate resolution in the R-region.

REFERENCES
1. Read by S. P. McGlynn; this work was supported by contract
    between the United States Atomic Energy Commission, Biology
    Branch and the Louisiana State University.
2. A molecular example is the low-temperature gaseous spectrum

of H_2: G. Hertzberg and C. H. Jungen, J. Mol. Spectry., 41, 425(1972).

3. A common statement is that R-transitions are "intense, narrow bands."

4. Or resolved only in certain regions, or completely resolved but not analysable.

5. These weighting factors depress the influence of $\nu_{\alpha,n}$ values which are experimentally uncertain.

6. A.B.F. Duncan, "Rydberg Series in Atoms and Molecules," Acad. Press, N. Y., 1971.

7. A. Sommerfeld, "Atombau und Spektralinien," Vol. II, Vieweg und Sohn, 1951.

8. W. C. Price, J. Chem. Phys., 4, 539(1936).

9. R. A. Boschi, D. R. Salahub, Mol. Phys., 24, 289(1972).

10. W. C. Price, Proc. Roy. Soc., (London), A 167, 216(1939).

ON MOLECULAR RYDBERG TERM VALUES[1]

S. Foster, P. Hochmann, W.S. Felps, H.t. Wang and S.P. McGlynn.

The Coates Chemical Laboratories, The Louisiana State University, Baton Rouge, Louisiana 70803.

INTRODUCTION

If the basis set of atomic orbitals for the $>C = 0$ group is restricted to the two $2p$ AO's x_π and x_0, which lie perpendicular to the sp^2 hybridization plane defined on the carbon center, and to the non-bonding AO of oxygen, n_0, which lies in the hybridization plane and is perpendicular to the $C = 0$ axis, it is not difficult to show that the ionization potential of the π-electron is related to the energy of the $\pi^* \leftarrow \pi$ transition by

$$I_\pi = 0.5h\nu_{\pi\pi}* - K(\alpha_c + \alpha_0)/2. \qquad \ldots\ldots 1$$

No simple relation between the ionization potential of the n-electron and the energy of the $\pi^* \leftarrow n$ transition is formulable; however, I_π and $h\nu_{n\pi}*$ can be connected via.

$$I_\pi = h\nu_{n\pi}* - (3/2)K\alpha_c, \qquad \ldots\ldots 2$$

where α_c and α_0 are Coulomb integrals for $2p_\pi$ electrons on carbon and oxygen, respectively, and K is a scaling factor. Granted the constancy of α_c and α_0 throughout a series of carbonyl-containing molecules, it follows that Eqs. 1 and 2 define very simple linearities which are subject to experimental test.

On the other hand, the energy of a Rydberg transition is given by

$$I = h\nu + R/(n \mp \delta)^2. \qquad \ldots\ldots 3$$

The energies of two electronic absorption bands of oxygen-containing molecules are plotted[3] against I_n in Figure 1. The

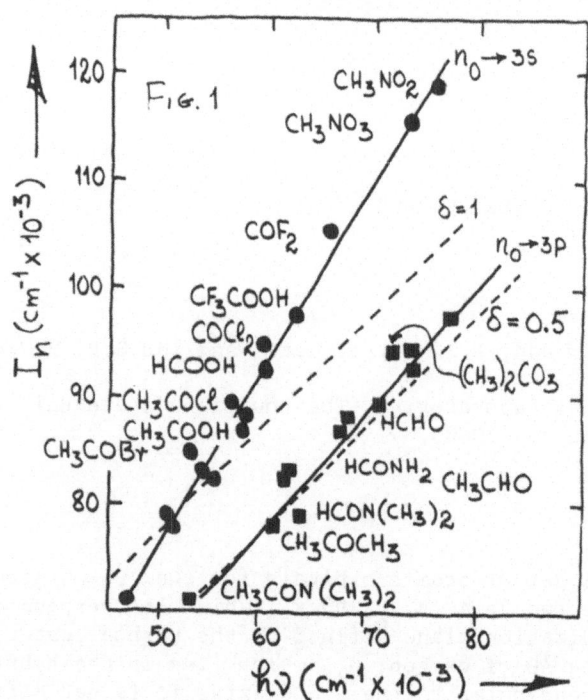

rightmost of the two linearities thus obtained fits Eq. 3 and is unambiguously identifiable as the $n_0 \to 3p(R)$ transition. Indeed, the deviation from the theoretical line for an $n_0 \to 3p(R)$ transition with $\delta = 0.5$, as shown by the dashed line, is quite negligible. One may conclude that the $n_0 \to 3p(R)$ transition is a perfectly normal member of the $n_0 \to np(R)$ series defined by the Rydberg formula.

The leftmost linearity of Fig. 1 is identified with the $n_0 \to 3s(R)$ transition for the following reasons: the theoretical line, for which $\delta = 1$, bisects the experimental line at acetone, the one point where a reasonably unambiguous $n_0 \to 3s(R)$ identification has been obtained by quite independent means. Secondly, the 3s orbital is a penetrating orbital and we expect it to be sensitive to nuclear field in the following way: in a highly electronegative molecule, it will sense the larger nuclear field and exhibit a larger δ (i.e., a lower energy) whereas in a highly electropositive molecule it will behave oppositively. We conclude, then, that the observed behavior is reasonable, δ being ≈ 1.2 for carbonyl fluoride with its heavily electron-withdrawing fluorine atoms and ≈ 0.9 for dimethylacetamide with its donor $-N(CH_3)_2$ and $-CH_3$ groups.

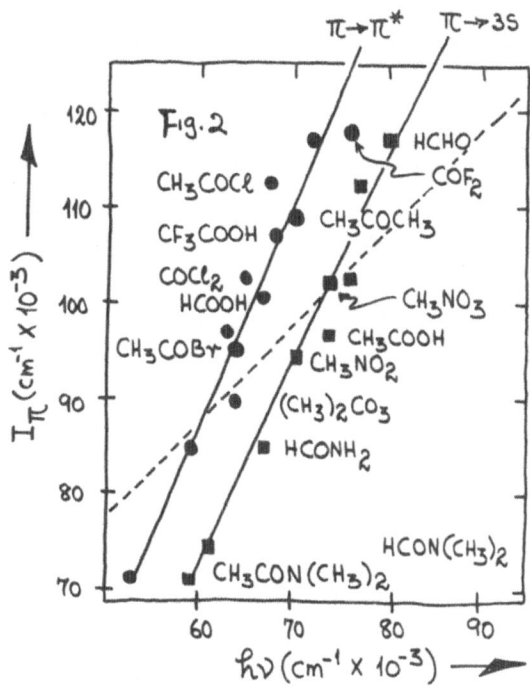

Fig. 2

The energies of two other absorption bands of the same set of molecules are linearized when plotted versus I_π. These plots[3] are shown in Figure 2. The slopes of both lines are identical and ≈ 2; they are also identical within experimental error with the slope of the leftmost line of Figure 1. The rightmost line of Figure 2 is thought to correlate the $\pi \rightarrow 3s(R)$ transition and the leftmost line of Figure 2 is thought to correlate the $\pi \rightarrow \pi^*$ transitions for reasons already outlined.

In any event, the utility of these ideas are evident in Figures 1 and 2 and they have also shown their value in discussions[4] of the R-transitions in both OsO_4 and RuO_4.

It was at this time, namely when we had finished the RuO_4/OsO_4 work, that we ran across a preprint of a paper by Robin.[5] This preprint stressed the importance of term values, T_n, as opposed to transition energies, $h\nu_n$; and we thought this to be a valuable point. It also introduced the idea that the 1st member of an R-series (in compounds containing an R-chromophore, which consisted of 2nd row elements, and an alkyl substituent) possessed a considerable chromophore to alkyl group charge transfer character. We found this idea very puzzling and, consequently, started a correlative game based on the term values of R-states and obtained the results which we now outline.

RESULTS

The alkyl iodides were studied first, simply because data were available to us. The result of our analysis is given in Table 1, where the Rydberg term values for the s-series which terminate at the quoted ionization limits are listed in serial order. As in Robin's analysis[5], the $(n+1)s(R)$ terms appear to approach a limiting energy of $\sim 24000 cm^{-1}$ as the alkyl substituent is enlarged. The T_{n+2} values, in contrast, vary but little; and the T_{n+3} terms, even less so. The insensitivity of T_{n+2} to alkyl substitution is, in our opinion, inconsistent with the idea that the T_{n+1} refers to an R-orbital which, when the alkyl group is large, is alkyl-localized. One might query, if such were correct, why T_{n+2} should suddenly revert to a clearly alkyl-independent behavior. More specifically, it is noted that in HI and Xe, where alkyl groups are absent, the term values of both T_{n+1} and T_{n+2} are quite similar to those for alkyl-containing compounds.

TABLE 1.

TERM VALUES AND LOWEST IONIZATION POTENTIALS FOR COMPOUNDS OF IODINE (in cm^{-1})

Molecule	Iodine Atom	Xe	HI	CH_3I	C_2H_5I	C_3H_7I	C_4H_9I	I_2	ICN
I	84340	97834	84128	77030	75417	74691	74449	75814	87675
T_{n+1}	29707	30766	27075	27316	25447	24674	24534	~24500	28790
T_{n+2}	12437	12645	12519	12192	12206	12113	12047	12276	12561
T_{n+3}	6890	7029	6910	6955					7095
T_{n+4}			4419	4515					4126
T_{n+5}			3047	3055					3138

We next decided to investigate trends in the s-series which might be independent of the R-chromophore. The results of such an investigation of atomic systems is shown in Figure 3. The T_{n+1} terms are clearly not isoenergetic, yet they do form a band centered at $\sim 30,000 cm^{-1}$, in suprising proximity to that found previously by Robin and by us (see Table 1) for molecules. The T_{n+j} band widths become increasingly narrow at higher j and, even at j = 2, the T_{n+2} values may be considered to be approximately constant.

It now became mandatory to explore as many molecular systems as possible and to determine what trends, if any, exist across a broad spectrum of R-chromophores. The results are shown in Table 2. The important content of this table is that all $T_{n+2,s}$ values are more or less identical, and all $T_{n+j, s}$, j > 2, virtually so. The variations of the T_{n+1} terms may reflect effects due to

TABLE 2.

TERM VALUES, T_n OF s-RYDBERG STATES, LOWEST IONIZATION POTENTIALS, AND QUANTUM DEFECTS FOR A VARIETY OF SYSTEMS (energies in cm^{-1})

MOLECULE	H_2O	H_2S	H_2Se	H_2Te	NH_3	HCl	HBr	HI	H-atom
I	101780	84420	79703	73705	82150	~102680	~94130	83815	109100
T_{n+1}	(36000)	(33000)	(28942)	(23700)	36020	(27546)	27046	27075	27420
T_{n+2}	12610	12651	12605	12146	12392			12519	12186
T_{n+3}	7000	6983	6999	6689	6938			6910	6855
T_{n+4}	4460	4422	4475	4297				4419	4387
T_{n+5}	3090	3070	3095	3002				3047	3046
T_{n+6}	2280	2279	2270	2208					2238
T_{n+7}	1740	1728	1737	1693					1713
T_{n+8}		1352	1365	1355					1354
T_{n+9}		1095	1099	1084					1097
T_{n+10}		920	917						906
T_{n+11}		766							761
δ	1.05	2.04	3.05	3.95	1.02	---	---	---	---

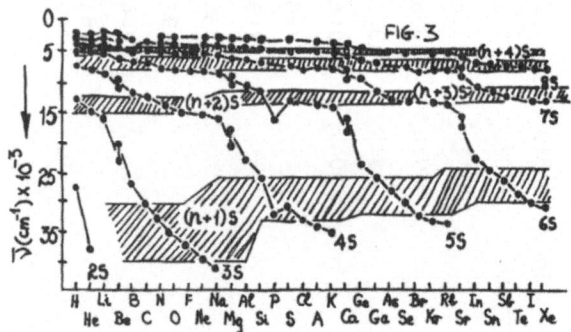

incomplete shielding and/or imperfect core/Rydberg orbital separability. The term values are also seen to be more or less hydrogenic (see last column of Table 2), indicating that the hole in the core behaves as a proton for $j \geq 2$. In addition, the study of other ionization limits in the same molecules produces a set of term values of similar type to that in Table 2 and indicates that the orbital nature of the hole has little effect on T_{n+j} values.

The constancy of s-term values in molecules, as compared to atoms, is probably a result of charge compensation. In other words, the $(n+1)s$ electron in an alkyl iodide sees a smaller effective charge when the alkyl group is large than when it is small. And it is this inductive effect which is, we believe, responsible for Robin's observations.

The results outlined facilitate the assignment of $N \rightarrow (n+j)s$ Rydberg transitions $(j \geq 2)$, and the $N \rightarrow (n+1)s(R)$ assignments, clearly, are hardly any more difficult. Our own attempts, (i.e., in I_2, CF_3I, CH_3Br, CH_3Cl, OsO_4 and RuO_4, etc.,) indicate that it requires no more than a few minutes. Thus, in conjunction with the recent availability of PES data, s-series assignments should become routine.

It now remains to investigate the p-series. Our efforts in this area are hampered by a lack, not only of reliable data, but even of tentative assignments. Nonetheless, it does seem clear to us presently that many regularities corresponding to those for the s-series will result.

CONCLUSIONS
Two rather simple methods for the assignment of Rydberg s-series

members have been outlined. One rather simple method for the assignment of Rydberg p-series has been illustrated. The d-series have not yet been investigated. We disagree with the alkyl-group nature of the $(n+1)s(R)$ state advocated by Robin; we interpret alkyl group effects on $T_{n+1~s}$ as a straightforward charge compensation effect: As the alkyl group becomes larger, the "Rydberg Hole" tends to delocalize away from the chromophore and on to the alkyl moiety.

REFERENCES
1. Read by S. P. McGlynn; this research was supported by contract between the United States Atomic Energy Commission, Biology Branch and The Louisiana State University.
2. H. J. Maria, J. L. Meeks, P. Hochmann, J. F. Arnett and S. P. McGlynn, Chem. Phys. Letters, 19, 309(1973).
3. These plots, as a result of the generation of better data, are somewhat different to, and supercede those given in reference #2 above.
4. S. Foster, S. Felps, D. Larson, L. Johnson and S. P. McGlynn, J. Amer. Chem. Soc., in press.
5. M. B. Robin, proceedings of : International Symposium on Atomic, Molecular, and Solid-State Theory and Quantum Biology, Sanibel Island, Florida, 1972.

NOTE ADDED IN PROOF: The implications of the text concerning the interpretation of the data in Figure 2 is probably wrong. The only statement that should be made concerning Figure 2 is that the transition energies listed thereon probably correspond to some one of the following transitions: $\pi \to \pi^*$; $\pi \to 3s(R)$; or $n_o \to 3d(R)$.

CORRELATION OF MOLECULAR AND RARE GAS TERM VALUES*

John D. Scott, G. C. Causley and B. R. Russell

Chemistry Department
North Texas State University
Denton, Texas 76203

The assignment of absorption bands for polyatomic molecules in the vacuum ultraviolet spectral region is often quite difficult; however, the observation of trends in the spectra of a series of compounds can be helpful in aiding the spectral assignments. The specific details of many of the complexities of the absorption bands require the use of high resolution data; however, low resolution spectral studies can provide insight into the basic electronic features. Therefore, the energetics of the interactions of substituents with a specific chromophore of interest can be inferred from this data. It should be noted that the chromophore can be "perturbed" either by varying the substituents adjacent to it, or by maintaining the substituents constant and replacing the atom(s) of the chromophoric unit with electronically similar species. The later technique was utilized in a study herein presented.

One of the trends that has been observed is that in many molecules the Rydberg transitions demonstrate certain similarities. Certain information can be gained by comparing the molecular values with the atomic transitions. Robin has shown that the initial members of a molecular Rydberg series can be related for many compounds when this transition is considered with respect to the term value (T_n),[1] where

---

*Research sponsored by The Robert A. Welch Foundation Grant B-470 and Faculty Research Funds of North Texas State University.

$$T_n = IP - \nu_n$$

Robin has developed a relationship of the term values which aids in assigning absorptions which are difficult to attribute to valency or Rydberg transitions. The term values are effectively constant for many compounds depending on the type of Rydberg transitions (ns, np, or nd). McGlynn and co-workers[2] observed similar trends of term values and also extended to the higher Rydberg members and emphasized the relation to hydrogen term values. These methods have been quite useful in treating data. We have noted another analogy which has aided in the assignment of not only the specific type of Rydberg but the principal quantum number as well.

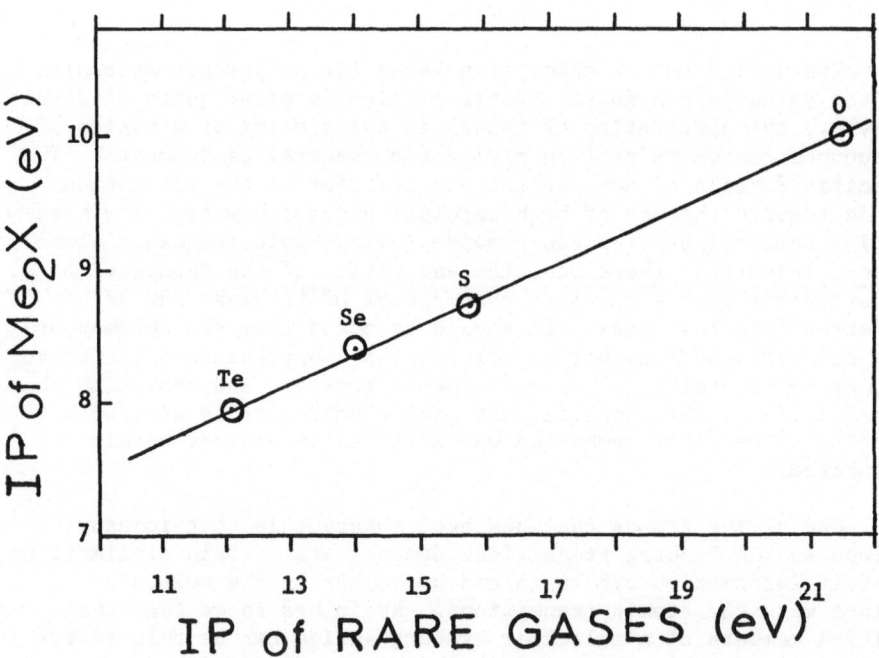

Figure. The first ionization potentials (IP) of Me_2X (X=O,S,Se and Te) vs the first IP's of the rare gases, where the IP of Me_2X is plotted vs the IP of the rare gas in the corresponding row of X.

Recently this group has investigated the spectra of three dimethyl compounds of group VI A and noted a correlation with the rare gases. Rydberg series have been assigned in the spectra of dimethylsulfide, dimethylselenide, and dimethyltelluride and converge on ionization potential (IP) values of 8.71 eV, 8.40 eV and 7.93 eV, respectively.[3] In the spectrum of each compound three series (s, p, and d), containing as many as eight members, were assigned. A plot of the IP's of the compounds, including the IP of dimethylether[4], verses the IP's of the respective inert gases, produced a linear relationship (see figure). This result, which has been observed by Price and co-workers[5,6] for other compounds, supports the assignment of the first IP's of all three compounds, and also that of dimethylether, to the non-bonding p orbital which is perpendicular to the C-M-C plane in each case. The results of this relationship were extended by considering the term values for the compounds. The atomic term values were taken from Moore's _Atomic Energy Levels_[7] and examples of the resulting values are given in the table. It should be noted that the initial members of the series for the molecules are expected to be the most divergent due to the interactions of the lower lying Rydberg members with the molecular core. The comparison of the term values was made initially with emphasis on the fitting of the

Table. Term values of dimethylsulfide (Me_2S), dimethylselenide (Me_2Se) and dimethyltelluride (Me_2Te) np molecular Rydberg series[a] to the corresponding atomic series[b] of argon, krypton and xenon. (units of cm^{-1})

	Me_2S	Ar		Me_2Se	Kr		Me_2Te	Xe
n	T_n	T_n	n	T_n	T_n	n	T_n	T_n
4	19063	20056	5	18511	18821	6	17690	17715
5	9462	9547	6	9184	9153	7	8855	8991
6	5681	5640	7	5591	5504	8	5392	5278
7	3759	3725	8	3671	3618	9	3667	3548
8	2672	2671	9	2640	2606	10	2576	2547
9	1990	1987	10	1998	1958	11	1952	1918
10	1549	1548	11	1550	1524	12	1523	1496
						13	1215	1199

a. The values of δ from the Rydberg formulas are Me_2S (δ=1.59), Me_2Se (δ=2.56) and Me_2Te (δ=3.50).

b. The particular atomic transition designated by Moore[7] for all the rare gases is $mp^5(^2P_{3/2})np(1/2)$, J=0.

third or fourth members and then extended for the lower and higher members. Corresponding relations were found for the ns and nd series for each spectra and although the agreement is not as good as for the np series, the correlation is still obvious. Initially, the first member of each Rydberg series was arbitrarily assigned an $\underline{n}$ of 3; however, the quantum defect values were changed by adding integers after the correlations with the atomic series were found. The integers were added to produce a correspondence between the already-correlated term values. For example, the δ of the dimethylsulfide series having a δ of 1.59 in the table was originally 0.59; however after the correlation with the argon $mp^5(^2P_{3/2})np(1/2)$, J=0 series (where the first value of n is 4) was established, addition of the integer $\underline{1}$ to δ allowed the comparable term values to have the same $\underline{n}$ values.

This correlation of molecular and rare gas term values has several significant aspects. The relation of atomic series to molecular Rydbergs can be used to aid spectral assignments, particularly if Rydberg members are obscured by other spectral features. Once fit into series, the specific manifold and excited state can be assigned. The quantum defect for each series of transitions can be assigned in terms of atomic notations. Finally, the magnitudes of the variations between the term values of the molecular series and the atomic series can be determined and thereby be helpful in providing an understanding of the molecular involvement in the lower Rydberg states.

References

1. M.B. Robin, "Higher Electronic States of Polyatomic Molecules," Vol. 1 in press.

2. S.P. McGlynn, "Optical and Photoelectron Spectra of Inorganic Molecules," paper presented at NATO Advanced Study Institute 1973, contained in proceedings.

3. John D. Scott, G.C. Causley and B.R. Russell, J. Chem. Phys., in press.

4. K. Watanabe, Toshio Nukayama and Joseph Mottl, J. Quant. Spectrosc. Radiat. Transfer., 2, 369 (1962).

5. W.C. Price, J.P. Teegan and A.D. Walsh, Proc. Roy. Soc., 201A, 600 (1950).

6. H.J. Lemka, T.R. Passmore and W.C. Price, Proc. Roy. Soc., 304A, 53 (1968).

7. C.E. Moore, "Atomic Energy Levels," Vol. I, NBS, 1949.

ON THE POSSIBILITIES OF STUDYING THE ELECTRONIC STRUCTURE OF ORGANIC MOLECULES THROUGH THE ANALYSIS OF THE WAVE FUNCTION

O. CHALVET and R. CONSTANCIEL

Centre de Mécanique Ondulatoire Appliquée
23, rue du Maroc, 75019-PARIS

Molecular diagrams are useful tools for studying the electronic structure of various states of a molecule.

By molecular diagram one means generally the ensemble of local indices as charges, bond orders or other properties which may be derived from one particle density matrix (1).

Such indices have played an important role in the study of chemical reactivity of the molecules in their ground state.

In the first part of this paper we recall some applicability conditions which limits in fact, the use of such indices.

In the second part we will show how the analysis of the wave function in term of two particles density can lead to another kind of indices. In a perturbational treatment of the molecular interactions we can say that the static indices are zero order quantities. According to the usual development of the perturbation theory the knowledge of zero order quantities is sufficient to obtain the energy to first order and then can give some informations about the orientation of the reaction at its beginning.

The most usual static indices (charges - free valence and so on) are quantities derived more or less

directly from the one-particle density matrix, whose
diagonal elements give the one electron density distribution of the system. In what follows we will em-
phazise two main points. The first is that the use of
quantities derived from the one particle density matrix
alone can be justified only by use of a very particular
model of reactant. The second point is that these indices are "correct" in the framework of a given
type of approximate method (e.g. SCF) only.

These two points are related to the applicability
of Hellman-Feynman's theorem (2). It is well known that
some change in the total hamiltonian can be related to
change in the total energy according to the formula

$$\frac{\partial < \Psi | H | \Psi >}{\partial \alpha} = < \Psi | \frac{\partial H}{\partial \alpha} | \Psi >$$

Where α is a parameter related to a given modification
of the system and ψ the exact wave function.

If we develop the total hamiltonian as is usual

$$H(1...n) = \sum_i H_1(i) + \sum_{i>j} H_2(i,j) \tag{1}$$

we have

$$\frac{\partial H}{\partial \alpha}(1...n) = \sum_i \frac{\partial H_1}{\partial \alpha}(i) + \sum_{i>j} \frac{\partial H_2}{\partial \alpha}(i,j)$$

and then if we consider that ψ is the exact wave
function we obtain

$$\frac{\partial}{\partial \alpha}<\Psi| H |\Psi> = \frac{\partial E}{\partial \alpha} = Tr \frac{\partial H_1}{\partial \alpha} P_1 + Tr \frac{\partial H_2}{\partial \alpha} P_2$$

where P_1 is the one particle density matrix and P_2
two particles density matrix.
This shows that even if we use the particular formula
(1) the change of energy is known not only in terms
of P_1 but also in terms of P_2.

It is clear from this expression that P_1 can
give some information about the possibility of the
reaction only if $\partial H_2/\partial \alpha$ is equal to zero. This is
the case for the point charge model of reactant by

which no two-electron perturbation is introduced.

On the other hand when we use as ψ function an approximate one (which is always the case) the Hellman Feynman theorem takes the form

$$\frac{\partial}{\partial \alpha} <\psi \mid H \mid \psi> = \frac{\partial}{\partial \alpha} \text{Tr } HP_n = \text{Tr } \frac{\partial H}{\partial \alpha} P_n + \text{Tr } H \frac{\partial P_n}{\partial \alpha}$$

which involves a correction term. In the framework of the perturbation theory it means that the change in energy is no longer determined by the knowledge of the wave function but also depends on the change of the wave function itself. There exists however a particular case where the corrective term vanish : that is in the Hartree-Fock approximation of the wave function. Then we have for a one electron perturbation

$$\frac{\partial}{\partial \alpha} \text{Tr } HP_n = \frac{\partial}{\partial \alpha} \text{Tr } H_1 P_1 = \text{Tr } \frac{\partial H_1}{\partial \alpha} P_1$$

One important example where the corrective term must be taken into account is when the excited states are represented by a function which is a superposition of a limited number of configurations (3).

A very demonstrative case is that of the protona- tion of molecules in their excited states (4) when we use one configuration only : here the density matrices P_1^S and P_1^T are rigorously equal although an energetical calculation shows that the energy variation under the influence of the proton is different in each states : consequently all this difference is accounted for the corrective term alone. It is important to observe that, in considering the two particle density matrix P_2, the term $\frac{\partial P_2}{\partial \alpha}$ is obviously different in singlet and triplet states. This provides a simple explanation of the behaviour of the molecules in this states.

On the other hand if we use an SCF procedure to calculate the wave function of excited singlet or trip- let states we obtain different P_1 for this states. For example for phenanthrene (table 1) (5) we have

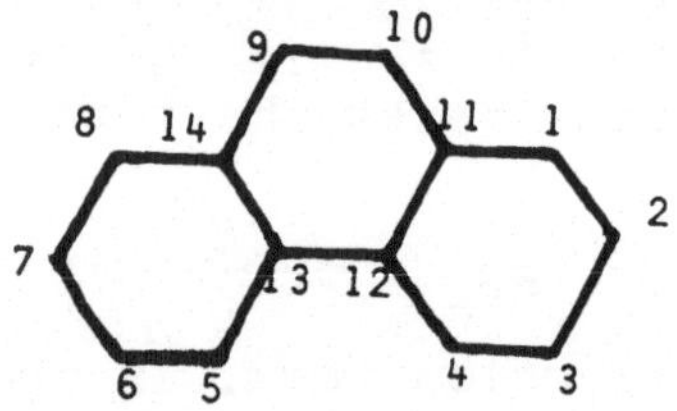

Bond Order

Bond	Ground state	Singlet state	Triplet state
1 - 11	0.570	0.452	0.559
1 - 2	0.719	0.703	0.657
2 - 3	0.612	0.638	0.675
3 - 4	0.716	0.572	0.615
4 - 12	0.580	0.684	0.686
9 - 10	0.816	0.447	0.171
10 - 11	0.458	0.625	0.557
11 - 12	0.580	0.482	0.516
12 - 13	0.421	0.319	0.301

Table 1

In the second part of this paper, we wish to describe a method of analyzing (6) the wave function with particular examples for the molecules in their excited states. The aim of such an analysis is to provide a picture of the electron distribution in a way which allows physical interpretation.

Let us first consider a wave function Ψ describing a n electron system. We may then talk about an independant subsystem of p electron if Ψ can be decomposed into the product

$$\Psi (1...n) = \Psi_A (1...p)\Psi_B (p + 1...n)$$

For such a case the p-particle reduced density matrix is idempotent. In reverse, knowing the p-particle density matrix P_p of a system and the projector P on a given functionnal subspace E_p the projection $P P_p P$ may be or not idempotent : in other words the idempotency property is caracteristic of P_p and of the subspace E_p . Then the study of the idempotency of

P P_p P can lead to define the best subspace adapted to
this property and we are able to precise where the p
electrons are better "localized" when some region of
space can be associated to the functionnal subspace
E_p. The "localizability" can be measured by means of
the dispersion of the diagonal elements of the matrix
P P_p P : this is achieved by using the missing information function I   (7).

We report here the details of such investigations
for the case of the two particle density matrix in the
framework of the closed shell SCF approximation of the
wave function. The two particle density matrix is
derived from the one particle one through the well
known formula in Löwdin's normalization (1).

$$(P_2)_{ab,cd} = \frac{1}{2} \left[(P_1)_{ac} (P_1)_{bd} - \frac{1}{2} (P_1)_{ad} (P_1)_{bc} \right] \qquad (2)$$

The subspace E_p is spanned by all the products
of the two atomic orbitals r and s defining a "bond".
It can be shown that the matrix P_2 is diagonal in
the basis of geminals constructed from two localized
bonding and antibonding orbitals L and A defined by

$$L = \frac{1}{\sqrt{1+\lambda^2}} (r + \lambda s)$$

$$A = \frac{1}{\sqrt{1+\lambda^2}} (\lambda r - s)$$

with an optimized value λ_0 of the polarity parameter
λ which is the positive value of the equation

$$(\lambda^2 - 1) P_{rs} + \lambda (q_r - q_s) = 0$$

The missing information is then computed from the four
diagonal elements. This procedure can be extended to
the case of excited states. The formula (2) is replaced
by the convenient formula for triplet or singlet state
corresponding to an open shell configuration.
For example we show in table 2 the results of the
application of this type of analysis for the ground
state of butadiene.

BUTADIENE, GROUND STATE

Bond	p_{rs}	n_1	n_2	n_3	n_4	I
12	0.9745	0.9747	0.0125	0.0377	0.0002	0.1482
23	0.2245	0.3873	0.2351	0.7052	0.1427	0.9768
13	0	0.2500	0.2500	0.7500	0.2500	1.0000
14	-0.2245	0.1503	0.2373	0.7122	0.3748	0.9808

Where p_{rs} is the usual bond order, n_1 the electron pair density in the local ground state, n_2, n_3, n_4 the pair densities in singlet state, triplet state, biexcited state respectively and I the missing information.
For the excited singlet and triplet states the following preliminary results were obtained.

BUTADIENE, EXCITED SINGLET STATE

Bond	p_{rs}	n_1	n_2	n_3	n_4	I
1-2	0.4976	0.5	0.3746	0.3786	0.0016	0.7956
2-3	0.5488	0.5488	0.3274	0.3714	0.0	0.7814
1-3	0.0	0.5	0.25	0.75	0.25	0.9755
1-4	0.4512	0.4512	0.4250	0.3714	0.0	0.7939

BUTADIENE, EXCITED TRIPLET STATE

Bond	p_{rs}	n_1	n_2	n_3	n_4	I
1-2	0.4383	0.4603	0.1358	0.6539	0.0	0.8498
2-3	0.7405	0.7405	0.0961	0.3555	0.0	0.6626
1-3	0.0	0.25	0.25	0.75	0.0	0.8982
1-4	0.2595	0.2595	0.0961	0.8366	0.0	0.8675

REFERENCES

(1) Löwdin P.O., J. Chem. Phys. 18, 365 (1950).
(2) Landau L., Lifchitz E., Mécanique Quantique, Ed. Miz (Moscou) 1966.
(3) Constanciel R., Theoret. Chim. Acta 26, 249 (1972).

(4) Bertran J., Chalvet O., Daudel R., Theor. Chim.
     Acta 14, 1 (1969).
     Chalvet O., Constanciel R., Rayez J.C., 6th
     Jerusalem Symposium. To be published "Chemical and
     Biochemical Reactivity".
(5) Bessis G., and Chalvet O., unpublished results.
(6) Constanciel R., Chem. Phys. Letters 16, 432 (1972).
(7) Aslangul C., C.R. Acad. Sc. (Paris) 272B, 1 (1971).

THE EARLY YEARS OF PHOTOCHEMISTRY IN THE VACUUM ULTRAVIOLET

Wilhelm Groth

Institut fuer Physikalische Chemie
der Universitaet Bonn

In 1934 Paul Harteck came to Hamburg to become director of the Institute for Physical Chemistry. He came from Cambridge, England, where he had worked with Rutherford on the production and the first nuclear reactions of deuterium, while I was a young assistant in the institute with Otto Stern, who emigrated in 1933. Harteck told me that he, together with a post-graduate student, F. Oppenheimer, had constructed a new light source for the extreme ultraviolet; this was in 1932 in the Kaiser Wilhelm Institute for Physical Chemistry in Berlin-Dahlem under Fritz Haber. They had never used the lamp for photochemical investigations.

I was very enthusiastic about the idea of starting a new field of research, and was anxiously looking forward to the arrival from Berlin of a box containing the lamp. When it arrived the disappointment was great; all that we found in the box were some fragments of quartz together with some metal pieces. So my work on the photochemistry in the Schumann ultraviolet started with the reconstruction and development of the xenon lamp, which I shall describe later.

The wavelength region between 1850 Å and about 1250 Å was opened to spectroscopy around 1900 through the pioneering work of V. Schumann, who introduced three essential experimental improvements: first, the use of CaF_2 (fluorite) windows and optics instead of the quartz apparatus which had been used up to that time in ultraviolet spectroscopy; second, the construction of vacuum spectrographs to eliminate the absorption of the atmosphere; third, the invention of the Schumann plate for the photographic registration of radiation of extremely short wavelengths.

Photochemistry, which is the younger sister of spectroscopy and is more related to chemistry, reached the Schumann region only after 1930. The reasons are that the use of fluorite (CaF_2) windows and the elimination of the atmospheric absorption caused experimental difficulties, and that suitable light sources with sufficient intensity in this wavelength region were not available. The difficulties were overcome in the thirties, and a considerable number of papers were published after 1930 by several institutes in the U.S.A., England, Germany, and Russia.

The relevance of this field of research for physics was that it was almost impossible - especially in the case of polyatomic molecules - to define all excitation levels of the molecule and its potential dissociation products by spectroscopic investigation only. As far as diatomic molecules are concerned, the situation was rather simple: according to the nature of the spectrum (lines, bands, predissociation or continuum), the physical process (excitation of electronic, vibrational, or rotational energy, dissociation into normal or excited atoms) could be explained very accurately. This was different in the case of polyatomic molecules; at about 10-12 atoms a limit of the molecule size was reached where no vibrational structure at all could be observed in the absorption spectrum. The larger the molecule, the more difficult was the interpretation of the spectroscopic data.

In its beginning, photochemistry was primarily concerned with the demonstration of Einstein's equivalence law. This changed when it became possible - by improvement of the analytical tools, by the construction of suitable light sources, and by exact measurement of light intensities - to determine quantum efficiencies with a reliability of a few percent; secondary reactions, which in most cases include free atoms or radicals, became more interesting. As is well known, the importance of photochemistry for the kinetics of thermal reactions originates here, since the secondary reactions are independent of the light. This independence is due to the high reactivity of the particles formed in the primary process, which causes a very low concentration and therefore a normally negligible light absorption by the primary dissociation products.

Of special importance was the mechanism of chain reactions. By the photochemical decomposition of suitable organic compounds, it was possible to start chain reactions with an exactly known number of primary radicals, which is given by the number of absorbed quanta, and to state directly the chain length by the quantum efficiency of the total reaction. Furthermore, it was possible to dissociate inorganic and organic molecules by light quanta of sufficient energy into partially excited radicals which could be detected directly by their emission spectrum.

These examples show that photochemical reactions - even though their activation mechanism is entirely different from that of thermal reactions - were an independent and therefore informative method for the investigation of the qualities and reactions of radicals.

The direct photochemical decomposition of simple hydrocarbons such as methane, ethane, etc., as well as of many important simple inorganic substances such as oxygen, nitrogen, carbon monoxide, carbon dioxide, water vapour, etc., is only possible in the Schumann ultraviolet, since all these elements and compounds are transparent for light of longer wavelengths.

The first spectroscopic investigations in the Schumann ultraviolet were carried out with the light of condensed sparks between metal electrodes and with quartz optic. Crystalline quartz is transparent down to about 1500 Å, while molten quartz in normal thicknesses already absorbs strongly at 1850 Å. For photochemical purposes in the Schumann ultraviolet, it is therefore usable only in thicknesses of some 1/10 mm and below in the form of quartz bubbles, etc.

Calcium fluoride and lithium fluoride are essentially the only useful optical materials, the absorption coefficients of which were known. The optical qualities of both materials are strongly dependent on the smallest impurities. In earlier years one had to find sufficiently large pieces of pure natural lithium fluoride, but later large fluoride crystals with excellent optical qualities could be synthesized. Though lithium fluoride has the highest transparency in the vacuum ultraviolet of all solid materials, difficulties arise when it is used for photochemical purposes. Electron bombardment as well as the contact with a low voltage discharge decrease the transparency of lithium fluoride after a short time, as is evidenced by a visible discoloration.

Absolute energy measurements in the Schumann ultraviolet with methods used in the visible or the near ultraviolet (bolometers, thermopiles) are impossible due to the low intensity after spectral decomposition. Photoelectric methods have been applied only in few cases, since spectral energy distribution curves of the photoelectric effect in this wavelength region were not well known. An important and generally usable method is photography. For photographic purposes in the Schumann ultraviolet, one uses either Schumann plates, which are low in gelatine content, or normal photographic plates which are sensitized for the short wavelengths by a thin layer of a substance which fluoresces in the visible (f.i. a thin oil layer). For the measurement of the relative energy distribution by this fluorescence photography, sodium salicylate is especially appropriate; the intensity of the fluorescence spectrum of this substance is proportional to the number of the incident quanta, independent of the wavelength of the exciting light.

The only absolute method for energy measurements in the Schumann ultraviolet is the photochemical one. It is possible – as we shall see later – to determine spectral energy distributions with an accuracy of a few percent by quantitative determination of the dissociation products of photochemical gas reactions.

The most important problem was the development of suitable light sources. Condensed sparks between metal electrodes were of some use, but they have several disadvantages: relatively small intensity, electrical perturbations which are caused by their operation, and large intensity fluctuations which produce large variations in the concentration of the formed primary products. In order to produce a constant intensity, electrodes in the form of slowly rotating disks were used. As electrode material, aluminium (1720 Å and 1850 Å), tin (1300 to 1600 Å), cadmium (between Al and Sn), silver solder, and tungsten carbide were used.

The hydrogen continuum which extends from the visible to about 1500 Å was also used for photochemical purposes. Since the emitted light intensity is proportional to the current, the energy in water-cooled quartz tubes was increased to >10 KVA.

In order to investigate photochemical gas reactions in the wavelength region below 1500 Å, P. Harteck and F. Oppenheimer had designed the low voltage xenon lamp with the resonance wavelengths at 1469 Å and 1296 Å which I mentioned in the beginning, and which I had the opportunity to develop. Harteck's first lamp was made of quartz, except for the fluorite window. The low voltage electrodes were melted into hard glass and connected to the lamp by ground glass joints; their top ends were situated in the centre of a sphere of 5 cm diameter blown out of the quartz tube. They emitted electrons very easily – if they were heated to about 1000°C – so the lamp could be operated by 220 V a.c. and a suitable resistance. The constant glow potential was about 50 volts. At first, the lamp was filled with a few torr of pure xenon. Later, it proved more practical to fill the lamp essentially with neon or argon and to add only a few percent of xenon. This had several advantages: first, neon and argon have a higher heat conductivity than xenon, so that the electrodes could be loaded to a higher degree; second, the electrodes are less sputtered in a mixture with a lighter rare gas; third, neon and argon are less expensive. Since the excitation energy of xenon is much smaller than that of neon and argon, practically only the xenon atoms were excited. The part of the tube near the fluorite window was reduced to about 7 mm in order to increase the potential gradient. Here the quartz tube was surrounded by a water mantle to keep it from melting. The lamp could be operated at up to 50 amps. Then the current density in the narrow part of the tube was about 140 amps/cm^2. The resonance lines are strongly broadened at this large current density and high temperature so that no significant self-absorption takes place.

The fluorite window was sealed into a metal mounting which was water-cooled and fitted the quartz ground joint of the lamp. In order to investigate photochemical gas reactions, a metal chamber could be attached to the metal mounting of the fluorite window.

In order to determine the sum of the line intensities at 1469 Å and 1296 Å, the photolyses of ammonia and of carbon dioxide were used. The quantum yield of NH_3 in the wavelength region below 2000 Å had been measured by Kistiakowsky (1932) to be 0.17. Under the assumption that the quantum yield for the xenon resonance lines was also 0.17, an intensity of 3×10^{17} quanta/sec followed from the amount of hydrogen and nitrogen produced. The intensity distribution between the two resonance lines could not be measured. Following the decomposition of CO_2 into CO and $1/2 \ O_2$, it was supposed that part of the O atoms recombined with CO to give CO_2; then again an intensity of 3×10^{17} quanta/sec was achieved.

Fig. 1. High-current low-voltage xenon lamp.

Improvements were made in different ways: (1) the heads of the electrodes were water-cooled to allow currents of up to 60 amps; (2) a circuit apparatus was used for high purification of the filling gases; (3) a magnetic field of 4000–5000 Gauss was operated on the narrow part of the lamp near the fluorite window, which has two effects: (a) the potential gradient and therefore the energy

which is consumed in this region is strongly increased, the constant
glow potential being raised from 50 V to about 90 V; (b) the self-
absorption of the resonance lines is essentially decreased by the
Zeeman effect produced by the magnetic field. These effects increased the light intensity by a factor of 10,

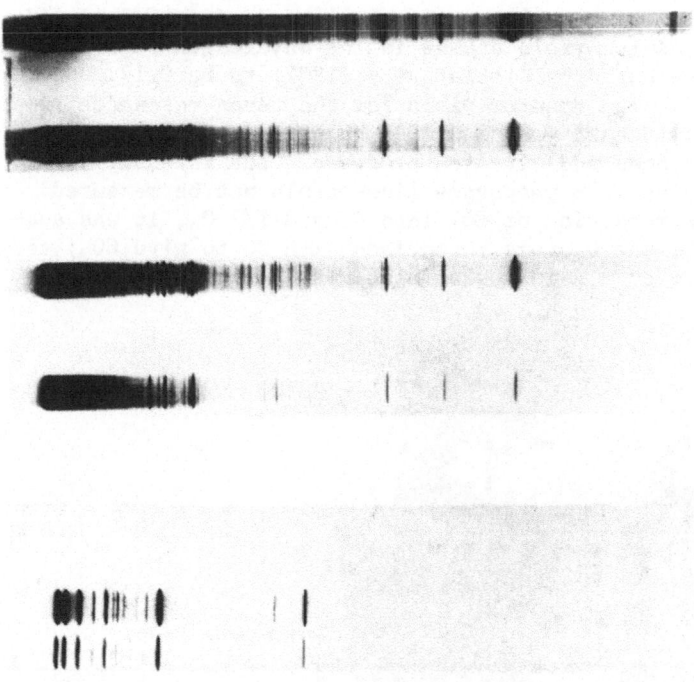

Fig. 2 Xenon resonance lines with and without magnetic field.

as fig. 2 shows. The two spectra of each plate, which were produced by a very small self-made lithium fluoride prism spectrograph, were taken with the same time of illumination, the upper
without, the lower with magnetic field. Plate b was tenfold longer
illuminated than plate a. This factor of 10 was compensated by the
magnetic field.

The intensity of the lamp was again determined by comparison
of the photolysis of NH_3, CO_2, and O_2 to give O_3, taking $\gamma(NH_3)=$
0.17 ± 0.02, $\gamma(CO_2)=0.98$, $\gamma(O_3)=1.90\pm0.05$. The distribution of
the total intensity in the extreme ultraviolet was determined by
using the very strong absorption of oxygen for $\lambda=1469$ Å. At an
absorption length of 30 mm, a pressure of 2 torr oxygen was sufficient to absorb 97% of $\lambda=1469$ Å, while only 23% of $\lambda=1296$ Å was
absorbed. The measurements - taking into account the absorption of
fluorite - gave for the ratio of $I_{1469\text{ Å}}$ to I (all wavelengths

absorbed by oxygen) about 2.7:1.

After a long interruption, the next step in the development
of lamps was achieved during my stay in George Kistiakowsky's laboratory at Harvard University in 1953. The intention was to develop
a krypton lamp with resonance wavelengths at 1235 Å and 1165 Å in
order to decompose molecules with higher dissociation energy. For
this purpose, lithium fluoride windows were of course necessary.
In addition, the high current of 50-60 amps and the magnetic field
had to be avoided since they caused a slow sputtering of the electrodes and dusting of the windows, and an attack on the narrow
quartz tubes, in spite of heavy water cooling. In order to decrease
the self-absorption of the krypton lines, the lamp was filled with
neon of 20 torr, and krypton was added only to such an amount that
the krypton lines in the visible could just be recognized. Under
these conditions the neon was also excited, and the discharge was
red, except for a blue fringe at the walls of the lamp. With currents of only 3-5 amps and without a magnetic field, light intensities could be attained in the extreme ultraviolet which were not
too far from the intensities of the high-current lamps which were
formerly used.

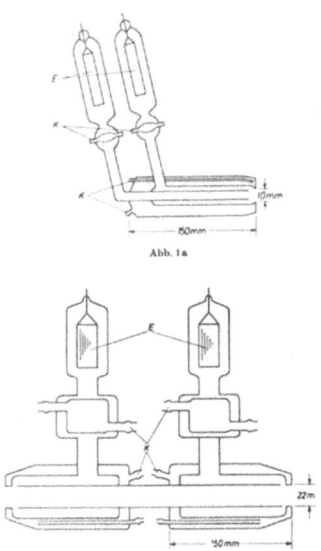

Abb. 1a

Fig. 3 Glow discharge lamps.

An important improvement was made in my Bonn institute in

1958/59 by Comes and Schlag independently of Lossing and Tanaka,
who had used a similar lamp in 1956 without giving its details.
They used an a.c. glow discharge between nickel electrodes in a
lamp made of Pyrex glass with LiF windows. The total pressure
(Ne+0.8% Xe or Ne+0.8% Kr) was 0.8-1.0 torr, the current 100-300
mA. The light intensity was 2-5 x 10^{15} quanta/sec. The absolute
intensity was measured by the ozone formation in flowing oxygen;
relative measurements were possible with a NO ionization chamber
which was calibrated by the ozone formation. In 1960, Schlag and
Comes replaced the a.c. discharge by a microwave discharge, following the work of Dieke and Cunningham, Tanaka and Zelikoff, Wilkinson a.o., using at first a quartz body and later a forced air-
cooled Pyrex body. The intensity of the lamp was again detected
by a NO ionization chamber and by the photochemical production of
ozone.

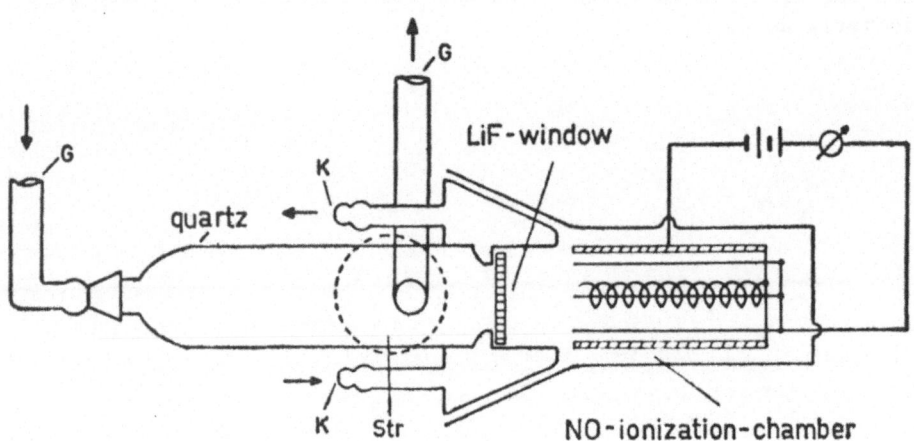

Fig. 4 Microwave discharge lamp.

        The emission maxima were found at 0.6 torr Xe and at about
1.0 torr Kr, corresponding to 1-3 x 10^{16} quanta/ sec; that means
the total excitation is from five to ten times as intense as
attained with the a.c. discharge. The microwave lamps, which at
present are generally used, also have the advantage of containing
no internal electrodes which require baking out, and hence an impurity source is avoided.

The next part will be concerned with the absorption and photo-chemical primary and total reactions of some gases in the Schumann ultraviolet, as far as they were known up to 1940.

OXYGEN

Oxygen is one of the most important and most effectively in-vestigated substances absorbing in the Schumann ultraviolet. Herz-berg had pointed out that the continuum between 1300 Å and 1750 Å with a sharp maximum at 1450 Å corresponded to a dissociation of the oxygen molecule into one normal $O(^3P)$ and one excited $O(^1D)$ atom; at longer wavelengths the Schumann-Runge-Fuchtbauer-bands are connected to this continuum. The absorption near the maximum is so strong, that in layers of only some 1/100 mm at atmospheric pres-sures practically all the incident light is absorbed. Added mole-cules (up to 80% N_2) do not influence the absorption coefficient in this region. The absorption in the banded region (above 1750 Å) is much smaller; it seemed that predissociation of the O_2 molecule in the Schumann-Runge band region takes place, that is, a radiation-less transition into two normal atoms between 1950 Å and 1750 Å.

The ozone formation in pure oxygen or in mixtures of oxygen with inert gases had been investigated by many authors and seemed to be completely elucidated. The formation of ozone by oxygen atoms and molecules in a triple collision was shown to proceed practically without activation energy, since it took place even at the tempera-ture of liquid hydrogen. The efficiency of inert gases as triple collision partners had been investigated by several authors. The temperature dependence of the photochemical ozone formation, which was measured at pressures between 20 and 50 torr up to equilibrium, gave for the activation energy of the reaction $O+O_3=2O_2$ a value of 6.1 ± 0.1 Kcal/mol.

The ozone formation has a quantum efficiency of ~2 independent of the wavelength. Therefore it seemed to be one of the most re-liable and convenient methods to determine the light intensity of unknown light sources in the Schumann-UV.

NITROGEN

Nitrogen is much more transparent in the Schumann-UV than oxy-gen. Its absorption was known to consist of a system of sharp small bands beginning at 1450 Å and reaching a continuum at 990 Å. The value of the dissociation energy of N_2 was not exactly known; the most probable value was given by 7.34 ± 0.02 eV, which was much too low. There existed no photochemical experiments in which N atoms, produced by the dissociation of N_2 molecules, were detected.

HYDROGEN

Hydrogen is even more transparent in the Schumann-UV than nitrogen. The absorption starts at 1115 Å in a band system; at 850 Å a continuum sets in, where dissociation into a normal and an excited atom is to be expected. Photochemical reactions with hydrogen as primary component were not known. Sensitized reactions with CO and Xe as sensitizers were investigated in the Schumann-UV. In both cases, the formation of mercury hydride, which interfered with the mercury sensitization, was avoided.

AMMONIA

The absorption and the photochemical decomposition of ammonia had been investigated by numerous authors; nevertheless the mechanism of the photodecomposition was not elucidated satisfactorily.

The absorption of gaseous ammonia was known to consist of diffuse bands starting at 2260 Å, which had been interpreted by Bonhoeffer and Farkas as predissociation bands. In the Schumann-UV three systems of sharp absorption bands with rotational structure between 1675 and 1150 Å were found; a continuum follows at about 1200 Å.

The quantum yield of the ammonia decomposition in the 2000 Å wavelength region was determined by several authors to be of the order of 0.25, whereas below 1500 Å, 0.17 was found, under stationary conditions independent of the pressure. As a primary step, $NH_3 + h\nu \rightarrow NH_2 + H$ was assumed, and unconventional assumptions, e.g. the formation of the radical NH_4 in equilibrium with NH_3 and H atoms, were necessary to explain the low quantum yield.

Only more than 25 years later was it shown by the work at the N.B.S. that at wavelengths below 1550 Å a second primary dissociation process, producing $NH + H_2$, sets in, and by us that in flowing systems, depending on the flow rate, the formation of N_2H_4 is an essential secondary process leading to much higher yields of the ammonia decomposition than in stationary experiments.

CARBON DIOXIDE

The absorption spectrum of CO_2 was not well known before World War II, nor did the absorption spectrum allow a decision whether predissociation processes play a role. At atmospheric pressure, the absorption was complete at 15 mm layer thickness. The direct photolysis of $CO_2 + h\nu \rightarrow CO + O$ was used as a comparison measurement for the determination of the intensities of light sources in the Schumann-UV. Since a few percent of the produced O-atoms will recombine with CO molecules to reform CO_2, the quantum yield was assumed to be 0.98, in good agreement with the photolysis of other gases.

CARBON MONOXIDE

The CO photolysis in its early stages is of special interest. The absorption was known to be weak between 1760 and 1820 Å; measurable band systems were found at a pressure of 750 torr and a layer thickness of 15 mm only below 1545 Å. The main difficulty was that the dissociation energy of the CO molecule – in spite of many published theoretical and experimental works – was an open question: values between 6.89 and 9.097 eV were discussed. Only many years later a value of 11.11 eV was determined by many different methods.

The direct photochemical decomposition of CO by the xenon resonance wavelengths at 1470 and 1295 Å produced CO_2 and C_3O_2. Since the quantum energy at 1295 Å (219 Kcal/mol) is higher than 9.097 eV (209 Kcal/mol), whereas that at 1470 Å (193 Kcal/mol) is lower, it was assumed that only the 1295 Å wavelength was effective and decomposed the CO molecule according to

$$CO + h\nu \longrightarrow C + O \text{ , followed by}$$

$$CO + O(+M) \longrightarrow CO_2 + M$$

$$2CO + C \longrightarrow C_3O_2$$

About 30 years later these experiments were repeated using modern methods and – in addition to the xenon resonance wavelengths – also those of krypton at 1236 and 1165 Å. Since the quantum energy of all four lines is not sufficient for a primary CO dissociation, a mechanism was assumed which had already been suggested by Gaydon:

$$CO + h\nu \longrightarrow CO^*$$

$$CO^* + CO^* \longrightarrow CO_2 + C$$

$$C + CO + M \longrightarrow C_2O + M$$

$$C_2O + CO(+M) \longrightarrow C_3O_2 (+M)$$

Shortly afterwards, Harteck and coworkers showed that even the iodine line at 2062 Å provides a specific excitation of CO to the a^3 level and that the above mechanism for forming CO_2 and C_3O_2 follows with high quantum efficiency.

Very recently we detected the C atoms produced in the cophotolysis by the attenuation of the carbon resonance wavelengths emitted by a carbon plasma lamp.

WATER VAPOUR

Until 1940 very little was known about the absorption spectrum and photochemistry of water vapour as primary component. Continuous absorption had been found at a pressure of 20 torr and a layer thickness of 20 mm from 1830 Å. Predissociated bands are superimposed. It was assumed that light below 1800 Å induced dissociation into normal H atoms and OH radicals, whereas in the region of the short wavelength band below 1392 Å excited OH radicals are formed.

I would like to add that at that time there existed many experimental and theoretical arguments for the assumption that in the extreme ultraviolet the radiation of the sun does not correspond to that of a black body of about 6000°K over its whole surface, but that it can be greater by many orders of magnitude in the sunspots. These assumptions prompted a discussion of the problem of the origin of free oxygen in the primitive atmosphere of the earth by Hans Suess and myself in 1938, long before the famous experiments of Miller.

We irradiated mixtures of water vapour and carbon dioxide by the xenon resonance wavelengths, and showed that molecular oxygen and aldehydes are formed by the processes:

$$H_2O + h\nu \rightarrow H + OH, \quad CO_2 + h\nu \rightarrow CO + O, \quad O + O + M \rightarrow O_2 + M,$$

$$H + CO + M \rightarrow HCO + M, \quad HCO + HCO \rightarrow H_2CO + CO \ .$$

This process, which may have played an important role in the earth's atmosphere, is similar to the assimilation of plants. Under the influence of extreme ultraviolet light, CO_2 and H_2O can produce O_2 and carbon compounds. This may explain the initial formation of free oxygen and of certain carbon compounds which may have been prerequisite for the development of organic life.

ORGANIC COMPOUNDS

In 1934 Bonhoeffer gave a report on the photochemistry of simple hydrocarbons. It contained the known experimental results of the absorption and the elementary photochemical reactions. In addition, he gave some hypotheses for decomposition processes which were still unknown. A few years later it was shown that some of these hypotheses were mistaken, but they have initiated many investigations on the changes of organic compounds by radiation.

I shall confine this report to the three simple hydrocarbons which were investigated in the following years in detail but with the use of very primitive analytical tools, compared to modern methods: methane, ethane, and propane. In order to demonstrate how difficult the analysis of the reaction products was at a time when neither mass spectrometers nor gas chromatography nor infrared spectroscopy were available, the next slide shows the apparatus used for the analysis.

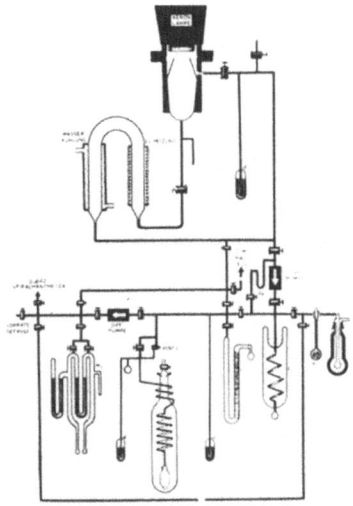

Fig. 5 Apparatus for analyzing reaction products of propane photodissociation.

The gas is pumped through the reaction vessel by a thermo-syphon pump. The quantum efficiency of the lamp is measured by the ozone formation in flowing oxygen. The analysis apparatus consists of two diffusion pumps, the Dewar vessel A, which can be pumped out for achieving low temperatures with a flask with active charcoal in it, two low temperature rectification columns, a U-tube with active charcoal, a cooling trap, the differential manometer M, and apparatus for the determination of unsaturated hydrocarbons, manometers, and storage vessels.

In addition to hydrogen, the following hydrocarbons were expected as reaction products in the photolysis of propane: methane, ethane, ethylene, acetylene, propylene, allene, allylene, saturated and unsaturated C_4 - and higher hydrocarbons.

Without going into the details of the very complicated analytical method, I would like to give a short resume of the results. The reaction products were, in the case of methane: hydrogen, ethane, ethylene, and acetylene; in the case of ethane: hydrogen, ethylene, acetylene, in minor amounts methane, propane, and butane; in the case of propane: hydrogen, propylene, no methane, and small amounts of ethane, ethylene, acetylene, and higher hydrocarbons.

The following primary processes were assumed:

Methane: $CH_4 + h\nu \rightarrow CH_3 + H$; secondary reactions under this assumption were only possible between the CH_3 radical and the H atoms. We know now, again from the work done at the N.B.S., that another primary process is involved: $CH_4 + h\nu \rightarrow CH_2 + H_2$.

Ethane: $C_2H_6 + h\nu \rightarrow 2CH_3$; Faltings rejected the process $C_2H_6 + h\nu \rightarrow C_2H_5 + H$ for the following reason: in the case of methane, addition of CO led to the formation of formaldehyde via $H + CO (+M) \rightarrow HCO(+M)$, which was not found in the case of ethane, but the H atoms would react with C_2H_6 much faster than with CH_4; therefore his argument was certainly unsubstantiated.

Propane: $C_3H_8 + h\nu \rightarrow C_3H_7 + H$ (1); due to the large energy of the absorbed light quanta, the fragments are primarily formed in thermally highly excited states, so that recombination reactions of the radicals are improbable, and radical reactions with high activation energy and decomposition reactions of the radicals become possible.

The following secondary reactions were assumed:

$$C_3H_7 + C_3H_7 \quad \rightarrow \quad C_3H_8 + C_3H_6 \qquad\qquad (2)$$

$$C_3H_7^* \quad\quad\quad\quad \rightarrow \quad C_3H_6 + H \qquad\qquad\qquad (3)$$

$$C_3H_7 \quad\quad\quad\quad\; \rightarrow \quad C_2H_4 + CH_3 \qquad\qquad\;\; (4)$$

$$H + C_3H_8 \quad\quad\; \rightarrow \quad C_3H_7 + H_2 \qquad\qquad\;\; (5)$$

$$H + C_3H_6 \quad\quad\; \rightarrow \quad C_3H_7 \qquad\qquad\qquad\quad (6)$$

$$CH_3 + CH_3 \quad\;\; \rightarrow \quad C_2H_6 \qquad\qquad\qquad\quad (7)$$

$$2C_3H_7 \quad\quad\quad\;\; \rightarrow \quad C_6H_{14} \qquad\qquad\qquad\quad (8)$$

According to this mechanism, hydrogen and propylene are the main products, but due to reaction (6) with a quantum yield < 1, by-products are ethane, ethylene, hexane, perhaps acetylene, but no methane.

The following quantum efficiencies were found:

H_2 - 0.53; C_3H_6 - 0.41; CH_4 - 0; C_2H_6 - 0.05; C_2H_4 - 0.09;

C_2H_2 - 0.02; higher hydrocarbons - 0.09.

Finally I would like to talk about some special photochemical reactions in the Schumann ultraviolet, which were carried out in the 1930's with a xenon lamp. I mentioned already that oxygen at 1496 Å, the strongest xenon resonance wavelength, has an extremely high absorption coefficient ($\alpha \sim 450$). Therefore it is possible to investigate in mixtures of oxygen with other gases at very low oxygen partial pressures (of the order of 0/00) the reactions of photochemically produced oxygen atoms. The advantage of this method is that even at these very low oxygen partial pressures nearly the total amount of incident energy is absorbed, and the O atoms undergo 10^5-10^6 collisions with other molecules (that means 10^2-10^3 triple collisions) before they react with molecular oxygen to form ozone. A comparison of the results of these experiments with those of thermal reactions has elucidated the qualities of atomic oxygen.

The reaction between O atoms and H_2 molecules had been investigated by many methods without clarifying the mechanisms of the reaction completely. Three primary reactions had been suggested:

(1) $O + O_2 + M_2 \rightarrow O_3 + M_2$

(2) $O + H_2 + O_2 \rightarrow OH + HO_2$

(3) $O + H_2 \rightarrow OH + H$

It was shown that even at the extremely small oxygen partial pressures used in the photochemical experiments, the ozone formation in a triple collision predominates the primary attack on hydrogen molecules. In a triple collision the resulting molecule is stabilized as soon as a small part of the binding energy is removed. Therefore the ozone molecules are formed with considerable inner energy. It is assumed that the primarily formed energy rich ozone at low O atom concentrations reacts with hydrogen to form water and hydrogen peroxide before it loses its energy by collisions, whereas it is rapidly destroyed by O atoms at the high atom concentrations of the gas discharge experiments before it can react with H_2. This assumption could be confirmed by experiments on the thermal reaction between hydrogen and ozonized oxygen, which showed that ozone molecules, under experimental conditions which prevent a thermal ozone dissociation, can attack hydrogen molecules and form water and hydrogen peroxide according to $O_3 + H_2 \rightarrow OH + HO_2$.

The simple experimental arrangement is shown in the next slide. Oxygen is purified, dried and ozonized, and mixed either with argon, nitrogen, or with very pure hydrogen; the mixture is flowed through a reaction vessel which could be heated to temperatures between 100°C and 360°C.

The results are shown in the next slide. Whereas with argon or nitrogen as carrier gas the decomposition of ozone sets in only at about 180°C and is complete at 306°C, already at 100°C 11% of the ozone is decomposed with hydrogen as the carrier gas, and the decomposition is complete at 210°C. The reaction products are water and hydrogen peroxide in amounts which correspond to the amount of decomposed ozone.

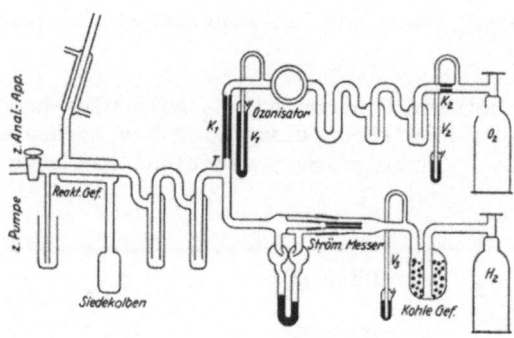

Fig. 6a Thermal reaction between hydrogen and ozonized oxygen.

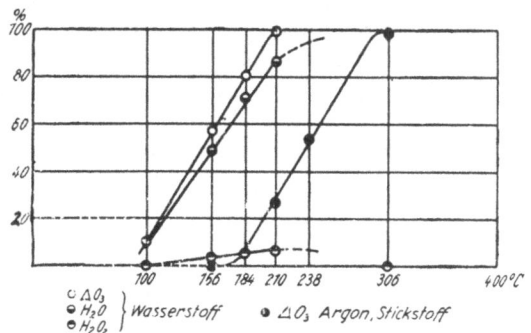

Fig. 6b. Experimental results.

Similar results were obtained for the CO oxidation. Photolysis of mixtures of CO and very small amounts of oxygen by the xenon resonance wavelengths produced O_3 and CO_2. Under the assumption that a triple collision formed CO_2 from CO and O atoms, a collision yield of 1/40 had to be assumed (if it is 1 for the ozone formation). The experiments on the reaction of thermally excited ozone with CO showed that this figure is too high; also in this case the primarily formed energy rich ozone molecules had a decisive effect.

An analysis of the reaction products of the photolysis of CO-H_2 mixtures by the xenon lamp showed that in addition to CO_2 also HCOH, H_2O, and, unexpectedly, H_2O_2 are formed, while the formation of C_3O_2, which had been detected in the photolysis of pure CO, was quantitatively suppressed. The formation of H_2O_2 goes via H atoms and HO_2 radicals, which are formed either by sensitization of the H_2 molecule by excited CO molecules or by the reaction of O atoms, primarily formed by the dissociation of CO, with H_2. Without CO, the H atoms would recombine. Therefore it was assumed that the H atoms react with CO to form COH and are stabilized in this form, and this radical reacts with traces of O_2, according to $HCO + O_2 \rightarrow HO_2 + CO$. The HO_2 radicals will produce H_2O and H_2O_2. This assumption was confirmed in mercury sensitized experiments with H_2-CO mixtures with few 0/00 O_2.

Several years later the nitrogen oxidation was investigated in the light of the xenon lamp.

Mass-spectrometrically a small amount of N_2O could be detected after about 30 hours illumination. This means that about 10^{-4} of the O atoms formed N_2O(dinitrogenmonoxide). The activation energy of the N_2O formation had been estimated to be 12-13 Kcal/mol; this amount of energy can be provided by the kinetic energy of the O atoms primarily formed in the ^3P ground state or in the excited ^1D state. In any case, the small amounts of N_2O which are found in the earth's atmosphere (2×10^{19} molecules per 1 cm^2) can at least partially be produced by this photochemical reaction.

Thus we end with one of the many problems for which the photochemistry in the Schumann ultraviolet has become so very important: reactions in the lower and upper earth atmosphere and in the atmosphere of other planets, after having started with the photolysis of very simple molecules in laboratory experiments.

PHOTOIONIZATION AND FRAGMENTATION OF POLYATOMIC MOLECULES

William A. Chupka

Argonne National Laboratory
Argonne, Illinois 60439

EXPERIMENTAL

The most detailed studies of photoionization and fragmentation have been carried out with apparatus which consists of a
vacuum ultraviolet monochromator with various light sources producing continua or near-continua and a mass spectrometer for ion
detection and identification. Several such instruments have been
used and are described or referred to in recent reviews.[1,2]
One particularly advantageous combination is that of a 1-meter
near normal incidence monochromator equipped for windowless operation with rare gas continua lamps and a magnetic-deflection mass
spectrometer. Photon band widths as narrow as 0.04 Å have been
used but 1.0 Å (~0.01 eV at 1100 Å) is more typical. Such an
apparatus has been described in the literature.[3] Of particular
interest in the present context are certain temporal characteristics of this apparatus. Between the instant of formation in the
photoionization chamber and final detection after mass analysis
an ion of typical mass (~100 amu) spends several microseconds in
each of the following regions of the instrument: (1) the ionization chamber, (2) acceleration and focusing regions, (3) a first
field-free region, (4) the magnetic field, (5) a second field free
region. Ions which dissociate in region (1) are detected as fragment ions. Those which dissociate in regions (2) or (4) are
effectively lost from the mass spectrum. Ions which do not dissociate at all or dissociate in region (5) are detected as parent
ions. Of special interest here are those ions which decompose in
region (3). They are known as "metastable ions" in mass spectrometry and appear at a generally non-integral mass position, m^*
given by the ratio $(m_f)^2/(m_p)$ where m_f and m_p are the masses of
fragment and parent ions respectively. The detection character-

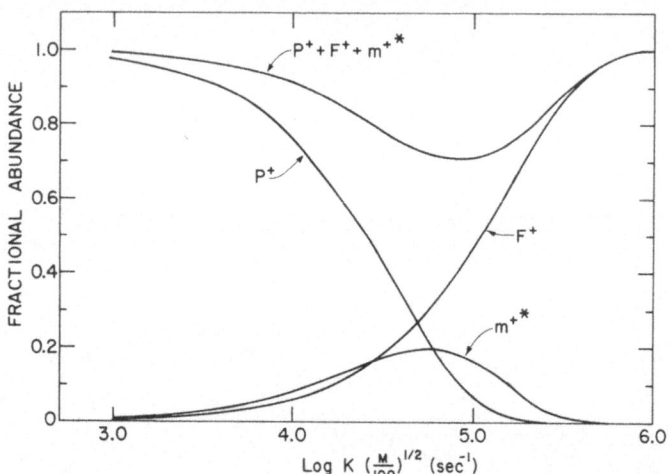

Fig. 1.   Fractional abundance of ions detected as parent, fragment
          and metastable ions as a function of k the rate constant
          for unimolecular decomposition of a parent ion of mass
          M in amu.   This applies to the apparatus described in
          ref. 3.

istics of the apparatus described in ref. (3) for parent, fragment
and metastable ions have been calculated as described previously[4]
for another instrument and are shown in Fig. 1.   Of particular
interest here is the fact that metastable ions are observed predominantly for decompositions with rate constants in the range
10^4 to 10^5 sec^{-1} and that fragment ions are not observed significantly until the rate constant is of comparable magnitude.   It is
assumed in these calculations that radiative de-excitation is not
a significant competitive process.   This point has important consequences as will be discussed later.

        The shape and width of the metastable ion peak in the mass
spectrum is also a very sensitive measure of the kinetic energy
released in dissociation.   For a decomposition which is isotropic
and monoenergetic in kinetic energy release, it has been shown[5]
that, to a good approximation, the metastable peak will be rectangular and of width proportional to the quantity $(eV \cdot T)^{1/2}$ where
eV is the kinetic energy to which the parent ions were accelerated
(typically of the order of kilovolts) and T is the kinetic energy
released in the center of mass system.   Neglected effects of components of velocity normal to the ion-optic axis and certain
losses of ions from the beam can produce some relatively small
distortion from the expected simple rectangular form.   However, a
more complete analysis and proper instrumental design and opera-

tion can yield highly precise absolute measurements of the distribution of kinetic energy released in the decomposition process.
Present instruments are capable of measuring values of T less than
one millivolt.[6] This high sensitivity is the result of the
large "amplification" effect of the transformation from the c.m.
to the laboratory system and is the precise opposite of the "de-
amplification" effect achieved in "merging-beam" experiments in
which the initial collisions are essentially the reverse of the
metastable decomposition process. Such amplification and de-
amplification effects are common and very important in a wide
variety of collision and dissociation phenomena.[7] Various techniques have also been used to measure the kinetic energy distribution of fragment ions formed in the ionization chamber.[8] These
latter techniques are particularly valuable for decompositions
which are not observable as metastable ions. However, the metastable ion measurements have the great advantage that they refer
to ions with a narrow range of dissociation rate constants which
implies a narrow range of internal energy.[8] This latter condition makes comparisons between theory and experiment much easier
and more direct.

GENERAL CONSIDERATIONS

     Absorption by molecules of photons of wavelengths shorter
than the first ionization limit of the molecule can lead to a
number of competitive and consecutive processes which may be
described briefly as follows. In this description the molecule
AB may be polyatomic and A and B may be atoms or groups of atoms.
Direct excitation to continua may occur:

$$AB + h\nu \rightarrow AB^+ + e^- \quad \text{(Direct Ionization)}$$
$$\rightarrow A^* + B \quad \text{(Direct Dissociation)}$$
$$\rightarrow A^+ + B^- \quad \text{(Direct Dissociation to an Ion Pair)}$$

Excitation may also occur to quasi-discrete ("superexcited")
states AB^* which may interact strongly with each other and with
one or more of the continua described above. The resulting indirect processes of autoionization (preionization) and predissociation are prominent and important in the photoionization spectral
region. The resulting fragments of both direct and indirect
processes may be sufficiently excited to undergo further decomposition. The most commonly observed such decomposition is that of
the parent ion, $AB^{+*} \rightarrow A^+ + B$. In the case of a polyatomic parent
ion, several competing modes of decomposition may occur and the
resulting fragments may have sufficient internal energy to undergo
still further competitive and successive decompositions. Another
possible decomposition process is that of autoionization of an

excited neutral fragment produced by direct dissociation or predissociation. This latter process has never been positively
identified and is not expected to account for anything but a minor
fraction of observed ionic fragments.

The above description may be too simplistic in some cases.
A more rigorous description would recognize the possibility that
the observed quasi-discrete states may interact strongly not only
with each other and with the continua described above, but also
with other optically weak or forbidden quasi-discrete manifolds
(i.e. internal conversion or intersystem crossing) and also
directly with three-body continua in a way such that it may not
be a good approximation to consider the processes of ionization
and dissociation to occur successively as described above. There
is very little experimental evidence bearing on the importance of
these latter processes. At present, the simple model of ionization followed by dissociation is capable of explaining practically
all experimental observations. It should be noted that this model
<u>need</u> not fail for transitions to a three-body continuum. For ex-
ample, the direct ionization of H_2 to produce H_2^+ in the repulsive
$^2\Sigma_u^+$ state is a process for which application of the Franck-Condon
principle predicts well the kinetic energy distribution of the
fragments, i.e. the Born-Oppenheimer separation of electronic and
nuclear motions is a good approximation and the ionization and
dissociation can be considered to occur consecutively. However,
the possibility of serious failure in some cases of decay of quasi-
discrete states should be kept in mind.

PHOTOIONIZATION AND ENERGY DEPOSITION

The general features of the photoabsorption curve of a molecule above its first ionization limit are readily understood at
least qualitatively from the photoelectron spectrum of the molecule
taken at a wavelength at which negligible autoionization occurs.
The absorption consists of direct ionization continua corresponding to the formation of the states of the ion displayed in the
photoelectron spectrum and of one or more Rydberg series converging to each of these states of the ion. In some cases an excited
valence state (or at least a state which cannot be unambiguously
assigned as a Rydberg state) may be above the first ionization
limit also. Thus the $^1A(\pi \rightarrow \pi^*)$ valence state of formaldehyde is
estimated to lie above the first I.P. and to be strongly auto-
ionized.[9] Some such valence states can be considered to be
formed by excitation of one of the more strongly bound electrons
(corresponding to one of the higher I.P.'s of the photoelectron
spectrum) into one of the normally unfilled valence orbitals. A
crude estimate of the excitation energy of such a state may be
given by the difference of the higher I.P. and the term value of
the unfilled valence orbital as determined from other transitions

to the latter orbital observed at longer wavelengths. Identification of such valence states is expected to be difficult since they will usually be very short-lived with respect to autoionization or predissociation and have a broadened band envelope not readily recognizable from other data in contrast to the situation for Rydberg states.

The vibrational structure of the Rydberg transitions, particularly higher members, approach those of the photoelectron spectrum and the interaction of these quasi-discrete states with the underlying continua often produces recognizable Beutler-Fano profiles.[10] However, especially for the larger polyatomic molecules, the overlapping and mutual interaction of rich vibrational structure together with rotational and lifetime broadening often produce an absorption curve with little assignable structure. The well known continuity of oscillator strength density through ionization limits[10] ensures that the thresholds for formation of various states of the ion are not apparent in the absorption spectrum.

If all absorption above the ionization limit led to ionization, the ionization and absorption cross section curves would, of course, coincide in this region. Except for atoms, this is rarely the case due to the competitive decay of some superexcited states via processes which do not produce charged particles, most importantly dissociation or predissociation into neutral particles. In the extreme case of strong predominance of predissociation of all intense Rydberg states converging to a particular limit, the resulting photoionization cross section curve will show a step-like rise at the limit, the height of the step being approximately proportional to the corresponding feature of the photoelectron spectrum. This behavior is most often observed just above the first ionization threshold and can lead to the display of a series of steps which correlate with the vibrational structure of the ground state photoelectron band. Some molecules which display this behavior fairly clearly are NH_3, NO, C_2H_2 and C_2H_4 although careful measurements still show remnant autoionization structure in this region.[11,12]

While the presence of autoionization has the disadvantage of preventing the determination of excited state thresholds in most cases from simple inspection of ionization cross section curves, it has an important advantage in the study of fragmentation thresholds, in that it can result in the population of excited states of the parent ion which are not accessible by direct ionization as indicated by the photoelectron spectrum taken at a non-autoionizing wavelength (the usual case for the HeI and HeII lines most frequently used). In the case of autoionization of a particular vibronic state, the vibrational distribution in the final ionic state is governed by the Franck-Condon factors between the

initial and final state. Since the ion in the excited vibronic
state often has several quanta of vibrational energy and a geometry significantly different from that of the ground state, the
resulting distribution of internal energy in the ion produced is
broader than that produced by direct ionization. Furthermore,
for larger polyatomic molecules, the autoionizing Rydberg bands
are often so overlapped that at any given wavelength many autoionizing bands may be excited so that this broadening of the internal energy distribution beyond that of the photoelectron spectrum is further enhanced. Several examples of marked changes in
the vibrational envelope of a photoelectron band due to excitation
of a strong autoionizing level are known in photoelectron spectroscopy,[13] and many more may be inferred from photoionization
efficiency curves for certain fragment ions which have thresholds
at energies at which the photoelectron spectra show no detectable
intensity for formation of excited ions.

Within the framework of the simple atomic and molecular
orbital description in which configuration interaction is neglected, only single electron excitation or ejection is permitted
by photon absorption and very nearly all photoelectron bands have
been assigned on this basis. However, many examples of two-
electron excitation processes, otherwise allowed by selection
rules, are known in atomic spectroscopy and photoionization[10,14,
15] although the intensities are typically one to two orders of
magnitude lower than single-electron allowed transitions.
Recently a few examples of photoelectron bands which can be
ascribed to formation of ionic states by simultaneous ejection of
one electron and excitation of another have been identified in the
photoelectron spectra of O_2 [16,17], N_2O[18] and I_2[19]. Undoubtedly many more such two-electron excitation bands are present
in photoelectron spectra but are in most cases obscured by the
stronger single-electron bands or buried in the noise of the
typical photoelectron spectrum. Such states would, of course, be
accompanied by associated Rydberg series converging to their
limit and would also be formed (usually weakly) by autoionization
of singly-excited states. The formation of such excited ionic
states can have significant effects in photoionization mass spectrometric studies in spite of the low intensity. Thus, if such a
state occurs in the energy region of a threshold for fragmentation
and this region is devoid of states produced by single electron
excitation, the intensity of the fragment ion can be due entirely
to the multiple excitation process. It is very commonly the case
that ion fragment intensities lower than parent ion intensities
by factors of 10^{-3}—10^{-5} are readily detectable by mass spectrometry. It is fairly common to observe a fragmentation threshold
in an apparently "blank" region of the photoelectron spectrum at,
or very near, the calculated thermochemical threshold. In many
cases, it is obvious from the structure in the ionization efficiency curve that autoionization is the major process responsible

for populating the "blank" region. In other cases an underlying continuum appears to be present and the mechanism is not clear.

DISSOCIATION PROCESSES AND THE DETERMINATION OF THERMOCHEMICAL QUANTITIES

Ions, as well as neutral molecules, may fragment by direct dissociation or by predissociation. Predissociation may be divided into three cases:(20) (I) electronic predissociation in which the molecular ion makes a radiationless transition into the dissociation continuum of another electronic state; (II) vibrational predissociation, in which the molecular ion remains in the same state and only vibrational motion is involved; (III) rotational predissociation, in which only the higher rotational levels of a particular vibronic state may predissociate. Rotational predissociation can be significant in the precise determination of dissociation thresholds especially for smaller ions.(21) However, the process accounts for only a very small fraction of observed fragmentation and is not significant to the main thrust of this paper. It will not be discussed further. It should also be noted that predissociation can involve a radiationless transition followed by vibrational predissociation. This is probably common for larger polyatomic molecules.

The observation and determination of thresholds for fragment ion formation by photoionization mass spectrometry is a very powerful and widely applicable technique for the determination of heats of formation of charged and neutral molecules, free radicals and atoms. From these quantities values of bond dissociation energies, ionization potentials and electron affinities can be determined reliably and accurately. The technique is not without its pitfalls, but by careful experiment and interpretation these can be almost always either eliminated or at least questionable value recognized.

The power and scope of the technique derives from several characteristics. The onset of fragmentation is unambiguously recognized by the appearance of the fragment ion in the mass spectrum. The minimum rate constant for observation of fragmentation is very low, - of the order of 10^4 sec^{-1}. This minimum can, and has been, decreased by the use of well-developed ion storage techniques or of other types of mass spectrometers such as the Ion-Cyclotron-Resonance type, but difficulties associated with secondary collision processes limit the decrease to a few orders of magnitude. Detection of such slow dissociations minimizes the effects of selection rules or other inhibiting kinetic factors. The identity of the fragment ion is usually unambiguously determined by its mass. In the case of possible isomers, experiments on several compounds often settle the question and can give heats

of formation of different isomers. Fragmentation processes of
very low intensity can be determined (typically 10^{-3}—10^{-4} and
sometimes as low as 10^{-6} of parent ion intensity). Parent ions
can be formed readily in electronic states spanning such large
ranges of energy that very often all values of internal energy
are effectively accessible. A large number of fragmentation
processes are typically observed, although many are not suitable
for thermochemical determinations as will be discussed later. The
kinetic energy released in fragmentation, especially of "metastable" ions, is measurable, often with high sensitivity and precision.

There are two general approaches used to determine the heat
of formation of an ion. In the first and more accurate approach,
one judiciously chooses a parent molecule and fragmentation process
such that the observed threshold is likely to be equal to the
thermochemical one. Various characteristics of the data will
usually indicate whether this circumstance is true or not. In the
other approach, one investigates a process, the threshold for
which is significantly higher than the "true thermochemical"
threshold and then applies a correction derived from other experimental data (such as kinetic energy released in fragmentation)
or by application of theory based on a plausible model (e.g. the
RRKM theory of unimolecular decomposition of an isolated molecule).
In any case, an observed threshold always yields at least an upper
limit to the value of ΔE for the process.

The major factors which can cause fragmentation to appear at
a higher threshold than that calculated thermochemically are
listed below.

(1) There may be a negligibly small probability for forming
the parent ion in the internal energy range of the thermochemical
threshold, in which case the observed threshold often coincides
with the threshold of the next highest band in the photoelectron
spectrum. The presence of significant band intensity in the photoelectron spectrum at and below the observed fragmentation
threshold is sufficient to rule out this possible source of error.
However, the above is not at all a necessary condition since an
apparently "blank" region may be populated by processes mentioned
earlier.

(2) There may be a potential barrier in every potential
curve or surface leading to the product fragments. This seems to
be relatively rare for simple bond rupture processes but a few
cases for small parent ions seem to be established. Thus adiabatic
correlation diagrams, theoretical calculations and some experiments
indicate that the fragments of lowest energy from NO_2^+ ($NO^+ + O$ in
their ground states) and from N_2O^+ ($NO^+ + N$ in their ground
states),[22] upon approaching to form the triatomic ion, do so

upon potential surfaces which are all initially repulsive (excluding the small ion-induced-dipole attraction at long distances). In such cases dissociation of the parent ion cannot occur at the thermochemical threshold other than by tunnelling which is far too slow for heavy atoms and thick barriers. This circumstance of large potential barriers is far more common for decomposition processes involving rearrangement where it should always be suspected.

(3) Spin conservation or correlation rules may require potential surface hopping (intersystem crossing or internal conversion) in order to yield ground state products at the thermochemical threshold from the electronic state in which the parent ion is initially formed. Such selection rules are considerably less effective than is found to be the case in ordinary optical spectroscopy in great part due to the far longer time ($\sim 10^{-4}$sec) available for observable predissociation and the high detection sensitivity of mass spectrometry. Thus, more cases are known in which such rules are broken than in which they are sufficiently effective as to prevent observable dissociation. Such rules should be more effective when the electronically excited parent ion has a fast radiative decay which reduces the available time.

(4) In the case of polyatomic molecules with a sufficiently large number of internal degrees of freedom, there may be a significantly large "kinetic shift".[4] That is, dissociation of the molecular ion may occur at the thermochemical threshold but with such a very low rate constant as to be practically undetectable and a significant amount of excess internal energy (the "kinetic shift") may be required to increase the dissociation rate constant to the readily detectable range of 10^4—$10^5 \mathrm{sec}^{-1}$. The observation of a corresponding intense metastable peak is a reliable warning of this possible source of error, since it indicates a significant fraction of ions with rate constants in the 10^4—$10^5 \mathrm{sec}^{-1}$ range and thus the strong possibility of a significant fraction with still lower rate constants. This behavior, expected for dissociation processes which are statistically controlled, e.g. describable by the RRKM theory, will be discussed in more detail later.

(5) In the case of those larger polyatomic molecules whose decompositions are describable statistically, only the decompositions having the lowest threshold energy or those within a small fraction of an eV of the lowest threshold are usually useful for the determination of accurate thermochemical thresholds. This limitation is due to the fact that decomposition processes occurring at higher energies must compete with those having lower thresholds resulting almost invariably in ionization efficiency curves for the higher processes which rise asymptotically from zero, have poorly defined thresholds and effectively larger

kinetic shifts. The same behavior occurs for processes in which three or more fragments are formed overall, since these generally occur successively and the energy carried off by the neutral fragment of the initial decomposition can have a wide distribution with a vanishingly small probability of zero energy loss.

In spite of the problems enumerated above the major fraction of simple bond scission decompositions of lower thresholds studied by photoionization mass spectrometry have yielded accurate and reliable thermochemical quantities. Unreliable thresholds are usually readily recognized and questionable heats of formation confirmed by investigating several processes in different molecules.

DIATOMIC MOLECULES

The processes leading to formation of fragment ions by photoionization of diatomic molecules are relatively well understood due to the wealth of optical spectroscopic information on dissociation and predissociation processes of neutral molecules.[23] The major experimental difference is the nearly universal observation of fragmentation due to the far greater sensitivity of mass spectrometric detection. For very nearly all molecules investigated by photoionization mass spectrometry, fragmentation has been observed at or at least very near the lowest thermochemical threshold. The shape and structure of the ionization efficiency curves for the fragment ions vary greatly from case to case depending on the characteristics and relative amounts of direct ionization and autoionization, on the shapes and relative positions of the potential curves for the ground state molecule, intermediate superexcited states and the various states of the ion, and on the interactions among the latter states.

SMALL POLYATOMIC MOLECULES

Optical spectroscopy provides relatively little information on dissocation and predissociation processes of polyatomic molecules due to the inherently greater complexity of the spectra. Nevertheless, analyses of spectra for several cases have provided useful examples of different decomposition mechanisms. Direct dissociation (i.e. within the time of a vibrational period) of neutral molecules is apparently well established for several cases from the appearance of true continua [20]. It is much more difficult to establish the occurrence of direct dissociation of a polyatomic ion formed by photoionization. The observation of an apparent continuum in the photoelectron spectrum of a polyatomic molecule taken with present limits of resolution ($\sim$10 mV) is not convincing evidence for direct dissociation unless one can exclude the possibility of a quasi-continuum due to crowded vibrational structure. Such quasi-continua have been observed in a number of instances in which dissociation is energetically impossible (e.g. the lowest photoelectron band of NO_2).[24] While no well established case of direct dissociation following photoionization of a polyatomic ion at or near threshold is known, such cases very probably occur. For example, direct dissociation seems likely for that class of molecules (e.g. CF_4, SF_6) for which only fragment ions are observed at the lowest ionization threshold.

Predissociation of small polyatomic molecules is well established for a number of cases by optical spectroscopy and is often a very obvious and common occurrence for polyatomic ions. Predissociation is obvious in the latter case from the appearance in the photoelectron spectrum of well defined vibrational structure in bands which correspond to the formation of fragment ions as observed by photoionization mass spectrometry, by the charge transfer technique developed by Lindhom,[25] and by the recently developed photoelectron-photoion coincidence techniques.[26-30]

The selection rules governing predissociation processes are discussed in detail by Herzberg.[20] The rigorous selection rules (conservation of total angular momentum and of overall symmetry) are not expected to be significant in excluding predissocation since appropriate final states are practically always available. However, the operation of various approximate selection rules (e.g. conservation of spin) can greatly affect rates of predissociation which may then depend on the strength of spin-orbit, vibronic or electronic-rotational interactions.

A number of examples of fragmentation of triatomic ions occurring from excited states which show distinct vibrational structure in the photoelectron spectrum have been discussed by Turner et al.[13] More extensive data of higher quality obtained with a variety of techniques have since become available for these and other triatomic molecules and a few examples will be discussed here.

The photoionization and ionic fragmentation of H_2O[13] provides an interesting example of predissociation processes in small polyatomic ions. Two fragmentation processes are observed to occur with high intensity at their thermochemical thresholds. The products (1) $OH^+(^3\Sigma^-) + H(^2S)$ appear at about 684.0Å (18.13 eV) and the products (2) $OH(^2\Pi) + H^+(^1S)$ appear at about 661.1Å (18.75 eV). (These values have been measured in this laboratory, are corrected for thermal energy effects and are somewhat more accurate than previously reported data.[31]) The shapes of the ionization efficiency curves indicate that process (2) becomes quite intense and possibly dominant within a few tenths of an eV above its threshold. In the energy range of both thresholds the parent ion must be initially formed in the $^2B_2$ second electronic-ally excited state which has its adiabatic threshold at 17.22eV[13] and a photoelectron band with vibrational structure which persists well above both dissociation hresholds although some broadening appears to set in at about the first threshold and to become progressively more pronounced at higher energies. The $^2B_2$ state does not correlate with either set of products in their ground states. Thus this is a clear-cut case of two predissocations occurring from the same excited state of the ion. Process (1) has been attributed to predissocation by a $^4A''$ repulsive state by spin-orbit coupling and process (2) to predissociation by a $^2A''$ repulsive state by rotational coupling.[32,33] Some other possi-bilities exist which cannot be discussed in this space. Never-theless, in summary, all possible predissociation processes of the $^2B_2$ state of H_2O^+ are forbidden by some (non-rigorous) selection rule, yet both decompositions are rapid (no metastable ions are detected) and occur readily at the thermochemical threshold. This is a common circumstance in photoionization mass spectrometry.

Photoionization and fragmentation of CO_2 into (1) $O^+(^4S)$ + $CO(^1\Sigma^+)$ and (2) $O(^3P)$ + CO^+ $(^2\Sigma^+)$ are both observed to occur at their thermochemical thresholds.[34,35] Process (2) is relatively low in intensity and will not be discussed further. The photo-ionization efficiency curve for O^+ and CO_2^+ from CO_2 is shown in Fig. 2. Process (1) is seen to have its threshold at about

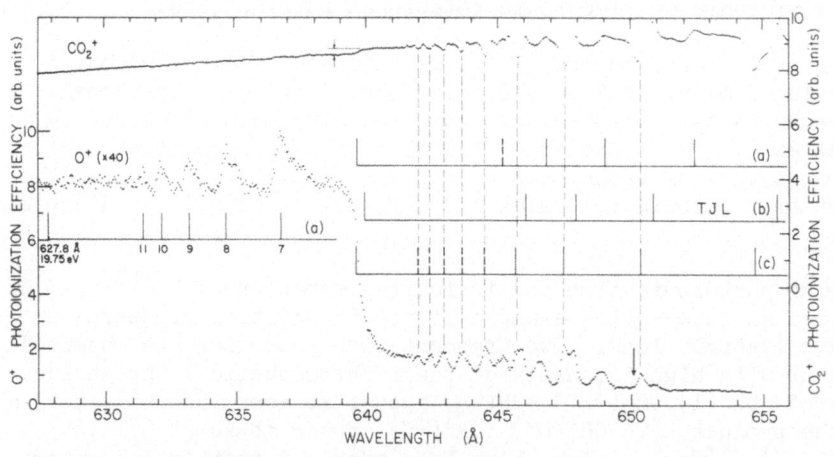

Fig. 2. Photoionization of CO_2 showing formation of O^+ at threshold (650 Å) and also by predissociation of C state of CO_2^+.

650.2 Å (19.07 eV) which corresponds to a gap in the photoelectron spectrum. In the region immediately above threshold, the data show clearly that the process occurs via initial formation of Rydberg states converging to the C $^2\Sigma_g^+$ excited state of the parent ion. The Rydberg states predominantly autoionize to form CO_2^+ in one or more of the several electronic states of lower energy ($X^2\Pi_g$, $A^2\Pi_u$, $B^2\Sigma_u^+$). The formation of O^+ probably occurs by predissociation of the small fraction of sufficiently highly vibrationally excited CO_2^+ ions in either the A or B states. At about 19.40 eV the CO_2^+ ion is formed in the $C^2\Sigma_g^+$ state. The photoionization efficiency curves for parent and fragment ions, the absence of fluorescence from the C state [36] and, most conclusively, a photoelectron-photoion coincidence study[28] show that the C state is completely predissociated, predominantly by process (1). Since the products can only form quartet states, predissociation of the C state as well as the lower energy fragmentation processes occur in violation of spin conservation.

Fragmentation of NO_2^+ has several unusual and interesting features (see Fig. 3). The energetically lowest possible fragmentation producing NO^+ + O in their ground states is observed to occur together with an unexpected associated metastable ion.

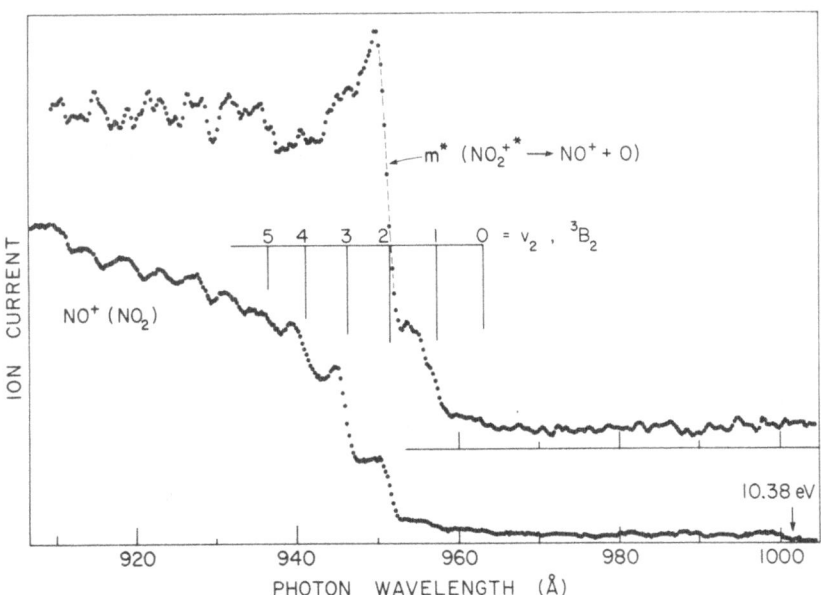

Fig. 3. Photoionization of NO_2 showing formation of NO^+ fragment fragment and associated metastable ion from various vibrational levels of the $^3B_2$ state of NO_2^+.

The thermochemical threshold is at 1001.5 Å (12.38 eV). There is some evidence for a threshold in this region although repeated experiments could not verify it and it is very doubtful at best. At about 964.0 Å (12.86 eV) the ion is formed in the first electronically excited $^3B_2$ state and the photoelectron spectrum[24] shows a strong progression in the bending vibrational mode. The photoionization data show that this predissociates with a rate that is very small and possibly zero for $v = 0$ but increases very rapidly with vibrational quantum number. Metastable ions corresponding to the decomposition are seen to be formed from the $v = 1$ and 2 (and possibly $v = 0$) vibrational states. By comparing the relative intensities of the metastable ions formed from these states in the photoelectron spectrum and considering the flight-time characteristics of the mass spectrometer, the following dissociative lifetimes were deduced:

$$v = 0, \quad \tau \geq 150 \ \mu sec$$
$$v = 1, \quad \tau = 55 \pm 10 \ \mu sec$$
$$v = 2, \quad \tau = 15 \pm 5 \ \mu sec$$
$$v = 3, \quad \tau \leq 5 \ \mu sec$$

Measurements of the metastable peak width show that nearly all of the excess energy available in the decomposition is released as kinetic energy.

The characteristics of the above decomposition process are very unusual. The $^3B_2$ state of the ion should correlate adiabatically to the ground state products but obviously must do so over a potential barrier at least in most of its configurations.

A survey of the available data shows that most fragmentation of small polyatomic molecular ions produced by valence shell photoionization proceeds by predissociation. Electronic predissociation is involved in most cases but it is usually not clear whether the process is purely case I or also involves vibrational predissociation of the final electronic state.

LARGE POLYATOMIC MOLECULES

While the decomposition of small polyatomic ions can be usefully discussed and explained in terms of details of potential surfaces, some features of which can be determined from experimental data, adiabatic correlation diagrams and theoretical calculations, such an approach for larger polyatomic ions is usually not fruitful due to the far greater complexity of the systems. The most commonly used and most successful approach which provides quantitative results with modest computational effort involves the use of statistical theories. The RRKM theory[37] in microcanonical form appropriate to isolated molecules of specific internal energy has been most useful. It is based on the assumption of effective randomization of internal energy such that the rate of reaction is proportional to the ratio of volume of phase space available to the molecule in the critical configuration for reaction to that available to the activated molecule. The theory has been tested most extensively and critically for neutral molecules decomposing on a single potential surface. It has been extensively successful[38] although evidence of failure of energy randomization has been found for some cases of very non-random chemical activation.[39,40]

The basic equation of RRKM theory for $k_a(E^*)$, the unimolecular rate constant for decomposition of a molecule of internal excitation energy E^*, is

$$(1) \quad k_a(E^*) = C \frac{1}{hN^*(E^*)} \sum_{E^\dagger}^{E^\dagger} P(E_{vr}^\dagger)$$

where $E^\dagger = E^* - E_a$, E_a is the activation energy for reaction, $N^*(E^*)$ the density of states (number of states per unit energy) of the activated molecule, $P(E_{vr}^\dagger)$ is the degeneracy of the state of the activated complex having vibrational and internal rotational energy $E_{vr}^\dagger$, and C is a dimensionless number (of the order of magnitude of unity) which includes ratios of symmetry numbers and moments of inertia as discussed in detail elsewhere.[37,41] Many methods for evaluating $N^*(E^*)$ and the summation in equation (1) have been described.[37,42] It is worth noting that the usual methods for evaluating k_a and especially the summation in equation (1) do not specifically conserve total angular momentum. This does not usually lead to significant errors except for certain cases of loose activated complexes in which so-called "Gorin-tumbling" rotational degrees of freedom of the separating fragments are active and are summed over without regard to conservation of angular momentum. Conditions for which the standard statistical calculations for dissociation of a molecule into an atom plus another molecule[43] are very bad approximations were first pointed out by Nikitin[43] who also derived more valid approximations. Klots[44] has considered this problem for a wide variety of fragment products and gives convenient approximate formulae.

At threshold $(E^* = E_a)$, equation (1) reduces to $k_a \approx 1/hN^*(E^*)$, that is it predicts a minimum rate[45] which is readily calculable. This minimum rate is inversely proportional to N^* which is a

rapidly increasing function of E^* and of the number of internal
degrees of freedom. Above threshold the calculated rate is
practically a monotonically increasing function of E^*. If RRKM
theory is applicable to decomposition of molecular ions formed by
photoionization, we expect that metastable ions will be undetectable for molecules with sufficiently few internal degrees of
freedom and/or such low activation energies that the minimum rate
is $\geqslant 10^6$ sec^{-1}. For sufficiently large molecules and large activation energies, metastable ions are expected to be observable and
the kinetic shift is expected to increase with activation energy
and number of degrees of freedom. These expectations are amply
borne out by a large amount of data with relatively few exceptions.

The RRKM theory (or one of its variant forms) can also be
extended to calculate the distribution of various forms of energy
in the fragment products. A number of such extensions have been
made, some specifically for ionic decompositions with their
characteristic long range forces between fragments (see references
40, 43, 44 and references therein). Of particular importance here
are calculations of the distribution of relative kinetic energy
of the products since this is readily measured experimentally.
The validity of straightforward applications of statistical theory
to fragment energy distributions depends on the characteristics
of the activated complex and particularly on the interaction
between fragments after passing through the critical configuration.
For simple bond dissociation of a molecular ion with no excess
activation energy, the theory would be expected to be adequate and
predicts a quasi-Maxwellian kinetic energy distribution which is
a measure of the total excess energy available. For reactions involving potential barriers and excess activation energy, as is
often the case with rearrangements, the observed kinetic energy
distribution is very often higher than the simple statistical prediction and can be interpreted to give information on the characteristics of the potential barrier.

Statistical theory of mass spectra.[46,47]

From its considerable success for neutral molecules the RRKM
theory would be expected to be applicable to the decomposition of
molecular ions occurring on a single potential surface. There
are a number of well known cases in which the lowest energetically
possible dissociations of polyatomic molecular ions occur from
vibrationally excited ions in the ground electronic state (e.g.,
n-alkyl amines and many alkanes). However, in many other cases,
fragmentation thresholds occur at energies corresponding to the
production of ions in electronically excited states. In any case,
at higher photon energies it is clear that the molecular ion is
initially prepared in a large number of electronically excited
states as shown by the photoelectron spectrum at that photon

energy. Photoionization mass spectrometry, charge-exchange mass spectrometry and the limited data available for larger molecules by the photoelectron-photoion coincidence technique show that most, if not all, ions with internal energies above the first fragmentation threshold dissociate. Strong arguments, based on theory and experiment, support the assumption that normally a negligible fraction of parent ions undergo direct dissociation (see reference 46). The dominant process of predissociation can occur in many ways. At one extreme, the ions may predissociate by vibration, remaining on the initial electronic potential surface, in which case RRKM theory is applicable with appropriate parameters. At the other extreme, the ions could undergo a series of rapid (compared to intervening dissociation) radiationless transitions, forming all isoenergetic states with equal probability, predominantly highly vibrationally excited ions in the ground electronic state. Again, RRKM theory is applicable with an appropriate set of structural and energy parameters. The latter case is that assumed in the usual form of the Statistical Theory of Mass Spectra (STMS) or, as it is now more commonly called, Quasi-Equilibrium Theory (QET). Predissociation (predominantly vibrational and describable by RRKM theory) is then assumed to occur along competitive reaction paths each with an activation energy and activated complex appropriate to the specific fragmentation process. For larger values of E^* many of the primary product ions will in turn possess sufficiently large (and statistically calculable) internal energies such that they can undergo still further competitive and successive decompositions until the resulting product ions have insufficient internal energy to decompose further. The result of a QET calculation is conveniently displayed as a "breakdown graph" showing the relative abundance of various ions (after a time of $\approx 10^{-5}$ sec) as a function of internal energy of the initial parent ion.

The general model of competitive and successive decomposition occurring on a time scale much larger than 10^{-14} sec is abundantly supported by a vast body of data (46,47) on ion abundances, ionization efficiency curves, measurements of metastable ions and of the kinetic energy released in formation of fragment and metastable ions. However, accurate and unambiguous quantitative tests of the theory are not readily made and probably the major uncertainty concerns the degree to which the assumption of complete electronic relaxation before dissociation is valid. This uncertainty should not be considered a fatal flaw of the model. It is similar to the uncertainty which accompanies the application of statistical thermodynamics in that it concerns the regions of energetically available phase space which are effectively excluded to the system during some time interval. At present, comparison of theory and experiment can be very helpful in inferring the degree of electronic relaxation and its correlation with pertinent parameters such as energy gaps between electronic states.

Radiationless transitions

Enormous advances in this area have been made in the past
decade as a result of much experimental and theoretical work on
fluorescence and phosphorescence phenomena. A number of good
reviews are available[48-51] and a very recent one[52] is particularly helpful in removing earlier confusion in the theory regarding proper choice of basis sets. Only the briefest outline of
those aspects of this topic most relevant to the present subject
can be presented here.

Practically all of the experimental and theoretical work on
non-dissociative radiationless transitions deal with molecular
systems whose ground and relevant excited state potential surfaces
are "nested", i.e. do not intersect in the region accessible by
photon absorption. Two major experimental generalizations, which
apply to gaseous molecules under collisionless conditions as well
as to molecules in the condensed phase are (1) the "energy gap
law" according to which the rate of radiationless transition decreases approximately exponentially with increasing energy gap
between the two potential surfaces, and (2) the isotope effect
whereby perdeuteration results in a strong decrease in the rate.
The simpler theoretical treatments which have explained these
rules have employed a model of (not too greatly) displaced and
distorted oscillators. The theoretical expression for the transition rate can be written in a very crude approximation as a
product of an electronic and a Franck-Condon factor. Nearly all
quantitative calculations have been concerned with the latter
factor which is capable of explaining the energy gap and isotope
effects. Accurate theoretical calculations of electronic factors
have not yet been achieved although the perturbations responsible
have been formally considered in some detail (spin-orbit coupling
for intersystem crossing and vibronic coupling for internal conversion within the same spin-multiplicity manifold) and crude
estimates made. It has been estimated [53] that variation of the
vibronic coupling term for different molecules may introduce an
uncertainty of only about two orders of magnitude in the rate so
that it is useful to note its magnitude for a large number of
aromatic molecules (see reference 48, chapter 5). If k_{nr}, the
rate of non-radiative transition is written as

$$k_{nr} = A \cdot F$$

where A is the elctronic factor and F the Franck-Condon factor,
A has been given the value of $\sim 4 \times 10^{12}$ sec^{-1} for internal con-

version and $\sim 4 \times 10^4$ sec^{-1} for intersystem crossing.[48] The value of F can be taken from Fig. 5.2, p. 148 of reference 48 or estimated for the perprotonated compound from the equation for the line of that figure:

$$- \log F \cong + 0.11 \ (\Delta E - E_o)/\eta + 0.7$$

where $(\Delta E - E_o)/\eta$ is the normalized energy gap in units of $10^3 cm^{-1}$, i.e. ΔE is the energy gap between zero-point levels of the two electronic states, $E_o \cong 4000$ cm^{-1} for the fully protonated compound and η is the relative number of hydrogen atoms as given in the formula $C_{1-\eta}H_\eta$. Englman and Jortner[53] give still simpler expressions with an estimated error of two orders of magnitude.

It should be emphasized that the internal conversion rates discussed above, which can be of the order of 10^{12} sec^{-1} (see table 5.7, p. 187 of reference 48) apply to what Englman and Jortner [53] call the "weak coupling case" for which energy gaps are larger than typical vibrational frequencies and displacements between potential surfaces are small ("nested" surfaces). These authors also treat a "strong coupling case" characterized by a large displacement of the potential energy surfaces along at least one normal coordinate so that the surfaces intersect in the vicinity of the minimum of the upper surface. The corresponding rate equation has the Arrhenius form with activation energy corresponding to the energy difference between the zero point level of the upper state and the minimum intersection energy.

Recently the variation of radiationless transition rates as a function of vibronic energy and even of specific vibronic states has been investigated[51-55] experimentally and theoretically. The experimental data show the rates to increase, often quite rapidly, with increasing vibronic energy although the theory indicates that the increase may sometimes be small or even negative, depending strongly on the specific vibrational modes excited.

A survey of photoelectron spectra[13] of larger molecules for which QET is expected to be applicable shows that energy gaps larger than ~ 2.0 eV are relatively rare (and this observation does not consider states which do not appear in the photoelectron spectra but may participate in electronic relaxation). Smaller gaps and overlapped bands which indicate a good possibility of the faster strong-coupling case are very common. From the above considerations we may then estimate that radiationless transition rates are usually much greater than 10^9 sec^{-1}. Thus, it is very likely that extensive if not virtually complete electronic relaxation occurs before the lowest energy decompositions accompanied by metastable ions indicating dissociation rate constants of the order of 10^5 sec^{-1}. Less reliable predictions can

be made regarding the extent of electronic relaxation before
dissociation for the more highly excited states of ions well
above one or more dissociation limits. From the energy gap law,
one expects the relaxation to occur stepwise and, for equal
energy gaps, with usually increasing rate for successive steps due
to the effect of increasing vibrational energy and to the greater
possibility of surface crossings and strong-coupling behavior.
Estimates of relaxation rates could be made for individual cases
but the estimation of the rates of competing dissociations of
excited electronic states would usually suffer from more uncertainty in the appropriate activation energies as compared with the
case of the ground state of the ion. For certain classes of compounds, such as the alkanes for instance, for which the photoelectron spectra indicate only broad and strongly overlapped
bands, extensive electronic relaxation seems very likely and
these compounds have generally shown very good agreement between
experiment and the predictions of QET based on complete electronic
relaxation.

Recent tests of QET

For a comprehensive presentation of experimental data relating to tests of QET, the reader is referred to review articles
[46,47] which may be summarized fairly as showing that QET accounts
well for the general features of mass spectra and decomposition
behavior of large molecular ions, but that there is evidence for
its failure in some instances. The theory is sufficiently successful as to become useful in practical applications of mass spectrometry to molecular structure determination.[56] This section will
emphasize recent tests of QET, particularly those employing more
powerful techniques.

Many earlier tests of QET have involved the experimental
measurement of breakdown curves by photoionization[4,57] techniques. The interpretation of these experiments required assumptions regarding cross section behavior. Charge transfer[25]
techniques which yield data points at widely separated energies
and sometimes suffer from difficulties of interpretation have also
been used. The recently developed technique[26-30] of detecting
mass-analyzed ions produced in coincidence with energy-analyzed
photoelectrons by monochromatic photons yields breakdown curves
unambiguously. Stockbauer[30] has recently published breakdown
curves of very high quality for methane and ethane. His data for
ethane are shown in Fig. 4, together with the results of an
earlier theoretical calculation. The small discrepancy is readily
attributed to the use of somewhat inaccurate thermochemical data
in the calculation. The complete absence of parent ions above
the first fragmentation threshold, the behavior of the competitive
decompositions producing $C_2H_5^+$ and $C_2H_4^+$ and of the succeeding

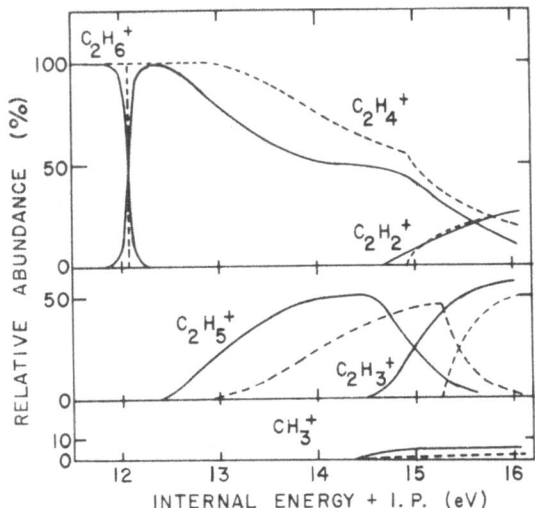

Fig. 4.   Comparison of experimental (solid lines) breakdown
curves of Stockbauer (ref. 30) with results of QET
calculations (dashed lines).

decomposition of these ions by elimination of H_2 are as predicted
by the theory. Of particular importance is the smooth behavior
of the curves (except at thresholds as expected) although the
photoelectron spectrum shows at least three electronic bands in
this region. The data show that the decomposition behavior is a
smooth function of the internal energy of the parent ion independent of initial electronic state. Such data for a number of molecules will provide excellent tests of theory. Lack of electronic
relaxation can evidence itself by persistence of parent ions well
above the first dissociation limit and by abrupt changes in breakdown curves at thresholds for formation of excited states.

Another series of important developments have been made by
Osberghaus, Ottinger and their co-workers.[58-60] By producing
initial ionization along a thin equi-potential plane in a strong
electric field and measuring the kinetic energy distribution of
fragment ions, they were able to measure decomposition times over
a range from 5×10^{-9} sec to 5×10^{-6} sec. The earlier work
employed electron impact ionization which produced ions with a
range of internal energies and they showed that the resulting ions
decayed with a distribution of rate constants as predicted by QET.
In the most sophisticated version of their experiment, Andlauer
and Ottinger[61] initiated ionization by charge-transfer by which

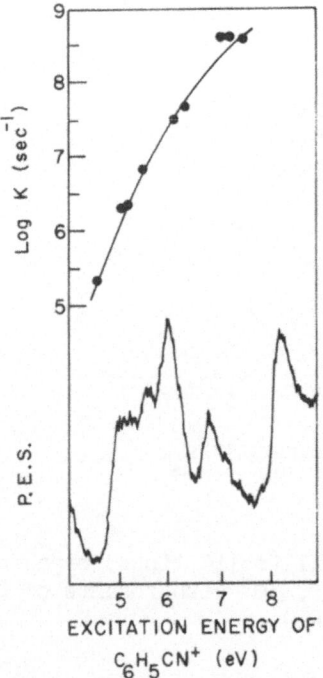

Fig. 5.   Comparison of the rate constant for decomposition of
          benzonitrile ion into $C_6H_4^+$ + HCN with the photo-
          electron spectrum.

parent ions were formed with specific amounts of internal energy.
For two molecules, they were able to measure the rate constant as
a function of excitation energy of the parent ion. The first
case was that of benzonitrile for which the experimental data, together with the corresponding region of the photoelectron
spectrum[13] is shown in Fig. 5. The data of Fig. 5 show that k
is a smoothly and monotonically increasing function of E^* even
though the photoelectron spectrum shows at least four different
electronic bands in the energy region covered. A theoretical
rate curve calculated by Klots[44] has the same general shape but
the quantitative agreement is poor, in great part due to the use
of an incorrect heat of formation for the fragmentation. Of far
more significance is the experimental evidence for electronic
relaxation as indicated by the monotonic behavior of k.

       The other molecule investigated by Andlauer and Ottinger is

benzene. Four decompositions occur with roughly the same thresholds and these workers were able to measure the decomposition rate by monitoring two fragments, $C_4H_4^+$ and $C_6H_5^+$. If these two fragmentation processes are competitive, the decay rates of the parent ion inferred by measuring the formation of each of the ions should be identical. The results are shown in Fig. 6 where it is seen that for E^* above 5.0 eV the two ion fragments yield different curves. Unfortunately the apparent discrepancy depends on only two sets of points but if correct, the data clearly indicate lack of complete randomization and apparently require that the two fragments come from different states or forms of the parent ion. Also shown in Fig. 6 is a dashed line obtained by photoionization as will be described later. The latter results indicate that for values of E^* below 5.0 eV all four decompositions are in competition. The facile conclusion from the above data is that randomization occurs very near the fragmentation threshold but fails at some higher energies.

Benzene and some derivatives

The fragmentation of the benzene ion has a long history in connection with QET. It was once proposed[46] as an obvious case of failure of QET since four metastable ions produced by decomposition of the parent ion were observed although the appearance potentials of the corresponding fragments measured by electron impact differed by more than 1.0 eV whereas QET requires that they be very nearly the same. Vestal[47] pointed out some reasons for the apparent discrepancy and recent photoionization mass spectrometric studies[62,63] have cleared up this problem entirely. Some of the relevant data and the photoelectron spectrum are shown sketched semi-quantitatively in Fig. 7 for both C_6H_6 and C_6D_6. In one photoionization study[63] the ionization efficiency curves of the four metastable ions were also measured and found to be very nearly identical (except for intensity) as required by QET for competitive decompositions. The metastable curves have the form of moderately sharp steps located in the threshold region for fragment production. The curvature and asymptotic form of the curves for the fragment ion make "threshold" an inappropriate description. The shape and position of the metastable curves are far more meaningful as will be described shortly. The fragment and metastable curves for C_6D_6 appear very similar to those of C_6H_6 except that they are shifted upward in energy by about 0.35 eV. The normalized derivative curve for an ion yields its breakdown curve[4] to an approximation which is especially good for a metastable ion. This experimental breakdown curve can be compared with the calculated one of Fig. 1 to determine a short segment of the curve of log k vs. E^* as illustrated in Fig. 8 for

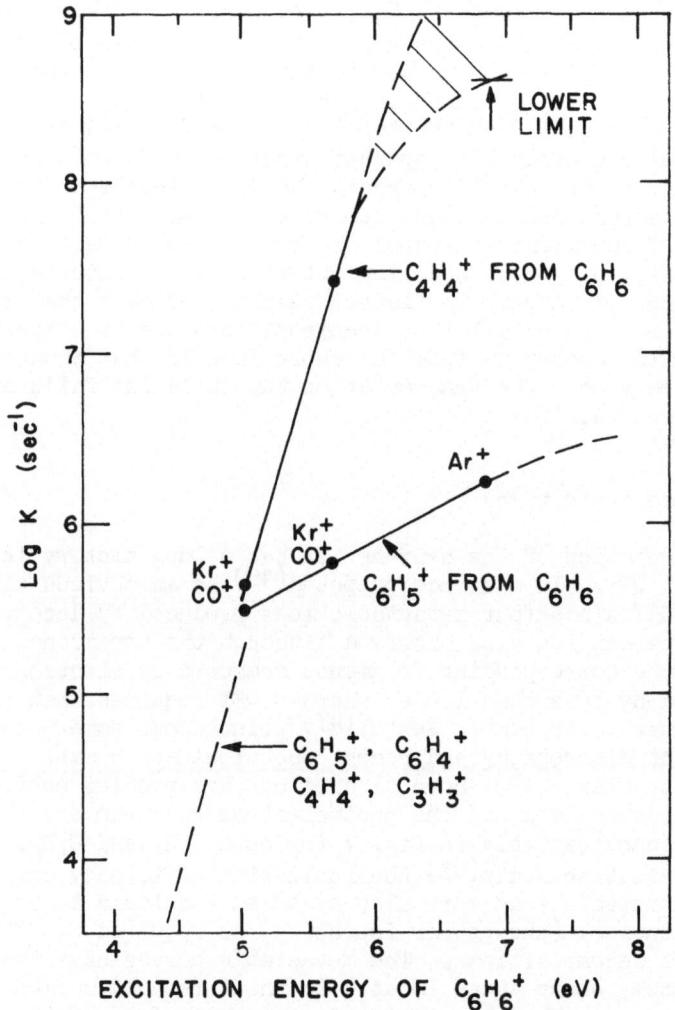

Fig. 6.   Rate constants as a function of internal energy for
          decomposition of the $C_6H_6^+$ ion as determined by
          charge-transfer excitation (points and solid lines,
          ref. 61) and by photoionization measurements of the
          corresponding metastable ions (dashed line, ref. 63).

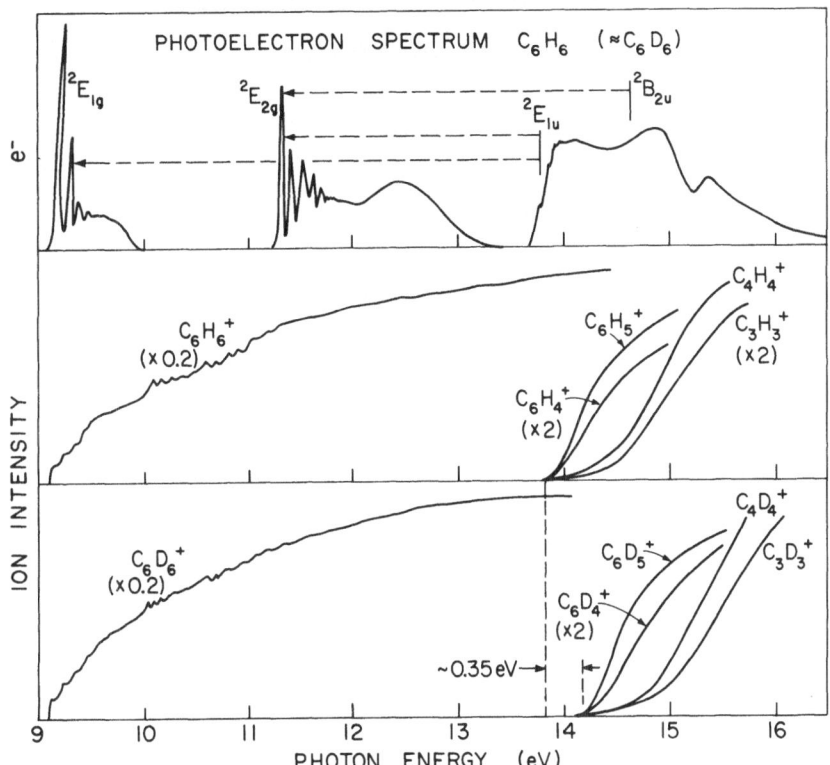

Fig. 7. Photoelectron spectrum of benzene (top) and photoionization efficiency curves for parent and fragment
ions of benzene and perdeuterobenzene.

C_6D_6. Thus $k = 6.0 \times 10^4$ sec^{-1} at 859 Å (14.43 ev) which
corresponds to E* = 14.43 - 9.25 = 5.18 eV and the slope is
2.87 eV^{-1} as shown. The dashed line of Fig. 6 was so obtained
for C_6H_6. The analogous line for C_6D_6 has very nearly the same
slope but is displaced to higher energies by about 0.35 eV.

The data of Fig. 7 show that fragmentation at threshold
occurs from parent ions initially formed in the $^2E_{1u}$ excited
state and at somewhat higher energies in the $^2B_{2u}$ state as well.
At threshold, i.e. in the region of metastable ion formation, it
is almost certain that a significant amount of electronic relaxation occurs for the following reasons. Both of the excited
electronic states mentioned above have one or more optically
allowed transitions to lower states as shown by the dashed lines

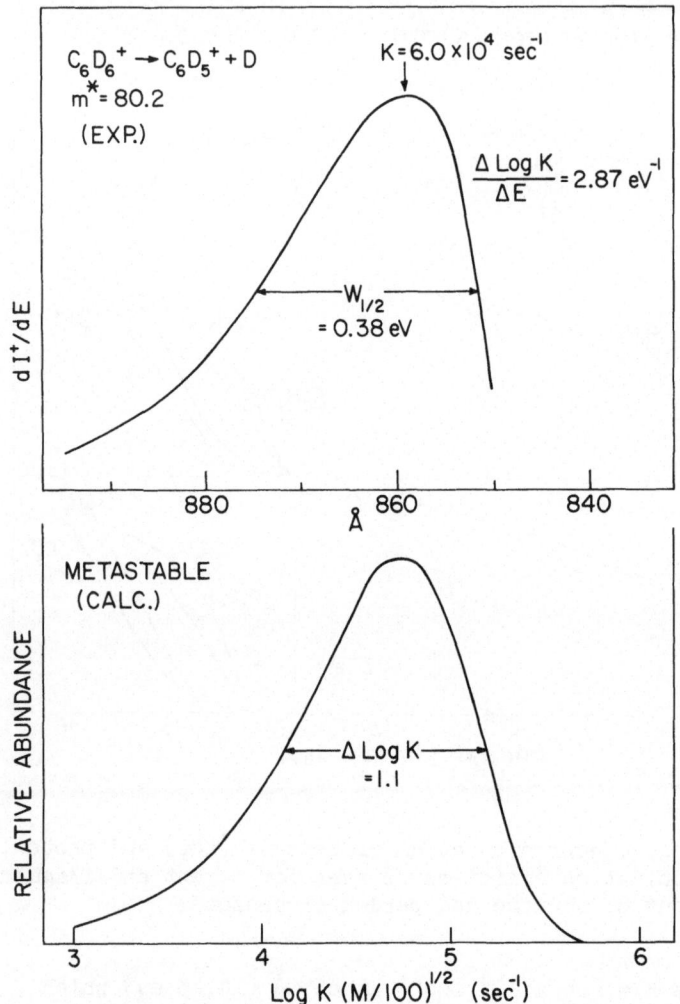

Fig. 8. Comparison of experimental breakdown curve as a
         function of energy with calculated curve as a
         function of rate constant in order to determine
         dependence of log k on E^* for $C_6D_6^+$.

of Fig. 7, yet no significant deactivation by radiation is observed in 10^{-4} sec as shown by the presence of intense metastable ions. Several attempts have been made to observe fluorescence of the benzene ion without success. Studies[64,65] of isotopically substituted benzenes have shown that complete isotopic scrambling of all carbon and hydrogen atoms occurs independently prior to metastable decomposition and nearly complete scrambling before at least some of the faster fragmentations. The thermochemical threshold for the reaction producing $C_6H_5^+$ is independently known[62] to be at about 12.95 ev or $E_a = 3.70$ eV. Figures 6 and 7 show a large kinetic shift of about 1.9 eV which implies, in the framework of RRKM theory, a very high density of states for the active molecular ion which could only be attained in one or more of the lowest electronic states. The shift of metastable and fragment curves upward in energy by about 0.35 eV for C_6D_6 is approximately as expected from the increased density of states of the perdeuterated ion while the positions of the $^2E_{1u}$ and $^2B_{2u}$ states in the photoelectron spectrum remain very nearly unchanged. A very crude estimate of the radiationless transition rate from the $^2E_{1u}$ state to the $^2A_{2u}$ state just below at about 12.2 eV gives a value of $10^{10} - 10^{11}$ sec^{-1}. However, the larger gap (about 2.2 eV) between the ground and first excited states suggests that complete relaxation to the ground state may not occur. Indeed, Rosenstock et al.[62] interpret their data in terms of dissociation from the ground and first excited states independently.

Extensive photoionization mass spectrometric measurements of fragmentation of the phenyl halides have been made.[63] The lowest energy process for the chloride, bromide and iodide is the simple bond fission forming $C_6H_5^+$. No other fragmentation process occurs until much higher energies. In all cases, intense metastable ions are seen and measurements of k vs. E^* similar to the dashed line of Fig. 6 were made. The parent ions at threshold are in all cases initially prepared in highly excited electronic states but the energy dependence of k and the values of the kinetic shift which increase with activation energy all indicate extensive and possibly complete electronic relaxation at the fragmentation threshold. The photoelectron spectra in all cases show energy gaps which are smaller than the larger ones in benzene.

Kinetic energy release in fragmentation

Early measurements [66,67] of kinetic energy release in fragmentation of polyatomic ions were made for ions prepared with a rather large and inaccurately known internal energy distribution. The experimental kinetic energy distributions were found to be predominantly quasi-Maxwellian and approximately as predicted by application of QET. The application of QET requires an additional assumption regarding disposition of kinetic energy in the decomposition of the activated complex. It has been assumed[68] that the kinetic energy distribution is equal to that along the reaction coordinate plus any due to activation energy for the reverse reaction. Other formulations[44] have employed microscopic reversibility and the assumption of a Langevin cross section (centrifugal barrier only) for the reverse reaction which is thus assumed to have no activation energy. Fortunately, many decompositions (e.g. practically all simple bond fission) are expected to have zero reverse activation energy and measured kinetic energies of such decompositions have usually been in reasonable agreement with theory. This agreement has encouraged the use of measured kinetic energies in a semi-empirical method of correction for excess internal energy in the calculation of thermochemical thersholds from measured values.[69] More important as test of the applicability of QET to the calculation of kinetic energy distributions are measurements on metastable ions which are produced from parent ions with a narrow range of internal energies and which have kinetic energy distributions measurable with very high precision. Such high precision measurements are only recently being made.[8] Together with precise determinations of E^* by photoionization techniques, such data should provide excellent tests of the applicability of QET to simple bond fission processes and may give valuable information on the characteristics of activated complexes for those rearrangement decompositions which occur with excess activation energy.

CONCLUSION

Fragmentation processes of diatomic and small polyatomic ions are fairly well understood in principle and a fruitful interplay of experiment and theory is in progress. For larger polyatomic molecules, progress requires much more experimental information on the degree of electronic and vibrational relaxation and the validity of RRKM theory for preparations of varying randomness, particularly at the higher interval energies (up to about 10 eV or so) produced in valence shell ionization. Nevertheless, QET has provided a fairly successful description of a wide range of phenomena en-encountered and there is reasonable hope that tractable extensions or completely new theoretical treatments which specifically include the effects of incomplete electronic and vibrational relaxation can be developed. The latter types of approach are being developed particularly in connection with studies of photochemical decompositions[70] which are formally the same as the dissociation processes considered here, although a few significant differences exist. The energy gaps between electronic states, particularly that between the ground and first excited state, are generally much

smaller for ions and hence electronic relaxation should be more
rapid. The polyatomic ions are practically always doublet radicals characterized by unusual lability and subject to a wide
variety of rearrangements. Simple bond dissociation nearly always occurs without excess activation energy and the long range
ion-induced-dipole potential for the departing fragments yields a
well known rotational barrier for the reaction. Most of these
factors support the expectation that simple statistical theories
will be valid more generally for ionic decompositions than for
photochemical ones. A serious disadvantage in the case of polyatomic ions is the relative lack of knowledge of the geometry and
other properties of parent and fragment ions. This gap in knowledge is slowly being filled by a variety of techniques but at
present remains a significant difficulty.

REFERENCES

(1) N. W. Reid, Int. J. Mass Spectrom. Ion Phys. 6, 1 (1971)
(2) W. A. Chupka in "Ion-Molecule Reactions" Vol. 1, Ed. J.L.
     Franklin (Plenum Press, New York 1972) pp. 33-76
(3) J. Berkowitz and W. A. Chupka, J. Chem. Phys. 45, 1287 (1966)
(4) W. A. Chupka, J. Chem. Phys. 30, 191 (1959)
(5) J. H. Beynon, R. A. Saunders and A. E. Williams, Z. Natur-
     forsch. 20a, 180 (1965)
(6) J. H. Beynon, R. M. Caprioli and T. Ast, Org. Mass Spectro-
     metry 5, 229 (1971)
(7) R. B. Bernstein, Comm. Atom. Mol. Phys. (in press)
(8) E. G. Jones, J. H. Beynon and R. G. Cooks, J. Chem. Phys. 57,
     2652 (1972). References are given for other types of kinetic
     energy measurements.
(9) J. E. Mentall, E. P. Gentieu, M. Krauss and D. Neumann, J.
     Chem. Phys. 55, 5471 (1971)
(10) U. Fano and J. W. Cooper, Rev. Mod. Phys. 40, 441 (1968)
(11) V. H. Dibeler and J. A. Walker, Int. J. Mass Spectrom. Ion
     Phys. 11, 49 (1973)
(12) P. C. Killgoar, Jr., G. E. Leroi, J. Berkowitz and W.
     Chupka, J. Chem. Phys. 58, 803 (1973)
(13) D. W. Turner, C. Baker, A. D. Baker and C. R. Brundle,
     "Molecular Photoelectron Spectroscopy" (Wiley-Interscience,
     New York, 1970)
(14) R. B. Cairns, H. Harrison, and R. I. Schoen, Phil. Trans.
     Roy. Soc. Lond. A 268, 163 (1970)
(15) J. A. Carlson, M. O. Krause and W. E. Moddeman, J. Physique
     32 Ce, 76 (1971)
(16) R. N. Dixon and W. E. Hull, Chem. Phys. Letters 3, 367(1969)
(17) O. Edquist, E. Lindhom, L. E. Selin and L. Åsbrink, Physica
     Scripta 1, 25 (1970)
(18) J. C. Lorquet and C. Cadet, Chem. Phys. Letters 6, 198 (1970)
(19) M. Jungen, Theoret. Chim. Acta (Berl.) 27, 33 (1972)

(20) G. Herzberg, "Molecular Spectra and Molecular Structure.
      III. Electronic Spectra and Electronic Structure of Poly-
      atomic Molecules" (D. Van Nostrand Co., Inc., Princeton,
      N.J., 1966)

(21) W. A. Chupka, J. Chem. Phys. 48, 2337 (1968)

(22) A. Pipano and J. J. Kaufman, J. Chem. Phys. 56, 5258 (1972)

(23) G. Herzberg, "Molecular Spectra and Molecular Structure. I.
      Spectra of Diatomic Molecules" (D. Van Nostrand Co., Inc.,
      Princeton, N.J., 1960), 2nd ed.

(24) O. Edqvist, E. Lindholm, L. E. Selin, L. Åsbrink, Physica
      Scripta 1, 172 (1970)

(25) E. Lindholm in "Ion Molecule Reactions in the Gas Phase",
      Adv. in Chemistry Ser. No. 58, (Amer. Chem. Soc., Washington,
      D.C., 1966) p. 1.

(26) J. H. D. Eland, Int. J. Mass Spectrom. Ion Phys. 8, 143
      (1972)

(27) C. J. Danby and J. H. D. Eland, Int. J. Mass Spectrom. Ion
      Phys. 8, 153 (1972)

(28) J. H. D. Eland, Int. J. Mass Spectrom. Ion Phys. 9, 397
      (1972)

(29) R. Stockbauer and M. G. Inghram, J. Chem. Phys. 54, 2242
      (1971)

(30) R. Stockbauer, J. Chem. Phys. 58, 3800 (1973)

(31) V. H. Debeler, J. A. Walker and H. M. Rosenstock, J. Res.
      Nat. Bur. Std. A 70, 459 (1966)

(32) F. Fiquet-Fayard and P. M. Guyon, Mol. Phys. 11, 17 (1966)

(33) J. Appell and J. Durup, Int. J. Mass Spectrom. Ion Phys.
      10, 247 (1973)

(34) K. E. McCulloh, J. Chem. Phys. (in press)

(35) W. A. Chupka, J. Berkowitz, E. Darland, G. E. Leroi (to be
      published)

(36) J. A. R. Samson and J. L. Gardner, J. Geophys. Res. 78,
      3663 (1973)

(37) R. J. Robinson and K. A. Holbrook, "Unimolecular Reaction"
      (Wiley-Interscience, New York, 1972)

(38) L. D. Spicer and B. S. Rabinovitch, Ann. Rev. Phys. Chem.
      21, 349 (1970)

(39) J. D. Rynbrandt and B. S. Rabinovitch, J. Phys. Chem. 75,
      2164 (1971)

(40) K. Shobatake, M. M. Parson, Y. T. Lee and S. A. Rice, J.
      Chem. Phys. 59, 1416 (1973); 59, 1427 (1973); 59, 1435(1973).

(41) E. V. Waage and B. S. Rabinovitch, Chem. Rev., 70, 377 (1970)

(42) W. Forst, Chem. Rev. 71, 339 (1971)

(43) E. E. Nikitin, Theor. Exp. Chem. 1, 144 (1965)

(44) C. E. Klots, Z. Naturforsch. 27a, 553 (1972)

(45) M. Wolfsberg, J. Chem. Phys. 36, 1072 (1962)

(46) H. M. Rosenstock and M. Krauss, Adv. Mass Spectrom., 2, 251
      (1962); H. M. Rosenstock, Adv. Mass Spectrom. 4, 523 (1968)

(47) M. L. Vestal in "Fundamental Processes in Radiation Chemistry,"
      ed. P. Ausloos (Interscience Publishers, New York, 1968),
      Chapter 2.

(48) J. B. Birks, "Photophysics of Aromatic Molecules", (Wiley-Interscience, New York, 1970).

(49) J. Jortner, S. A. Rice, R. M. Hochstrasser, Advan. Photochem. $\underline{7}$, 149 (1969)

(50) J. Chim. Phys., Special Issue "Transitions Non Radiatives dans les Molecules", (1970)

(51) E. W. Schlag, S. Schneider and S. F. Fischer, Ann. Rev. Phys. Chem. $\underline{22}$, 465 (1971)

(52) K. F. Freed, Topics Current Chem. $\underline{31}$, 105 (1972)

(53) R. Englman and J. Jortner, Mol. Phys. $\underline{18}$, 145 (1970)

(54) A. A. Abramson, K. G. Spears and S. A. Rice, J. Chem. Phys. $\underline{56}$, 2291 (1972)

(55) D. F. Hellyer, K. F. Freed and W. M. Gelbart, J. Chem. Phys. $\underline{56}$, 2309 (1972)

(56) F. W. McLafferty, "Interpretation of Mass Spectra" (Addison-Wesley, Reading, Mass., 1973), 2nd ed. chapter 8.

(57) W. A. Chupka and J. Berkowitz, J. Chem. Phys. $\underline{47}$, 2921(1967)

(58) O. Osberghaus and Ch. Ottinger, Phys. Lett. $\underline{16}$, 121 (1965)

(59) Ch. Ottinger, Z. Naturforsch. $\underline{22a}$, 20 (1967)

(60) I. Hertel and Ch. Ottinger, Z. Naturforsch. $\underline{22a}$, 40 (1967); $\underline{22a}$, 1141 (1967)

(61) B. Andlauer and Ch. Ottinger, Z. Naturforsch. $\underline{27a}$, 293 (1972)

(62) H. M. Rosenstock, J. T. Larkins and J. A. Walker, Int. J. Mass Spectrom. Ion Phys. $\underline{11}$, 309 (1973)

(63) W. A. Chupka, S. I. Miller and J. Berkowitz (to be published)

(64) J. H. Beynon, R. M. Caprioli, W. O. Perry and W. E. Baitinger, J. Amer. Chem. Soc. $\underline{94}$, 6828 (1972)

(65) R. J. Dickinson and D. Williams, J. Chem. Soc. (B), 249 (1971)

(66) R. Taubert, Z. Naturforsch. $\underline{19a}$, 484, 911 (1964)

(67) R. Fuchs and R. Taubert, Z. Naturforsch. $\underline{19a}$, 494 (1964)

(68) C. E. Klots, J. Chem. Phys. $\underline{41}$, 117 (1964)

(69) M. A. Haney and J. L. Franklin, Trans. Faraday Soc. $\underline{65}$, 1794 (1969) and references therein.

(70) S. A. Rice in "Advances in Electronic Excitation and Relaxation", ed. E. C. Lim (Academic Press, New York) vol. 2 (in press).

FAR ULTRAVIOLET PHOTOCHEMISTRY OF ORGANIC COMPOUNDS[*]

P. Ausloos and S. G. Lias

National Bureau of Standards
Radiation Chemistry Section
Washington, D. C. 20234

This brief survey will be concerned with reviewing what is known about the fates—particularly the modes of decomposition—of organic molecules excited by absorption of photons having energies above about 8 eV. Since the dissociation energies of most chemical bonds in organic molecules lie in the energy range of 3-5 eV, it is to be expected that for most compounds, the major mode of energy dissipation in the vacuum ultraviolet photolysis will be dissociation. With the notable exceptions of aromatic and acetylenic molecules, molecules excited to neutral states in this energy region usually have a quantum yield of fragmentation close to unity. Of course, other modes of energy dissipation such as fluorescence may have to be considered under conditions where molecules may have long dissociative lifetimes—such as high molecular weight molecules at high densities[1/].

It should be remembered that the ionization potentials of most organic compounds lie in the energy range 8-13 eV, and so in our discussion we will need to be concerned at times with the presence of ions in the system. Although we will confine our attention to the fates of neutral excited molecules, the role of ions in any given system must be understood for an adequate appraisal of the overall photochemical process. At the photoionization threshold and in the region just above the onset, the quantum yield of ionization (Φ_+) is usually small (i.e., the quantum yield of excited molecule formation is large). At energies above the ionization threshold, it is not extremely difficult to study the modes of decomposition of the neutral excited molecule in regions where the quantum yield of ionization is relatively low, particularly if

---
* Work supported in part by the U.S. Atomic Energy Commission.

ionic decomposition is unimportant or the ion-molecule reaction
mechanism is well known[2]. The ionization quantum yield generally
increases in importance with increasing energy (although plots of
Φ_+ versus λ may show structure), and for the great majority of organic compounds for which ionization quantum yield determinations
at high photon energies (> 15 eV) are available, it is seen that
Φ_+ generally is unity or close to unity (Table I). In discussing

TABLE I

Photoionization Quantum Yields and Ionization Potentials of Organic
Molecules

Compound	Ionization Potential eV	Ionization Quantum Yields (Φ_+)		
		106.7 nm[a] 11.6 eV	14.4-73.6 nm[b] 16.66-16.84 eV	58.4 nm 21.2 eV
CH_4	12.5	0.00	1.0	0.96[b]
C_2H_2	11.4	0.772	0.92	1.0[c]
C_2H_4	10.5	0.225	0.98	0.98[c]
C_2H_6	11.5	0.017	1.0	0.98[c]
C_3H_8	11.1	0.240	1.0	0.94[b]
$n-C_4H_{10}$	10.6	0.363	1.0	0.91[b]
C_6H_6	9.2	~ 0.47[d]	1.0	n.d.
$c-C_6H_{12}$	9.8	~ 0.49[d]	1.0	n.d.
C_2H_5OH	10.5	~ 0.20[d]	0.92	0.98[c]
CH_3COCH_3	10.0	~ 0.21[d]	1.0	n.d.

a/ C. E. Klots, J. Chem. Phys. 56, 124 (1972).
b/ R. E. Rebbert and P. Ausloos, NBS J. Res. 75A, 481 (1971).
c/ S. W. Bennett, J. B. Tellinghuisen and L. F. Phillips, J. Phys.
   Chem. 75, 719 (1971).
d/ P. Ausloos and S. G. Lias, Rad. Res. Rev. 1, 75 (1968).

the modes of decomposition of molecules excited to energies above
the ionization onset, it is useful to express the probability for
the occurrence of a particular dissociative process, or the probability for the formation of a particular fragment, relative to the
total number of excited molecules in the system which do not ionize
rather than relative to the number of photons absorbed, since in
this way more meaningful comparisons with the events occurring below the ionization threshold can be made. Whenever appropriate,
therefore, we will express product yields or probabilities of the
occurrence of primary processes using the notation $M(X)/N_{ex}$, (that
is, either molecules of product X or the probability of process X
formed per neutral excited molecule, N_{ex}, in the system). Of course,
at energies below the ionization potential, this is simply equivalent to the quantum yield. The relationship of the quantity $M(X)/$

N_{ex} to the quantum yield is given by the expression:
$\Phi = [M(X)/N_{ex}](1-\Phi_+)$.

The fragments formed by dissociation of an ion are simply and directly monitored by examining the mass spectrum of the compound. Thus, it is not a difficult task to derive the primary modes of dissociation of a molecule which has been ionized. Establishing the modes of fragmentation of an electrically neutral molecule, however, is much more difficult in that there is no single, direct method of measuring the "spectrum" of molecules, radicals, and atoms formed through the various dissociative processes. Before going on to discuss what is known about the modes of decomposition of various types of compounds in the far ultraviolet photolysis, we will first briefly describe experimental approaches which have been used to obtain information about photochemical fragmentation processes.

DIRECT METHODS OF DETERMINING PHOTOFRAGMENTS

Direct methods of determining the fragment species which are formed in the decomposition of electronically excited molecules include absorption spectroscopy, electron spin resonance, and the monitoring of emission from excited fragments. While the former two methods have given valuable information about the kinetics of the secondary reactions of the photofragments, and about their structure, these approaches are usually not adequate to the task of determining the overall picture of the modes of fragmentation of an excited molecule.

Information which is of direct relevance to an understanding of the primary dissociative processes can be obtained from an examination of the emission from fragments which originate with excess electronic energy, but of course this technique does not yield any information about fragmentation processes which lead to product species in their ground states. The value, as well as the limitations, of this type of study is illustrated by the results of Judge and Welch[2] on methane (Figure 1). First, little emission is seen in the low energy region for the obvious reason that at low energies photofragments will be produced mainly in their ground electronic states. Even in the region where the emission cross section of a fragment such as CH is particularly high, other ground state fragments (including CH) will be formed, and this technique will give us no information about these simultaneously-occurring processes. It should also be pointed out that at high energies there is a region where no emission is seen due to the fact that the photoionization quantum yield of methane is equal to unity in this energy range (Table I). It has actually recently been shown that the photochemical transformations occurring when CH_4 is irradiated with neon resonance light source are entirely

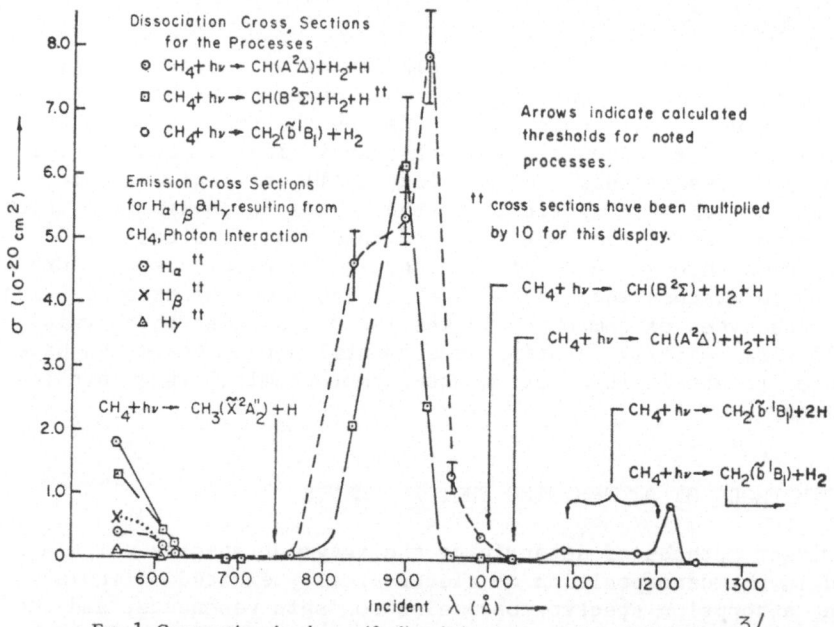

Fig. 1 Cross sections for the specific dissociation processes observed in methane.[3]

accounted for by ionic processes[4-5].

INDIRECT METHODS OF DETERMINING PHOTOFRAGMENTS

Because of the limitations of direct methods of determining photolytic fragments, the vast majority of photochemical investigations on organic molecules in the far ultraviolet have approached the determination of the "spectrum" of fragments through various techniques involving the chemical analysis of the final products formed in the compound after photolysis. Of course, the identities of these final products do not in themselves give much information about the modes of dissociation of the excited molecule, since a given product molecule might be formed, for instance, directly in a primary dissociation process, as a result of a reaction of a radical or atom fragment, or even, depending on the experimental conditions, as a result of a secondary fragmentation of one of the primary photochemical products[6]. Thus, one must resort to the use of various diagnostic techniques in order to obtain more specific information about the identities of the primary photofragments. These diagnostic techniques include the use of isotopic labeling, observing the effects on the product yields when various experimental parameters such as density or wavelength are varied, and experiments in which foreign compounds are added to the system

which are expected to interact in specific ways with certain reactive intermediates. For instance, it is common to add compounds
to a system which will interfere with the reaction mechanism of the
free radicals by reacting with these radicals; identities and
yields of the radicals present in the system are derived from the
variations in the yields of products observed when the "scavenger"
is added, or, in some cases, from the appearance of a specific
product which can be attributed to a known reaction between a given
radical and the added compound. This brief description of the
devices used by photochemists in order to wring information about
photochemical processes from analyses of chemical products is by
no means a complete list, but it should at least give an impression
that obtaining answers about primary photochemical processes by
analyzing final products is not simple and straightforward. A
fully-defined, completed picture of the photochemistry of a given
compound is difficult to obtain, and this ideal goal can only be
approached by combining the information gained from different diagnostic techniques. Nevertheless, as some of the results discussed below will demonstrate, extensive information about the
primary photochemical processes in certain systems has been obtained from such experiments.

At this point, it is in order to point out that accurate
quantitative analysis of the final products formed in a photochemical system requires, in practical terms, that within a reasonably
short length of time it must be possible to accumulate quantities
of product molecules large enough to be detected by current analytical techniques. This means, in turn, that the light sources used
for such photochemical experiments must be of relatively high intensity, that is, must deliver more than about 10^{11} photons per
second. Continuum light sources, used in conjunction with a monochromator in order to select photons of a given energy, do not
give light of sufficient intensity for end-product-type experiments,
although future improvements in the light sources and analytical
techniques will make this a feasible approach. At the present time,
the most useful light sources for photochemical experiments in the
far ultraviolet are enclosed lamps, fitted with appropriate windows,
in which a gas excited by microwave discharge emits its characteristic resonance radiation; such lamps are directly attached to the
reaction vessel, and deliver up to 10^{16} photons per second. Table
II lists the resonance light sources most commonly used for far
ultraviolet photolysis experiments, the energies which they deliver
and the appropriate window materials for each. Actually, the fact
that most current studies are practically limited to these few
selected energies is not really an important limitation at the
present time. After all, a detailed quantitative determination of
the modes of photodecomposition is a goal which has not really been
successfully achieved yet, even at these wavelengths. As an illustration, two papers published within the last year disagree about
the primary decomposition modes of a simple molecule, propane, at

TABLE II

Microwave Light Sources Most Commonly Used[a-b]

	Optimum Pressure Torr	Wavelength λ nm	Photon Energies eV	Recommended Window
Bromine	0.1% in 1 torr He	163.4*	7.6	Suprasil quartz
X_e	~ 0.5	147.0	8.4	Saphire
Kr	~ 0.5	123.6	10.0	CaF_2
H (Lyα)	2% H_2 in 1 torr He	121.6	10.2	LiF (~ 1 mm thick)
Ar	~ 1.0	104.8 (~50%) 106.7 (~50%)	11.6 11.8	LiF (~ 0.3 mm thick)
Ne	~ 2.0	74.4-73.6**	16.7-16.8	Al (1,500Å thick)
He	~ 2.0	58.4	21.2	Al (1,500Å thick)

* Depending on thickness of window 158.2 nm and 157.65 nm radiation will be transmitted as well.
** Ratio of two resonance lines depends on thickness of Al layer.
a/ J. R. McNesby, W. Braun and J. Ball, Chapter 11 in "Creation and Detection of the Excited State", A. A. Lamola, Editor, M. Dekker, Inc., N. Y. (1971).
b/ R. Gorden, Jr., R. E. Rebbert and P. Ausloos, NBS Tech. Note No. 496, 55 pages (Oct. 1969).

one chosen wavelength (123.6 nm)[7-8].

ALKANES AND CYCLOALKANES

Since saturated hydrocarbons absorb energy only in the far ultraviolet region of the spectrum[9], their photochemistry has a relatively short history, dating back only to the beginnings of far ultraviolet photochemistry as an active field of research[10]. In the period 1961-1962, it was reported that elimination of a molecule of hydrogen was the most important primary process occurring in the gas phase photolysis of methane, with 10.0 eV photons[11], and ethane[12-13] and n-butane[13] with 8.4 eV photons. Of particular interest was the observation reported by Okabe and McNesby[12] in 1961, that in the photolysis of ethane, the molecule of hydrogen eliminated originated almost entirely from a single carbon atom. This conclusion was based on experiments in which CH_3CD_3 was photolyzed; the molecular hydrogen consisted mainly of H_2 and D_2:

$$CH_3CD_3 + h\nu \rightarrow C_2H_3D + D_2 \tag{1}$$

$$CH_3CD_3 + h\nu \rightarrow C_2D_3H + H_2 \qquad\qquad (2)$$

This unanticipated result prompted a number of investigations on partially deuterium labeled alkanes (propane, isobutane, and n-butane)[10]. In addition to hydrogen elimination, other possible primary processes in the alkanes are direct elimination of an alkane product molecule, elimination of an H-atom, and simple C-C bond cleavage processes leading to the formation of two radicals. The results of these studies showed that alkane elimination processes occurred mainly through a concerted mechanism in which a hydrogen species moves across a C-C bond which is undergoing cleavage. For example, in propane[14], most of the molecular methane is formed through the process:

$$CD_3CH_2CD_3 + h\nu \rightarrow C_2D_3H + CD_3H \qquad\qquad (3)$$

Four-centered molecular elimination processes such as:

$$CD_3CH_2CD_3 + h\nu \rightarrow C_2H_2D_2 + CD_4 \qquad\qquad (4)$$

also occur, but are less important. Evidence was also obtained for the occurrence of simple C-H and C-C bond cleavage processes, but because such processes lead to the formation of radical products, which in the gas phase may also originate from secondary decompositions of other products, it was more difficult to establish the occurrence of simple bond-cleavage primary processes or to estimate their importance[7-8].

Although in the early work on the photochemistry of alkanes, no quantum yield determinations were made, and accurate determinations of the relative yields of radical fragments were difficult to obtain, much painstaking effort did lead to fairly detailed estimates of the relative importances of the major primary processes in the lower alkanes, especially ethane and propane. Current efforts in our laboratory are directed to determinations of accurate quantum yields of all the fragments formed in the photolysis of linear and branched alkanes up to about $C_{11}H_{24}$, and thereby to an eventual derivation of the quantum yields of all the primary processes. It is hoped that such an accumulation of quantitative information will make possible a broader understanding of the photochemistry of alkanes in the near future. Some preliminary descriptions of systematic trends which have been observed as a function of structure, chain length, and photon energy will be presented here, although at the present time a comprehensive interpretation (which might make it clear to what degree information obtained from optical absorption studies are relevant to photochemistry) is premature.

It is obvious that hydrogen and alkane products formed through direct elimination from an excited alkane are fairly easy to determine when radical scavengers such as O_2 or NO (which prevent the radicals from forming hydrocarbon products) are added to the system; through the use of partially deuterated compounds, the exact mechanism for the formation of these products can be established without difficulty, as shown by the examples given in reactions 3 and 4. Until recently, the quantitative determination of radical products was more difficult, and the results less certain. It has now been established[8], however, that small amounts of HI added to alkanes do not interfere with the photolytic mechanism, but do quantitatively intercept hydrocarbon radicals and H-atoms to form the corresponding RH compound:

$$R + HI \rightarrow RH + I \tag{5}$$

(where R is a radical or H-atom). Figure 2 shows the yields of RH compounds formed in the photolysis of propane at 8.4 eV as a function of the concentration of added HI; it is seen that very low concentrations of HI ($\sim$0.2%) quantitatively intercept the slow reacting radicals (although species such as CH_2 are not quantitatively intercepted, it should be mentioned). Thus, the yield of a given radical may be based on the difference in the quantum yields of the RH compounds formed when HI is added, compared to an experiment in which NO or O_2 scavenges the radicals. A more elegant technique involves the photolysis of the fully deuterated alkane in the presence of HI, and a mass spectrometric analysis of the isotopic composition of the photolytic products; products formed by molecular elimination will be fully deuterated, while products with a radical precursor will contain an H-atom.

Such experiments lead to the determination of a "spectrum" of most of the fragments formed by dissociation of an excited alkane. Typical fragmentation spectra, representing the species formed in the dissociation of propane and n-butane, are shown in Table III. These results on the photolysis of n-butane, along with supplementary information derived from determining the isotopic structures of the products formed in the photolysis of $CD_3CH_2CH_2CD_3$, and $CD_3CD_2CH_2CH_3$ in the presence of radical scavengers, led to estimates[15] of the quantum yields of all the primary processes occurring in the photolysis of n-butane at 8.4 eV and 10.0 eV; these are shown in Table IV.

One general observation illustrated by these results is that the hydrogen (or deuterium) elimination process which predominates at low energies becomes less important as the photon energy is increased. In ethane[6], the quantum yield of the hydrogen-elimination process diminishes from 0.85 at 8.4 eV to 0.4 at 10.0 eV; similar estimates can be made about the variation with energy of the quantum yield of the hydrogen-elimination process in propane[8]. The drop in the importance of the hydrogen elimination process is compensated by increases in the other primary processes, (particu-

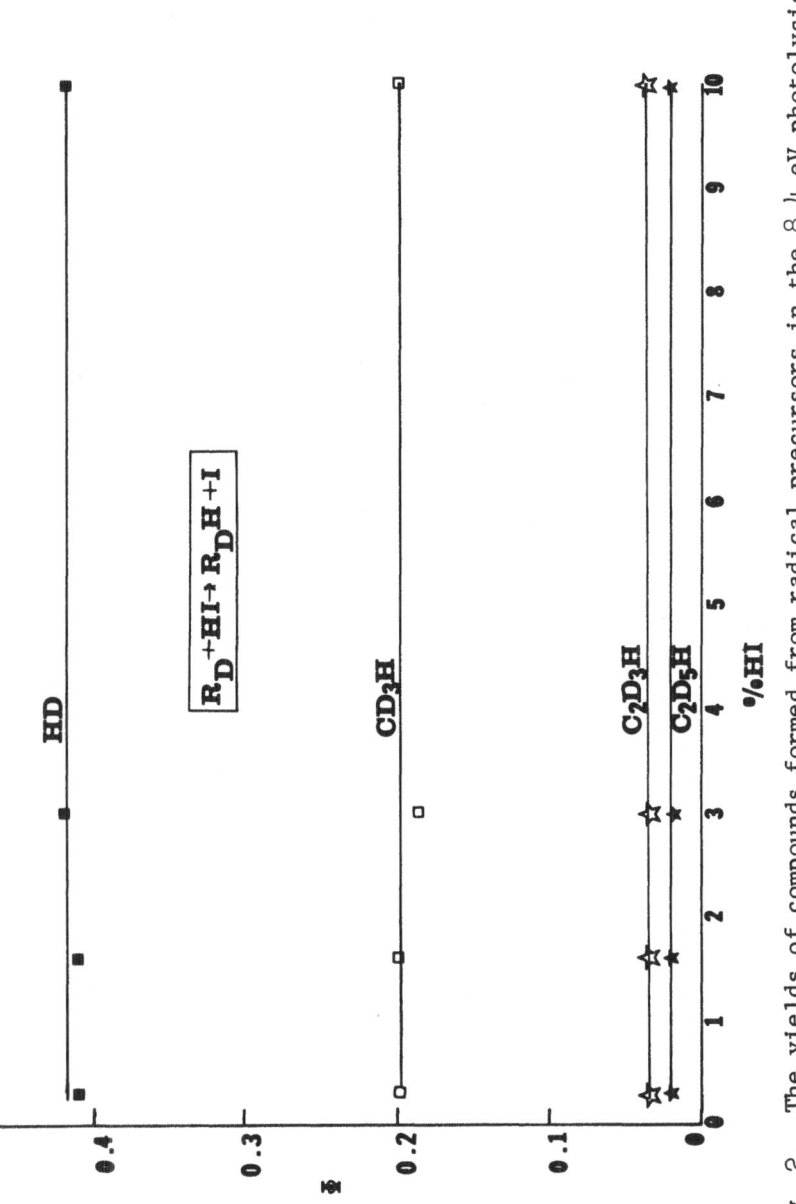

Fig. 2. The yields of compounds formed from radical precursors in the 8.4 eV photolysis of C_3D_8-HI mixtures, as a function of HI concentration.

TABLE III

Quantum Yields of Fragments Formed in the Photolysis of Alkanes
(Gas Phase, 15 torr)

	Propane[8]		n-Butane[15]	
	8.4 eV	10.0 eV	8.4 eV	10.0 eV
D	0.40	0.92	0.37	0.64
D_2	0.82	0.55	0.62	0.45
CD_3	0.20	0.43	0.29	0.40
CD_4	0.07	0.21	0.0089	0.040
C_2D_2	0.018	0.10	0.060	0.12
C_2D_3	0.032	0.082	0.024	0.070
C_2D_4	0.17	0.35	0.38	0.48
C_2D_5	0.019	0.060	0.20	0.24
C_2D_6	0.021	0.088	0.15	0.20
C_3D_5	0.080	0.040	0.046	0.065
C_3D_6	0.50	0.14	0.061	0.093
C_3D_7			0.005	0.013
C_3D_8			0.006	0.019

TABLE IV

Quantum Yields of Primary Processes in n-Butane (Gas Phase,
46 torr)[15]

	8.4 eV	10 eV
$n-C_4D_{10} + h\nu \rightarrow CD_4 + C_3D_6$	0.009	0.04
$\rightarrow CD_2 + C_3D_8$	0.005	0.018
$\rightarrow C_2D_4 + C_2D_6$	0.21	0.28
$\rightarrow C_2D_5 + C_2D_5$	0.15±0.01	0.17±0.02
$\rightarrow C_3D_7 + CD_3$	0.04	0.07
$\rightarrow C_4D_9 + D$	0.05	0.06
$\rightarrow C_4D_8 + D_2$	0.55±0.05	0.40±0.05

larly alkane-elimination processes in the case of linear molecules as illustrated by the interpretations of the butane system represented in Table IV). This is illustrated in another way by the results given in Table V, where the total quantum yield which can be attributed to alkane elimination processes in a series of linear alkanes is listed for experiments at 8.4 eV and 10.0 eV. Indeed, in each case, the importance of alkane elimination processes is greater at the higher energy.

Table V

Total Quantum Yields of Alkane Elimination Processes in Linear Alkanes

	8.4 eV	10.0 eV
C_2H_6[6/]	0.020	0.26
C_3H_8[8/]	0.092	0.30
$n\text{-}C_4H_{10}$[15/]	0.17	0.26
$n\text{-}C_5H_{12}$	0.11	0.21
$n\text{-}C_6H_{14}$	0.069	0.19
$n\text{-}C_7H_{16}$	0.060	0.15
$n\text{-}C_8H_{18}$	0.048	0.13
$n\text{-}C_9H_{20}$	0.023	0.098

It is also interesting to examine the trends in the total quantum yield of the alkane elimination processes (and of the compensating process, H_2-elimination) as a function of molecular structure. The results given in Table V, for instance show that for the homologous series $n\text{-}C_4H_{10}$ through $n\text{-}C_9H_{20}$, at a given photon energy in the gas phase the importance of alkane elimination decreases (i.e., the importance of H_2 elimination increases) as the chain is lengthened. For the lower members of the series, C_2H_6 to $n\text{-}C_4H_{10}$, an increase in the importance of alkane elimination with molecular size is seen, so that alkane elimination is of most importance (i.e., H_2-elimination is least important) for $n\text{-}C_4H_{10}$ at 8.4 eV or for C_3H_8 at 10.0 eV.

The effect of branching and of cyclization on the relative importances of H_2-elimination and the other competing processes is suggested by the results given in Table VI. This Table lists the quantum yields of the hydrogen formed at 8.4 eV in the gas phase by elimination as a molecule from various linear, branched, and cyclic alkanes. The results show that the probability of H_2-elimination in a given set of isomers is strongly reduced by branching. In the C_5H_{12} isomers, the isomer with maximum branching, neopentane, essentially does not eliminate hydrogen. As in the linear alkanes,

TABLE VI

Quantum Yields of H_2-Elimination in Alkanes in the Gas Phase

	8.4 eV		8.4 eV
CH_3CH_3	0.85	$(CH_3)_4C$	< 0.05
$CH_3CH_2CH_3$	0.70	$c-C_3H_6$	< 0.02
$CH_3CH_2CH_2CH_3$	0.59	$c-C_4H_8$	< 0.02
$CH_3CH_2CH_2CH_2CH_3$	0.65	$c-C_5H_{10}$	0.7 ± 0.1
$(CH_3)_2CHCH_3$	~ 0.22	$c-C_6H_{12}$	0.7 ± 0.1
$(CH_3)_2CHCH_2CH_3$	~ 0.24		

in the branched isomers a diminution in the probability of H_2-elimination is compensated by an increase in the quantum yields of other primary processes especially direct C-C cleavage and alkane elimination processes.

In the case of the cycloalkanes, it is seen that the $c-C_5H_{10}$ and $c-C_6H_{12}$ species exhibit probabilities for H_2-elimination which are similar to the straight chain C_5 and C_6 compounds. However, $c-C_3H_6$ and $c-C_4H_8$ undergo H_2-elimination to a negligible extent. Apparently, the ring strain in cyclopropane favors the C-C cleavage process[16-17]

$$c-C_3H_6 + h\nu \rightarrow C_2H_4 + CH_2 \qquad\qquad (6)$$

Having considered the variations in the quantum yields of H_2-elimination processes and the compensating alkane-eliminations or C-C bond cleavage processes as a function of energy, molecular structure, and in the case of straight-chain alkanes, as a function of chain length, it is now of interest to examine the effect of a change of phase on the importance of these processes. Table VII shows the quantum yields of the H_2-elimination processes and the alkane-elimination processes in n-pentane and 2,2,4-trimethylpentane in both the gas phase and liquid phase. In the case of the linear alkane, the change of phase makes little or no difference in the quantum yields of the various processes. For 2,2,4-trimethylpentane, H_2-elimination is of negligible importance in both the gas and liquid phases; the quantum yields of alkane elimination processes are greater in the liquid phase than the gas, but the relative probabilities of these processes are about the same in the two phases.

Information is also available on the mechanisms of alkane elimination processes in the alkanes. Table VIII for example, shows the relative importances of the various alkane-elimination processes

TABLE VII

Comparison Between Gas and Liquid Phase Photolysis at 147 nm

	Gas Phase[a]	Liquid Phase[b]
$n\text{-}C_5H_{12} \xrightarrow{h\nu} CH_4$	0.01	0.01
$\rightarrow C_2H_6$	0.048	0.05
$\rightarrow C_3H_8$	0.053	0.06
$\rightarrow H_2$	0.65 ± 0.1	~ 0.70
$iso\text{-}C_8H_{18} \rightarrow CH_4$	0.063	0.17
$\rightarrow i\text{-}C_4H_{10}$	0.190	0.36
$\rightarrow neo\text{-}C_5H_{12}$	0.029	0.05
$\rightarrow H_2$	~ 0.1	0.08

a/ R. Lesclaux and P. Ausloos (unpublished results).
b/ R. A. Holroyd, J. Amer. Chem. Soc. 91, 2708 (1969).

TABLE VIII

Modes of Alkane Elimination in n-Hexane

	C-C Bond Cleaved	Percent 8.4 eV	10.0 eV
$n\text{-}C_6H_{14}{}^* \rightarrow CH_4 + C_5H_{10}$	1-2	6	11
$\rightarrow n\text{-}C_5H_{12} + CH_2$		1	3
$\rightarrow C_2H_6 + C_4H_8$	2-3	29	28
$\rightarrow n\text{-}C_4H_{10} + C_2H_4$		31	32
$\rightarrow C_3H_8 + C_3H_6$	3-4	33	26

in n-hexane. In this molecule, if the 1-2 C-C bond is broken with
the concerted movement of a hydrogen species across the bond in one
direction or the other, CH_4 or $n\text{-}C_5H_{12}$ will be formed; such a mechanism at the 2-3 C-C bond leads to the formation of C_2H_6 or $n\text{-}C_4H_{10}$,
and at the 3-4 C-C bond, C_3H_8. The results given in Table VIII
lead to several generalizations, which seem to hold true for all
the larger ($C_5\text{-}C_{11}$) linear alkanes. (1) The probability for alkane
elimination processes involving primary C-C bonds (i.e., the formation of CH_4 and of $n\text{-}C_5H_{12}$, in this case) is always much lower than
for such processes at other sites in the molecule. Furthermore, the
process at this site in which a secondary H-atom is transferred (to

form methane) is always strongly favored over the competing process in which a primary H-atom moves in the concerted process. In our discussion of alkane-elimination processes in branched alkanes, below, we shall see further evidence that for competing processes, the breaking of a weaker C-H bond is favored. (2) The alkane-elimination processes at the 2-3 C-C bond, C_2H_6 and $n-C_4H_{10}$ formation occur with about equal probability. Here, the C-H bonds being broken in the two processes are of approximately equal strength. (3) The probability of alkane elimination at the 3-4 C-C bond (C_3H_8 formation) is about half that at the 2-3 C-C bond. Since there is only one 3-4 C-C bond, but 2 bonds equivalent to the 2-3 bond (4-5 and 2-3 are the same), this result can be interpreted to mean that, as might be expected, the energy of excitation is randomly distributed in the carbon chain among bonds of equal strength.

In the alkane-elimination processes in branched alkanes, interesting observations about the relative probabilities of various alkane-elimination processes can be made. Partial deuteration experiments lead to the results given in Table IX, which show the relative probabilities of various equivalent alkane elimination processes involving transfer of tertiary, secondary, or primary H-atoms. These results show that, when statistical factors are taken into account the elimination of a given alkane molecule through the cleavage of a given bond will proceed preferably through transfer of a tertiary H-atom, but transfer of a secondary H-atom occurs readily; transfer of a primary H-atom occurs with a low probability.

TABLE IX

Modes of Alkane Elimination in Branched Alkanes at 8.4 eV.

$\overset{\displaystyle CZ_3}{\underset{\displaystyle |}{}}$	
$CH_3CX_2CY\text{--}CH_3 \rightarrow CH_3Y$	1.0
$\rightarrow CH_3X$	1.3
$\rightarrow CH_3Z$	0.15
$\overset{\displaystyle CZ_3}{\underset{\displaystyle |}{}}$	
$CH_3CX_2CY\text{---}CH_2CH_3 \rightarrow CH_3CH_2Y$	1.0
$\rightarrow CH_3CH_2X$	1.0
$\rightarrow CH_3CH_2Z$	0.17

X, Y, or Z represent H and/or D atoms in various partially deuterated compounds. Corrections for isotope effects have not been made.

ALKENES

Of the alkenes the photolysis of ethylene has been investigated most extensively. At photon energies from 6.1 eV to 11.8 eV,

H_2-elimination is an important process[10,18-19] ($\Phi \sim 0.45\pm0.05$ in the range 8.4 eV to 6.4 eV)[19]. On the basis of CH_2CD_2 experiments Okabe and McNesby[20] suggested that at 8.4 eV and 10.0 eV terminal elimination of H_2 and D_2 is more probable than 1,2 elimination of HD. This has later been shown to be also the case for the super-excited molecule (11.6-11.8 eV)[18]. Furthermore, the relative probability for the formation of H_2, D_2 and HD was within experimental error, found to be independent of wavelength and phase[21]. There are, however, indications that the competing process

$$C_2H_4 + h\nu \rightarrow C_2H_3 \ (C_2H_2 + H) + H \tag{7}$$

does increase in relative importance with an increase in photon energy from 8.4 to 11.6-11.8 eV. Stable vinyl radicals are observed in appreciable yields at lower photon energies[19] and there are also clear indications that because of collisional stabilization at any given wavelength, vinyl radicals are more abundant in the solid phase photolysis[17] than in the gas phase. In addition to the photofragmentation processes mentioned above there is also concrete evidence[21] for the occurrence of the process

$$C_2H_4 + h\nu \rightarrow 2CH_2 \tag{8}$$

It has been suggested that at least three different electronic states of ethylene must be invoked to explain the photofragmentation processes in the 6.4 eV to 8.4 eV region[18]. It is doubtful however that the present experimental observations on this simple molecule allow correlation of primary processes with distinct electronic states.

Higher molecular weight alkenes have been investigated less extensively and the _primary_ processes in the far ultraviolet are poorly understood because of secondary fragmentation of vibrationally excited species formed in the initial fragmentation. This is for instance illustrated by a study of the photolysis of isobutene at 8.4 eV[22], where the overall process

$$iso-C_4H_8 + h\nu \rightarrow C_3H_4 + H + CH_3 \tag{9}$$

accounts for nearly 90 percent of the photodecomposition. There is, however, no easy way to establish the quantum yields of the two primary processes:

$$iso-C_4H_8 + h\nu \rightarrow C_3H_5^* + CH_3 \tag{10}$$
$$\rightarrow C_4H_7^* + H \tag{11}$$

which precede the formation of C_3H_4. Longer wavelength studies are more revealing because the primary fragments dissociate only to a limited extent.

For instance, recent studies carried out on olefins around 6.7 eV have revealed that cleavage of the C-C bond β to the double bond is strongly favored even in cases where this is not the weakest bond in the molecule[23-24]. For instance, in a recent flash photolysis study of 2-methylbut-1-ene, monitoring of the β methyl-alkyl radicals by absorption spectroscopy led to the conclusion that the β(C-C) cleavage process

$$CH_3CH_2C(CH_3)=CH_2 + h\nu \rightarrow CH_3 + CH_2C(CH_3)=CH_2 \qquad (12)$$

accounts for nearly 80 percent of the primary decomposition modes. The question remains as to what extent this information obtained at the long wavelength end of the absorption spectrum can be applied to shorter wavelength studies where secondary dissociative decompositions mask the initial cleavage mechanisms. Besides the simple C-C and C-H cleavage processes, molecular elimination of alkanes from the electronically excited alkenes does occur as well. As expected, the quantum yield of such processes is, however, lower than for the corresponding alkanes[25], and does depend on the position of the double bond.

The importance of H_2-elimination is also lower at a given photon energy in an alkene than in the corresponding alkane. For instance, as shown in Table VI, the major process in the photolysis of cyclohexane at 8.4 eV is H_2-elimination. By contrast, in the photolysis of cyclohexene[26], the quantum yield of H_2-elimination is negligible (< 0.02); instead, elimination of a molecule of ethylene is the major process ($\Phi = 0.84$).

CARBONYL COMPOUNDS

While carbonyl compounds have been investigated extensively in the near ultraviolet[27], relatively few studies have been carried out at shorter wavelengths[28-33] where absorption cross sections are much higher. In the photolysis of both straight-chain ketones[28-31] and cycloketones[32], cleavage of the C-C bond adjacent to the CO group predominates. However, in addition, new primary processes have been observed which had not been reported to occur in the near ultraviolet. For instance, in the case of acetone, the following primary decomposition processes have been suggested[28] to occur in the 8.4-11.8 eV region

$$CH_3COCH_3 + h\nu \rightarrow H_2 + C_3H_4O \qquad (13)$$
$$\rightarrow CH_4 + CH_2CO \qquad (14)$$
$$\rightarrow C_3H_4 + H_2O \text{ (or } H_2 + O) \qquad (15)$$

The M/N_{ex} values for these processes which occur both below and above the ionization region are given in Table X. In view of the fact that with the exception of process 13 a monotonous increase

TABLE X

The Primary Processes in the Photolysis of Acetone

Photon Energy eV	M/N_{ex}		
	$C_3H_6O \rightarrow H_2 + R\cdot$	$C_3H_6O \rightarrow CH_4 + CH_2CO$	$C_3H_6O \rightarrow C_3H_4 + H_2O$
4.0		< 0.0002	
4.9		~ 0.0015	~ 0.0003
8.4	0.03	0.020	0.015
10.0	0.35	0.037	0.025
11.6-11.8	0.38	0.054	0.029

of M/N_{ex} is seen, it would be difficult to associate these processes with a particular electronic transition. Actually, recent experiments carried out in our laboratory[28] indicate photolysis at 4.9 eV (253.7 nm) in the absorption region characteristic of the singlet $n_o \rightarrow \pi^*$ transition leads to a very small but finite quantum yield of process 14 ($\Phi \sim 0.0015$). However, at 4.0 eV (313.0 nm) where the singlet $n_o \rightarrow \pi^*$ absorption is followed by intersystem crossing to the triplet state, no evidence for process 14 was obtained ($\Phi < 2 \times 10^{-4}$).

REFERENCES

1. W. Rothman, F. Hirayama and S. Lipsky, J. Chem. Phys. 58, 1300 (1973), and references cited therein.
2. P. Ausloos, R. E. Rebbert and S. G. Lias, J. Phys. Chem. 72, 3904 (1968).
3. A. R. Welch and D. L. Judge, J. Chem. Phys. 57, 286 (1972).
4. R. E. Rebbert and P. Ausloos, J. Photochem. 1, 171 (1973).
5. R. E. Rebbert and P. Ausloos, NBS J. Res. 77A, 101 (1973).
6. S. G. Lias, G. J. Collin, R. E. Rebbert and P. Ausloos, J. Chem. Phys. 52, 1841 (1970).
7. J. H. Vorachek and R. D. Koob, Can. J. Chem. 51, 344 (1973).
8. P. Ausloos, R. E. Rebbert and S. G. Lias, J. Photochem. 2, 267 (1973).
9. J. W. Raymonda and W. T. Simpson, J. Chem. Phys. 47, 430 (1967).
10. J. R. McNesby and H. Okabe, Advan. in Photochem. 3, 157 (1964).
11. B. H. Mahan and R. Mandal, J. Chem. Phys. 37, 207 (1962).
12. H. Okabe and J. R. McNesby, J. Chem. Phys. 34, 668 (1961).
13. M. C. Sauer, Jr. and L. M. Dorfman, Ibid. 35, 497 (1961).
14. P. Ausloos, S. G. Lias and I. B. Sandoval, Disc. Faraday Soc. 36, 66 (1963).

15. J. A. Jackson and S. G. Lias, to be published.
16. A. A. Scala and P. Ausloos, J. Chem. Phys. 49, 2282 (1968).
17. A. K. Dhingra and R. D. Koob, J. Phys. Chem. 74, 4490 (1971).
18. R. Gorden, Jr. and P. Ausloos, J. Chem. Phys. 47, 1799 (1967).
19. P. Potzinger, L. C. Glasgow and G. von Bunau, Z. Naturforsch. 27a, 628 (1972).
20. H. Okabe and J. R. McNesby, J. Chem. Phys. 36, 601 (1962).
21. R. Gorden, Jr. and P. Ausloos, NBS J. Res. 75A, 141 (1971).
22. J. Herman, K. Herman, and P. Ausloos, J. Chem. Phys. 52, 28 (1970).
23. P. Borrell, P. Cashmore, A. Cervenka and F. C. James, J. Chim. Phys. 229 (1970).
24. F. Bayrakceken, J. H. Brophy, R. D. Finks and J. E. Nicholas, J. Chem. Soc. Faraday Trans. I, 69, 228 (1973).
25. P. M. Perrin and G. J. Collin, Can. J. Chem. 51, 724 (1973).
26. R. Lesclaux, S. Searles, L. W. Sieck, and P. Ausloos, J. Chem. Phys. 53, 3336 (1970).
27. J. G. Calvert and J. N. Pitts, Jr., Photochemistry, Wiley, New York (1966).
28. L. J-T Lin and P. Ausloos, J. Photochem. 1, 453 (1973).
29. A. G. Leiga and H. A. Taylor, J. Chem. Phys. 41, 1247 (1964).
30. A. A. Scala and P. Ausloos, J. Phys. Chem. 70, 260 (1966).
31. A. A. Scala, Ibid. 74, 2639 (1970).
32. A. A. Scala and P. G. Ballau, Ibid. 76, 615 (1972).
33. S. Glicker and L. J. Stief, J. Chem. Phys. 54, 2852 (1971).

ENERGY PARTIONING IN PHOTOCHEMISTRY OF ALKANES

R. D. Koob

Department of Chemistry
North Dakota State University
Fargo, North Dakota   58102

The factors which control the partioning of energy
to fragments in the photodissociation of alkanes are
poorly understood. This is not surprising as there are
no reliable estimates of alkane photofragment energies
now available. I am convinced, however, that the pur-
suit of such estimates will provide insight into the
detailed mechanisms of alkane photochemistry. This
belief is based on the success now being enjoyed by the
variety of methods used to determine energy distribution
aspects of unimolecular and bimolecular reactions of
simple alkyl halides.[1,2,3]

In this presentation, I will try to provide a view
of the problem measuring alkane photofragment energies,
a review of methods which have found success in other
systems and may hold potential for the alkane problem,
a summary of available data pertinent to the energy
partioning problem, and alternate models for interpreting available data.

PHOTOPHYSICAL AND PHOTOCHEMICAL DESCRIPTION.
The absorption spectra of acyclic hydrocarbons are
typically broad band continua with onset of absorption
near or below 170 nm.[4] The absorptions rise rapidly
and remain intense into the ionization region. With
the exception of ethane, little structure is to be
observed in any of the spectra. Fluorescence is not
observed for small alkanes and is weak in higher members
of the series.[5] Molecular orbital calculations indicate

a relatively high density of electronic states. The lowest excited singlet state of methane, for instance, is probably triply degenerate or nearly so.[6] While the degeneracy is clearly removed with substitution of methyls for hydrogens, ethane would be doubly degenerate and propane singly degenerate, spacing between adjacent states is probably small compared to the energy of the transition from the ground state. The oscillator strengths for some of the low-lying states have been calculated.[6a,b] Unambiguous assignments of states corresponding to the absorption spectra are not available.

The photochemistry of the lower alkanes is characterized by simultaneous breaking of two bonds. This is illustrated for propane by equations (1) - (3).

$$C_3H_8 \overset{h\nu}{=} C_3H_6 + H_2 \text{ (2H)} \tag{1}$$

$$C_3H_8 \overset{h\nu}{=} C_2H_4 + CH_4 \text{ (CH}_3 + H) \tag{2}$$

$$C_3H_8 \overset{h\nu}{=} C_2H_6 + CH_2 \tag{3}$$

Some evidence also exists for a process such as reaction (4)[7] although the pressure dependence of products ascribed to the ethyl radical appear anomolous.[8]

$$C_3H_8 \overset{h\nu}{-} C_2H_5 + CH_3 \tag{4}$$

If present, reaction (4) amounts to less than 5°/o of the overall quantum yield for disappearance of propane. Similarly, reaction (5) is close to twenty times more important than reaction (6) in the photolysis of methane.[9]

$$CH_4 \overset{h\nu}{=} CH_2 + H_2 \text{ (2H)} \tag{5}$$

$$CH_4 = CH_3 + H \tag{6}$$

The lack of any appreciable pressure dependence of photodissociation quantum yield for the various small hydrocarbons indicates a relatively short excited state lifetime ($\leq 10^{-10}$ sec). This is consistent with the diffuse nature of the absorption spectrum and the lack of significant fluorescence.

The apparent rapidity of the dissociation process argues against the occurance of intersystem crossing, and one may assume with confidence that only singlet

excited states are precursors to products. Some
chemical evidence to support this assumption has been
obtained in the photolysis of 1,2-dimethylcyclopropane.[10]
Since the quantum yields for the photodissociation of
the lower hydrocarbons are, within experimental error,
all unity, it is highly probable that it is the dissociation rate which controls the excited state lifetime.[11]

     Lifetimes as short as 10^{-12} sec are not necessarily
restrictive to statistical redistribution of available
energies.[12] Speaking formalistically, a Franck-Condon
excitation will generate, in a collection of molecules,
a distribution of states with angular momenta and
nuclear geometries reminiscent of the vibrational-
rotational distributions of the ground state molecule.
It is highly improbable that any of the ground state
configurations will be the same as any "equilibrium"
geometries of the excited state. The system will
evolve in time through a series of Born-Oppenheimer
states. The transition probabilities between these
states determine the final disposition of energy among
the photofragments.

PHYSICAL METHODS FOR DETERMINING ENERGY PARTIONING.
     Potentially, kinetic spectroscopy offers a method
for observing alkane photofragments directly. Technical-
ly, however, instrumentation is not now available which
incorporates light sources of sufficient intensity and
chromaticity required for alkane flash photolysis
coupled with detector systems with response times short
enough to view nascent photofragments.

     Photofragment spectrometry was developed recently
by Busch and Wilson.[13] The photolyzing light is a
polarized beam from a laser. The laser intersects the
sample as it emerges from a nozzle as a molecular beam.
Fragments recoiling from the photolysis region are
detected, on a time resolved basis, by a quadrapole mass
analyzer as a function of both distance and angle from
the reaction region. From the measurement of translational energy and angular distribution of fragments,
internal energies are estimated. The entire experiment
is carried out at high vacuum to assure that the observed
energies represent initial events and not subsequent
collisions. Photofragment spectrometry has the
advantage of not requiring very fast electronic
detection. Although the technique has not yet been
applied to alkane photolysis, the availability of a
laser in the region of 160 nm or lower would appear to

make it feasible.

In experiments with alkyl iodides, data indicated the presence of both low lying electronic states of iodine atom for methyl and ethyl iodide, but resolution grew worse as the number of degrees of freedom in the alkyl fragment increased.[2] Data for propyl iodides could not be interpreted unambigously.

Two important methods which have been successful in directly identifying vibrational level populations, flash photolysis of sample confined in a laser cavity[1] and IR fluorescence techniques[3,14] are less likely to be able to be applied directly to the alkane problem but may offer useful information if applied to related compounds. Although perturbed, photon absorption in partially fluorinated alkanes is likely to be related to alkane processes. For example, the absorption spectrum of methyl fluoride is very similar to that of methane.[15] If the photolysis of alkyl fluorides yields hydrogen fluoride, and there is some evidence that it does,[16] then techniques mentioned above would perhaps be more readily applied. HF lasers and IR fluorescence[3] have both been generated in other systems. One might hope that information obtained for alkyl fluorides would be applicable to the alkanes.

CHEMICAL METHODS

The absence of a physical method for determining energy partioning in alkane photolysis forces us, for the present time, to turn to chemical methods. The behavior of reaction rates as a function of various reaction parameters, typically pressure and photolysis energy, must reflect the history of the reactants. Unfortunately, the sampling time in a steady state photolysis system is long with respect to the processes of interest. One cannot sample the system for the primary distribution of photofragments, but sees only the end result. A multitude of other processes including secondary reactions and collisions perturb the initial spectrum.

Interpretation of the chemical data is heavily dependent upon kinetic models. The only kinetic theory which has been developed to the sophistication to handle routinely molecules as complex as alkanes is the RRKM formulation. In its most commonly applied form, it may be conveniently expressed as

$$D/S = k_a/\omega = \frac{\Sigma \; k_\epsilon/k_\epsilon + \omega) \; f(E)}{\Sigma \; \omega/(k_\epsilon + \omega) \; f(E)}$$

Where S is stabilized product, D is dissociated product, k_a is the apparent rate constant for the reaction, ω is a collision rate constant times total pressure, k_ϵ is an energy dependent microscopic rate constant, and $f(E)$ is the distribution function for energies greater than E_0, the minimum energy required for reaction to occur. ω is typically calculated using simple collision theory with suitably chosen collision cross sections. k_ϵ is calculated following absolute rate theory. This expression is built on a number of assumptions, the most important for our discussion is that it assumes that deactivation of an excited species to energies below E_0 occurs in a single collision (strong collider assumption). One must be especially cognizant of this assumption when working with vacuum ultraviolet photolysis because large energies in excess of E_0 may be imparted to the reacting fragments.

If one wishes to take into account failure of an excited species to be deactivated below E_0 on a single collision (weak collider), application of the theory becomes somewhat more complex. A stationary state treatment centered about some state, i, is carried out in which noncollisional and collisional population of i is equated with collisional transfer out of i plus unimolecular reaction from i. The concepts expressed may be written conveniently in matrix form. To apply any such equations it is necessary to model the transition probabilities between the various states i.[35] In practice these are usually chosen to best fit available data. The parameter most often varied to obtain "best fit" is the average energy transferred per collision in a "step-ladder" model.[18]

In the strong collision model, a plot of $(D/S)^{-1}$ vs ω^{-1} should provide a straight line for sufficiently narrow energy distribution functions, $f(E)$. As $f(E)$ broadens, this plot will curve upward at the high pressure end of the plot. The weak collider model also will cause such a plot to become concave upward. But this is the result of a turn up in the low pressure region. The less energy removed per collision, on the average, the more dramatic the low pressure turn-up.

EXPERIMENTAL RESULTS AND DISCUSSION

In order to apply an RRKM model to a particular system it is necessary to have a well-defined mechanism for the formation of the species of interest. The appropriate data is relative yield of dissociation product to stabilized reactant as a function of total sample pressure. For alkanes, the most numerous data which fit both of these criteria are those which report the fate of photolytically produced methylene. It is well established that methylene is the primary product in the photolysis of methane.[19] Deuterium labeling has demonstrated that of the two possible sources of methylene from propane, elimination of CH_2 from a terminal methyl is the only actual source.[20] Methylene is also an important product in the photolysis of cyclopropane and 1,2-dimethylcyclopropane.[10]

Since methylene itself is not a stable end product, it is necessary to examine it by means of its insertion products. This seriously complicates interpretation from the viewpoint of energy partioning as methylene may be collisionally deactivated prior to insertion.

Rebbert, Lias and Ausloos have photolysed methane and deuteromethane at 123.6 and 104.8-106.7 nm and fail to find any wavelength dependence on the dissociation rate constants of the insertion produced ethanes.[21] Similar results are obtained for butanes produced in the photolysis of propane at 147 and 123.6 nm.[11] The failure to observe any wavelength dependence argues strongly that collisional deactivation of methylene occurs prior to insertion. Such a conclusion would also be consistent with the similarity of dissociation rates found by Dhingra, Vorachek and Koob for butanes produced by insertion into propane of methylene produced by the photolysis of propane at 123.6 nm and the photolysis of cyclopropane at 165 nm, respectively.[22] It is in contrast, however, to the findings of Topor and Carr.[23] In this work, ketene was photolyzed in the presence of oxygen and propylene at wavelengths ranging from 355 to 268 nm. This corresponds to an excess photon energy range of 0 to 25 kcal/mole, a narrower energy range than represented by the different wavelengths used in either methane of propane photolysis by Ausloos and co-workers. Although the energy retained by the methylene at the shortest wavelength used by Torpor and Carr is relatively small, 3 kcal/mole, a clear distinction between the data taken at 355 and 268 nm is evident.

There are at least two ways the results of Topor and Carr can be reconciled with those of Ausloos and co-workers, and both are consistent with what is presently known about methylene. The first possibility is that the sampling times are different in the two systems. A more rapid addition of methylene to the propylene double bond than insertion of methylene into a C-H bond[23] would allow less time for collisional deactivation of the methylene with a possible concomitant rise in the average energy of the reacting methylene. Such an explanation would not, however, explain the dependence upon photolysing wavelength of the dissociation rate of isobutane formed by methylene insertion into propane, where the singlet methylene was generated from diazomethane as observed by Johnson, Hase and Simons.[24]

An alternative explanation would be that deactivation is occuring within a different electronic manifold in the alkane photolysis than in the photolysis of either ketene or diazomethane. If this were the case, neither energy transfer or insertion rates would necessarily be similar and it would be unwarranted to expect identical behavior.

Molecular orbital calculations examining the process

$$CH_4 = {}^1CH_2 + H_2$$

have suggested that the methylene formed should be in a ${}^1B_1$ rather than the lower ${}^1A_1$ state.[25] Herzberg reports ${}^1B_1$ as 7100 cm^{-1} (20.3 kcal/mole) above ${}^1A_1$.[26] The photolyses of the alkanes occur at energies sufficiently high to populate the upper state, while only the shortest wavelengths used by Topor and Carr are energetic enough. Energies available to methylene in the photolysis of diazomethane by Johnson, Hase, and Simons are, within the uncertainties of the thermochemistry, able to populate the ${}^1B_1$ state if all available energy goes to the methylene.

Hase has used the p 1/2 value of 770 torr obtained by Rebbert, Lias and Ausloos for the stabilization of ethane in the methane photolysis system to calculate an average excitation energy of 117 kcal/mole.[27] Using values for the heats of formation of methane, ethane and methylene from NBS Tech. Note 270-3,[28] this corresponds to 20.9 kcal/mole brought to the reaction by the inserting

methylene. Using a p1/2 of 1.3 torr for butane from
methylene insertion into propane and the model of
Johnson, Hase and Simons, a similar calculation yields
approximately 25 kcal/mole as the energy of the inserting methylene. These rough estimates demonstrate that
methylene produced in the photolysis of these alkanes
could be in a $^1B_1$ state. The argument cannot be considered closed, however. Methylene produced by vacuum
ultraviolet photolysis of propane at 123.6 nm has the
same selectivity for secondary over primary C-H bonds
and the same intersystem crossing rates as methylene
produced in the photolysis of ketene at 366 nm.[10]
Further, yield vs pressure plots on both methane and
methane-d_4 show considerable curvature. This may be
indicative of step-wise deactivation of the excited
ethanes. Topor and Carr have argued convincingly of
the necessity of taking this aspect of the model into
account when attempting to estimate average energies.
To date, no calculations have been done using the data
of Ausloos and co-workers which take this into account.

While the photogeneration of methylene from alkanes
is of interest and offers interesting comparisions with
methylene generated from other sources, the apparent
loss of vibrational energy before insertion makes it of
less use in determining energy partitioning characteristics. Cofragments generated when methylene is produced
are often of unique origin in a given photochemical
system. The only source of ethane, for example, in the
photolysis of propane, in the presence of a radical
scavenger, is the primary process which also produces
methylene.[20] The quantum yield of ethane is found to be
independent of pressure down to 2 torr at 123.6 nm.[8]
There is approximately 133 kcal/mole of energy in excess
of the endothermicity of the reaction at this wavelength
(113 kcal/mole if CH_2 is produced as $^1B_1$). Clearly all
this energy is not carried from the primary process by
ethane as the critical energy for dissociation of ethane
is only 85 kcal/mole.[21] Not even statistical partioning[30]
of 133 kcal to vibrational, rotational and translational
degrees of freedom of methylene and ethane is consistent
with the pressure independence of the quantum yield of
ethane at these pressures. Statistical partioning of
113 kcal/mole is consistent with these observations,
however. Lack of an observable pressure dependence in
ethane yield provides no real estimate of the energy
partitiones to ethane, but provides only an upper limit.

Another system in which more information is potentially available is the primary process observed in the photolysis of cis-1,2-dimethylcyclopropane (CDMC) which yields methylene plus cis-2-butene (C2B).[32] Photolysis of CDMC at 165 and 147 nm in the presence of oxygen yields no trans-2-butene (T2B) even at pressures as low as 1 torr. Photolysis at 123.6 nm, however, gives rise to a pressure dependent yield of T2B which has been interpreted as arising from the isomerization of C2B. The ratio C2B/T2B was found to vary from near 2 to greater than 10 over the pressure range 0.1-105 torr. No variation in the quantum yield of total butene was found. These observations place a number of restraints upon the energy partitioned to C2B in this photochemical reaction. (Numbers in parentheses refer to available energies if $^1B_1$ CH_2 is produced instead of $^1A_1$). Of the 105 kcal/mole (85 kcal/mole) available after photolysis at 147 nm, not enough is retained by C2B to cause it to isomerize. Activation energy for the isomerization is 65 kcal/mole.[33] Of the 142 kcal/mole (122 kcal/mole) excess energy available after photolysis at 123.6, sufficient energy is retained by C2B to cause isomerization, but not enough to initiate an observable dissociation. The activation energy for the lowest energy decomposition is 85 kcal/mole.[34] Finally, a significant fraction of C2B must be produced with energies small enough that it is incapable of isomerizing even at pressures as low as 0.1 torr. An RRKM treatment, which included an investigation of possible stepwise deactivation of C2B and T2B suggested a broad energy distribution.[32] An alternative which could not be ruled out is that an excited electronic state of C2B is populated by photolysis of CDMC at 123.6 nm. Distribution of available energy among two states can be made to fit the available data.

The secondary dissociation of ethylene produced in the photolysis of ethane also has been discussed as an alternative between ethylenes produced in two different processes versus a broad energy distribution.[35] Although no clear distinction has been made, the former interpretation is currently favored.[11]

CONCLUSION

The absence of physical methods for determining energy partioning features in alkanes forces reliance, at present, on chemical methods. While attempts to apply chemical methods have uncovered a number of interesting aspects of alkane photochemistry, it cannot yet be claimed that an unambiguous assignment of nascent energy of any photofragments has been made.

REFERENCES

1. M.J. Berry and G.C. Pimentel, J. Chem. Phys., 53, 3453 (1970).
2. S.J. Riley and K.R. Wilson, Faraday Discuss of the Chem. Soc., 53, 132 (1972).
3. H.W. Chang and D.W. Setser, J. Chem. Phys., 58, 2298 (1973). This paper provides an excellent entrance to the IR luminescence literature.
4. (a) B.A. Lombos, P. Savageau, and C. Sandorfy, J. Molec. Spectr., 24, 253 (1967), Chem. Phys. Letters, 1, 42 (1967). (b) J.W. Raymonda and W.T. Simpson, J. Chem. Phys., 47, 430 (1967).
5. F. Hirayama and S. Lisky, J. Chem. Phys., 51, 3616, (1969).
6. (a) D.R. Salahub and C. Sandorfy, TCA, 20, 227 (1971). (b) P. Saatzer, M.S. Gordon, and R.D. Koob, Fall ACS Meeting, C Paper 61, Chicago, (1973), (c) J.E. Del Bene, R. Ditchfield and J.A. Pople, J. Chem. Phys., 55, 2236 (1971), (d) S. Katagiri and C. Sandorfy, Theor. Chim. Acta, 4, 204 (1966), (e) R. Hoffman, J. Chem. Phys., 39, 1397 (1963).
7. P. Ausloos and S.G. Lias, J. Chem. Phys., 44, 521 (1966).
8. J.H. Vorachek and R.D. Koob, Can. J. Chem., 51, 344 (1973).
9. J.G. Calvert and J.N. Pitts, "Photochemistry", J. Wiley and Sons, Inc., New York, (1966), citations on p. 498.
10. A.K. Dhingra and R.D. Koob, J. Phys. Chem., 74, 4490 (1970).
11. P. Ausloos, Mol. Photochem., 4, 39 (1972).
12. I. Oref, P. Schuetzle, and B.S. Rabinovitch, J. Chem. Phys., 54, 575 (1971).
13. G.E. Busch and K.R. Wilson, ibid., 56, 3626 (1972).
14. J.C. Polanyi and K.B. Woodall, ibid., 57, 1574, 1547, (1972).
15. L. Edwards and J.W. Raymonda, J. Amer. Chem. Soc., 91, 5937 (1969).
16. E. Tschuikow-Roux and S. Kodama, J. Chem. Phys., 50, 5297 (1969).
17. P.J. Robinson and K.A. Holbrook, "Unimolecular Reactions", Wiley-Interscience, New York (1972).
18. J.H. Georgakekos, B.S. Rabinovitch, and E.J. McAlduff, J. Chem. Phys., 52, 2143 (1970).
19. (a) B.H. Mahan and R. Mandel, ibid., 37, 207 (1962), (b) P. Ausloos, R. Gorden, and S.G. Lias, ibid., 40, 1854 (1964), (c) R. Gorden, P. Ausloos, ibid., 46, 4823 (1967).
20. P. Ausloos, S.G. Lias, and I.B. Sandoval, Discuss. Faraday Soc., 36, 66 (1963).

21. R.E. Rebbert, S.G. Lias, and P. Ausloos, Chem. Phys. Letters, 12, 323 (1971).
22. A.K. Dhingra, J.H. Vorachek, and R.D. Koob, ibid., 9, 17 (1971).
23. M.G. Torpor and R.W. Carr, J. Chem. Phys., 58, 757 (1973).
24. R.L. Johnson, W.L. Hase, and J.W. Simons, ibid., 52, 3911 (1970).
25. (a) E. Lindholm, Arkiv Fysik, 37, 37 (1968), (b) S. Karplus and R. Bersohn, J. Chem. Phys., 51, 2040 (1969).
26. G. Herzberg, "Electronic Spectra and Electronic Structure of Polyatomic Molecules", D. Van Nostrand Company, Inc., Princeton, N.J., (1967), p. 583.
27. W.L. Hase, J. Chem. Phys., 57, 730 (1972).
28. D.D. Wagman, W.H. Evans, V.B. Parker, I. Halow, S.M. Bailey, and R.H. Schumm, NBS Technical Note 270-3, U.S. Govt. Printing Office, 1968.
29. R.A. Marcus, J. Chem. Phys., 20, 363 (1952).
30. R.J. Campbell and E.W. Schlag, J. Amer. Chem. Soc., 89, 5103 (1967).
31. J.H. Vorachek and R.D. Koob, J. Phys. Chem., 76, 9 (1972).
32. K. Dees and R.D. Koob, ibid., 77, 759 (1973).
33. (a) M.C. Lin and K.J. Laidler, Trans. Faraday, Soc., 64, 94 (1968), (b) G.M. Wieder and R.A. Marcus, J. Chem. Phys., 37, 1835 (1962).
34. E. Jakubowski, H.S. Sandu, and O.P. Strausz, J. Amer. Chem. Soc., 93, 2610 (1971).
35. A.W. Kirk and E. Tschuikow-Roux, J. Chem. Phys., 51, 2247 (1969).

RECENT STUDIES OF THE FLUORESCENCE FROM SOME HYDROCARBON MOLECULES[*]

Sanford Lipsky
Chemistry Department
University of Minnesota
Minneapolis, Minnesota 55455

I. INTRODUCTION

We present here three short reports each based on the emission characteristics of some recently investigated class of molecules. All the molecules we consider are hydrocarbons, all are excited at relatively short wavelengths and all emit with rather low quantum yield. Aside from these rather weakly binding constraints, the reports are otherwise independent and no attempt will be made to further connect them in any way. We begin with an experimental section which summarizes most of the pertinent techniques that we employ and then follow this with the three separate sections on i) Saturated Hydrocarbons, ii) Aromatics and iii) Simple Olefins.

II. EXPERIMENTAL

For most of the work we describe below, a variety of microwave powered lamps have been used all generally constructed as suggested by Gordon, Rebbert and Ausloos.[1] The lamp radiation is passed through a 1/2 meter Seya-Namioka monochromater and then focussed sharply via a LiF lens onto the front face of the sample cell. The sample emission is collected at an angle of 45° from the exciting beam and focussed by two Suprasil quartz lenses onto the entrance slit of an analyzing monochromater. Both 1/3 meter McPherson and 1/2 meter Bausch and Lomb monochromaters have been used, at various times, for this purpose. The detector is a dry-ice cooled EMI 6256S photomultiplier whose output is fed to an electrometer and then to a potentiometric recorder. For the spectral region in which we have so far worked ($\lambda > 1200$ Å), nitrogen flushing of the entire optical system is adequate although occasionally the excitation system is pumped to $\approx 10^{-6}$ torr to check the effectiveness of the nitrogen flush.

In Sections IV and V we discuss the properties of electronic states with fluorescence quantum yields of the order of 10^{-6}. At this level of intensity, most window materials noticeably fluoresce and appropriate caution must therefore be exercised in selecting a window with low emission in the spectral region of interest. For various purposes we have used LiF, MgF_2 or sapphire attached to a cell body of either pyrex or quartz with Pt-Ag/AgCl. Although both pyrex and quartz are strongly fluorescent, they nevertheless can be employed for construction of a cell body for samples of sufficiently high optical density. Otherwise we have found it useful to cover the inside wall of the cell with an opaque coat of platinum. The window fluorescence is determined by filling the cell with liquid CCl_4 which exhibits negligible fluorescence even at the 10^{-6} level.

To measure a very low level of fluorescence, it is imperative to remove all sources of stray light that can contribute a background signal in the pertinent spectral region. The stray light has always a variety of origins and often, therefore, even with the most careful baffling of the optical system may remain a serious source of difficulty. With a resonance lamp, it is usually possible to avoid, at least in the excitation beam, stray light that lies in the frequency interval of the expected emission. This, together with the generally high intensity of the resonance lamp makes its use almost mandatory for low level emission work. Nevertheless, the excitation light itself after scattering from the cell window interfaces is a potential source of trouble. If the quartz analyzing optics cannot block this scattered light, then additional filters are required to prevent its appearance as stray light at the analyzer's exit slit. The filters must be chosen to have sharp cut-off at the exciting frequency in order to avoid obscuration of the short wavelength edge of the sample emission. Thus, for example, for the work described in Section IV we have found satisfactory a 2 cm path length of H_2O and a 1 cm path length of 4 volume % thiophene in isooctane for 184.9 nm and 253.7 nm respectively.

When measuring fluorescence from upper states of aromatic liquids (Section IV), a very serious stray light problem arises from scattering within the analyzing monochromater of the relatively intense emission from the sample's S_1 state. This emission can usually be removed with chemical filters each appropriately chosen for the compound being studied to block S_1 and transmit S_n emission. For example, for measuring the S_3 emission of naphthalene we have employed a 2 cm path length of Cl_2 gas at 1 atm. Combinations of this with aqueous solutions of $NiSO_4 \cdot 6H_2O$ and $CoSO_4 \cdot 7H_2O$ have proven to be generally adequate for S_3 emissions from most of the other aromatic compounds thus far studied. In the case that S_n emission lies very close to the S_1 emission, as happens for the S_2 emission from alkyl benzenes, appropriate chemical filters are difficult to find and we then have relied exclusively on O_2 saturation to quench S_1, while leaving undistrubed the much shorter-lived S_2 states. Confirmation of the reality of S_n emission can be obtained by demonstrating that it can itself be removed with a filter that blocks S_n and transmits S_1 emission. For the alkyl benzenes we have used 1 cm of liquid CCl_4 for this purpose.

When the exciting frequency lies close to the sample emission band, Raman scattering may obscure a weak fluorescence. Although we usually, in solution measurements, work at solute concentrations sufficiently high to reduce solvent Raman scattering, nevertheless resonance Raman scattering from the solute can be appreciable. If the sample emission is very short-lived, Raman scattering and fluorescence cannot be simply distinguished by quenching and the most reliable alternative (although not always possible) is then to move the exciting light to higher frequency. However, it is to be remembered that to avoid distortion of the sample emission spectrum by reabsorption it is required that the sample absorptivity at the exciting frequency be significantly larger than the absorptivities at the frequencies of the emission spectrum. Otherwise tedious, albeit standard, corrections are required.

Recorded spectra are corrected for the spectral response of the analyzing system. The technique employed for this correction has been fully described elsewhere.[2] All fluorescence quantum yields are based on a value of 0.35 for a degassed dilute solution of p-xylene in cyclohexane excited at 253.7 nm[3] which in turn derives from Berlman's suggested value of 1.00 for a degassed dilute solution of 9,10-diphenylanthracene in cyclohexane also excited at 253.7 nm.[4] A comparison of lamp intensity at 253.7 nm and at the shorter wavelength excitations was achieved through use of neat oxygenated p-xylene whose emission yield has been found to be independent of excitation wavelength from λ_{ex} = 253.7 nm to λ_{ex} = 147.0 nm.[2,5] The pertinent equations for calculating reflection-loss corrections at the window interfaces have been described elsewhere.[5] For the saturated hydrocarbons (Section III), since refractive indices are not generally available at the excitation frequencies employed, these corrections could not be made. However, the error caused by their neglect is estimated not to exceed 3%.[2] In all other cases, unless otherwise stated, the appropriate corrections have been applied.

For an isotropically emitting point within a solution and at a distance from a solution/vacuum interface which is small compared to the distance a detector lies above this interface, it is easily derived[6] that the number of photons of frequency ν emitted into the solid angle $d\Omega (= \sin\alpha d\alpha d\beta)$ subtended by the detector is proportional to $d\Omega/n\sqrt{n^2-\sin^2\alpha}$ where n is the solution refractive index at the frequency ν and $\alpha = 45°$ in our geometry. To obtain therefore the emission quantum yield of some liquid by comparison with the yield of p-xylene in cyclohexane it is necessary to know the refractive indices of cyclohexane at the p-xylene emission frequencies and those of the liquid at its emission frequencies. For most saturated hydrocarbons, the necessary refractive indices are not reliably known and therefore the appropriate corrections have not been made although it is estimated that the correction factor should not exceed $\approx$ 13%.[2] As a result our reported yields may be too low by approximately this amount.[7] For all other aromatic and olefinic molecules whose quantum yields are reported here, these "geometrical" corrections, where necessary, have been applied unless otherwise stated.

When comparing emission quantum yields of some substance at two wavelengths, one close to the absorption edge and the other where absorption is much stronger, it is important to consider that differences in the measured yields may be attributable, at least in part, to quenching by photolysis products which will be present at higher concentrations for the less penetrating excitation. In the usual case, the same state emits for the two excitation wavelengths and a diagnostic test for the significance of this "penetration depth" effect is to add some quencher and determine to what extent the quenching efficiencies differ at the two wavelengths. At sufficiently high quencher concentrations, the ratio of measured yields should approach a constant which may be identified as the true internal conversion efficiency from the higher to lower state. A detailed treatment of this effect has been presented elsewhere for the case of benzene.[8] For saturated hydrocarbons, with their much shorter lifetimes, the effect has been found not to be too significant at least down to excitation wavelengths of 147.0 nm.[2] Although the "penetration depth" effect tends to somewhat complicate emission measurements for excitation into intensely absorbing regions of the spectrum, there is the following compensation. As the extinction becomes increasingly large, so does the reabsorption of emission become increasingly negligible, thus making possible, even in condensed phases, reliable observations of the short wavelength edge of the emission spectrum. An interesting example of this has been presented for the S_1 emission from liquid benzene where not only the 0-0 but also hot transitions are clearly revealed upon excitation into the third strong absorption system at 184.9 nm.[5]

III. SATURATED HYDROCARBONS

For many years, it had been commonly believed that excited states of saturated hydrocarbons lived too short a time not only for the development of an observable fluorescence, but indeed even for the appearance of a resolvable vibrational structure in

the absorption spectrum. Direct dissociation or a strong predissociation were considered to occur, on these time scales, much too
rapidly. The discovery in 1967 of vibrational structure in the
ethane spectrum[9] and also in cyclopropane[9a] and cyclohexane[9a] came
therefore as some surprise. Clearly, at least for these saturated
hydrocarbons, excited states existed for a sufficiently long time
for vibrational quantization to develop.

In 1965 Laor and Weinreb[10] reported that cyclohexane liquid,
optically excited, could sensitize the fluorescence of a solute
present at low concentrations. Hexane was shown to behave similarly. Two years later, Holroyd demonstrated a sensitization of
N_2O decomposition by liquid cyclohexane excited at 147.0 nm[11] and
found too a strong reduction in the hydrogen yield from cyclohexane
upon addition of low concentrations of benzene or cyclohexene.[12]
From subsequent extensions of this work,[13,14] it was possible to
estimate that the cyclohexane excited state lifetime was at least
of the order of magnitude of 1 nsec if its bimolecular reactions
were not faster than diffusion limited. The implication, though
not fully realized at the time, was that unless the radiative constant of this excited state was abnormally low (eg <10^4 sec^{-1}) a
measurable fluorescence should be seen.

In 1969 observation of the emission of liquid cyclohexane was
finally reported together with observations of similar emissions
from a number of other saturated hydrocarbons.[15] During the past
three years, this work has been extended and the emission spectra
and emission quantum yields are now known for about 150 compounds.
[16,17] In the following paragraphs, some of the more salient features of this work are summarized.

Due to the weakness of the emission from saturated hydrocarbons, spectra have not yet been obtained for an analyzer band pass
less than about 60 cm^{-1}. However, at this value the spectra are
all completely featureless with a spectral distribution in photons/
cm^{-1} which is approximately Gaussian in its dependence on wavenumber $\tilde{\nu}$. As a consequence, the emission may be characterized by
the wavelength λ_f of the maximum of the distribution and the full
width at half height, σ_f.

At 25°C all of the n-alkane liquids from pentane to heptadecane have been studied.[17] The spectral distribution shows slight
but systematic increase in σ_f from 8400 cm^{-1} for pentane to 7700
cm^{-1} for heptadecane with λ_f remaining essentially constant at
about 207 $\pm$ 1 nm. For excitation at 165 nm the fluorescence quantum yields ϕ_f increase with increasing number of carbon atoms (N_c)
from 0.35 x 10^{-3} for pentane to 7.4 x 10^{-3} for heptadecane. Defining the absorption edge, λ_a, as the wavelength at which the
molar decadic extinction coefficient is 5 liter/mole cm, we find
λ_a to increase from 169 nm for pentane to 176 nm for heptadecane

so that the 165 nm excitation lies within $\approx$ 1500 cm^{-1} - 3500 cm^{-1} of the absorption onsets. For excitation at 147 nm, the emission spectrum is unchanged but the fluorescence quantum yields, ϕ_f, strongly decline. The extent of this decline monotonically de-$\cdot$ creases as N_C gets larger from $\beta_f = \phi_f(147)/\phi_f(165) = 0.18$ for pentane to 0.62 for heptadecane. From this we conclude that were we able to excite at the absorption edge, $\phi_f(165)$ would tend to increase slightly and by about the same factor β_0^{-1} for all alkanes. Defining k_f and k_{nr} as the radiative and non-radiative constants of the emitting state, then also be definition

$$\phi_f(165) = \beta_0 \, k_f/(k_f + k_{nr}) \quad .$$

The lifetime of the emitting state $\tau_f = (k_f + k_{nr})^{-1}$ has been determined for a number of n-alkanes by Henry and Helman[18] and more recently by Lyke and Ware.[19] Taking the ratio of our $\phi_f(165)$ to their τ_f gives a remarkably constant value for $\beta_0 k_f$ of 1.7 $\pm$ 0.2 x 10^6 sec^{-1} for all of the n-alkanes studied by both groups (ie from pentane to heptadecane). Since, as previously indicated, β_0 should be close to unity and only weakly dependent on the nature of the alkane, we conclude that k_f must be essentially constant for all n-alkanes at a value $\geq$ 1.7 x 10^6 sec^{-1}. Thus the molecular variability of the fluorescence quantum yield is attributed exclusively to a strong decline in k_{nr} as N_C increases.

To understand this effect, we first comment on the extraordinarily large frequency shift between absorption onset λ_a^{-1} and the frequency of maximum emission λ_f^{-1}. This spectral shift $\Delta = \lambda_a^{-1} - \lambda_f^{-1}$ varies monotonically from 10.6 x 10^3 cm^{-1} to 9.0 x 10^3 cm^{-1} on increasing N_C from 5 to 17. Similarly defined shifts for aromatic compounds usually are about an order of magnitude lower (ie $\approx$ 10^3 cm^{-1}).[4] When the emission originates in the same electronic state as is the terminal state of the first absorption, Δ is proportional to the so-called "Stokes-shift" and is a measure of the difference in equilibrium geometries of ground and excited states. On this basis we would then conclude that the first electronic excitation of a saturated hydrocarbon must induce very significant alterations in the nuclear geometry. Alternatively, if the terminal state of the first absorption internally converts to a lower, optically forbidden state and it is this state which is seen in emission, then of course Δ is without significance and we can conclude nothing about the geometry change. However, when the geometries of ground and emitting states are similar, the absorption strength of the emitting state can usually be reasonably well estimated under the assumption that the electronic transition matrix elements are the same in absorption and emission. This leads to the relation[20]

$$k_f = 2.88 \times 10^{-9} \, n^2 {<\lambda_f^3>}_{av}^{-1} (g_\ell/g_u) \int \frac{\varepsilon d\tilde{\nu}}{\tilde{\nu}} \tag{1}$$

where n is the solvent refractive index, g_ℓ and g_u the degeneracies of upper and lower states and ε the decadic molar extinction coefficient. With $k_f > 1.7 \times 10^6$ sec^{-1}, n = 1.5, $g_\ell = 1$, $g_u > 1$ and assuming a Gaussian absorption spectrum shifted $\approx 10^3$ cm^{-1} from the emission spectrum but with equal width (ie $\sigma_f \leqslant 8 \times 10^3$ cm^{-1}), we estimate for the absorption an $\varepsilon_{max} > 10$ liter/mole cm. But there is observed no transition of this order of magnitude in the absorption spectra of n-alkanes in the region close to 200 nm. Indeed, as we have already remarked, ε first rises to 5 ℓ/mole cm only at $\lambda_a = 169 - 176$ nm.

Our conclusion then is that Δ truly measures a "Stokes" shift and that its magnitude for the n-alkanes signifies a large nuclear distortion in the excited state. Other arguments, based on energy transfer studies further support this view.[17] Also consistent is the width of the spectral distribution of $\approx 8 \times 10^3$ cm^{-1} which is larger (by about a factor of 2) than the widths of aromatic emissions and, of course, too it must be remembered that the promotion of a bonding or antibonding electron is expected generally to generate substantially more nuclear rearrangement than the promotion of a π or non-bonding electron.

As N_c increases, both Δ and σ_f decline and therefore, we are suggesting, so too must the disparity in equilibrium geometries of ground and excited state. The decrease in k_{nr} then follows directly from the view that k_{nr} is determined most importantly by radiationless transition to the ground state with a probability that depends crucially on the relative displacement of excited and ground state potential energy hypersurfaces.

The nuclei in a saturated hydrocarbon react to an altered distribution of the σ electrons by changing   their equilibrium positions. The less localized is the electronic distortion, as for example will result from spreading the excitation over a larger molecular frame, the smaller should be the change in nuclear geometry[21] and conversely the geometry change should be more severe for an electronic distortion which is strongly localized. Upon adding a methyl branch to a n-alkane, the emission spectrum red shifts.[17] Thus whereas all the n-alkanes emit at ~ 207 nm, for 2-methyl butane $\lambda_f = 233$ nm, for 2-methyl hexane $\lambda_f = 223$ nm, for 2-methyl octane $\lambda_f = 217$ nm, etc. This rather profound influence of the branch suggests a role akin to that of a low energy chromophore with a correspondingly large localized excitation amplitude. Accordingly, we would expect now considerable nuclear rearrangement manifesting itself in large Δ, large σ_f and decreased ϕ_f. Exactly these are found. In 2-methyl hexane, for example, Δ and σ_f increase by about 3×10^3 cm^{-1} and 1.5×10^3 cm^{-1} respectively whereas ϕ_f is about an order of magnitude reduced from that of heptane. These effects persist for other members of the series although the extent of these changes is generally reduced as N_c increases. Thus for

2-methylnonadecane, λ_f = 208 nm, σ_f = 7.7 x 10^3 cm^{-1}, ϕ_f = 2.0 x 10^{-3} and Δ = 8.9 x 10^3 cm^{-1} all very similar to the values obtained for the larger n-alkanes.

An even stronger localization of the electronic excitation, with its concomitant effect to increase k_{nr}, is suggested as an explanation for the absence of emission from geminally substituted alkanes.[17] Thus for 2,2-dimethylbutane, 3,3-dimethylheptane, etc., ϕ_f must be less than 10^{-5} since no emission could be detected within our experimental sensitivity. On the other hand, for branches which are separated by more than one CH_2 group (eg 2,4-dimethyl-pentane), emission characteristics quite similar to those of single branched compounds are observed. But whenever the two branches are placed vicinal to each other (eg 2,3-dimethylbutane) a remark-able change occurs. The emission is strongly red-shifted and the quantum yield strongly enhanced. For 13 such compounds that we have studied,[17] λ_f = 242 $\pm$ 2 nm with ϕ_f generally an order of magnitude larger than for other (but non-geminally) branched compounds. Un-fortunately lifetime measurements are not yet available for deter-mining the origin of the increase in ϕ_f although preliminary quen-ching studies indicate that, at least, in part, this is due to sub-stantial increase in k_f. What property of this particular config-uration of carbon atoms induces these effects is not yet known, but the influence is clearly a dominating one. Even in cyclic com-pounds, vicinal substitution leads to characteristic effects regard-less of how otherwise disparate the structures are.

Of the non-substituted cycloalkane liquids only cyclohexane has been observed to fluoresce.[16,17] The emission peaks at 201 nm and with quantum yields of 8.8 x 10^{-3} and 3.5 x 10^{-3} for excitation at 165 nm and 147 nm respectively (ie β_f = 0.40). The spectral width is similar to that of the n-alkanes (ie σ_f = 7.7 x 10^3 cm^{-1}) but with a somewhat reduced Stokes shift, Δ = 6.7 x 10^3 cm^{-1}. The absence of emission from cyclopentane, cycloheptane, cyclooctane and cyclodecane and its reappearance for solid cyclododecane has been explained in terms of ground state strain energies and the release of this strain upon excitation.[17] Thus with increasing strain in the ground state the equilibrium nuclear geometries of ground and excited states are expected to become increasingly dis-parate. Comparison of methylcyclohexane with the very weakly emit-ting methylcycloheptane tends to confirm this view. For methyl-cycloheptane, Δ is increased by about 40% and ϕ_f reduced by about a factor of 100.

Although the emission quantum yields of cycloalkanes show gen-erally less sensitivity to structural changes than are found for alkanes, other properties, however, behave rather similarly for the two classes of compounds. Thus adding a single branch red shifts the emission (λ_f $\approx$ 212 nm) and increases both σ_f and Δ. As the size of the branch increases from methyl to dodecyl, λ_f remains about constant whilst σ_f and Δ decline. Adding a second, geminal branch, as for example in 1,1-dimethylcyclohexane although not now extinguishing the emission, nevertheless reduces the quantum yield by about an order of magnitude and red-shifts the spectrum (λ_f = 225 nm). When the second branch is removed by at least one CH_2 group, spectral characteristics similar to those of a single branched compound are observed. But placing the two branches vicinal, as for example in 1,2-dimethylcyclohexane or isopropylcyclohexane or sec-butylcyclohexane strongly red-shifts the spectrum to λ_f = 232 nm. A more complete examination of these results has been presented elsewhere.[17]

Emission has also been observed from many saturated hydrocarbons in the vapor phase.[15,17] In general the emission is slightly blue-shifted but otherwise possesses similar spectral characteristics to those of the liquid. One striking difference, however, is that whereas in the liquid, emission is observed even at excitation wavelengths to 120 nm, in the vapor phase, the emission suddenly declines to unobservable levels for excitation below $\approx$ 160 nm. In the solid too, emission has been observed from several compounds.[22] In general the spectrum of the liquid shows little effect on cooling even below the freezing point although there is observed very large increase in fluorescence quantum yield. Thus for example the emission yield of n-pentane increases about 15 fold on cooling from 25°C to -78°C whereas λ_f and σ_f are unchanged.[17] In the case of cyclohexane, the spectrum remains unchanged to -87.1°C at which point there occurs a solid-solid phase transition from a face-centered cubic (I) to monoclinic (II). At the transition temperature both emission and absorption spectra have been observed to blue shift by $\sim$ 100 Å.[22] The modification in crystal structure is such as to substantially alter the closest nearest neighbor center-center distances from 6.1 Å in solid I to 5.2 Å in solid II[23] and it is perhaps this which is responsible for the blue shift. Raymonda has recently suggested an assignment of cyclohexane's first absorption system to a 3s Rydberg[24] and a blue-shift triggered by a sudden large change in intermolecular distances is then perhaps not unexpected.[25] A more complete discussion of this is developed elsewhere.[22]

The nature of the emitting state remains unknown. If it is indeed a Rydberg state then presumably there will be similarity between its geometry and that of the ion. An interesting consequence of this follows. Beck and Thomas[26] have found the positive ion of cyclohexane to be highly mobile in liquid cyclohexane with a rate constant for reaction with pyrene more than an order of magnitude larger than diffusion limited. If the process involves resonance exchange of charge between neighboring cyclohexane molecules we would predict that positive ion mobilities in hydrocarbons should be larger the more similar in geometry are the ground and ion states. Thus the fluorescence yield of the neutral,

which in large part we suggest is determined by such geometry differences, could be a significant indicator of whether or not there
will be observed high positive ion mobilities. When ϕ_f is large,
or more precisely when k_{nr} is small, the positive ion mobility is
correspondingly expected to be large. Consistent with this, Zador,
Warman and Hummel[27] have recently found reaction rate constants of
cyclohexane and methylcyclohexane positive ions with pyrene to be
$\sim > \pm 4 \times 10^{11}$ liter/mole sec whereas for the much weaker emitting
compounds such as n-hexane, cyclopentane, cyclooctane and 2,2,4-
trimethylpentane, rate constants more than an order of magnitude
lower are obtained. In this connection too, it is perhaps pertinent to note that similarly high rate constants have been obtained
for bimolecular reactions of excited neutral cyclohexane[26] and
decalin[18] with low concentrations of CCl_4 or benzene dissolved in
the neat liquids. The corresponding process to a resonance charge
exchange for the ion could be, for a Rydberg state, the propagation
of a Wannier exciton. However, the effect of dilution[17] and temperature[28] are not yet clearly consistent with this view.

In general, as the saturated hydrocarbon increases in size,
the efficiency of its internal conversion from an upper to a lower
electronic state increases. This has interesting implications for
the liquid phase photochemistry. For example cyclohexane excited
at 147 nm disappears with essentially unit photochemical quantum
yield.[14] However, since $\beta_f = 0.4$ for cyclohexane,[17] about 40% of
the molecules excited at 147 nm must cascade to the emitting state.
Therefore the probability for photochemistry from the levels achieved directly by 147 nm excitation cannot exceed 0.6 whereas the
emitting state must photochemically disappear with close to unit
yield. In the case of bicyclohexyl or decalin, since $\beta_f = 1$, all
the photochemistry must proceed from the emitting state regardless
of the excitation wavelength.

For those compounds for which $\beta_f < 1$, as λ_{ex} is further reduced below 147.0 nm, the fluorescence quantum yield continues to
decline until an excitation wavelength λ_t ($\approx 135.0 - 140.0$ nm) is
achieved below which the emission yield has been observed to
recover.[29] With N_2O present, which quenches the emission, we have
found the quenching efficiency to remain constant to λ_t and then
abruptly to increase. Since N_2O is known to be an efficient electron scavenger and λ_t is approximately where we might expect to see
a liquid photoionization threshold (≈ 1 ev below the gas phase
value) the following mechanism is suggested. As λ_{ex} decreases, the
emission yield also decreases due to a decreasing probability of
internal conversion to the emitting state. This continues until a
state S_t is excited at λ_t. From S_t an electron is ejected. Recombination of the electron with the positive ion follows rapidly producing a neutral excited state S_s. The probability for this we take
to be g (with 1-g being the probability for producing on recombination any non-fluorescent state). Clearly the emission yield must

increase if g exceeds the probability for internal conversion from S_t to S_s. In the case that this conversion efficiency is unity, the emission yield must decline at λ_t (for $g \neq 1$) and indeed exactly such behavior is observed for decalin.[29] Thus, with some refinement, measurements of emission intensities may ultimately provide thresholds for photoelectron ejection in condensed phases.

IV. AROMATIC MOLECULES

Fluorescence, of course, is always a possible decay channel for an excited state of a polyatomic molecule and therefore should, in principal, be observable from any state. The question then is never whether or not a state will emit but rather with what probability will it do so and is this of sufficient magnitude for detection by one's equipment. The fluorescence quantum yield, ϕ_f, experimentally defined as the number of photons emitted per photon absorbed is generally expressible as the ratio of the radiative (k_f) to the sum of radiative and non-radiative ($k_f + k_{nr}$) rate constants. In the case that the non-radiative decay is into a continuum (or pseudocontinuum) of states (as will usually be true for highly excited states of large molecules internally converting to lower states), the absorption or emission vibronic line is expected to be approximately Lorentzian with a full width at half height, σ_f, proportional to the sum of $k_f + k_{nr}$, ie $\sigma_f = \hbar(k_f + k_{nr})$.[30] From this, an upper bound of $\simeq 10^{15}$ sec^{-1} can be established for k_{nr} since clearly σ_f cannot exceed the entire electronic width of the absorption or emission band (usually $\lesssim 0.5 \times 10^4$ cm^{-1}). Therefore, for any excited state with a strongly allowed optical transition (ie $k_f \simeq 10^9$ sec^{-1}), the fluorescence quantum yield of such a state cannot be less than $\sim 10^{-6}$. The conclusion then is that with a capability for measuring yields at the 10^{-6} level, we should not be surprised to find emission from optically allowed states regardless of where they may lie. Indeed the surprise would be in not finding such emission. With this in mind we have searched and recently located fluorescence from upper electronic states of several aromatic molecules both in liquid and vapor phases.[31,32] Our progress to date is summarized below.

In isooctane solution, the third absorptive transitions (S_3) of naphthalene and pyrene are located at λ_{max} = 220.5 nm and 272.5 nm respectively. Their oscillator strengths (defined as f = 4.32 $\times 10^{-9} \int \epsilon d\tilde{\nu}$) are for naphthalene 1.3 and for pyrene 0.35. Weak S_3 emissions have been observed from both molecules for excitation into S_3 at λ_{ex} = 206.2 nm (naphthalene) and λ_{ex} = 253.7 nm (pyrene).[31] In both cases the emission lies about 1400 $\pm$ 100 cm^{-1} red shifted from the absorption (naphthalene λ_f = 227 nm; pyrene λ_f = 284 nm) with quantum yields of 22 $\times 10^{-6}$ for naphthalene and 5 $\times 10^{-6}$ for pyrene. The emission spectra are sufficiently close to mirror images of the absorptions to permit a reliable estimate of k_f to be obtained from the Strickler-Berg equation[20] (see

Section III, eq. 1). Using this together with the measured quantum yield gives for naphthalene, $k_{nr} = 1.5 \times 10^{14}$ sec^{-1} ($\sigma_f \cong 800$ cm^{-1}) and for pyrene $k_{nr} = 1.0 \times 10^{14}$ sec^{-1} ($\sigma_f \cong 500$ cm^{-1}).

Pyrene has also been excited into its fifth absorption system with $\lambda_{ex} = 184.9$ nm. Although quantum yields have not yet been obtained, distinct emissions have been observed from S_5 ($\lambda_f \cong 213$ nm), S_4 ($\lambda_f \cong 255$ nm) and S_3 ($\lambda_f = 284$ nm).[31] An S_2 emission is also seen both for $\lambda_{ex} = 184.9$ nm and 253.7 nm but its proximity to the intense S_1 emission makes somewhat difficult its analysis. Additionally, population of S_2 is known to derive, at least in part, from thermal equilibration with S_1[33] and therefore its quantum yield is not simply related to k_f and k_{nr}.

Benzene, toluene, p-xylene and mesitylene neat liquids when excited into S_3 at $\lambda_{ex} = 184.9$ nm exhibit an S_2 but no S_3 emission.[31] However, in dilute isooctane solution, and in the vapor phase the S_3 emission is indeed observed. Both pyrene and naphthalene behave similarly in this regard with S_3 emission disappearing for the pure melts. The oscillator strength of the S_3 transition in neat liquid benzene is $\cong 0.8$.[34] Using this, together with an S_3 emission yield that cannot exceed $\cong 1 \times 10^{-6}$, requires that $k_{nr}(S_3) \gtrsim 10^{15}$ sec^{-1} and therefore $\sigma_f \gtrsim 5500$ cm^{-1}. Consistent with this, Inagaki[34] reports the neat liquid S_3 absorption spectrum to be abnormally broad ($\cong 7500$ cm^{-1}) as compared to that in dilute solution ($\cong 3500$ cm^{-1}). The very rapid process that depletes S_3 is not yet known. However, the observation that the S_2 emission yield is not greatly changed between neat liquid and dilute solution suggests as a possible mechanism an intermolecular $S_3 \rightarrow S_2$ conversion involving reaction of S_3 with neighboring S_0 to generate vibrationally excited S_2.

The S_3 emission from naphthalene has also been observed in the vapor phase.[32] The spectrum is very similar to that observed in isooctane solution except for a 7 nm blue shift which parallels the solution to vapor shift of the S_3 absorption. From the emission quantum yield of $\cong 10 \times 10^{-6}$ [35] and the integrated vapor absorption, a value of $k_{nr} \cong 1.5 \times 10^{14}$ sec^{-1} [35] is obtained. The close agreement between this value and that obtained in solution indicates either that vibrational relaxation in the liquid does not occur within the time scale of S_3 decay or if it does (which is unlikely) that such relaxation does not effect the decay amplitude. Both S_2 and S_1 emissions have also been observed from naphthalene vapor for excitation into S_3. The proximity of the S_2 to the S_1 emission makes S_2 analysis difficult but its spectral position ($\lambda_f \cong 290$ nm) is close to that previously reported by Wannier, Rentzepis and Jortner.[36] The S_1 emission is significantly broader, red-shifted, weaker and shorter lived when generated at 184.9 nm than at $\lambda_{ex} = 253.7$ nm (excitation into S_2) and similar changes have been observed on comparing 253.7 nm with still longer wavelength excitation.[37]

S_3, S_2 and S_1 emissions have also been observed from a number of alkylbenzene vapors excited into S_3 at λ_{ex} = 184.9 nm.[32] For p-xylene, which is a representative example of the group, $k_{nr}(S_3) \approx$ 3 x 10^{14} sec^{-1} ($\sigma_f \approx$ 1500 cm^{-1}).[35] (Were a similar value, to obtain in the neat liquid, an emission quantum yield of $\approx$ 10 x 10^{-6} should have been obtained, again arguing then for the development, in the neat liquid, of a much more rapid decay channel.) The S_2 emission from p-xylene is clearly resolved from S_1 and appears with a quantum yield about 1.8 times larger than the S_3 yield. Since the S_2 state must be populated via internal conversion from S_3 , with a probability $\beta_{32} < 1$, it is simply calculated from the quantum yield ratio, the ratio $k_f(S_3)/k_f(S_2)$ and $k_{nr}(S_3)$ that $k_{nr}(S_2)$ has an upper bound of $\approx$ 2 x 10^{13} sec^{-1}. Assuming that the reaction of S_2 with O_2 occurs with a hard sphere radius of $\sim$ 5 Å, we estimate from the complete absence of S_2 quenching by 1 atm O_2 that $k_{nr}(S_2)$ cannot be less than $\approx$ 5 x 10^{11} sec^{-1}.

As for naphthalene, so too for p-xylene vapor, the S_1 emission spectrum is significantly altered by changing the excitation wavelength. Additionally there is observed for both molecules when excited at 184.9 nm, an effect of increasing density to blue-shift, narrow, intensify and lengthen the lifetime of the S_1 emission causing generally an approach of the emission to that characteristic for excitation into S_1. In the case of p-xylene, where this has been studied more carefully, large effects are observed from 4 to 100 torr yet the addition of even 1 atm of neohexane has been found to have absolutely no effect. The following picture seems to emerge. The high vibrational levels of S_1 (ie $S_1^\dagger$) that are generated via internal conversion from S_3 (directly or via S_2) must have available a non-radiative channel to some lower electronic state (perhaps S_0). Due to this coupling, $S_1^\dagger$ is short-lived, but with a k_{nr} (estimated from its quantum yield at 4 torr) not greater than $\approx$ 2 x 10^{12} sec^{-1}. Thus, in solution, vibrational relaxation easily competes with decay through this channel and S_1 emission in solution is always large regardless of excitation wavelength. In the vapor phase, 1 atm O_2 has very little effect on $S_1^\dagger$, suggesting (for a 5 Å hard sphere reaction) a lower bound for k_{nr} of $\approx$ 5 x 10^{11} sec^{-1}. The difference between p-xylene and neohexane is explained by postulating an intermolecular resonance mechanism for relaxation, ie $S_1^\dagger + S_0 \rightarrow S_0^\dagger + S_1$. Although the cross-section for this reaction must be larger than that involving neohexane as relaxer, it nevertheless need not be too large to generate an important effect since there will be at least a three order of magnitude gain in fluorescence intensity on converting $S_1^\dagger$ to S_1.

V. SIMPLE OLEFINS

The absorption spectrum of ethylene and its simple alkyl derivatives exhibit a strong transition ($V \leftarrow N$) at 170-190 nm which generally red-shifts with increasing alkyl substitution.[38] To the

red of this absorption, there appears, at least in the vapor phase,
a somewhat weaker transition (R ← N) with a separation from the
V ← N system that increases with increasing alkyl substitution.
Merer and Mullikan[39] have summarized the evidence for assigning
the R-N system to a Rydberg transition. A number of calculations
on ethylene have indicated an equilibrium geometry for the R state
(unlike the V state) not too dissimilar from that of the ground
state.[40] If this be similarly true for the alkyl derivatives then,
in general, the Einstein B coefficient in absorption should provide
a reasonable estimate of the radiative constant of the R state.
For many simple olefins this rate constant is about 10^7-10^8 sec^{-1}
which is comparable to that of the lowest singlet states of many
aromatic compounds. If then the R state is the lowest singlet
state of simple olefins, it becomes curious that whereas aromatics
emit strongly, no fluorescence has ever been reported from the
olefins.

With a capability for measuring emission quantum yields as
low as 10^{-6} (for selected excitation wavelengths), we have recently
searched for olefin fluorescence and have indeed now observed weak
emission in the region around 230-265 nm.[41] A brief survey of the
results are presented below. Since spectral response calibrations
are not yet available, the absolute wavelengths of maximum emission
intensity, λ_f, are uncertain to $\approx \pm 3$ nm. Differences, however,
are reliable to $\approx \pm 1$ nm. A more extensive report of this work
will be published elsewhere.

For excitation at 184.9 nm, weak emissions have been observed
from the gases at 1 atm of propylene ($\lambda_f \approx 223$ nm) and cis- and
trans-2-butene ($\lambda_f \approx 223$-227 nm) with spectral widths, $\sigma_f \approx 8$-9 x
10^3 cm^{-1}. No structure has yet been detected. Quantum yields are
not yet known absolutely but are of the order of ≈ 1-2 x 10^{-6} and
increase from propylene to cis-to trans-2-butene. No emission has
been seen from ethylene under these conditions.

Similarly at λ_{ex} = 184.9 nm, emissions have been observed from
the neat liquids (nitrogenated) of 1-hexene ($\lambda_f \approx 236$ nm), cis-2-
hexene ($\lambda_f \approx 238$ nm) and trans-2-hexene ($\lambda_f \approx 235$ nm) with spectral
widths of ≈ 8-9 x 10^3 cm^{-1} and quantum yields ≈ 5-7 x 10^{-6}. For
1-hexene a vapor emission has also been located at ≈ 231 nm.

With the exception of cis- and trans-2-hexene, the R ← N
absorptions of all the above compounds are insufficiently resolved
from the V ← N absorptions to determine the existence of a correlation with the observed emission spectra. Both V ← N and R ← N absorptions apparently shift, for these compounds, to sufficiently
similar extent that regardless of the emission assignment we expect
and indeed confirm that maximum frequencies of the V ← N absorption[38] when shifted by a constant interval (~ 13.3 x 10^3 cm^{-1}),
superimpose on the emission frequencies. The solution absorptions

of cis- and trans-2-hexene show the V ← N absorption maximum of
the cis compound about 600 cm^{-1} to the blue of the trans whereas
the emission spectra are exactly reversed in this order. This
would be consistent with an R → N emission assignment were the
V-R gap significantly larger for the cis compound. Although the
vapor spectra are not unequivocal on this point, there is, nevertheless, to some extent, a general indication of this and not only
for the hexenes but also for other cis, trans olefin pairs.[38]

For 2-methyl-2-pentene, as for other alkyl substituted olefins,
the R ← N absorption now begins to move out clearly from under the
V ← N absorption and we observe the emission spectrum to strongly
red shift (λ_f ≈ 253 nm) and the emission quantum yield to increase
(ϕ_f ≈ 10 x 10^{-6} for the neat liquid). These trends continue with
increasing substitution and we have examined thus far, in somewhat
more detail than for other compounds, the emission characteristics
of 2,3-dimethyl-2-butene (tetramethylethylene, TME). Emission
spectra and emission quantum yields have been obtained for excitation at 184.9 nm, 213.8 nm (Zn) and 228.8 nm (Cd) with TME as neat
liquid, dissolved in pentane, dissolved in perfluorinated hexane
and as a vapor. For any given sample condition, the emission spectrum remains essentially invarient to excitation wavelength although
in the vapor at 184.9 nm there is some slight broadening similar
to that observed for the S_1 emissions from naphthalene and p-xylene
when excited into higher states (see Section IV). Since at λ_{ex} =
228.8 nm, the absorption is exclusively into the R system, a V → N
emission assignment is clearly eliminated. For the neat liquid,
λ_f ≈ 262 nm with a quantum yield at λ_{ex} = 228.8 nm of ≈ 4 x 10^{-4},
which is now, for us, a relatively intense emission. In pentane
solution, λ_f blue shifts by about 2 nm and in perfluorinated hexane
is again blue-shifted by an additional 5 nm but ϕ_f remains approximately constant for λ_{ex} = 228.8 nm. At shorter λ_{ex}, ϕ_f declines
but less so for the neat liquid than for solution in perfluorinated
hexane, which is reminiscent of similar effects obtained for benzene
S_1 emission.[5] The emission spectrum of the vapor at λ_{ex} = 228.8 nm
is unshifted from that obtained in perfluorinated hexane (ie λ_f ≈
255 nm) and the quantum yield too is not very much altered (ie
ϕ_f ≈ 3 x 10^{-4}).

To within about 5%, no quenching of the emission of TME could
be observed on addition of 1 atm O$_2$. From this we estimate a lifetime of the emitting state (assuming hard sphere reaction with
d = 5 Å) not exceeding ≈ 5 x 10^{-12} sec. Using this together with
the vapor quantum yield of 3 x 10^{-4} gives for the radiative constant, k_f, a lower bound of ≈ 6 x 10^7 sec^{-1} which is within about
20% of the value of k_f ≈ 5 x 10^7 sec^{-1} obtained from the integrated
vapor absorption of the R ← N system. Therefore, unless the O$_2$
quenching reaction has an abnormally small cross-section, we must
conclude that the emission cannot arise from some weak transition
buried within the R ← N system. The assignment suggested therefore

is R → N with $k_{nr} \approx 2 \times 10^{11}$ sec^{-1} (from $k_f = 5 \times 10^7$ sec^{-1}).

The R state lives a very short time (for an S_1 state) which raises the interesting question as to the nature of the decay channel that so rapidly depopulates it. Since the TME vapor emission and its R ← N absorption spectra exhibit a relatively small Stokes shift, it appears unlikely that the R and N states differ greatly in geometry. In this regard the olefin emission is more similar to that of an S_1 emission from an aromatic than from a saturated hydrocarbon. However, for the π^* state of the aromatic, the singlet-triplet gap is rather large ($\sim$ 1 ev) whereas if the R state of the olefin is indeed a Rydberg, this splitting should be much reduced.[42] As a consequence an efficient Rydberg singlet to Rydberg triplet intersystem crossing could conceivably be responsible for the short emission lifetime. The small ϕ_f for unsubstituted olefins may in part arise from reduced internal conversion efficiency from the V to R state (all of these have so far only been excited into the V ← N system). However, there are also spectral indications (increased Stokes-shift and band width) of increased nuclear distortion of the R state for the unsubstituted compounds and therefore perhaps some enhancement of k_{nr}.

Although the R ← N absorption is very different in vapor and solution (flattened out and blue-shifted), the emission is relatively unchanged both in spectrum and yield. Since the excited state electronic wave functions in emission and absorption are not necessarily the same, a possible explanation is that the emitting, vibrationally relaxed molecule is in a state with electron density somewhat less spatially extended. It would be interesting, therefore, to study the olefin emission in a higher density solid phase where even further important changes in the R ← N absorption have been reported.[43]

ACKNOWLEDGEMENTS

All of the work reported here has been performed in close collaboration with my colleague Dr. Fumio Hirayama. Additionally Dr. Thomas Gregory, Dr. William Rothman and Dr. Craig Lawson have each contributed extensively and importantly to various aspects of the project.

REFERENCES AND FOOTNOTES

*   This research has been supported in part by the U. S. Atomic Energy Commission (Document No. COO-913-54).

1. R. Gordon, Jr., R. E. Rebbert and P. Ausloos, NBS Technical Note **496** (1969).

2. W. Rothman, F. Hirayama and S. Lipsky, J. Chem. Phys. **58**, 1300 (1973).

3. F. Hirayama and S. Lipsky, J. Chem. Phys. **51**, 1939 (1969).

4. I. B. Berlman, "Handbook of Fluorescence Spectra of Aromatic Molecules" (Academic Press, New York, 1971), 2nd ed.
5. C. W. Lawson, F. Hirayama and S. Lipsky, J. Chem. Phys. 51, 1590 (1969).
6. W. H. Melhuish, J. Opt. Soc. Amer. 51, 278 (1961) and references cited therein.
7. Professor J. B. Birks has recently suggested (Chem. Phys. Lett. 17, 370 (1972)) that the 9,10-diphenylanthracene yield should be 0.83 rather than 1. This would have the effect of making our reported yields too high by ~17% thus tending to cancel the error due to neglect of the refractive index correction.
8. C. W. Lawson, F. Hirayama and S. Lipsky, in "Molecular Luminescence", E. C. Lim, Ed. (W. A. Benjamin, Inc., New York, 1969).
9. a) J. W. Raymonda and W. T. Simpson, J. Chem. Phys. 47, 430 (1967); b) B. A. Lombos, P. Sauvageau, and C. Sandorfy, J. Mol. Spect. 24, 253 (1967); c) E. N. Lassettre, A. Skerbele and M. A. Dillon, J. Chem. Phys. 46, 4536 (1967); d) S. Lipsky and J. A. Simpson, Intern. Conf. Phys. Electron. At. Collisions, Vth, Nauka, Leningrad (1967) p. 575.
10. U. Laor and A. Weinreb, J. Chem. Phys. 43, 1565 (1965).
11. R. A. Holroyd, J. Am. Chem. Soc. 72, 759 (1968).
12. R. A. Holroyd, J. Y. Yang and F. M. Servedio, J. Chem. Phys. 46, 4540 (1967).
13. U. Laor and A. Weinreb, J. Chem. Phys. 50, 94 (1969).
14. J. Y. Yang, F. M. Servedio and R. A. Holroyd, J. Chem. Phys. 48, 1331 (1968).
15. F. Hirayama and S. Lipsky, J. Chem. Phys. 51, 3616 (1969).
16. F. Hirayama, W. Rothman and S. Lipsky, Chem. Phys. Lett. 5, 296 (1970).
17. W. Rothman, F. Hirayama and S. Lipsky, J. Chem. Phys. 58, 1300 (1973).
18. M. S. Henry and W. P. Helman, J. Chem. Phys. 56, 5734 (1972).
19. W.R. Ware and R. L. Lyke, Chem. Phys. Lett. (submitted for publication 1973).
20. S. J. Strickler and R. A. Berg, J. Chem. Phys. 37, 814 (1962).
21. J. Franck and E. Teller, J. Chem. Phys. 6, 861 (1938); J. L. Magee and K. Funabashi, J. Chem. Phys. 34, 1715 (1961); J. C. Lorquet, S. G. ElKomoss and J. L. Magee, J. Chem. Phys. 37, 1991 (1962).
22. F. Hirayama and S. Lipsky, Chem. Phys. Lett. (accepted for publication 1973).
23. R. Kahn, R. Fourme, D. André and M. Renaud, Acta Crystallogr. B29, 131 (1973).
24. J. W. Raymonda, J. Chem. Phys. 56, 3912 (1972).
25. M. B. Robin and M. A. Kuebler, J. Mol. Spectrosc. 33, 274 (1970); E. Miron, B. Raz and J. Jortner, J. Chem. Phys. 56, 5265 (1972) and references cited therein.
26. G. Beck and J. K. Thomas, J. Phys. Chem. 76, 3856 (1972).

27. E. Zader, J. M. Warman and A. Hummel, Chem. Phys. Lett. (submitted for publication 1973).
28. W. P. Helman, Chem. Phys. Lett. 17, 306 (1972).
29. F. Hirayama and S. Lipsky, to be published. A preliminary report of this work has been presented at the Notre Dame Conference on Elementary Processes in Radiation Chemistry, Notre Dame, Indiana, 1972.
30. J. Jortner, S. A. Rice and R. M. Hochstrasser, in Advan. Photochem. 7, 149 (1969) and references cited therein.
31. F. Hirayama, T. A. Gregory and S. Lipsky, J. Chem. Phys. 58, 4696 (1973).
32. T. A. Gregory, F. Hirayama and S. Lipsky, J. Chem. Phys. 58, 4697 (1973).
33. P. A. Geldorf, R. P. H. Rettschnick and G. J. Hoytink, Chem. Phys. Lett. 4, 59 (1969); A. Nakajima and H. Baba, Bull. Chem. Soc. Japan 43, 967 (1970); H. Baba, A. Nakajima, M. Aoi and K. Chihara, J. Chem. Phys. 55, 2433 (1970).
34. T. Inagaki, J. Chem. Phys. 57, 2526 (1972).
35. In ref. 32, quantum yields were not corrected for the effect of refractive index on the collection efficiency. Values presented here are appropriately corrected.
36. P. Wannier, P. M. Rentzepis and J. Jortner, Chem. Phys. Lett. 10, 193 (1971).
37. J. O. Uy and E. C. Lim, Chem. Phys. Lett. 7, 306 (1970); U. Laor and P. K. Ludwig, J. Chem. Phys. 54, 1054 (1971).
38. L. C. Jones, Jr. and L. W. Taylor, Anal. Chem. 27, 228 (1955).
39. A. J. Merer and R. S. Mullikan, Chem. Rev. 69, 639 (1969).
40. M. B. Robin, R. R. Hart and N. A. Kuebler, J. Chem. Phys. 44, 1805 (1966); A. J. Lorquet and J. C. Lorquet, J. Chem. Phys. 49, 4955 (1968).
41. F. Hirayama and S. Lipsky, A. E. C. Document No. COO-913-51 (1973).
42. R. J. Buenker, S. D. Peyerimhoff and W. E. Kammer, J. Chem. Phys. 55, 814 (1971).
43. W. J. Potts, Jr., J. Chem. Phys. 23, 65 (1955).

PRODUCTION OF ELECTRONICALLY EXCITED SPECIES IN PHOTODISSOCIATION OF SIMPLE MOLECULES IN THE VACUUM ULTRAVIOLET

H. Okabe

Physical Chemistry Division
National Bureau of Standards
Washington, DC 20234   U.S.A.

INTRODUCTION

The detailed understanding of the primary photochemical process in the vacuum ultraviolet is difficult even for simple triatomic molecules. It is known [1] for example that the photochemical primary process of N_2O at 1470 Å and 1236 Å involves transient products such as $N_2(B^3\Pi)$, $N_2(A^3\Sigma)$, $N(^2D)$, $O(^1S)$, $O(^1D)$, and $O(^3P)$. It is not possible to identify these intermediate products from the analysis of final products or from the absorption spectrum of N_2O. With an advancement of flash photolysis and sensitive measuring instruments, it has recently become possible to detect transient species directly, either by measuring absorption or emission of the transient species.

The photolysis of simple molecules in the vacuum ultraviolet often yields an electronically excited product which undergoes fluorescence before colliding with reactant molecules. In this case it is often possible to detect the fluorescence produced by an extremely weak light source from a monochromator. Studies of the intensity and the spectrum of the fluorescence as a function of incident light wavelength provide useful information on (1) the bond dissociation energy and the heat of formation, (2) the spin conservation rule in the primary photochemical process, (3) the nature of dissociation (i.e., whether the process is predissociative or a direct dissociation), and (4) the configuration of an upper state. In this review several examples will be given on these subjects.

APPARATUS [2]

Fig. 1 shows the apparatus for measuring the fluorescence intensity as a function of incident wavelength. The light source is a hydrogen-discharge lamp operated on a high-voltage AC power supply. The lamp is provided with a LiF window of 1 mm thickness. The beam of monochromatic light of desired wavelength in the region 1050 to 2000 Å can be obtained at the exit slit of the monochromator by a rotation of a concave grating. The fluorescence cell, made of monel, is provided with three windows. The fluorescence light emerging at right angles to the incident beam passes through a Suprasil window and is detected by a 13-stage photomultiplier. A glass filter is placed in front of the multiplier to select emission in the proper wavelength region. The intensity of the incident light over the region 1100 to 2000 Å is monitored by another photomultiplier coated with sodium salicylate.

For identifying the fluorescence spectrum, a high intensity Xe or Kr resonance lamp was used instead to irradiate the sample directly to yield the fluorescence, and its spectrum was taken by scanning the monochromator over the desired wavelength region.

APPLICATION OF THE THRESHOLD MEASUREMENT

Determination of the bond dissociation energy and related thermochemical data

The bond dissociation energy, $D_0(A-B)$, may be obtained from a relationship

$$D_0(A-B) \leq h\nu_1 - E_0(B^*) \qquad (1)$$

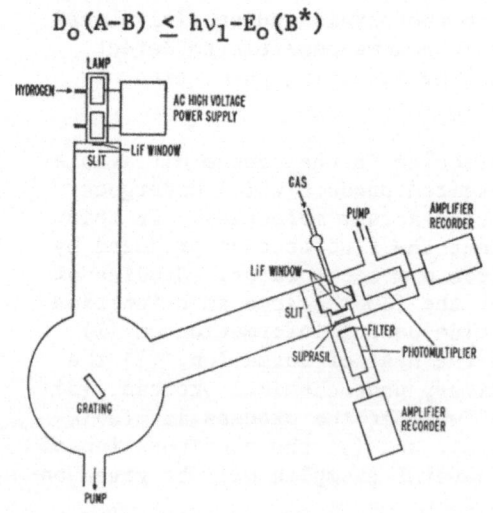

Fig. 1. Experimental arrangement for the measurement of fluorescence.

where $h\nu_1$ is the threshold energy of incident photons required to produce electronically excited species, B^*, and E_o, an electronic energy of B^*.

Table I shows the results obtained from photolysis, of some cyanogen compounds. [3,4]

Table I: Comparison of bond dissociation energies of cyanogen compounds obtained by photodissociation photoionization and electron-impact methods.

Molecule	Bond Dissociation Energies (ev) $D_o^{\cdot}(R-CN)$		
	Photo-dissociation	Photo-ionization	Electron Impact
HCN	5.20 ± 0.05	5.20 ± 0.02	5.1 ± 0.2
FCN	4.82 ± 0.05	5.01 ± 0.02	-----
ClCN	4.33 ± 0.05	4.31 ± 0.02	4.9 ± 0.1
BrCN	3.77 ± 0.05	3.68 ± 0.02	4.4 ± 0.1
ICN	3.11 ± 0.04	3.17 ± 0.02	3.9 ± 0.1
C_2N_2	5.62 ± 0.05	6.22 ± 0.02	6.2 ± 0.3
CH_3CN	5.32 ± 0.03	-----	4.8 ± 0.1
C_2HCN	6.21 ± 0.04	-----	-----

This method is similar to the measurement by a monochromator-mass spectrometer of the appearance threshold of fragment ions produced from photoionization of a molecule. In the latter case $D_o(A-B)$ is given by

$$D_o(A-B) \leq h\nu_2 - I.P.(A) \qquad (2)$$

where $h\nu_2$ is the threshold energy required to produce A^+ from AB by photoionization and I.P.(A) is the ionization potential of A. It is necessary to obtain I.P.(A) by some independent method. Very often, however, I.P.(A) is not known. By a combination of two methods (i.e., photodissociation and photoionization) it is possible to obtain I.P.(A).

$$I.P.(A) = h\nu_2 - h\nu_1 + E_o(B^*) \qquad (3)$$

We have recently studied [4] photodissociation and photoionization of C_2HCN. In the vacuum ultraviolet photolysis less than 1% of the total primary process results in the production of $CN(B^2\Sigma)$.

$$C_2HCN \xrightarrow{h\nu_1} C_2H + CN(B^2\Sigma)$$

The threshold energy of incident photons required for this process is found to be 9.41 ± 0.04 eV. We also studied the photoionization process to produce C_2H^+.

$$C_2HCN \xrightarrow{h\nu_2} C_2H^+ + CN + e$$

The threshold energy of incident photons for this process is 18.19 ± 0.04 eV. Since the electronic energy of $CN(B^2\Sigma)$ is 3.198 eV, $I.P.(C_2H) = 11.98 \pm 0.05$ eV is derived from (3). From $\Delta Hf_0^o(C_2H^+) = 17.47 \pm 0.01$ eV [4], $\Delta Hf_0^o(C_2H) = 5.50 \pm 0.04$ eV, or 127 ± 1 kcal mol^{-1} is obtained. This value is in excellent agreement with $\Delta Hf_0^o(C_2H) = 129 \pm 3$ kcal mol^{-1} obtained from high temperature reaction of graphite with hydrocarbons [5]. $D_0(C_2H-H) = 124 \pm 1$ kcal mol^{-1} is calculated from $\Delta Hf_0^o(C_2H)$, $\Delta Hf_0^o(H)$, and $\Delta Hf_0^o(C_2H_2)$. $D_0(C_2H-H) = 124$ kcal mol^{-1} is much larger than a value [6] of $D_0(C_2H_3-H) \sim 108$ kcal mol^{-1}.

Determination of electronic energies of Cl_2 yielding the 2580 and 3063 Å emission continua [7]

The photolysis of $OCCl_2$ in the vacuum ultraviolet produces Cl_2 emission continua similar to those produced in an electric discharge, as shown in Fig. 2.

The electronic energies of Cl_2 yielding these continua at 2580 Å and 3063 Å have not been determined since these emission continua do not exhibit fine structure which makes the spectral analysis possible. Since the photolysis of $OCCl_2$ in the vacuum ultraviolet produces these emission continua, it is possible to determine electronic energies of Cl_2 responsible for the continua from the measurement of threshold energies required to produce these continua

$$OCCl_2 \xrightarrow{h\nu} CO + Cl_2{}^*$$

where $Cl_2{}^*$ signifies electronically excited Cl_2 producing the 2580 (or 3063 Å) emission continuum. The threshold energy, $h\nu$, to produce the 2580 Å continuum is 8.28 ± 0.05 eV and that for 3063 Å is 9.00 ± 0.03 eV. Since the heat of reaction of the process

$$OCCl_2 \rightarrow CO + Cl_2$$

is 1.071 eV, the electronic energy of Cl_2 yielding the 2580 Å emission is 7.21 ± 0.05 eV and that yielding the 3063 Å emission is 7.93 ± 0.03 eV. Since no diffuse bands were observed in the 3063 Å continuum, lower states must be repulsive, producing two ^2P chlorine atoms. An energy level diagram for the upper and lower states and transitions involved for the production of the 2580 and 3063 Å continua is given in Fig. 3. The electronic energy of $NH(a^1\Delta)$ with respect to $NH(X^3\Sigma)$ is similarly determined [8].

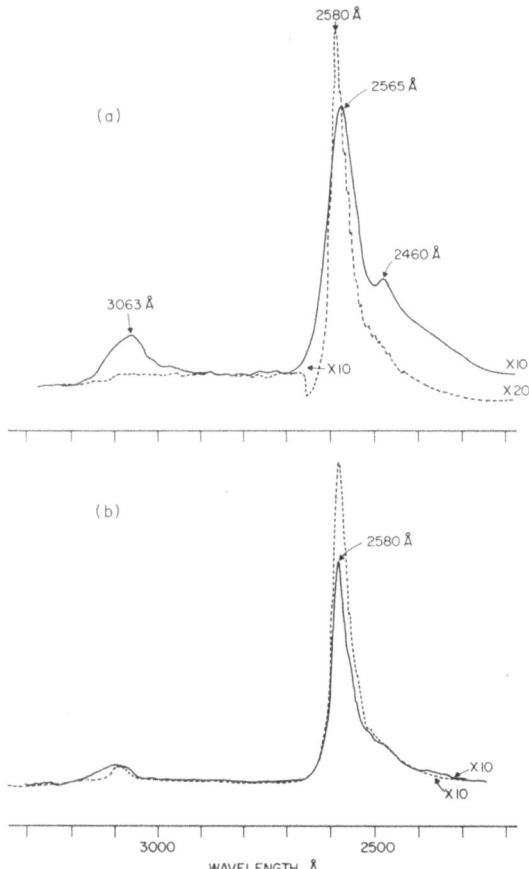

Fig. 2. (a) Full line: emission continua observed from 0.2 torr OCCl$_2$ excited by the Kr lamp. The spectrum was scanned with a speed of 100 Å per minute and a time constant of 1 sec. Resolution, 3 Å. Dashed line: emission spectrum when 30 torr N$_2$ was added to the above sample. (b) Solid line: emission continuum from 0.2 torr OCCl$_2$ excited by the Xe lamp. Scanning speed, 100 Å per minute; time constant, 1 sec; resolution, 6 Å. Dashed line: emission spectrum with 60 torr of Ar added to the sample.

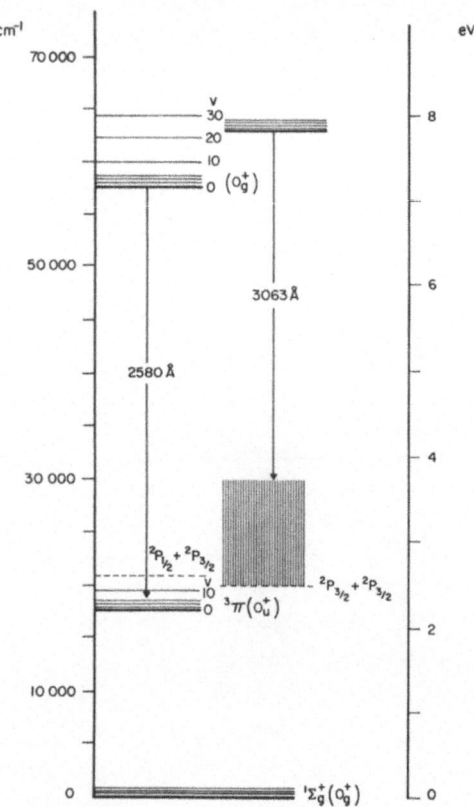

Fig. 3. Energy-level diagram of Cl$_2$; transitions producing
the 2580- and 3063-Å emissions are indicated. Vibrational levels
of the upper and lower states for the 2580-Å emission were drawn
after Briggs and Norrish (Ref. 9) and for the 3063-Å emission
after Venkateswarlu (Ref. 10) and this work (Ref. 7).

NATURE OF THE PRIMARY PROCESS

The spin conservation rule

It is assumed in many photochemical reactions that the spin
conservation rule is obeyed. However, the direct demonstration
that the spin is conserved in photolysis is few. Stuhl [11] has
shown that NH(a$^1\Delta$) is initially produced in the vacuum ultraviolet
flash photolysis of HN$_3$ in agreement with the spin rule, since HN$_3$
and N$_2$ are both in singlet state

$$HN_3 \xrightarrow{h\nu} NH(a^1\Delta) + N_2.$$

He observes further that as the $NH(a^1\Delta)$ concentration decreases with
time the $NH(X^3\Sigma)$ concentration increases, indicating the reaction

$$NH(a^1\Delta) + M \rightarrow NH(X^3\Sigma) + M$$

where M represents HN_3 or inert gas. Kroto [12] makes a similar
observation in flash photolysis of NCN_3 that $NCN(^1\Delta)$ is formed
shortly after the flash

$$NCN_3 \overset{h\nu}{\rightarrow} NCN(^1\Delta) + N_2.$$

Subsequently, the singlet starts to decay and the triplet to grow,
signifying the reaction

$$NCN(^1\Delta) + M \rightarrow NCN(X^3\Sigma) + M.$$

An observation of fluorescence in the vacuum ultraviolet photolysis of NH_3 and HN_3 also reveals that the spin is conserved in the
reaction. The photolysis of NH_3 at 1236 Å [13a] produces $NH(c^1\Pi)$

$$NH_3 \overset{h\nu}{\rightarrow} NH(c^1\Pi) + H_2$$

The photolysis of HN_3 [13b] at 1470 Å forms mostly $NH(c^1\Pi)$

$$HN_3 \rightarrow NH(c^1\Pi) + N_2$$

Fig. 4 shows the NH emission spectrum produced from photodissociation of HN_3 at 1470 Å and 1236 Å. The NH $A^3\Pi$ formed at the Kr
line is due to a secondary process. On the other hand the sensitized
reaction of HN_3 [14] by Xe^3P_2, equivalent in energy to photons of
wavelength 1470 Å, forms $NH(A^3\Pi)$ only and no $NH(c^1\Pi)$ emission is
observed

$$Xe(^3P_2) + HN_3 \rightarrow NH(A^3\Pi) + N_2$$

In the case of CS_2 photolysis in the vacuum ultraviolet it was found
that a main primary process is

$$CS_2 \overset{h\nu}{\rightarrow} CS(A^1\Pi) + S(^3P_2)$$

in apparent violation of the spin conservation rule. The photolysis
in the region 2300–1850 Å also results almost exclusively in the
production of $S(^3P)$ atoms [16].

$$CS_2 \overset{h\nu}{\rightarrow} CS(X^1\Sigma) + S(^3P)$$

in analogy with the photolysis in the vacuum ultraviolet.

Predissociation

In some cases it is possible to show that the primary photochemical process is predissociative; that is, dissociation takes
place from a stable excited state formed from initial light absorption. For example, the production of $CS(A^1\Pi)$ from the photolysis

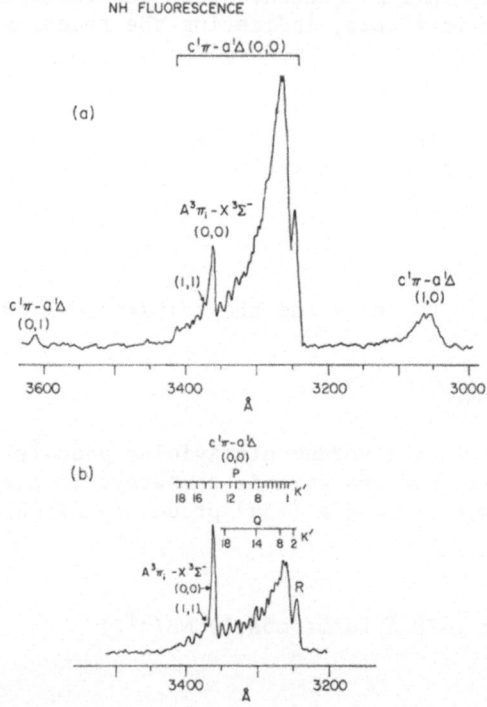

Fig. 4. (a) Fluorescence spectrum of the transitions NH $c^1\Pi$-$a^1\Delta$ and NH $A^3\Pi_i$-$X^3\Sigma^-$ in the photolysis of 0.2 torr HN_3 by the Xe lines with a resolution of 7 A. (b) The same spectrum of 0.3 torr HN_3 by the Kr lines with a resolution of 3 Å.

of CS_2

$$CS_2 \overset{h\nu}{\to} CS(A^1\Pi) + S(^3P_2)$$

has been shown [15] to occur from Rydberg states of CS_2 and not from direct dissociation via a repulsive state. This conclusion is drawn from the results that the $CS(A^1\Pi)$ fluorescence efficiency vs. incident wavelength curve shows the same Rydberg series I and II as those in the absorption spectra of CS_2, as shown in Fig. 5.

Configuration of an excited state

Welge [17] has shown that $OH(X^2\Pi)$ produced from photodissociation of H_2O in the vacuum ultraviolet has very little rotational (K" up to 5) and vibrational energy (v" up to 1) indicating that

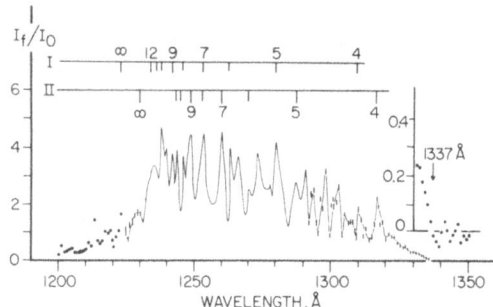

Fig. 5.   Fluorescence (CS $A^1\Pi$) excitation spectrum of CS_2 at
a resolution of 0.8 Å in the region 1200–1350 Å and 3 Å in the
region 1330–1350 Å; pressure ∿0.1 torr scanning speed, 2 Å/min.
with a time constant of 5 sec. I and II designate Rydberg series.

excess energy (incident photon energy-bond energy) must be trans-
formed into translational energy of the products, H and OH.   Since
OH acquires little rotational energy, the H–O–H angle must be
invariant in the ground and excited state.   The photolysis of HNCO
at the Xe or Kr lines produces $NCO(A^2\Sigma)$ [18].   It is estimated,
from the intensity of the emission bands, $NCO(A^2\Sigma-X^2\Pi)$, that bending levels $v_2' = 1,2,3$, and 4 are more populated than the ground
vibrational levels $v_2 = 0$ of the upper state.   The process of ac-
quiring the bending vibration may be understood from the following
consideration.   It is assumed that dissociation takes place almost
immediately after light absorption without making many vibrations.
In this case the vibrational energy distribution of $NCO(A^2\Sigma)$ may
be determined by an equilibrium configuration of the upper state
of HNCO.   The molecule in the ground state has a linear NCO structure with an HNC angle of 128°.   The $NCO(A^2\Sigma)$ is also linear with
almost the same internuclear distance as NCO of the ground state
HNCO.   In order to have bending vibrations, v_2', excited up to 4,
the NCO group of the molecule in its equilibrium configuration
must be considerably bent in the upper state produced by initial
light absorption.   As the excited molecule dissociates into H +
$NCO(A^2\Sigma)$, the NCO group changes its structure from bent to linear
resulting in bending vibrations.   Although no analysis has been
made of the absorption spectrum in the vacuum ultraviolet, an
analysis of near ultraviolet absorption spectrum [19] indicates a
bent NCO structure in the upper state.

Table II lists some thermochemical data obtained by the
photodissociation method.

Table II: Thermochemical Data Obtained from Photodissociation

Quantity Measured	eV	kcal mol^{-1}	Reference
$D_0(SC-S)$	4.463 ± 0.014	102.92 ± 0.32	[15]
$D_0(OS-O)$	5.64 ± 0.02	130 ± 0.4	a
$D_0(NC-N_3)$	4.2 ± 0.1	96 ± 2	b
$D_0(NCN-N_2)$	0.3 ± 0.1	7 ± 2	b
$D_0(H-NCO)$	4.90 ± 0.01	112.9 ± 0.2	[18]
$D_0(N-CO)$	2.14 ± 0.15	49 ± 3	[18]
$D_0(N-CN)$	4.3 ± 0.2	99 ± 5	b
$E_0(NH\ a^1\Delta)$	1.6 ± 0.1	36 ± 2	c
$\Delta Hf_0^0(C_2H)$	5.50 ± 0.04	127 ± 1	[4]
$\Delta Hf_0^0(CS)$	2.82 ± 0.02	64.96 ± 0.4	[15]
$\Delta Hf_0^0(SO)$	0.056 ± 0.03	1.3 ± 0.7	a
$\Delta Hf_0^0(CH_2N_2)$	2.22 ± 0.05	51.3 ± 1	d

[a] H. Okabe, J. Chem. Phys. 56, 3378 (1972); J. Am. Chem. Soc. 93, 7095 (1971).

[b] H. Okabe and A. Mele, J. Chem. Phys. 51, 2100 (1969).

[c] H. Okabe and M. Lenzi, J. Chem. Phys. 47, 5241 (1967).

[d] A. H. Laufer and H. Okabe, J. Am. Chem. Soc. 93, 4137 (1971).

REFERENCES

[1] R. A. Young, G. Black, and T. G. Slanger, J. Chem. Phys. 48, 2067 (1968); ibid 49, 4769 (1968).

[2] H. Okabe, J. Chem. Phys. 47, 101 (1967).

[3] D. D. Davis and H. Okabe, J. Chem. Phys. 49, 5526 (1968).

[4] H. Okabe and V. H. Dibeler, J. Chem. Phys., to be published.

[5] J. R. Wyatt and F. E. Stafford, J. Phys. Chem. 76, 1913 (1972).

[6] D. M. Golden and S. W. Benson, Chem. Rev. 69, 125 (1969).

[7] H. Okabe, A. H. Laufer, and J. J. Ball, J. Chem. Phys. 55, 373 (1971).

[8] H. Okabe and M. Lenzi, J. Chem. Phys. 47, 5241 (1967).

[9] A. G. Briggs and R. G. W. Norrish, Proc. Roy. Soc. (London) A276, 51 (1963).

[10] P. Venkateswarlu, Proc. Ind. Acad. Sci. 26, 22 (1947).

[11] F. Stuhl, Dissertation, The University of Bonn, Bonn, Germany, 1966.

[12] H. W. Kroto, J. Chem. Phys. 44, 831 (1966).

[13] a. K. H. Becker and K. H. Welge, Z. Naturforschg. 18a, 600 (1963).
     b. H. Okabe, J. Chem. Phys. 49, 2726 (1968).

[14] D. H. Stedman, J. Chem. Phys. 52, 3966 (1970).

[15] H. Okabe, J. Chem. Phys. 56, 4381 (1972).

[16] M. deSorgo, A. J. Yarwood, O. P. Strausz, and H. E. Gunning, Can. J. Chem. 43, 1866 (1965).

[17] K. H. Welge and F. Stuhl, J. Chem. Phys. 46, 2440 (1967).

[18] H. Okabe, J. Chem. Phys. 53, 3507 (1970).

[19] R. N. Dixon and G. H. Kirby, Trans. Faraday Soc. 64, 2002 (1968).

ENERGY PARTITIONING IN THE PHOTOFRAGMENTS

C. Vermeil

C. N. R. S. , E. S. P. C. I.
Paris, France

INTRODUCTION

For about one century photochemistry has been widely used to
produce short lived transients in order to study the physical properties and the chemical reactivities of atoms and molecules in
their ground electronic states. For the last ten years, it is the
excited states of these transients which are of interest and all
the work already done in the U.V. range is being repeated with V.
U.V. photons. So the energy deposited in a molecule greatly exceeds that needed to break any chemical bond and the forthcoming
question is that of the partition of this excess energy among the
photofragments. This is an old problem in chemical kinetics shar-
ing close analogy with thermal unimolecular reactions. However
the basic assumption of the R.R.K.M. theory of reaction rates is
that of a random distribution of energy over all the internal vibrational and rotational degrees of freedom of the molecule. If
application of R.R.K.M. theory to photodissociation has not been
very successful (1), one tentatively may ascribe this failure to
the fact that the excited molecule does not live long enough for
this randomization to occur. So, from the point of view of energy
partition, the first important point is the <u>time scale of the photophysical events before the molecule dissociates.</u>
The other theoretical approaches (2) to the problem appear as
developments of radiationless transition theories in an attempt to
connect the directly excited molecular state to a continuum of unbound states via for example a dense manifold of vibrational states.
As large densities of states are needed, application is mainly to
large molecules beyond the statistical limit. From the point of
view of fragmentation, the size of the molecule does not seem to be
an important point: there are more differences between diatomic

and triatomic molecules than between triatomic and very large ones
because of the appearance of vibrational and rotational degrees of
freedom in the photofragments. It is not to say that such theories
are of no value: they are or will be beneficial to understand competition between various processes, that is to explain or even predict quantum yields.   Indeed a good way to look at the conversion
of electronic to vibrational energy inside the molecule may be the
study of fluorescence quantum yields as a function of the exciting
wavelength (3). However the lack of fragmentation, although a con-
ceptual advantage, reduces the number of monitoring parameters of
the system under investigation.

        In the present review we will begin a little later in time
and assume that some intramolecular energy transfer (from electronic to vibrational excitation) has already led the excited molecule
to select its own reaction coordinate, a single dissociative channel being assumed for the sake of simplicity. Obviously enough,
for dissociation to occur, the location of energy in one preferen-
tial vibrational mode is more important than the electronic state
to which the excited molecule belongs. In a study of the luminescence of two excited singlet states of chromene derivatives, the
observation was made that "it appears as if internal conversion
may proceed only within a particular vibrational manifold" (4). I
will tentatively assume that a given photofragmentation pattern
proceeds only within a particular vibrational manifold irrespective
of the electronic states involved. We indeed know that electron
impact and atomic collisions may compare with photonic absorption
for the molecule to reach that stage of "prepared" fragmentation.
However it is not clear presently if the whole spectra of rotational and vibrational distributions are or are not identical in each
case and more experiments are needed in order to assert this working assumption.

        It is worthwhile to emphasize that the same problems are in-
volved each time large quantities of energy are transmitted to
molecules or ions, whatever the way chosen to do it; it follows
that photofragmentation, at least in the V.U.V. region, is more
likely to be connected to collision theories (5) than to unimolecular decomposition.

        So far only the so called primary photochemical processes have
been under consideration. On the other hand, secondary reactions
in photochemical systems, as long as they imply highly excited
species, may be ascribed to the same class of chemical systems
characterized by non-thermal equilibrium populations. Diatomic
systems, belonging obviously to a different group will not be considered here.

INTERNAL ENERGY OF PHOTOFRAGMENTS

        From the point of view of energy balance, the amount of energy
required for electronic excitation of one (or both) photofragment

in any electronic state is in principle the easiest to evaluate. Terenin in the thirties was the first to seek the threshold dissociation energies of molecules leading to light emission from photofragments such as OH, CN, NH_2, I (6). Unfortunately in too many cases the spacing of the various electronic states of radicals as simple as NH, CH_2 is not yet known with accuracy (7). However, the spectroscopy of excited radicals is such an active field that much progress is expected in the forthcoming years.

Work in absorption spectroscopy has mostly been performed using flash photolysis, that is excitation into an array of states. As far as vibrational and rotational structures of the photofragments are viewed as reflecting the geometrical properties of the dissociative molecule, at least one would like to know which state of the parent molecule is excited by photon absorption. Even for rather simple molecules the properties of upper electronic states are not known precisely. A double improvement is clearly needed: better spectroscopic assignments of absorption spectra in the far U.V. and still more powerful photon detectors.

One would like to know the energy partition between vibration and rotation for each photofragment electronic state as a function of the exciting wavelength at pressures low enough to avoid relaxation by collisions. In order to obtain such data, a factor of 10^3 is needed either in the intensities of the light sources or in the detector sensitivity. Once available, and it may be soon, undoubtedly such data will stimulate important theoretical work on the details of photofragmentation processes, but presently one is forced to say that the subject is still in its infancy.

An interesting semiclassical approach is that of Wilson and coworkers (8) who examined the photodissociation of the first excited state of ICN treated as a quasidiatomic molecule, since the length of the CN bond is nearly the same in the ground state ICN, in its first excited state and in the CN radical. Attempts to correlate experimental data on internal energy partition between vibration and rotation of fragments and the symetry properties of the excited states have been performed by Mitchell and Simons for halogenated compounds (XNO, XCN, XCH_3), RONO, CS_2 (9) and by Mele and Okabe for cyanogen halides XCN (10).

TRANSLATIONAL ENERGY

This non-quantized energy plays a less dignified part in the energy distribution, being thought of as carrying away the remainder of the energy balance. However two interesting points must be noticed: the velocity spectrum of the photofragments (before any collision) may be already quite large and, in special cases, the distribution of the trajectories may not be fully isotopic. Quite general aspects of anisotopic effects created in any collisions by a beam of incident particles (photons, electrons, neutrons, atoms, ions) directed onto other particles (nuclei, atoms, molecules, ions)

initially isotropically oriented may be found in the Bersohn and
Lin review (11). Anisotropy induced in photochemical systems by
light absorption and leading to the polarization of the light emitted by electronically excited photofragments as well as their angular recoiling distribution has been looked at by Zare (12). Jonah
has derived expressions for loss of anisotropy due to molecular
rotation and to the velocity of the dissociation (13).

Although experiments with polarized light are presently limited to the near U.V. part of the spectrum, I feel that the use of
polized light is a good way to get some insight into the photofragmentation processes, since only in that way may the law of conservation of energy and momentum be introduced. To do that two angles
must be defined: first, the angle between the electric vector of
the light beam and the electric dipole transition moment of the
molecule; and second, the angle between the flying fragment and the
transition moment. So one is forced to consider the dissociation
as an individualized event rather than from a statistical point.
In the diatomic case, once given the direction of the polarized
light beam and the characteristics of the molecule, everything is
known, almost in principle. The experimental determination of the
recoiling fragment distribution allows the orientation of the transition dipole moment to be known and if the symmetry species of
the ground state is known that of the excited dissociative state
may be inferred. (14).

Experimental determinations of anisotropic photofragmentation
distribution have been undertaken using two types of detectors:
the old Paneth method of detection of radicals by removal of a thin
metallic mirror and a more elaborate one using time of flight mass
spectrometry. The first one gives only the angular distribution
of the recoiling fragments: the polarized light enters a hemispheric cell whose internal surface has been covered by a thin
metallic deposit. The pressure of the gas to be photolysed is controlled so that the mean free path between collisions is larger
than the diameter of the cell. The atomic or radical product
attack to the mirror reflects the angular recoiling photofragmentation. This method of "photolysis mapping" has been applied to
aliphatic carbonyl compounds (H_2CO, C_2H_5CO, C_3H_7CO, CH_3COCH_3) (15),
unfortunately not with monochromatic light. A slightly different
form of the same method has been to photolyze an organometallic
compound, $(CH_3)_2Cd$, so that the cadmium atoms struck onto the clean
pyrex surface of the hemispheric cell (16). It was concluded that
the transition moment is approximatively perpendicular to the motion
of the Cd atom and also, as shown by another experiment, perpendicular to the motion of the CH_3 radicals. Assuming that the molecule
is linear it seems most likely to the authors that the fragments
exit along the bond axis through excitation of the antisymmetric
stretching vibration. Picturing the process classically Bersohn
et al wrote: "One Cd-C bond will be overextended and will break
while the other Cd-C bond will be highly compressed. Thus the
second methyl will leave (at least) one half period of vibration

later than the first".

In the second method a pulsed polarized laser light source is crossed with a molecular beam and the kinetic energy of recoil fragments is measured orthogonally at a selected value of m/e by a time of flight mass spectrometer. A half wave retardation plate rotates the electric vector of the light pulse. The peak intensity, measured at a fixed m/e value, when recorded as a function of time, gives the <u>kinetic energy</u> distribution of the fragments; recorded as a function of the angle of rotation of the half wave plate, it gives the <u>angular distribution</u> of the fragments. First successful application has been to Cl_2 (17). Application to NOCl and NO_2 has been performed by Wilson and coworkers (18): exciting molecular beams with laser light of 3.57 eV (347.1 nm) they measured the velocity and angular distributions of both fragments, NO and O, or NO and Cl. In the simplest case (NOCl) the velocity distribution gives direct information about the states of the fragments (NO up to v'=2 or even 3), while the angular distribution gives direct information about the excited state of the parent molecule. The triatomic model used to fit the data assumes a repulsive force between N and Cl atoms and NO as a rigid rotor; the dissociation is calculated to take place within less than 10^{-14} sec. The transition dipole moment lies in the plane of the molecule leading to an excited state of A' symmetry; it is most likely nearly parallel to the NCl bond. The NO_2 case is more complex: two peaks are found in the velocity distribution spectrum of the O atoms and may come from different excited states (of NO_2, of O or of NO). Wilson and Busch prefer to assume that the first and second peak correspond respectively to dissociation into NO v'=0.1 on the statement that "different electronic states would in general yield angular distributions for the two peaks which were more dissimilar than those observed". However their very thorough discussion leaves the reader with the impression that dissociation by two channels, one obtained directly and one by predissociation, is the authors' favorite conclusion.

The major interest of these studies and specially those of Wilson is due to the numerous physical models of energy partitioning in <u>triatomic</u> molecule photofragmentation which are put forward and compared to the experimental data by computation (Table 1). The limiting cases used to perform the calculations enlighten the intricacy one has to expect actually. For the angular distribution of the fragments it may be shown to depend on four angles: 1) angle between the electric vector of the polarized light and the direction of the recoiling fragments; 2) angle between the transition dipole moment and the direction of the bond to be broken; 3) angle of rotation of the recoiling fragments due to the rotation of the molecule; and 4) angle of rotation of the excited molecule during its lifetime. The first angle gives the peak location, the other three the shape of the angular distribution.

To go back to radiationless transition theories as compared to dynamical models as described above, one can say that, as far as

TABLE 1

Wilson's Models for Energy Partition in ABC + hν → A + BC

A. <u>Statistical models</u> → low values of translational energies
1. Statistics before dissociation → total equilibration (RRKM)
2. Statistics after dissociation → statistical vibrational, non-statistical rotational distributions (19)
B. <u>Direct models</u> ← no time for equilibration of internal energy
1. Interfragment (the changes in potential surface lead to forces between fragments)
a. pure impulsive: Steep repulsive potential along A-B, C being a spectator → vibrations in BC, high values of translational energies
b. modified impulsive: very stiff B-C bond, rovibrations in BC
2. Intrafragment (the changes in potential surfaces lead to intra-fragment forces) → high vibrations in BC.

the <u>pathway to the dissociative state</u> is concerned, the density of available levels is of paramount importance and as the <u>dissociation</u> is concerned, it is the <u>lifetime</u> of the excited molecule which is of interest. The two concepts, if linked together, nevertheless yield two different mathematical treatments.

HOT PHOTOFRAGMENTS

Although Welge and Gilpin made time-of-flight spectroscopy studies of the velocity of the O atoms such as those formed in the photodissociation of N_2O in the V.U.V. (20), studies like those of Wilson are not yet possible in V.U.V. photochemistry. So one must content himself with indirect chemical methods to get an idea of translational or internal energy released to the fragments. When the fragments are polyatomic it is rather difficult to know the energy partition between them: for instance "hot" CH_3 has been shown to form CH_4 by H-atom abstraction rather than recombining to give C_2H_6, but one does not know if this change is effected by vibrational or translational energy (21). This problem does not exist when the fragment is an atom, and so-called hot atom chemistry is an active field of theoretical and experimental investigation in which both photoreactions and nuclear reactions are used for hot atom generation (22). The cross section of an exchange chemical reaction of the kind A + BC = AB + C, A being an atom, is an increasing function of the kinetic energy of A from a threshold value not very different from the activation energy of the "thermal" reaction (A, BC being that time in thermal equilibrium) up to about 10 eV, above which it decreases. In the energy range of photo-chemically produced hot atoms (mainly H or D atoms between 0.2 and 4 eV) the cross section increases almost linearly with the kinetic

energy. Linear relationships for a set of hot atom reactions once
obtained may be used as calibration curves to deduce the mean
kinetic energy of an atomic photofragment produced by photolysis
of a parent molecule at a known photon energy. As crude as they
are, these evaluations give valuable insights into the primary
processes; when two processes are energetically possible their
relative probabilities may be estimated from the energy balance.
We have applied this method to the V.U.V. photochemistry of H_2O,
CH_4 and NH_3 (23); other applications have been to H_2S (24), D_2S (25)
methanethiol and ethane thiol (26). We used as a monitor for calibration the ratio of the rate constants of the two reactions
$H + D_2 = HD + D$ and $D + H_2 = HD + H$. The following reactions which
have been demonstrated to occur with hot H atoms have been or could
be used for calibration (27-28)
$$H + N_2O = N_2 + OH$$
$$H + CO_2 = CO + OH$$
$$H + COS = CO + SH$$
At kinetic energies above 1.7 eV, the reaction with N_2O takes
another path (24)
$$H + N_2O = NO + NH$$
The HI and HBr systems are still the most frequently investigated
and also used for calibration. These studies have so far yielded
the following interesting results: 1) the part of the excess energy
which goes into translation is often rather high ($\simeq$ 75%); 2) small
variation in kinetic energy of the H atoms yield large variations
of the secondary reaction rate constants; and 3) moderation by collisions is not as efficient a process as would be expected from
simple considerations. References prior to 1970 may be found in a
review (29) and others in the latest papers (30-31). Hot H atom
chemistry being in itself a whole field of research is far beyond
the scope of the present review.

OTHER METHODS OF EXCITATION

    At several places in this review has been put forward the idea
that the way in which the dissociating state is created is of no
concern to the dissociation itself. Inversely, if the same dissociative pattern is obtained from two different excitation devices,
does this mean that the same dissociating state is involved in the
two cases? If this is true, does this common unbound state come
necessarily from the same bound excited state? Obviously enough
for the comparison to be valid, both experiments have to be made
at exactly the same energy under similar experimental conditions
of pressure and detection.
    Much work has been done along this line on diatomic molecules
with comparison of photoionization and Penning ionization, using
for example He $2^{1,3}S$ collisions and 584 Å photons. The energies of
the ejected electrons are analyzed with the same experimental setup in both cases. From a careful study of NO, McDowell and co-

workers (32) were able to show identical distributions of vibrational levels of NO^+ in both cases and to conclude that the Penning ionization of NO to the ground state of NO^+ is vertical within experimental error.

Such a comparison is not yet easy to do for polyatomic molecules but there are indications that excitation by metastable argon atoms leads to the same energy partition as does photoexcitation. The HCN photolysis (but at the krypton lines) (10) has been shown to yield the same type of energy pattern as that obtained by argon and xenon collisions (33). The rotational and vibrational population of OH ($A^2\Sigma^+$) has been obtained from H_2O dissociated by collision with Ar ($^3P_{2,0}$) (34), 1236 and 1216 Å photons (35) and electron impact (36). All of them are characterized by abnormal rotational distributions (with strong modifications due to relaxation). Valuable comparisons of these distributions are not yet possible because of no data on photolysis at 1048-1066 Å, although Sokabe (34) has calculated a value of 9.4 eV for the electronic energy transferred from the argon atom to the water molecule implying a non-resonnant energy transfer.

In our laboratory, work is in progress on NH_3, a molecule which affords the case of multichannel dissociation processes. It has been photolyzed at the argon lines (37): the emission spectrum of NH $c^1\pi$ - $a^1\Delta$ obtained shows rotational structure in the o-o vibration band up to K = 6. Metastable argon collisions performed at Boulder, Colorado, by J. Fournier and F. C. Fehsenfeld (38) with the afterglow method give an identical emission spectrum of NH $c^1\pi$ (same rotational structure). Additionally, argon collisions produce the triplet NH $A^3\pi$ in the v' = 0.1 level with K up to at least 12. Pulse radiolysis of NH_3 gives identical emission spectra of both triplet and singlet NH, as do argon collisions (Febetron 706 gun, 560 keV electrons) (39). Further the photoionization yield and the Penning ionization yield are identical within the limits of experimental error, 0.49±0.02. From these results one can draw the following points: 1) At the same energy above the ionization potential the relative probabilities of ionization and excitation are the same for photon impact and argon collisions. 2) For one dissociative channel, that is $NH_3 \rightarrow NH\ c^1\pi + H_2$ the energy partition among the fragments does not seem to depend on the mode of excitation. As surprising as it may appear, the same is true for the dissociation into NH $A^3\pi$. $NH_3 \rightarrow NH\ A^3\pi + H_2$ produced by argon collision or by pulsed radiolysis. 3) Optically forbidden transitions may be allowed in the case of argon collisions (40), as they are known to be in pulse radiolysis. The points given above must not be taken as stated yet; they are mainly given here to illustrate the connection between collision processes and V. U. V. photochemistry.

REFERENCES

1a. G. B. Porter, B. T. Connelly, J. Chem. Phys. 33, 81 (1960).
 b. O. K. Rice, J. Chem. Phys. 55, 439 (1971).
2a. S. A. Rice, I. McLaughlin, J. Jortner, J. Chem. Phys. 49, 2756 (1968).
 b. K. G. Kay, S. A. Rice, J. Chem. Phys. 58, 4852 (1973).
 c. W. Rhodes, J. Chem. Phys. 50, 2885 (1969).
3a. N. A. Borisevich, V. A. Tolkachev, Opt. Spect. XXI, 18 (1966).
 b. K. Evans, R. Scheps, S. A. Rice, D. Hellner, Faraday Trans. II 69, 856 (1973).
4. R. S. Becker, E. Dolan, D. E. Balke, J. Chem. Phys. 50, 239 (1969).
5. Disc. Faraday Soc. No. 44, 1967 "Molecular dynamics of the chemical reaction of gases."
6a. A. Terenin, H. Neujmin, Nature 135, 543 (1935); J. Chem. Phys. 3, 436 (1935).
 b. A. Terenin, Zeits. f. Physik 44, 713 (1927).
7. G. Herzberg, Electronic spectra of polyatomic molecules (1966).
8. K. E. Holdy, L. C. Klotz, K. R. Wilson, J. Chem. Phys. 52, 4588 (1970).
9. R. C. Mitchell, J. P. Simons, Disc. Faraday Soc. 44, 208 (1967).
10. A. Mele, H. Okabe, J. Chem. Phys. 51, 4798 (1969).
11. R. Bersohn, S. H. Lin, Advan. Chem. Phys. 16, 80 (1969).
12. R. J. van Brunt, R. N. Zare, J. Chem. Phys. 48, 4304 (1968).
13. C. Jonah, J. Chem. Phys. 55, 1915 (1971).
14. K. G. Kay, S. A. Rice, J. Chem. Phys. 57, 3041 (1972).
15. J. Solomon, C. Jonah, P. Chandra, R. Bersohn, J. Chem. Phys. 55, 1908 (1971).
16. C. Jonah, P. Chandra, R. Bersohn, J. Chem. Phys. 55, 1903 (1971).
17. R. W. Diesen, J. C. Wahr, S. E. Adler, J. Chem. Phys. 50, 3635 (1969).
18. G. E. Busch, K. R. Wilson, J. Chem. Phys. 56, 3655, (1972); ibid. 3626, 3638.
19. P. Pechukas, J. C. Light, C. Rankin, J. Chem. Phys. 44, 794 (1966).
20. K. H. Welge, R. Gilpin, J. Chem. Phys. 54, 4224 (1971).
21. J. K. Rice, F. K. Truby, Chem. Phys. Lett. 19, 440 (1973).
22. F. S. Rowland, in M.T.P. International review of Science, Physical Chemistry series I, Vol. 9, Page 109, Butterworths (1972).
23a. M. Cottin, J. Masanet, C. Vermeil, J. Chim. Phys. 63, 959 (1966).
 b. J. Masanet, C. Vermeil, ibid. 66, 1248 (1969).
 c. L. Hellner, J. Masanet, C. Vermeil, J. Chem. Phys. 55, 1022 (1971).
 d. J. Masanet, J. Fournier, C. Vermeil, Can. J. Chem. (1973).
24. G. A. Oldershaw, D. A. Porter, A. Smith, Faraday Trans. I, 68 2218 (1972).
25. L. E. Compton, R. M. Martin, J. Chem. Phys. 52, 1613 (1970).
26. J. M. White, R. L. Johnson, J. Chem. Phys. 56, 3787 (1972).
27. R. E. Tomalesky, J. E. Sturm, Faraday Trans. I 68, 1241 (1972);

J. Chem. Phys. 55, 4299 (1971).

28. G. A. Oldershaw, D. A. Porter, Faraday Trans. I 68, 709 (1972).
29. C. Vermeil, Israel J. Chem. 8, 147 (1970).
30. D. K. Jardine, N. M. Ballash, D. A. Armstrong, Can. J. Chem. 51, 656 (1973).
31. K. Y. Hong, G. J. Mains, J. Phys. Chem. 76, 3337 (1972).
32. C. E. Brion, C. A. McDowell, W. B. Stewart, Chem. Phys. Lett. 13, 79 (1972).
33. T. Ursu, K. Kuchitsu, Chem. Phys. Lett. 18, 337 (1973).
34a. M. A. A. Clyne, J. A. Coxon, D. W. Setser, D. H. Stedman, Trans. Faraday Soc. 65, 1177 (1969).
  b. N. Sokabe, J. Phys. Soc. Japan 33, 473 (1972).
35a. T. Tanaka, T. Carrington, H. P. Broida, J. Chem. Phys. 35, 750 (1961).
  b. T. Carrington, J. Chem. Phys. 41, 2021 (1964).
36. T. Horie, T. Nagura, M. Otsuka, J. Phys. Soc. Japan, II, 1157 (1956).
37. J. Masanet, A. Gilles, unpublished work.
38. J. Fournier, F. C. Fehsenfeld, unpublished work.
39. M. Clerc, M. Schmidt, J. Hagege-Temman, J. Belloni, J. Phys. Chem. 75, 2908 (1971).
40. J. A. Coxon, D. W. Setser, W. H. Duewer, J. Chem. Phys. 58, 2244 (1973).

GENERATION OF COHERENT LIGHT IN THE VACUUM ULTRA-VIOLET

Stephen C. Wallace

IBM, Thomas J. Watson Research Center
P.O. Box 218
Yorktown Heights, N.Y. 10598

Remarkable progress has been made in the past few years in the development of sources of coherent light in the vacuum ultra-violet spectral region. This review will discuss recent experimental work and relevant basic physical principles of both vacuum ultra-violet lasers and harmonic generation into the vacuum ultra-violet, utilizing visible wavelength lasers. In exciting electronic transitions which would be made to lase in the vacuum ultra-violet, two problems immediately arise. The first is the severe frequency dependence of the pumping power required to attain significant inversion densities, and hence optical gain at these short wavelengths. For example, with a Doppler broadened line, simple considerations based on the Schawlow-Townes equations show that the pump power necessary to achieve some arbitrary gain scales as ν^4. The second difficulty has been more of an operational problem, that of finding a pumping source capable of efficiently exciting electronic levels above 7 eV.

Stimulated emission in the Lyman band of hydrogen ($B'\Sigma_u^+$, see Figure 1(a)) was first reported by Hodgson[1] in 1970, exciting the gas by means of a fast rise time, high current electric discharge from a Blumlein device. Subsequently, using a high current, relativistic electron beam for excitation, stimulated emission has been attained[2] in the Werner band transitions ($C'\pi_u$, see Figure 1(a)) yielding the shortest "laser" wavelength so far observed, 109.8 nm in para -H_2. Because of the difficulty in obtaining high densities of electrons ~ 50 eV (where the cross section for electronic excitation becomes significant) in a discharge, only in a Blumlein circuit of extremely high performance[3] has it been possible to make hydrogen lase[4] in the Werner bands. In lasers of this type using travelling wave excitation, there is no optical cavity. The photon

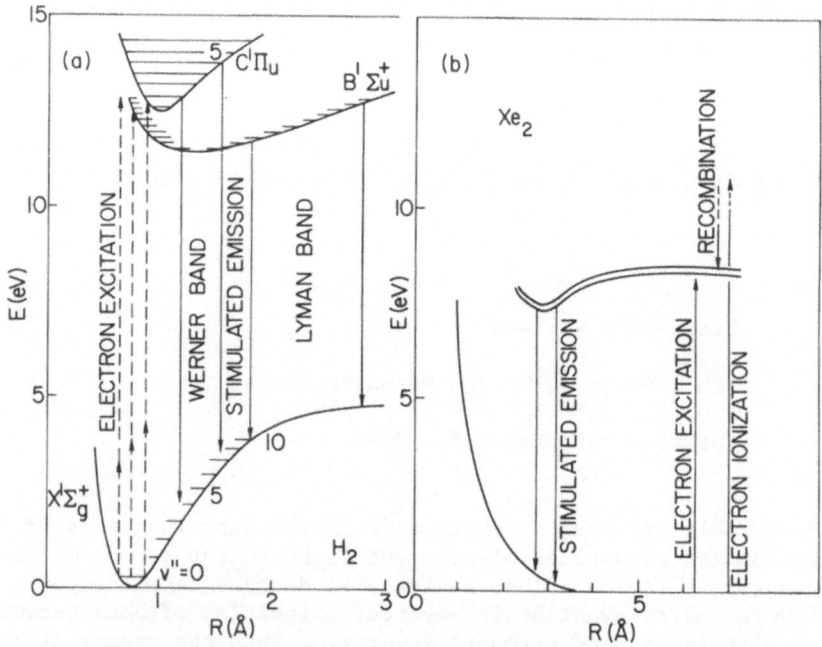

Figure 1 (a) Potential energy curves for H_2
(b) Potential energy curves for Xe_2

output is thus amplified spontaneous emission and lacks discrete
mode structure. Confirmation of laser action[1,2,4-7] comes from an
analysis of the large perturbation on individual line intensities
which results from stimulated emission, considering first the
Franck-Condon factors and then the relative line strengths of the
individual rotational transitions. For example, in H_2, the ratio
of the P(3) to R(1) lines of the 6-13 band of the $(B'\Sigma_U^+)$ state is
3/2 in spontaneous emission, but in stimulated emission the P(3)
line is $> 10^3$ times as intense as the R(1)[5]. As shown in Figure 2
the P(3) lines predominate in the Lyman band of H_2. Peak powers
of 100 kW.cm^{-2} and 5 kW.cm^{-2} in a few nanoseconds have been reported for the Lyman[6,8] and Werner[4,8] bands respectively. Full
spectral analyses have been made for H_2, D_2, HD, and para -H_2[8].
The linewidths in the hydrogen laser are all extremely narrow
(~ 0.5 cm^{-1}) because at the low pressures (20 torr) in this laser
the lines are only Doppler broadened. A detailed discussion of
the theory of the hydrogen laser may be found in reference 9.

Recently, stimulated emission on two carbon IV lines at 157.82
nm and 155.08 nm has been reported[10] in a Blumlein device[3]. This

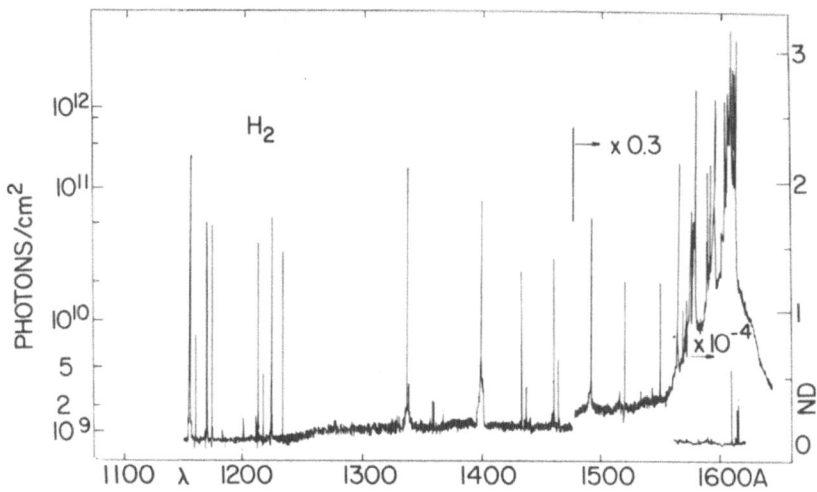

Figure 2   Laser spectrum from H_2

is the first observation of vacuum ultra-violet amplification of a resonance transition in an atomic ion. The output obtained so far is low ($\sim$ 1 kW.cm^{-2} in a 10^{-9} sec pulse), and as these lines arise from some impurity hydrocarbon in the filler gas, the physics is yet to be understood.

The xenon molecular dissociation laser at 172.9 nm is the first vacuum ultra-violet laser to use an optical cavity. The lasing transition (see Figure 1(b)) is from the lowest excited state of the excited xenon diatomic molecule (Xe$_2^*$) to the dissociative ground state. These xenon diatomic molecules are formed very rapidly and with high efficiency ($\sim$ 20%) by exciting xenon at pressures greater than 10 atm with relativistic electron beams. However in order to observe lasing action, large inversion densities must be attained because of the very broad fluorescence bandwidth and the rather long radiative lifetime[11] ($\tau \sim$100 nsec) of the principal excited state, which is produced in high pressure xenon by high energy electrons. Whether the upper laser level is the $^3\Sigma_u^+$ or $^1\Sigma_u^+$ state is presently not known. These molecular states of the rare gases are formed from the lowest neutral atomic excited states, in three body collisions with two ground state atoms. Since these excited atomic states are radiatively trapped at such high number densities, the conversion of atomic excitation into the excited molecular species is quite efficient.

Conclusive evidence[12] for laser action at 173 nm in high pressure xenon was first obtained following excitation by high energy (400J) 50 nsec pulses of relativistic electrons. The laser

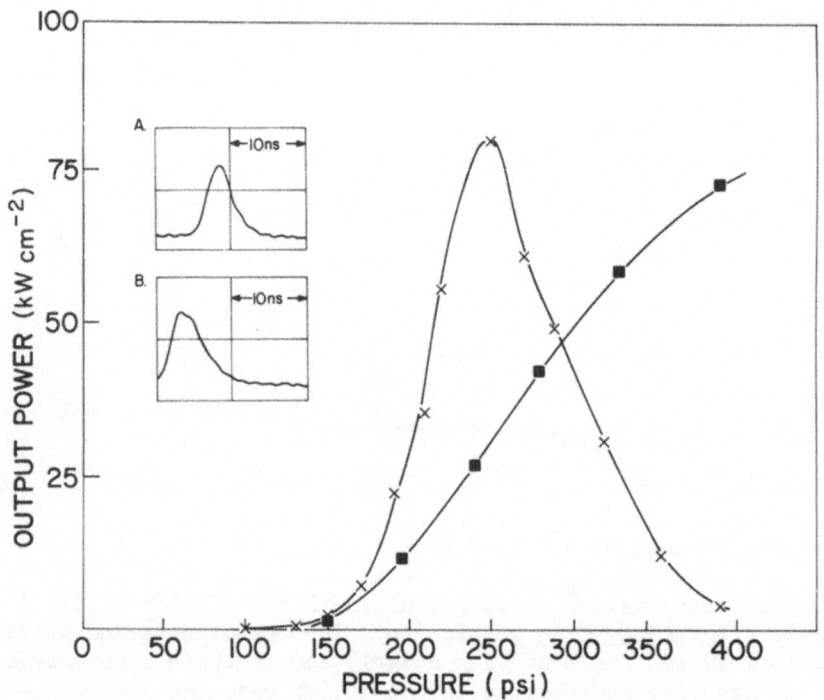

Figure 3 Pressure dependence of the output power in the
Xenon laser.

output has a very large linewidth, ~ 2 nm, as a consequence of the
breadth ~ 15 nm of the spontaneous emission from $Xe_2{}^*$. This linewidth might be used to advantage, allowing tunability over a range
of ~ 10 nm. Output powers ~ 100 kW.cm^{-2} in a 20 nsec pulse have
been reported. Subsequent experiments[13] using essentially the
same pumping scheme as above have also demonstrated laser action
in high pressure krypton at 145.7 nm and here the linewidth is of
the order of 1 nm.

The xenon laser described above is not a convenient laboratory device for two reasons: a large electron accelerator (Febetron 705[14]) is required and the aluminized mirrors forming the
optical cavity rapidly degrade as a consequence of the significant
absorptivity of Al coatings at these wavelengths. By careful optimization of the pumping efficiency, a xenon laser which is
excited by a relatively small (10J stored energy) electron accelerator (Febetron 706[14]) has now been developed[15], and produces
~ 100 kW.cm^{-2} in a 5 nsec pulse. Since these output pulses are
significantly shorter, the heating of the Al coatings due to absorption of the laser light is considerably reduced and no mirror

damage has been observed over several hundred pulses.[15]

The energy efficiency of the xenon lasers is strongly pressure dependent. As shown in Figure 3, the output power in xenon drops off very rapidly above 250 psi. Diluting xenon in another rare gas (argon or krypton), while maintaining the same number density of excited Xe_2^* enhances[13] the output power for long pulse (50 nsec) excitation, and yet has no effect[15] on the output power for short pulse (2.5 nsec) excitation. These observations are interpreted[15] to be a result of quenching of the excited state xenon molecules by ground state xenon atoms, for which a rate constant of $\sim 5 \times 10^{-13} cm^3 sec^{-1} mole^{-1}$ has been measured[11], and thus there is an upper limit to the fraction of the input beam energy which can be usefully absorbed in xenon.

A viable alternative to vacuum ultra-violet lasers for the production of coherent light in the vacuum ultra-violet is harmonic generation in atomic vapours, using currently available visible lasers. Recent work by Harris and co-workers has demonstrated that very short wavelengths[16] (88.7 nm) and good conversion efficiencies,[17] such as $\sim 3\%$ from 354.7 nm to 118.2 nm, can be achieved in atomic gases. Apart from the obvious requirement of transparency in the vacuum ultra-violet (solid state materials such as ADP and KDP have their absorption edges[18] ~ 200 nm), gases come in arbitrary lengths, have wide regions of low absorption, and are self-healing after breakdown. Note that since gases are isotropic media, only the odd order harmonics are generated.

Two phenomena are important for efficient harmonic generation, resonant enhancement of the nonlinear susceptibility and phase matching. Consider first the expression[19] for the third order nonlinear susceptibility $\chi^{(3)}$

$$\chi_{(3\omega)}^{(3)} = \frac{8\pi^3}{h^3} \sum_{a,b,c} \mu_{ga}\mu_{ab}\mu_{bc}\mu_{cg} A_{abc} \qquad (1)$$

where,

$$A_{abc} = \frac{1}{(\Omega_{ag}-3\omega)(\Omega_{bg}-2\omega)(\Omega_{cg}-\omega)} + \frac{1}{(\Omega_{ag}+\omega)(\Omega_{bg}+2\omega)(\Omega_{cg}+3\omega)} \qquad (2)$$

$$+ \frac{1}{(\Omega_{ag}+\omega)(\Omega_{bg}+2\omega)(\Omega_{cg}-\omega)} + \frac{1}{(\Omega_{ag}+\omega)(\Omega_{bg}-2\omega)(\Omega_{cg}-\omega)}$$

and $\mu_{ij} = e \langle i/z/j \rangle$. With appropriate choice of the driving frequency, ω, combined with suitable atomic energy levels, (i.e., Ω_{ij}) the possibility exists for a large resonant enhancement of $\chi^{(3)}$ by minimizing the frequency difference terms in A_{abc}. Nevertheless

even a very high value of $\chi^{(3)}$ would lead to a low efficiency of
third harmonic generation without phase-matching. This may be
seen from Equation (3)[20],

$$\frac{W_{3\omega}}{W_\omega} = \frac{5.134 \cdot 10^{-3}}{\lambda^2} L^2 N^2 [\chi_{(3\omega)}^{(3)}]^2 \left\{\frac{\sin\frac{\Delta k L}{2}}{\frac{\Delta k L}{2}}\right\}^2 \cdot (\frac{W_\omega}{A})^2 \qquad (3)$$

where W_ω and $W_{3\omega}$ are respectively the powers in the fundamental
and third harmonic beam, n_1 and n_3 are the indices of refraction
at ω and 3ω, $\Delta k = k_3 - 3k_1 = 6\pi/\lambda(n_3 - n_1)$, W_ω/A is the power density
of the fundamental beam, N is the number density, L is the length
of the medium and λ the wavelength of the driving wave. Unless
$\Delta k = 0$, the third harmonic polarization moves out of phase with
the generated wave at ω, modifying the term within the curly
bracket in (3), in which case the conversion efficiency is small.
Phase-matching is in general possible if an atomic system is chosen such that the driving frequency is below a strong resonance
line while the third harmonic lies above it; that is, the system
is anomalously dispersive at 3ω. It is then possible to add a
buffer gas which is normally dispersive at both ω and 3ω until
$n_1 = n_3$. One may anticipate that ultimate conversion efficiencies
of up to $\sim 50\%$ may be attained.[20]

In the experiments of Harris and co-workers a single, picosecond pulse from a mode-locked Nd:YAG laser propagates through
conventional nonlinear crystals to produce the second, third, or
fourth harmonic of the fundamental 1.06μ output. The resultant
pulse is then used as the driving frequency for third harmonic
generation in the atomic vapour, and in this way wavelengths of
177.3 nm[21], 118.2 nm[17], 152 nm[21] and 88.7 nm[16] have been generated
with powers ranging from ~ 1 kW up to 300 kW for 118.2 nm. It
should be noted that the properties of the fundamental laser beam
are also very important in these nonlinear processes. High efficiencies can only be obtained using a diffraction-limited, single-
mode laser.

With several nanosecond lasers as well as the harmonics now
available in picosecond pulses for vacuum ultra-violet studies,
there are many possibilities for direct measurements of dynamic
photophysical properties of atoms and molecules for the first
time. If as expected, the molecular dissociation lasers, such as
xenon or krypton become fully tunable over their linewidths, or
the harmonic generation using tunable driving frequencies is
developed, the possibilities for high resolution studies in this
region become very attractive. As the refinement of laser beam
characteristics occurs to obtain narrow linewidths, mode selection,
and low divergence, the study of nonlinear optical phenomena such
as multiphoton absorption will become feasible, thus further extending the presently accessible frequency domain. For example,

two photon absorption spectroscopy will reveal excited states resulting from even parity transitions normally forbidden in one photon spectroscopy. It may be the case however, that the production of coherent light at wavelengths less than 100 nm will be more practicable by harmonic generation than with vacuum ultraviolet lasers, primarily because of the difficulties of obtaining optical components with high transmission or reflectivity at these short wavelengths.

REFERENCES

1. R.T. Hodgson, Phys. Rev. Lett., 25, 464 (1970).

2. R.T. Hodgson and R.W. Dreyfus, Phys. Rev. Lett., 28, 536 (1972).

3. J.D. Shipman, Jr., Appl. Phys. Lett., 10, 3 (1967).

4. R.W. Waynant, Phys. Rev. Lett., 28, 533 (1972).

5. R.T. Hodgson and R.W. Dreyfus, Physics Lett., 38A, 213(1972).

6. R.W. Waynant, J.D. Shipman, R.C. Elton and A.W. Ali, Appl. Phys. Lett., 17, 383 (1970).

7. R.W. Waynant, J.D. Shipman, R.C. Elton and A.W. Ali, Proc. IEEE, 59, 679 (1971).

8. R.W. Dreyfus and R.T. Hodgson, to be published.

9. A.W. Ali and P.C. Kepple, Appl. Optics, 11, 2591 (1972).

10. R.W. Waynant, Appl. Phys. Lett., 22, 419 (1973).

11. S.C. Wallace, R.T. Hodgson and R.W. Dreyfus, Appl. Phys. Lett., 23, 22 (1973).

12. J.B. Gerardo and A. Wayne Johnson, IEEE J QE-9, 748 (1973).

13. P.W. Hoff, J.C. Swingle, and C.K. Rhodes, to be published Appl. Phys. Lett.

14. Field Emission Corporation, McMinnville, Oregon.

15. S.C. Wallace, R.T. Hodgson and R.W. Dreyfus, submitted to Appl. Phys. Lett.

16. S.E. Harris, to be published.

17. A.H. Kung, J.F. Young, and S.E. Harris, Appl. Phys. Lett., 22, 301 (1973).

SYNCHROTRON RADIATION AS A LIGHT SOURCE

James W. Taylor

Department of Chemistry
University of Wisconsin
Madison, Wisconsin 53706

INTRODUCTION

The electromagnetic radiation resulting from the high velocity centripetal acceleration of electrons is an intense continuous
spectrum covering the x-ray to the infra-red region. Theories
concerning the production of light from the synchrotron process
were placed on a sound basis between 1944-1949 [1] and experimental
observation of the light was performed in 1947.[2] Since that time,
the number of experiments involving wavelength scanning has grown
enormously.[3-13] At the present time, there are major radiation
facilities located in England, France, Germany, Italy, Japan, USSR,
and the United States. Most of these facilities were originally
constructed for the study of problems in high energy physics with
the spectroscopic radiation experiments functioning in a parasitic
mode. More recently, new facilities (primarily storage rings) are
being proposed as dedicated light sources. The reason for this
increased interest lies in the characteristics of the light as it
is produced from synchrotron radiation. These characteristics are:
1) the continuous nature of the spectrum; 2) the intensity of the
photon flux, particularly from the x-ray region to 600 Å; 3) the
polarization of the light; 4) the available range of the short time
pulses; and 5) the high vacuum from which the light originates. In
this paper the discussion will be focused on the way in which the
light is produced, the characteristics of the various machines,
expected new developments, and finally a few examples of experiments which utilize the unique characteristics of synchrotron
radiation for spectroscopic purposes.

MACHINES FOR THE PRODUCTION OF LIGHT

Basically there are two types of machines presently used by
chemists and physicists as light sources. These are the synchrotron and the storage ring. The basic synchrotron produces a pulse
of high energy electrons at a fixed repetition rate on the order
of 50-60 pulses/sec and accelerates the electrons to the desired
energy. For machines of this type, each pulse produces a related
number of photons at an experimental port dependent on the number
of electrons in the pulse. Variations between pulses do occur.
As a consequence of producing a pulse at one energy and accelerating it to a higher energy, it is necessary that the magnetic
field of the machine be reset during each cycle. This leads to a
wavelength dependent duty factor of less than one for the light
production.[3] A further consequence of the large number of electron pulses produced in the synchrotron is the necessity to eliminate a pulse before the new one is injected. A great deal of
radiation is produced from this process, and extensive shielding
of the machine is required. For example, the experimental bunker
at DESY,[6-7] the German synchrotron located at Hamburg, is located
40 meters away from the source because of the shielding requirements. To an observer of the photon radiation at one point along
the electron orbit, then, the photon flux has a repetition rate
dependent on the frequency of the synchrotron pulse, an intensity
variation depending on the number of electrons per pulse, and a
duration dependent on the electron pulse width in time. The pulse
duration depends on the particular machine producing the pulse.

A storage ring functions in somewhat the same fashion as a
synchrotron, using a device called an injector to produce an initial pulse of electrons. In this type of machine, however, a
pulse is captured, usually further accelerated to the final
desired energy, and stored for a long-time period in a periodic
structure which completes a closed loop. To the photon observer
the light flux again has pulse character, but in this case the
frequency is dependent on the revolution frequency of the electron
bunch in the ring. The light pulse duration is determined by the
design and operating characteristics of the storage ring. For
example, the PSL-1 ring at Wisconsin [12,14-15] is constructed on
1.5 meter average radius with a repetition frequency of 32 MHz.
This ring produces a light pulse every 32 nsec with a duration of
1.6 nsec. As with synchrotrons, the photon flux in a storage ring
is directly related to the number of electrons radiating. The
loss of the stored bunch, however, is slow and arises primarily
from scattering of the electrons with gases desorbed from the
chamber walls, the inevitable imperfections in perfect focusing
and timing, and the electron-electron repulsion which expands the
beam size and leads to defocusing. The loss of photon flux in a
storage ring, then, is described as a log function whose slope
generally depends on the number of electrons circulating. At ACO,

the French 550 MeV storage ring, the light flux for a 100 mA circulating current has a half-life of 18 hours.[16] At 30 mA the half-life is 35 hours. The beam lifetime also depends on the pressure. The beam currents produce gas molecules from the chamber walls; and thus, scattering becomes a dominant mechanism for electron loss. For this reason, the ring pressure is kept low, typically in the 10^{-11} to 10^{-10} torr range. Thus, the light is produced from a source operating at such a low pressure that surface and sample contamination from the source are quite small. Because the pulse once captured stays in the machine, the duty cycle is essentially unity over short time periods as compared to a synchrotron, but a time dependent compensation for the light intensity must be provided.

In Table I are listed the known synchrotrons and storage rings where experimental programs have been established to utilize the synchrotron radiation for spectroscopic purposes. Also listed are the machines which are planned or soon to be completed. This table cannot be considered complete, but the data were compiled from information listed in the described references [17,18] and from solicited responses in anticipation of this report.[19] In addition to the energy, the radius of the electron orbit is listed along with the latest known operating current and the critical wavelength, λ_c, in angstroms. These listings are the important parameters for accelerator builders, but for use of the machines as light sources, other parameters such as light flux at specific wavelengths, usable range, and light emittance at an experimental port located some distance from the machine are critical. The following section describes how the listed machine information can be used to generate the maximum and usable intensity values for the various machines.

BASIC EQUATIONS RELATING TO PHOTON ENERGY AND INTENSITY

The intensity of radiation emitted by a given machine may be estimated by the equations describing the radiation from monoenergetic electrons.[1,3-9] For storage rings, the results for a single electron are multiplied by the number of electrons circulating and a factor somewhat less than one which corrects for the straight sections in the ring where no radiation occurs. For synchrotrons the single electron results are multiplied by the average number of electrons in a pulse times the duty factor of the machine and a similar correction for the straight sections. In NINA, for example, Codling [9] evaluates the factor for the straight sections as 0.59 from the knowledge that it takes an electron 7.37×10^{-7} sec for one revolution in the machine, whereas an electron traveling at the speed of light would require only 4.35×10^{-7} sec for one revolution if the machine were completely circular with its radius described by the radius in the

TABLE I. SYNCHROTRON RADIATION VACUUM ULTRAVIOLET LIGHT SOURCES

		E (GeV)	R (m)	λ_C (Å)	J, mA	1 Å[a]	10 Å[a]	100 Å[a]	1000 Å[a]
DESY	Hamburg, GERMANY	7.5	31.7	0.42	30	9000	600	30	1.6
ACU	Erevan, USSR	6.0	24.65	0.65	22	7000	760	43	2.2
NINA	Daresbury, ENGLAND	5.0	20.8	0.93	20	470	510	36	1.8
	Bonn, GERMANY	2.3	7.65	3.5	30				
	Moscow, USSR[b]	1.36	4.0	9.9	10				
INS-SOR	Tokyo, JAPAN	1.3		15	50				
	Frascati, ITALY	1.1	3.6	23	14				
VEPP-2'	Novosibirsk, USSR	0.68	2	35	10				
	Moscow, USSR	0.68		76					
	Bonn, GERMANY	0.50	1.7						
SURF	Gaithersburg, MD, USA	0.18	0.83	795	30				
PTB	Braunschweig, GERMANY	0.14	0.46	947	5				
DORIS	Hamburg, GERMANY[b,c]	3.5	12.12	1.58	200[b]	3600	760	41	2.2
CEA	Cambridge, USA	3.5	26.0	3.38	30	200	640	39	2.1
SPEAR	Stanford, USA	2.5	12.7	4.54	250				
VEPP-3	Novosibirsk, USSR	2.0			250[b]				
	Daresbury, ENGLAND[e]	2.0[e]	5.55[e]	3.88[e]	1000[e]	290[e]	440[e]	31[e]	1.6[e]
DCI	Orsay, FRANCE[b]	1.8	3.8	3.8	250[b]	270	460	27	1.4
PSL-II	Stoughton, WI, USA	1.76[e]	4.5[e]	4.61[e]	100[e]	130[e]	380[e]	28[e]	1.5[e]
DORIS	Hamburg, GERMANY[b,c]	1.75	12.12	12.7	1000[b]	0.11	230	36	1.9
ACO	Orsay, FRANCE	0.55	1.11	30	100		18	14	0.76
	Tokyo, JAPAN[b]	0.30	1.1	228	100[b]				0.72
PSL-I	Stoughton, WI, USA	0.24	0.64	260	20			1.8	0.61
NBS	Gaithersburg, MD, USA[f]	0.18	0.83	795	10			1.1	0.32

[a] Intensity expressed in photons/Å sec mA mR x 10^{-10} with 0.5 synchrotron duty factor. Available intensity = I x J x True Duty Factor x Acceptance x Reflectivity x Bandpass x Straight Section Factor (see Text); [b] Machines under construction and scheduled for completion during 1973 or 1974; [c] DORIS is the same machine at two energies; [d] Funding and operation status uncertain; [e] New Machines proposed but not presently funded; [f] Conversion from synchrotron to storage ring planned.

bending magnets of 20.77 meters. In addition to these two factors, there is a photon energy dependent factor which is important for some synchrotrons which corrects for the variation of the energy of the electrons during the acceleration cycle.[3-4] At long wavelengths compared to the peak wavelength, this latter factor may have a value around 0.25 for some modes of machine operation. At shorter wavelengths, it may decrease to 0.1 or less.[9]

For monoenergetic electrons the several reviews [7-9] on the theory of synchrotron radiation define the critical wavelength, λ_c, in practical units as

$$\lambda_c = \frac{4\pi R}{3} \gamma^{-3} = 5.59 \ R(\text{meters}) \ [E(\text{GeV})]^{-3} \qquad (1)$$

where $\gamma = E/m_0 c^2$; $m_0 c^2$ is the rest mass energy of the electron (= 0.511 MeV); R is the orbit radius in the bending magnet; and E is the electron energy. The radiation emitted by relativistic electrons at one point of a circular orbit is confined to a narrow cone tangential to the forward direction of motion. The cone half-angle is given by $\psi = (\lambda/\lambda_c)^{1/3} \gamma^{-1}$.[20] At λ_c the value of ψ at DESY with 7.5 GeV is 0.07 mrad; at NINA at 4 GeV is 0.12 mrad; at SPEAR at 2.5 GeV is 0.20 mrad; at ACO at 0.55 GeV is 1.0 mrad; and at PSL-1 at 0.24 GeV is 2.13 mrad.

According to Schwinger,[17] the energy $P(\lambda)$ radiated per second by a monoenergetic electron in a bandwidth $\Delta\lambda$ and integrated over all horizontal and vertical angles is given by

$$P(\lambda) = 0.0987 \ \frac{ce^2}{R^3} \ (\lambda)^7 \ G(y)\Delta\lambda \qquad (2)$$

where numerical values have been inserted for some of the constants.

In this expression the R is assumed to be the radius of the electron orbit in essentially a circular machine, and the function $G(y)$ can be evaluated numerically.[4,7-9,20-21] Suller [22] suggests that a more convenient expression would be

$$N(\lambda)d\lambda = \frac{3^{5/2} \ 10^{-33}}{16\pi^2} \ \frac{e}{hR^2} \ (\gamma)\lambda G(y) \ d\lambda \qquad (3)$$

where $N(\lambda)d\lambda$ is the number of photons per second at a wavelength λ in a bandwidth $d\lambda$ emitted into a horizontal acceptance angle of one milliradian for a circulating current of one milliamp. In this expression e is the electronic charge in coulombs, h is Planck's constant in joule seconds, λ is the wavelength in angstroms, c is the velocity of light in meters/sec, R is the orbit radius in the bending magnet in meters, and $y = \lambda_c/\lambda$. Converting to practical units gives $N(\lambda)$ in photons/Å sec mR mA as

$$N(\lambda) = 2.46 \times 10^{13} \; (\lambda/\lambda_c)^2 \; EG(y)\lambda^{-1} \qquad (4)$$

If the desired intensity is to be expressed in units of interest to those working in the photoelectron spectroscopy area, $N(\lambda)$ can be multiplied by $\lambda^2/12{,}398.5$ to give $N(\lambda)$ in photons/eV sec mR. The value of $G(y)$ as a function of the argument λ/λ_c may be listed [23] as follows

TABLE II.   Values of $G(y)$ as a Function of λ/λ_c

λ/λ_c	$G(y)$	λ/λ_c	$G(y)$
5.0-02[a]	8.5-04[a]	1.00	6.51-01
1.00-01	1.92-02	5.00	3.61-02
1.43-01	1.62-01	10.0	8.18-03
2.00-01	5.32-01	50.0	2.19-04
3.33-01	1.16	100	4.45-05
4.00-01	1.24	500	1.07-06
5.00-01	1.21	1000	2.14-07

[a]Graphical extrapolation value.   (5.0-02 also reads as 5.0×10^{-2}.)

By using equation 4 and the numerical solution for $G(y)$, or graphical interpolation for values not listed in Table II, the photon flux in terms of a beam current of one milliampere can be calculated. In order to evaluate operating machines with larger currents, the average circulating current, J, must be evaluated from n, the number of electrons present at a given time in the curved section of the machine. For this case, $n = 2\pi RJ/ec$.[8] The values listed in Table I are for J, and a straight section correction is *not* required.

     If this is done for a number of the light sources listed in Table I, assuming the operating data listed therein, assuming a duty cycle of 0.5 for the synchrotrons, using the $G(y)$ values of Table II, and making no other assumptions, the intensities for 1 Å, 10 Å, 100 Å, and 1000 Å can be determined on a per Å mA mR basis. The actual intensities observed at a given experimental port must take into account the various fractions listed in footnote a of Table I. For example, at 100 Å the DESY synchrotron at 30 ma is calculated to produce 9.0×10^{12} photons/Å sec mR whereas PSL-1 at 20 ma is calculated to produce 2.2×10^{11} photons/Å sec mR. At DESY, focusing optics with a 2 cm aperture located at a 40 meter

distance from the electron orbit can accept approximately 0.5 mR.
At PSL-1, because of the lower shielding requirements on the storage ring, it is possible to locate a 2 cm aperture within one
meter, giving 20 mrad of beam acceptance. The available intensities under these acceptance conditions are 4.5×10^{12} photons/Å sec
at DESY and 4.4×10^{12} photons/Å sec at PSL-1. (The limiting
aperture at PSL-1, in fact, actually turns out to give a maximum
of 70 mR. DESY is not limited to a 2 cm aperture either.) At
shorter wavelengths, however, the higher energy machines do produce
greater flux. The problem in using this radiation becomes one of
matching the emittance of the source with the acceptance of the
optics. The focusing of synchrotron radiation into an angle of
$\simeq \gamma^{-1}$ about the synchrotron plane causes its intensity to decrease
with D^{-1} (the inverse of distance, D) rather than the D^{-2} expected
for a normal point source. The necessary matching of optics with
monochromators has been discussed with reference to specific
machines,[24-27] and Kunz has given a comparative review of the
various monochromators adapted for synchrotron radiation.[28]

THE POLARIZATION OF SYNCHROTRON RADIATION

It has been shown that synchrotron radiation is elliptically
polarized and that the angular distribution components can be calculated.[5,9,29] That with the electric vector parallel to the
electron orbit is labeled $I_{||}$ and that component perpendicular, $I_{\perp}$.
The result of that calculation shows that $I_{\perp}$ goes to zero as the
observation angle relative to the plane of the orbit goes to zero.
As a result, the polarization, P, [defined as $= (I_{||}-I_{\perp})/(I_{||}+I_{\perp})$]
goes to 100% when the radiation is viewed directly on the orbital
plane. There is no simple relationship to describe the variation
in polarization with vertical observation angle except to note
that it is related to ψ, the cone half angle, and that ψ can be
calculated from $(\lambda/\lambda_c)^{1/3}\gamma^{-1}$. In general, as long as the observation angle does not exceed 0.3 ψ, the polarization is better than
80%. At angles of 0.5 ψ the polarization drops to 60-80% and at
0.7 ψ the value is in the range of 40-60%. Describing the polarization in this general way is helpful because it points out that the
polarization is *greater* at longer wavelengths for the same observation angle because ψ is larger at longer wavelengths. For example,
Codling shows that with a fixed 0.1 mR observation angle the calculated polarizations for 1, 10, 100 and 1000 Å radiation from NINA
at 4 GeV are 60, 75, 90 and 98% respectively.[9] Similar results
were previously reported for the NBS machine [5] and DESY.[6]

POSSIBLE MACHINE IMPROVEMENTS

Improvements have occurred over the past few yaars to make
the machines more amenable for spectroscopic studies and the new

facilities being built or proposed have incorporated them. Because
the new machines of spectroscopic interest are all storage rings,
the comments which follow are concerned with their operation.

    One problem which causes some difficulty in using a continuous
source is that of order overlap. With intense fluxes at 10 Å, it
is difficult to operate at 100 Å and exclude all second-order radiation at 50 Å, third-order at 33 Å, fourth-order at 25 Å, *etc.*
Reflectivity losses in optics operated at larger than grazing
angles are an aid [7,9-10,30-31] as are the various thin metal and
organic film window.materials.[7-8,10,30-34] With the machine
energy as a variable, however, it is possible to change the intensity distribution sufficiently to observe the effects of some of
the higher order contributions. This can be easily seen with the
two energy listings for DORIS in Table I. At 1 Å we find that the
flux drops from 3600 to 0.11 x 10^{10} photons/Å sec mA mR; at 10
Å the flux decreases from 760 to 230 x 10^{10}; whereas at 100 Å the
decrease is from 2.2 to 1.9 x 10^{10}. At longer wavelengths with
the high energy machines, the changes are not as dramatic but may
be calculated and thereby provide a valid technique for compensation. Figure 1 shows the variation in intensity from a lower

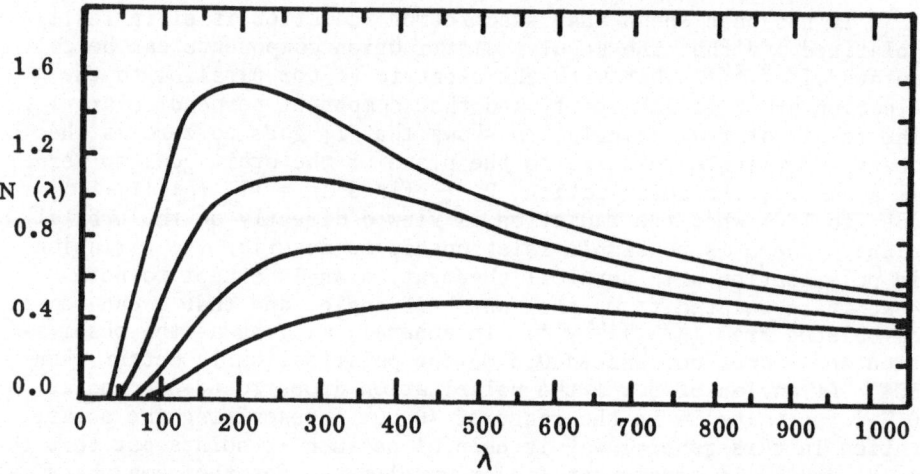

Fig. 1. Intensity distribution plotted as photons x 10^{-10}/Å sec
mR mA for PSL-1 as a function of wavelength in Å and energy in
GeV. Top curve, 0.240 GeV; second curve, 0.220 GeV; third curve,
0.200 GeV; bottom curve, 0.180 GeV.

energy storage ring, PSL-1, as a function of electron energy and
also shows the continuum nature of the emitted light.

    It is evident from the equations that the flux from the rings is directly proportional to the average circulating current in the ring. Because of various machine parameters, there are limits to the number of electrons that can be injected in any one given pulse. Multiple injection techniques are possible, however, with each new injection effectively adding to the previously stored bunch. The major penalty for this mode of operation is the decreased beam half-lifetime mentioned previously.

For certain experiments, to be described later, it is necessary to view the radiation as a pulse of fixed duration followed by a dark time. These two time periods are fixed by the geometry of the ring and the accelerating frequency. Although the positional beam stability in the rings can be held to better than 20 microns, it is possible to introduce periodic motion of the electron bunch with reference to an observation aperture. In this way the effective dark periods may be increased. Unfortunately, creating beam motion for one observation port in this manner causes the same effective flux loss around the periodic structure. Only by building deflection elements in the ring itself can local conditions be altered at one point.

One technique of making local perturbations on certain sections of the ring has been the incorporation of so-called wiggler magnets [7] to produce a shift in the peak wavelength.[35] The operation of this type of device can simply be described by reference to the three magnet array constructed for operation at PSL-1 as shown schematically in Figure 2.[35] The first and third

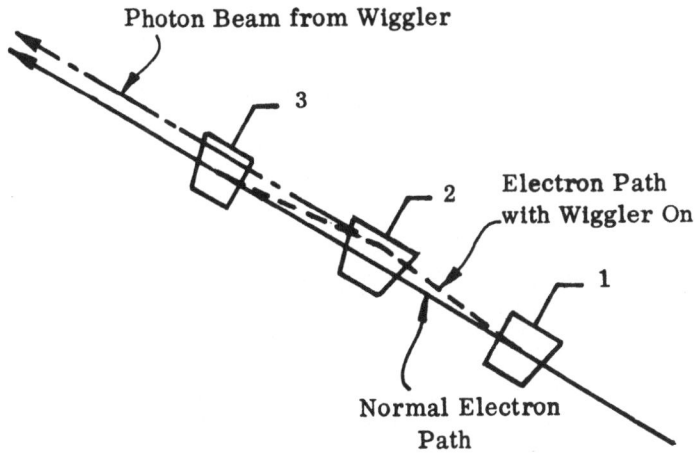

Fig. 2. Schematic of wiggler magnet system installed at PSL-1 consisting of two identical magnets #1 and #3 and one high field magnet #2. The photon beam of interest originates from magnet #2.

magnets, with 12.5 kG peak field, produce a 4° bend and deflect the electron bunch one cm radially away from the ring center. The second magnet in the array, with 25 kG peak field, turns the beam through 8°. In this way the array returns the beam to its original orbit position, a necessary condition for stable operation, but effectively presents the user with a port where the bending radius is determined by the second magnet. Because λ_c is directly proportional (Equation 1) to the radius in the magnet, the peak in the spectrum is shifted to shorter wavelengths with a corresponding increase in photon flux as given by Equation 4. Figure 3 illus-

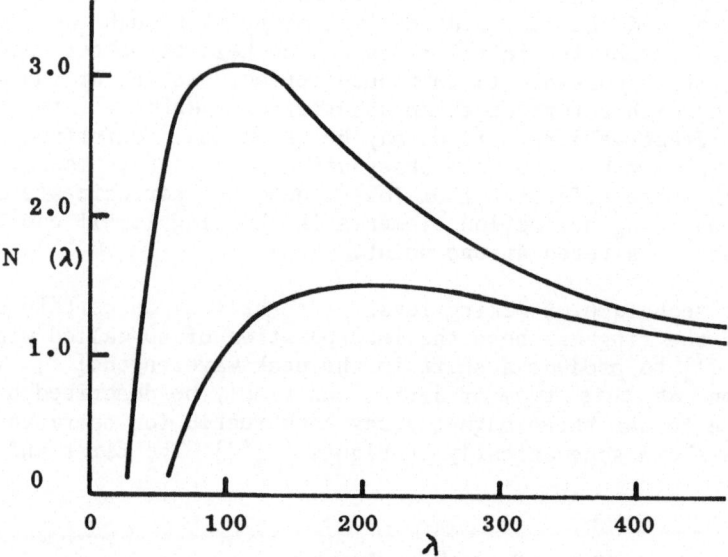

Fig. 3. Intensity distribution in photons x 10^{-10}/Å sec mR mA plotted as a function of wavelength in Å. Ring energy is 0.240 GeV. The top curve is from #2 magnet; the bottom, normal magnet.

trates the normal distribution and that expected for the wiggler. [21] Further extensions of this idea could permit energy selection or scanning of the peak maximum by control of the field current. It would also be possible to use superconducting magnets to decrease R to much smaller values than are possible with conventional magnets giving a small orbit bending radius and intense short wavelength photons. The use of multiple-pole wiggler magnet sections (which effectively increase the total radiated power per mR by the factor n R_W/R, where R_W is the wiggler magnet radius and R is that of the main ring system,[36]) would be another way to increase the available flux at a particular port without altering the basic energy of the machine or disturbing other experiments at longer wavelengths.

SOME EXPERIMENTS UTILIZING THE PULSE AND POLARIZATION
CHARACTERISTICS OF SYNCHROTRON RADIATION

Of all of the unique characteristics of synchrotron radiation,
the fact that the source produces a pure continuum is perhaps the
most important spectroscopically. Energy and bandpass selection
are presently limited by the dispersion optics and intensity. High
resolution studies of absorption, reflection, transmission, photoionization, fluorescence, photoelectron spectroscopy, photon
desorption, ESCA, x-ray microscopy, and medical diagnostics are all
possible and have been accomplished at various installations. The
applications are too broad to be covered in this review but leading references are available.[10-13] There are two characteristics, however, which merit further comment because their potential
for applications cannot presently be duplicated by other laboratory
sources. These are the pulse character and the polarization.

At the Orsay storage ring, ACO, the form of the light pulse
was determined, by a single photon counting technique, to be perfectly gaussian with a 1.4 nsec half-width and about a 0.5 nsec
decay time. Using this pulse, the fluorescence decay characteristics of 2-naphthol and fluorescein were determined. In this paper
[37] the ACO pulse was also compared with other nanosecond light
sources. These authors found that the ACO beam had a smaller FWHM
pulse than any of the N_2, air, D_2 and H_2 nanosecond discharge
sources and did not produce post-luminescence phenomena. The
pulse intensity below 2000 Å was also superior to any of these
discharge sources. For studies of this type, the r.f. acceleration time determines the pulse duration, and this time is maintained precise to one part in 10^6 to 10^8 by the master oscillator.
Very precise studies of lifetimes become possible, limited primarily by the detection electronics. Because the gate pulse can be
triggered before the light pulse reaches the experiment window
(trigger positioned a known distance along the electron orbit
ahead of the experimental port) compensation for electronics dead-
time can be provided and accurate delay times measured.

The use of the polarization of the light in conjunction with
photoelectron measurements has been employed with studies on the
rare gases at the Glascow 300 MeV synchrotron [38-39] and on
benzene at PSL-1.[40] These latter studies made use of the relationship derived by Samson [41-42] for describing the differential
cross-section as a function of the light polarization and the
assymmetry parameter, β, given by

$$\frac{d\sigma}{d\Omega} \propto 1 - \frac{B}{2} + \frac{3\beta}{2(g+1)} \left[g\cos^2 A + \cos^2 C \right] \qquad (5)$$

In this expression $g = I_x/I_z$, where I_x and I_z are the components of
the incident light in the x and z axes, respectively. The angles

A and C are defined in Figure 4. By collecting the electrons in a

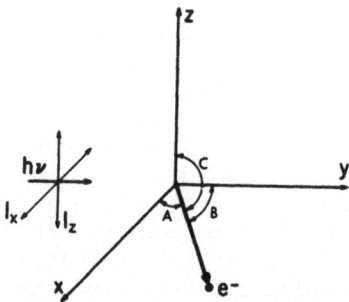

Fig. 4. Ejection of an electron by a photon, traveling along the
y-axis, polarized along I_x and I_z. The electron leaves the origin
in the direction indicated.

direction such that A = B = C = 54°44', the observed intensity data
are independent of the angular distribution and the degree of polarization.

    At other observation angles, β can be obtained if the polari-
zation is known. In addition, it is possible to tune the light
source energy to remove any ambiguity in the photoelectron intensity data arising from autoionization and to study autoionization
phenomena by photoelectron spectroscopy. This has been done with
O_2 using synchrotron radiation [43] and compared to other studies
using laboratory sources.[44] Price [45] has advocated the study
of the energy dependence of β and the orbital giving rise to the
photoejected electron. These studies become possible over a wide
wavelength range using the continuum synchrotron radiation.
Detailed studies of this type and of those with rare gases also
permit tests of the validity of the various calculations and
theories describing the photoelectron process.

ACKNOWLEDGMENT

    A number of colleagues responded to my request for the latest
information on facility operation and experiments in preparation
of this paper. I wish to thank E. Rowe, C. Pruett, D. Dexter and
W. Trzeciak of PSL-1; K. Codling, University of Reading; Y. Farge
and R. Lopez-Delgado, ACO; V. Mikhailin, Moscow State University;
Ian Munro, Daresbury; S. Doniach, SPEAR; W. Paul, Harvard; and
T. Sasaki, University of Tokyo. I wish to thank F. Mills,

Brookhaven National Laboratory and F. Brown, now at Xerox in Stanford, and E. Rowe and C. Pruett, Wisconsin, for their early encouragement into this research area. During the summer of 1972 visits to the University of Reading, ACO, and DESY were made possible through a grant by the Wisconsin Alumni Research Foundation and the hospitality of my hosts K. Codling, P. Guyon and R. Haensel. The research presently in progress is supported by the National Science Foundation under grant GP-36236X and was previously supported by the Air Force Office of Scientific Research under AFOSR-69-1725. The storage ring, PSL-1, is operated under contract to the Air Force Office of Scientific Research under Contract F44620-70-C-0029, and this support is gratefully acknowledged.

REFERENCES

1. J. Schwinger, *Phys. Rev. 70*:798, 1946; *Phys. Rev. 75*:1912, 1949.

2. F. R. Elder, A. M. Gurewitsch, R. V. Langmuir, and H. C. Pollock, *Phys. Rev. 71*:829, 1947.

3. D. H. Tomboulian and P. L. Hartman, *Phys. Rev. 102*:1423, 1956.

4. D. H. Tomboulian and D. E. Bedo, *J. Appl. Phys. 29*:804, 1958.

5. K. Codling and R. P. Madden, *J. Appl. Phys. 36*:380, 1965.

6. R. Haensel and C. Kunz, *Z. Angew. Phys. 23*:276, 1967.

7. R. P. Godwin in "Springer Tracts in Modern Physics," *51*, Springer-Verlag, Berlin, 1969.

8. C. Gähwiller, F. C. Brown, and H. Fujita, *Rev. Sci. Instrum. 41*:1275, 1970.

9. K. Codling, *Rep. Prog. Phys. 36*:541, 1973.

10. G. V. Marr, I. H. Munro, and J. C. C. Sharp, "Synchrotron Radiation: A Bibliography," DNPL/R24 Daresbury Nuclear Physics Laboratory, Daresbury, Nr. Warrington, Lancashire, 1972.

11. "Research Applications of Synchrotron Radiation," R. E. Watson and M. L. Perlman, Eds., BNL 50381, Brookhaven National Laboratory, Upton, New York, 1972.

12. "Notes on the Annual Synchrotron Users Conference," E. Rowe, Ed., Physical Sciences Laboratory, Stoughton, WI, 1968 to 1972.

13. "International Symposium for Synchrotron Radiation Users," G. V. Marr and I. H. Munro, Eds., DNPL/R26 Daresbury Nuclear Physics Laboratory, Daresbury, Nr. Warrington, Lancashire, 1973.

14. C. H. Pruett, R. A. Otte, E. M. Rowe, and J. D. Steben, *Bull. Amer. Phys. Soc. Series II 14, No. 1:*17, 1969.

15. E. M. Rowe, R. A. Otte, C. H. Pruett, and J. D. Steben, *IEEE Trans. Nucl. Sci. NS-16, No. 3:* Part I, 1969.

16. P. Dehez, J. Durup, Y. Farge, P. M. Guyon, P. Jaeglé, S. Leach, R. Lopez-Delgado, P. Martin and C. Vermeil in reference 13, p. 37.

17. R. Haensel in reference 13, p. 3.

18. E. M. Rowe, private communication and report on Third All Union Conference of Particle Accelerators, Moscow, USSR, October, 1972.

19. See Acknowledgments for other sources.

20. F. C. Brown, P. L. Hartman, P. G. Kruger, B. Lax, R. A. Smith, and G. H. Vineyard, "Synchrotron Radiation as a Source for the Spectroscopy of Solids," NRC Solid State Panel subcommittee report, March, 1966.

21. E. M. Rowe in reference 11, p. 1.

22. V. P. Suller, private communication and Science Research Council internal report, SRS/NS/7314, Daresbury Nuclear Physics Laboratory.

23. Values taken from computer output of $G(y)$ function courtesy J. A. Stevenson and H. Ellis, Georgia Institute of Technology.

24. C. Kunz, R. Haensel, and B. Sonntag, *J. Opt. Soc. Amer. 58:* 1415, 1968.

25. K. P. Miyake, R. Kato, and H. Yamashita, *Sci. of Light 18:*39, 1969.

26. C. H. Pruett, N. C. Lien and J. D. Steben, *Third Int. Conf. Vacuum Ultraviolet Radiation Physics,* Tokyo, 31a A2-5, 1971.

27. R. P. Madden and D. L. Ederer, *J. Opt. Soc. Amer. 62:*722, 1972.

28. C. Kunz in reference 13, p. 68.

29. K. C. Westfold, *Astrophys. J. 130:*241, 1959.

30. J. A. R. Samson, "Techniques in Vacuum Ultraviolet Spectroscopy," John Wiley and Sons, New York, N.Y., 1967.

31. A. N. Zaidel and F. Ya. Shreider, "Vacuum Ultraviolet Spectroscopy," Ann Arbor-Humphrey Science Pub., Ann Arbor, Mich., 1970.

32. W. C. Walker, O. P. Rustgi, and G. L. Weissler, *J. Opt. Soc. Amer. 49:*471, 1959.

33. R. Gordon, Jr., R. E. Rebbert, and P. Ausloos, *Nat. Bur. Stand. (U.S.) Tech. Note 496,* 1969.

34. J. A. Kinsinger, W. L. Stebbings, R. A. Valenzi, and J. W. Taylor, *Anal. Chem.* 44:773, 1972.

35. W. Trzeciak, *IEEE Trans. Nucl. Sci. NS-18, No. 3*:213, 1971.

36. H. Winick, private communication and internal report transmitted through the courtesy of W. Paul, Harvard University.

37. L. Lindqvist, R. Lopez-Delgado, M. M. Martin, and A. Tramer, in reference 13, p. 257.

38. P. Mitchell and K. Codling, *Phys. Lett. 38A*:31, 1972.

39. M. J. Lynch, A. B. Gardner and K. Codling, *ibid. 40A*:349, 1972; *43A*:213, 1973.

40. J. A. Kinsinger and J. W. Taylor, *Int. J. Mass Spectrom. Ion Phys. 10*:445, 1973.

41. J. A. R. Samson, *J. Opt. Soc. Amer. 59*:356, 1969.

42. J. A. R. Samson, *Phil. Trans. Roy. Soc. London A 268*:141, 1970.

43. J. A. Kinsinger and J. W. Taylor, *Int. J. Mass Spectrom. Ion Phys. 11*:461, 1973.

44. J. L. Bahr, A. F. Blake, J. H. Carver, J. L. Gardner, and V. Kumar, *J. Quant. Spectrosc. Radiat. Transfer 11*:1839, 1971; *11*:1853, 1971.

45. W. C. Price, A. W. Potts, and D. G. Streets, in D. A. Shirley, Ed., "Electron Spectroscopy," North-Holland Publ. Co., Amsterdam, 1972, p. 187.

OPTICAL STUDIES OF MOLECULAR CRYSTALS IN THE VACUUM ULTRAVIOLET
USING SYNCHROTRON RADIATION

E.E. Koch

Sektion Physik der Universität München,
8 München 40

I. INTRODUCTION

Several experimental problems encountered in studies of the
optical properties of gases and solids in the vacuum ultraviolet
(VUV) and soft X-ray region were essentially overcome during the
last few years. This was due to the development of new vacuum
spectrometers, ultrahigh vacuum sample chambers, new detectors
and improvements on light sources. The utilization of synchrotron
radiation[1] as an intense, highly polarized continous light source
was one of the major breakthroughs. Since 1953, when synchrotron
radiation was used for the first time for optical experiments in
the VUV by Hartman and Tomboulian[2], spectroscopic work using this
unique light source has matured in many ways. The techniques and
results obtained have been summarized in a number of review articles[3-9], which reflect the extend of recent activities as well as
the various lines along which research is carried out at the different synchrotron radiation laboratories.

In the following, the emphasis will be placed on those experiments which have been performed during the last three years at
the Deutsches Elektronen-Synchrotron DESY on molecular crystals
in the excitonic region of the spectrum. Experimental results for
molecular crystals formed by rare gases, atmospheric gases and
simple hydrocarbon molecules will be discussed.

Since, in molecular crystals, the covalent bonding within
the molecule is strong in comparison with the van der Waals binding between the molecules, electronic excitation lines of an individual molecule will appear in the optical results from the
crystalline solid as excitons, often shifted but little in energy.

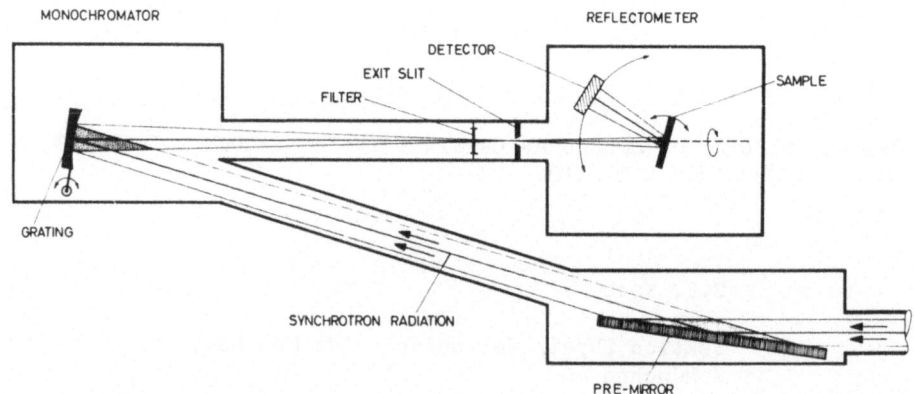

Fig. 1 Sketch of the experimental setup cut perpendicular to the synchrotron plane

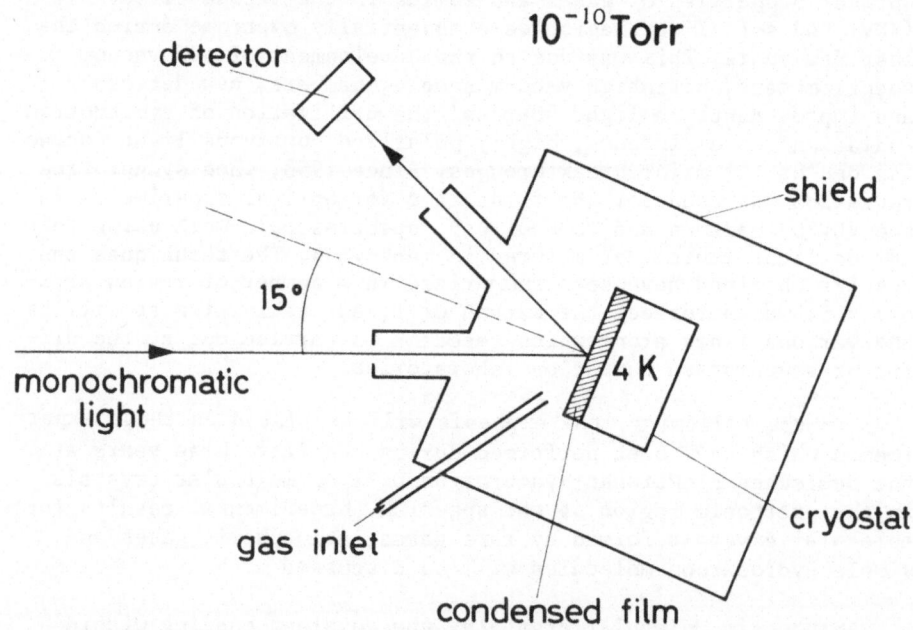

Fig. 2 Sketch of the reflection experiment on condensed films of molecular crystals

Comparison between results from the vapour and solid phase therefore gives valuable information about the properties of these excitons. In single crystals containing more than one molecule per
unit cell the crystal field splitting (Davydov-splitting) and
also the influence of longitudinal macroscopic fields can be
studied. Only recently have studies of secondary processes such
as fluorescence and photoemission using synchrotron radiation as
the primary VUV-radiation source been undertaken.

II. EXPERIMENTAL ASPECTS

As the large oscillator strength of singlet excitons makes
absorption measurements in the VUV for most molecular crystals
impossible, investigations of them has to be done by reflection
spectroscopy. Using synchrotron radiation, which makes it possible to cover a large range of frequencies, a fact important for
subsequent Kramer-Kronig analysis, this technique has been shown
to be a very useful tool. Figures 1 and 2 are sketches of the
monochromator and experimental setup presently used for such
studies at DESY. The reflectometer or cryostat and multiplier are
mounted'into an ultrahigh vacuum system attached to the exit slit
of the monochromator with a vertical dispersion plane especially
designed for use with synchrotron radiation[10],[11]. The best resolution obtained with this instrument, which uses the size of the
electron beam within the synchrotron as an entrance slit, is
0.5 Å. In future experiments, improved resolution will permit more
accurate determinations of line shapes and vibrational structure.
Further, experiments using internal modulation techniques such as
thermoreflectance or electroreflectance are in preparation. Thus
detailed information about exciton phonon coupling may be obtained. The high degree of polarization of the synchrotron light
makes it an excellent source for spectroscopic measurements on
anisotropic single crystals. Details of the evaluation of optical
constants from reflection spectroscopy with polarized light on
monoclinic crystals, which form the largest group of organic molecular crystals are given in Ref. 12.

The high intensity of synchrotron light in the VUV allows
the study of secondary processes such as photoemission[13] and the
fluorescence radiation from molecular crystals excited with primary VUV photons[14]. This promising field, the results from which
will contribute to a better understanding of dynamical processes,
e.g., exciton decay via various channels and energy transfer processes[15], will benefit from the availability of high current
storage rings. Presently, at DESY, a photon flux behind the exit
slit of a normal incidence monochromator (one grazing incidence
reflection at the premirror, at 4.5 GeV, 30 mA)[16] of 3×10^9
Photons/Å sec is obtained at 600 Å $\hat{=}$ 20.66 eV. This photon flux
will be increased by a factor of up to 100 when the high current

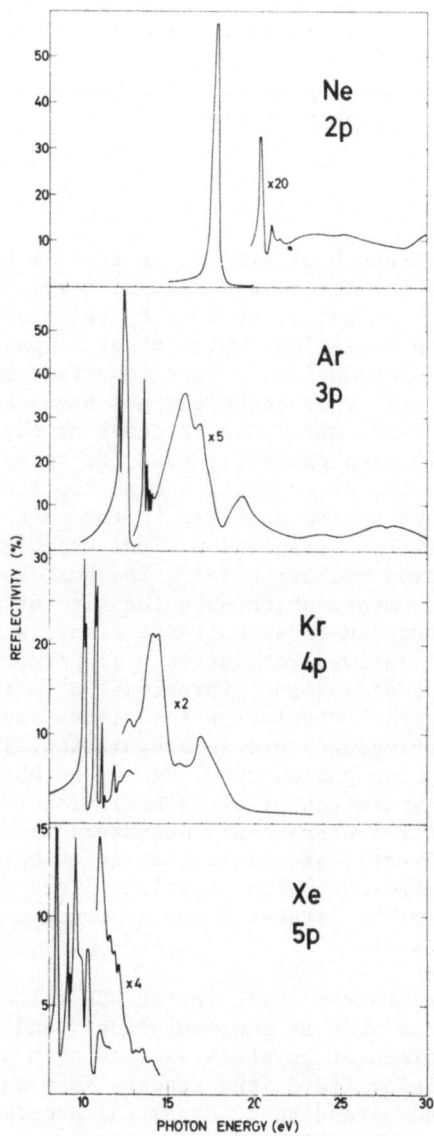

Fig. 3 Reflectance of solid Ne, Ar, Kr and Xe for an angle of incidence of 15°

storage ring DORIS, which is presently under construction will be in operation.

III. EXCITONIC EXCITATIONS IN SOLID RARE GASES

The absorption edge of solid rare gases, often considered to be the most simple molecular crystals, is characterized by a number of narrow lines followed by broader continuum-like structures at higher energies (Fig. 3). Comparison is made in Fig. 4 between the positions of resonance lines in the gas due to the excitation of those p-electrons forming the filled outer shells of the rare gas atoms, with the maxima in ϵ_2, the imaginary part of the dielectric function, for the solid. The latter was deduced via Kramers-Kronig analysis from the reflectance data[17-19]. Such an analysis is also necessary for a discussion of the line shapes. The excitation of roughness induced surface plasma waves may influence the high reflectivity of a sample in the reflectance band severely causing a dip or shoulder. The interpretation of these lines in the case of

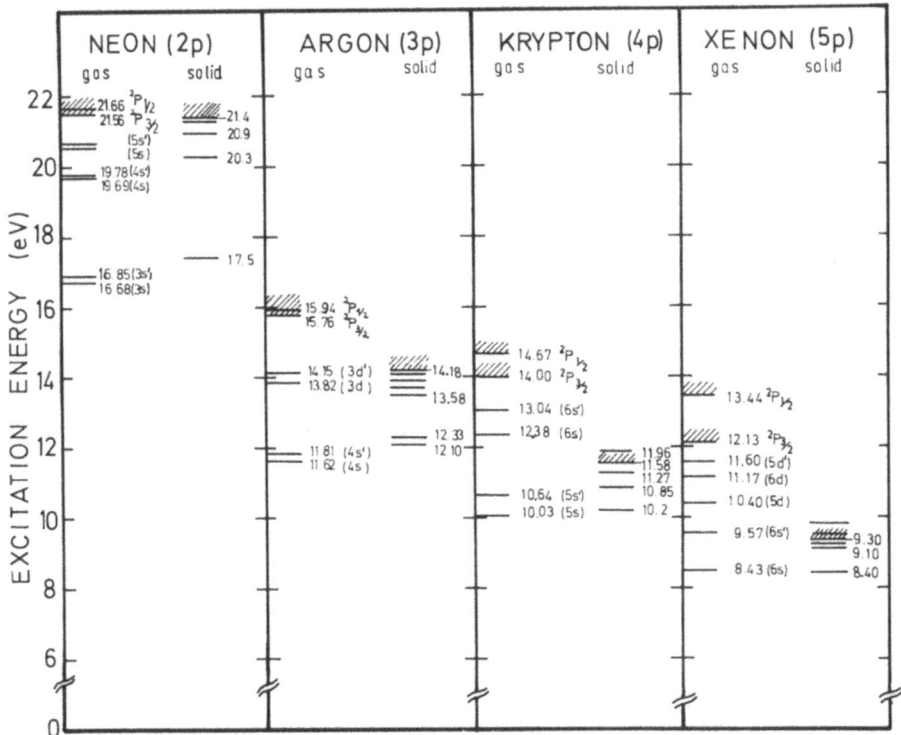

Fig. 4   Excitation energies of rare gas atoms and solids for the onset of outer p-shell electron transitions

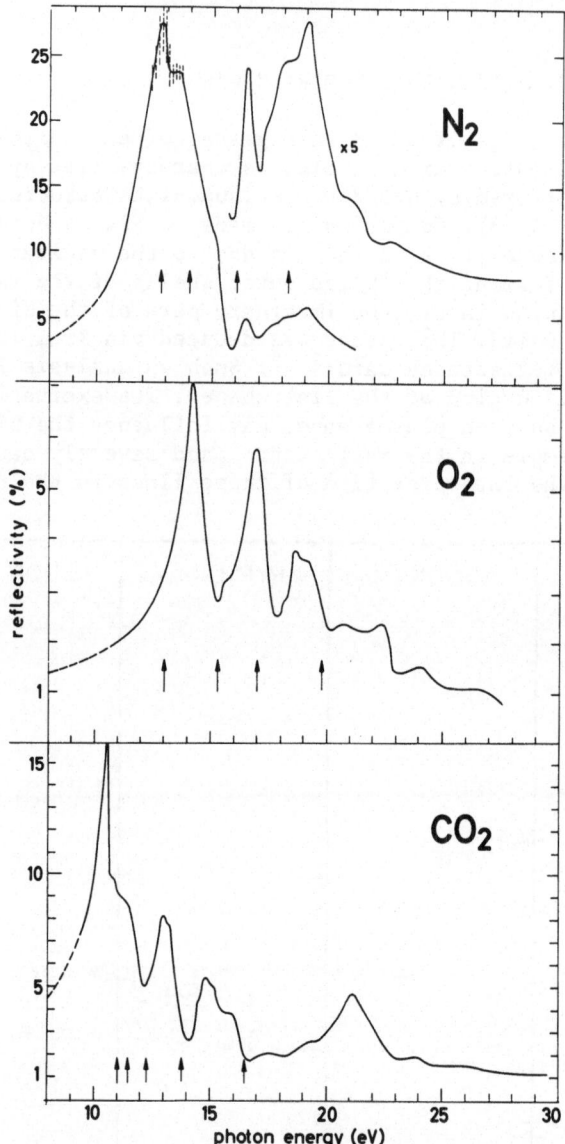

Fig. 5 Reflectance of solid N_2, O_2 and CO_2 at an angle of
        incidence of 15°. Arrows indicate the positions of
        the main absorption bands in the gaseous phase.

argon, krypton, and xenon in terms of Wannier excitons by Baldini[20] has been subsequently refined. For instance, in a recent analysis by Rössler and Schütz[21], in which the electron hole interaction was considered as a perturbation of the one-electron band model, the calculated binding energies, oscillator strengths and relative hydrogenic defects compare favorably with data deduced from experiment. Photoemission yield spectra from solid argon, krypton, and xenon[13] in the whole range of p-valence shell electron excitation show weak emission below the band gap caused by the decay of excitons. This range is followed by a steep increase of the photoemission yield to ~ 0.5 electrons per incidence photon.

For a recent discussion of the interband transitions as well as the inner shell excitations in solid rare gases[22] and solid rare gas mixtures[23], see, for instance, the paper by Kunz and Mickish[24].

IV. ATMOSPHERIC MOLECULES

Discrete structure in the absorption continua of N_2 and O_2 was found several years ago by Codling and Madden[25] using synchrotron radiation. In contrast to the gaseous phase only broad bands are generally observed in the reflection spectra of solid N_2, O_2[26] and CO_2[27] (Fig. 5) in the range 10 to 30 eV. There is no obvious correlation between the bands in the solid and the gas and in order to establish a rough correlation, a considerable chemical shift due to the crystal potential has to be assumed. There is one surprising exception in solid N_2, where one observes a progression of sharp reflectance bands in the solid at about 13 eV (Fig. 6). For this $b^1\Pi_u$ transition, 14 bands with a vibrational spacing of 0.1 eV are observed. Compared to the gas, they are no longer perturbed by Rydberg bands[28]. The solid bars in Fig. 6 indicate the positions and strengths of the corresponding bands in the gaseous phase[29]. The vibrational structure is more regular in the solid as regards the spacing as well as the intensity distribution. No such sharp vibrational structure could be observed in solid O_2 and CO_2. Since these molecular crystals are more complicated than the solid rare gases, comparativelly little theoretical work has been done in order to understand their electronic properties.

V. ORGANIC MOLECULAR CRYSTALS

In the photon energy range from 5 to 35 eV, the excitations of all but the C 1s-core electrons of organic materials can be studied. An example of a homolog series of molecules are the n-alkanes[10], which are pure σ-electron systems. The first pronounced reflectance maxima for solid methane and ethane (Fig. 7) are not much shifted from the absorption maxima for the gas phase[30].

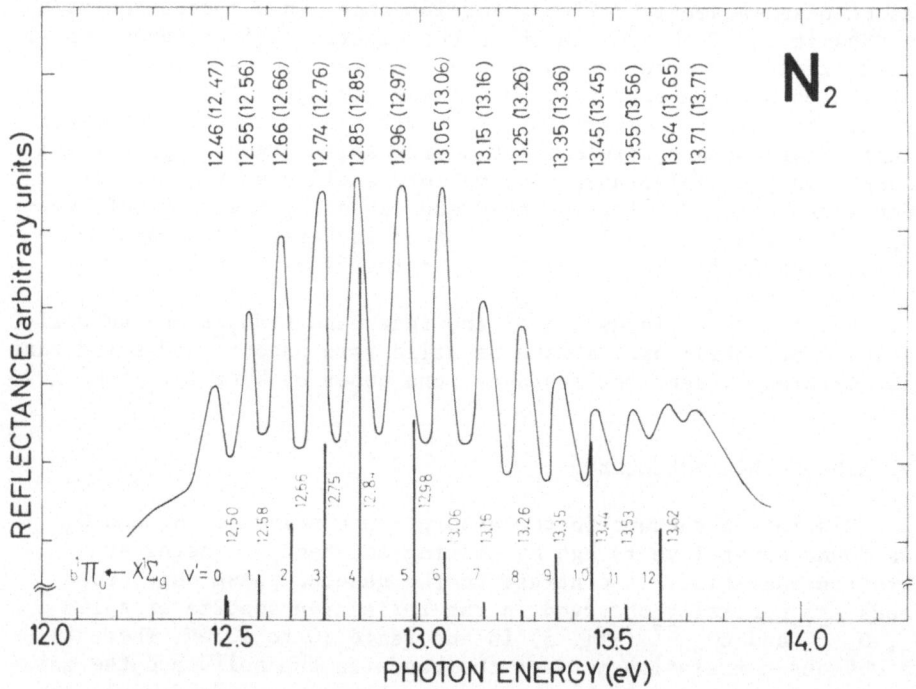

Fig. 6   Reflectance of solid N$_2$ at 10° K for an angle of incidence
of 15°, showing the b$^1\Pi_u$ vibrational levels. The values
given by Buxton and Duley[28] are added in brackets. Position
and strengths of vapour bands[29] are indicated as vertical
bars.

For solid ethane, the vibrational levels observed in the vapour are
smeared out. The fact that, also at higher energies, the main solid
reflectance maxima coincide with the principal absorption peaks,
indicates that the electronic structure of these molecules is not
much influenced by the solid environment.

For solid ethylene and benzene[11], the fine structure due to
Rydberg series[31] has disappeared in the solid. These extravalence
excitations are significantly distorted by scattering processes
involving the neighbouring molecules. Recent photoyield measurements on solid benzene show an onset of the yield at about 8.70
eV[32]. This shift of 0.55 eV from the first ionization potential
in the vapour at 9.25 eV, is nearly the same as the gas to solid
shift for the $^1E_{1u}$ band (6.69 eV to 6.26 eV).

An assignment of absorption bands of the anthracene molecule[33] has been obtained using information from studies of the
anisotropic reflectance from single crystals[34]. The optical con-

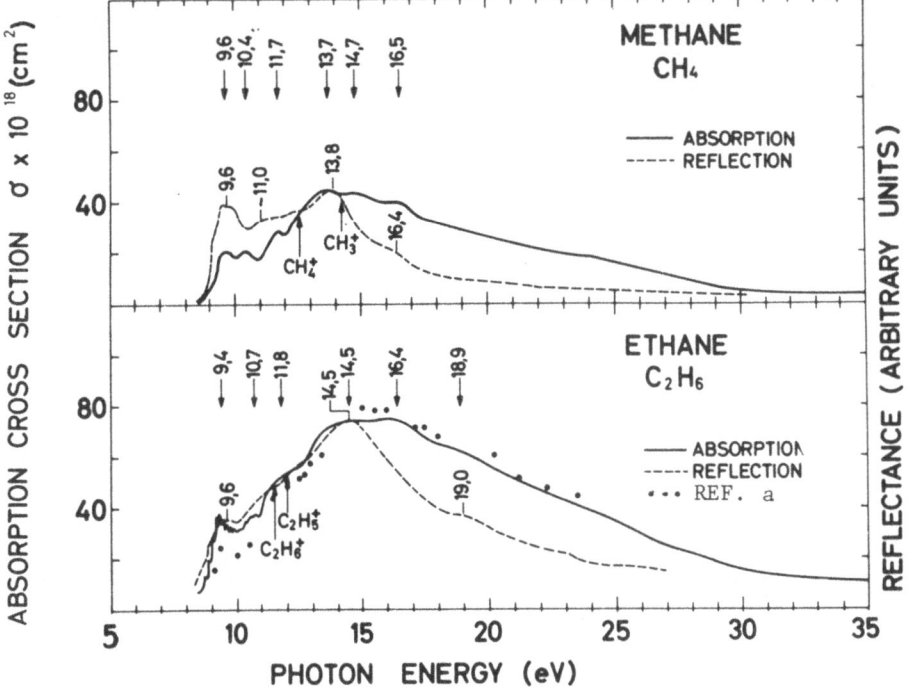

Fig. 7   Absolute absorption cross sections for methane and ethane
(solid line) and reflectance for solid films of methane
and ethane. Ref. a: R.J. Schoen, J.Chem.Phys. 37, 2032 (1962)

stants in the photon energy range between 4 and 10.5 eV could be
derived from the reflectance over a large range of frequencies for
the most important directions of polarization ($\underline{E} \parallel \underline{b}$ on (001) and $\underline{E} \parallel \underline{L}$,
$\underline{E} \perp \underline{L}$ on the (010) plane) (Fig. 8, see next page). In particular,
the influence of longitudinal fields on the exciton bands was
studied experimentally by means of non normal incidence reflection
and theoretically from the point of view of experimentally obtained
frequency dependent dielectric functions[35]. The results obtained
are in good quantitative agreement with recent theoretical results
based on the microscopic theory of exciton bands in molecular crystals[36-37].

ACKNOWLEDGEMENT

I would like to thank my colleagues at the synchrotron radiation group of DESY, with whom much of the work described above
was done, for their cooperation. This work was supported by Deutsche Forschungsgemeinschaft DFG and Deutsches Elektronen-Synchrotron DESY.

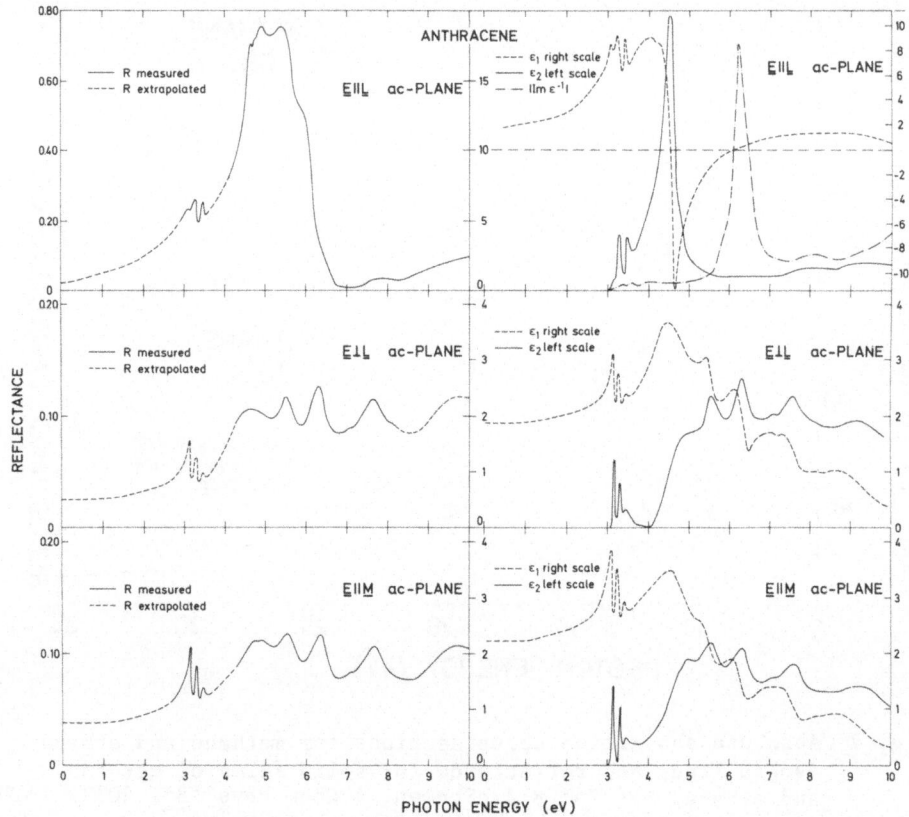

Fig. 8   Reflectance of anthracene single crystals for s-polarized
         light at near normal incidence (7.5°) from the (010)-plane
         for $\underline{E} \parallel \underline{L}$ and $\underline{E} \perp \underline{L}$ and $\underline{E} \parallel \underline{M}$ and dielectric functions
         deduced from it.

REFERENCES

1. A.A. Sokolov and I.M. Ternov Synchrotron Radiation, Oxford,
   Pergamon Press 1968, 207 p. (translated from the Russian)
2. P.L. Hartman and D.M. Tomboulian, Phys.Rev. 91, 1577 (1953)
3. R. Haensel and C. Kunz, Z.Angew.Phys. 23, 276 (1967)
4. R.P. Godwin, in Springer Tracts in Modern Physics, ed. by
   G. Höhler, Springer Verlag, Berlin, 1969, Vol. 51, p. 1
5. C. Gähwiller, F.C. Brown and H. Fujita, Rev.Sci.Instr. 41,
   1275 (1970)
6. W. Hayes, Contemp. Phys. 13, 441 (1972)
7. K. Codling, Rep.Prog.Phys. 36, 541 (1973)
8. F.C. Brown, in Solid State Physics ed. by H. Ehrenreich,
   F. Seitz and D. Turnbull, in press

9. See also: G.V. Marr, I.M. Munro and J.C.C. Sharp, Synchrotron Radiation: A Bibliography, Daresbury Nuclear Physics Laboratory 1972

10. E.E. Koch and M. Skibowski, Chem.Phys.Letters 9, 429 (1971)

11. E.E. Koch, Thesis München 1972, Verhandl. DPG 5, 307 (1971)

12. E.E. Koch, A. Otto and K.L. Kliewer, submitted to Chem.Phys.

13. N. Schwentner, M. Skibowski and W. Steinmann, Phys.Rev. B (1973) in press

14. Ch. Eitenmüller, R. Haensel, U. Nielsen and G. Zimmerer, Verhandl. DPG (VI) 8, 661 (1973)

15. e.g. O. Cheshnovsky, B. Raz and J. Jortner, J.Chem.Phys. 57, 4628 (1972)

16. U. Backhaus, Diplomarbeit Hamburg (1973)

17. R. Haensel, G. Keitel, E.E. Koch, M. Skibowski and P. Schreiber, Phys.Rev. Letters 23, 1160 (1969)

18. R. Haensel, G. Keitel, E.E. Koch, M. Skibowski and P. Schreiber, Optics Commun. 2, 59 (1970)

19. R. Haensel, G. Keitel, E.E. Koch, N. Kosuch and M. Skibowski, Phys.Rev. Letters 25, 1281 (1970)

20. G. Baldini, Phys.Rev. 128, 1562 (1962)

21. U. Rössler and O. Schütz, phys.stat.sol. (b) 56, 483 (1973)

22. R. Haensel, G. Keitel, C. Kunz, P. Schreiber and B. Sonntag, Extrait des Mémories de la Société Royale des Sciences de Liège, Tom 20, 169 (1970)

23. R. Haensel, N. Kosuch, U. Nielsen, U. Rössler and B. Sonntag, Phys.Rev. B7, 1577 (1973)

24. A.B. Kunz and D.J. Mickish, Phys.Rev. B (1973) in press

25. K. Codling and R.P. Madden, J.Chem.Phys. 42, 3935 (1965), K. Codling, Astrophys. J. 143, 552 (1966)

26. R. Haensel, E.E. Koch, N. Kosuch, U. Nielsen and M. Skibowski, Chem.Phys. Letters 9, 548 (1971)

27. E.E. Koch and M. Skibowski, Chem.Phys.Letters 14, 37 (1972)

28. the same observation was reported by E. Boursay and J.Y. Roncin, Phys.Rev.Letters 26, 308 (1971), see also: R.A.H. Buxton and W.W. Duley, Phys.Rev.Letters 25, 740 (1970)

29. P.K. Carroll and C.P. Collins, Can.J.Phys. 47, 563 (1969)

30. See also: B.A. Lombos, P. Sauvageau and C. Sandorfy, J.Mol. Spectr. 24, 253 (1967); Chem.Phys.Letters 1, 382 (1967)

31. E.E. Koch and A. Otto, Chem.Phys.Letters 12, 476 (1972)

32. E.E. Koch and N. Schwentner, unpublished

33. E.E. Koch, A. Otto and K. Radler, Chem.Phys.Letters (1973) in press

34. E.E. Koch and A. Otto, phys.stat.sol. (b) 51, 69 (1972)

35. E.E. Koch and A. Otto, submitted to Chem.Phys.

36. M.R. Philpott, J.Chem.Phys. 58, 588 (1973); Chem.Phys.Letters 17, 57 (1972)

37. See also: M.R. Philpott and J.W. Lee, J.Chem.Phys. 58, 595 (1973)

PHOTOCHEMISTRY OF PLANETARY ATMOSPHERES AND INTERSTELLAR MOLECULES

L. J. Stief

Astrochemistry Branch
Laboratory for Extraterrestrial Physics
NASA/Goddard Space Flight Center
Greenbelt, Maryland 20771

Photodissociation is the primary loss process for the major components of the atmospheres of planets and for interstellar molecules in unobscured regions of the galaxy. In the case of planetary atmospheres, a fundamental problem is the explanation for the apparent stability of, for example, the CO_2 atmosphere on Mars. It now appears that the ultimate solution to this problem is more likely to depend on new knowledge concerning the dynamics of atomic and free radical reactions occurring subsequent to photodissociation rather than on additional details concering the primary photochemical process. For the interstellar molecules, a quantitative discussion is given of the role of photochemistry in determining the lifetime of interstellar molecules. The results of the lifetime calculations have important implications concerning the conditions under which interstellar molecules are formed and the conditions required in order to protect them from photodissociation.

PLANETARY ATMOSPHERES

The scope of the subject the photochemistry of planetary atmospheres is obviously too broad to be adequately described in a limited presentation. I had initially intended to discuss two examples of photochemical problems relating to the atmospheres of planets, viz. the stability of the CO_2 atmosphere on Mars and the stability of the NH_3 and CH_4 in the atmosphere of Jupiter. However, the latter problem will not be discussed because the astronomical "facts" are not yet sufficiently firm and a choice between the two theories relating to this problem depends on these "facts" as well as laboratory data not presently available. The interested reader is referred to papers by McNesby[1] and Strobel[2] for details regarding the Jupiter problem. The present discussion will therefore be limited to the question of the stability of the Martian CO_2 atmosphere.

The composition of the lower atmosphere of Mars is shown in Table 1. The densities given are those at or near the surface where the temperature is about 150°K. There are some time variations observed as well as differences between the results of different observers, but the results for CO and O_2 are certainly good to within a factor of 2 and the CO_2 density is now known to an accuracy of $\pm$ 15%. The O_3 observations were made with the ultraviolet spectrometer aboard Mariner 7 and 9. No ozone is present in the equitorial region, but in the polar regions ozone appears in the fall, reaches a maximum in winter and decreases in the spring[3]. The inverse correlation of ozone with water vapor suggests that the presence of water surpresses the ozone formation.

The contents of Table 1 should be very disturbing to any laboratory photochemist who has used CO_2 as a chemical actinometer. It is well known that CO_2 is photodissociated in the laboratory and that CO is formed with a quantum yield Φ equal to or close to unity. It may be readily calculated from the known flux of solar radiation, the absorption cross section for CO_2 as a function of wavelength and assuming $\Phi = 1$ that the amount of CO now present in the atmosphere of Mars would be produced in only two years. Even if $\Phi = 0.5$, the essential problem remains since the observed CO would

Table 1:   MARS: ATMOSPHERIC COMPOSITION

Molecule	Concentration (cm^{-3})
CO_2	1.5×10^{17}
O_2	2.0×10^{14}
CO	1.2×10^{14}
H_2O	$\sim 10^{14}$
O_3	$< 10^{10a}$

a. estimated limit for planet outside the polar regions; ozone only observed over polar regions and reaches maximum in winter of $\sim 10^{11}$ cm^{-3}.

then be produced in four years. One is thus forced to conclude that either the absorption cross sections or quantum yields of dissociation are many orders of magnitude smaller than inferred from laboratory data or that there is an efficient process for converting CO and O_2 back to CO_2. Julienne, Neuman and Krauss[4] have drawn attention to the possible errors involved in using room temperature laboratory cross sections in modeling studies for CO_2 kinetics. Their theoretical calculations suggest that halving the temperature from 300° to 150°K produces only about a 10% decrease in CO_2 absorption for the region 1200 - 1750 Å. In addition, De More and Patapoff[5] have shown that cross sections for the region 1700 - 2000 Å measured at 200°K are about one-half the room temperature values. Thus, while this effect should be considered for accurate modeling, it appears that there are no order of magnitude effects on absorption cross sections. The situation is similar for the primary quantum yield. While there may be evidence for Φ as low as 0.5 (see for example the work of Inn[6]), order of magnitude effects appear to be definitely excluded. The task thus reduces to identifying the mechanism or mechanisms by which the photodissociation products CO and O_2 are converted back to CO_2.

The processes occurring in the upper and lower atmosphere leading to the formation of O_2 and CO may be summarized as follows:

Upper Atmosphere

$$CO_2 + h\nu \rightarrow CO + O(^1D)$$

$$O(^1D) + CO_2 \rightarrow O(^3P) + CO_2$$

Lower Atmosphere

$$CO_2 + h\nu \ (\lambda > 1650 \ \text{Å}) \rightarrow CO + O(^3P)$$

$$O(^3P) + O(^3P) + CO_2 \rightarrow O_2 + CO_2$$

Three body recombination of O and CO may be neglected based on the results of Slanger, Wood and Black[7].

The HO_2 Mechanisms

There are a number of mechanisms which maybe considered jointly since they all involve reaction and formation of the HO_2 free radical. The source of H is considered to be the photolysis of H_2O, known to be present in the atmosphere ($\sim$ 0.1% of CO_2). It is well to recall here the CO_2 photochemical equilibrium studies of Harteck et al.[8] using the 1633 Å bromine line. It was found that with as little as 0.1% water vapor present (the same level as for Mars), complete recombination to CO_2 occurred ($O_2/CO_2 < 0.005$). While these experiments do not point to a specific mechanism, they clearly show that the presence of water is important and that under conditions comparable to that of the Martian atmosphere, there is little net CO_2 dissociation.

The HO_2 mechanism may be summarized by the following cycle of reactions:

$$H + O_2 + CO_2 \rightarrow HO_2 + CO_2 \tag{1}$$

$$HO_2 + X \rightarrow OH + XO \tag{2}$$

$$OH + CO \rightarrow CO_2 + H \tag{3}$$

It is clear that the net effect of X is to convert less reactive HO_2 to OH which leads to recombination of CO and O_2 to CO_2 and regeneration of the chain carriers H and OH. We will briefly consider here mechanism for X = CO, NO, O_3, H, O and HO_2. Table 2 lists the rate constant for each $HO_2 + X$ reaction, the approximate steady state concentration required at or near the surface to compensate for the O_2 production rate. In the full model, many factors are included and concentrations are calculated as a function of altitude.

Table 2:   STEADY STATE CONCENTRATION REQUIRED
BY THE VARIOUS HO_2 MECHANISMS

X	$k (cm^{-3} s^{-1})$	$(X)_{ss} (cm^{-3})$ req.	$(X)_{ss} (cm^{-3})$ obs.
CO	$<10^{-20}$	10^{18}	10^{14}
NO	3×10^{-13}	10^{10}	--
O_3	3×10^{-15}	10^{12}	$<10^{10}$
H	$(\sim 10^{-11})$	10^{9}	10^{4a}
O	7×10^{-11}	10^{8}	10^{8a}
HO_2	3×10^{-12}	10^{9}	--

a. Calculated from Mariner observations at 250 km for H and 135 km for O. See ref. 9.

Despite early suggestions that reaction (4) should be fast[8,10] and the very indirect determination by Westenberg and de Haas[11] that $k_4 \cong 10^{-12} cm^3 molec^{-1} s^{-1}$, it is now relatively certain that the reaction

$$HO_2 + CO \rightarrow CO_2 + H \qquad (4)$$

is negligibly slow near 300°K. This was first suggested by the high temperature work of Baldwin[12] and subsequently verified by attempts to measure k_4 at 300°K.[13-16] The best estimate comes from the isotopic labelling-product analysis experiment[14] and this result of $k_4 < 10^{-20} cm^3 molec^{-1} sec^{-1}$ would require a CO density that exceeds the observed value by 10^4. Using the same isotopic labelling technique, an estimate has been made[17] of the rate constant for the reaction[18]

$$HO_2 + NO \rightarrow OH + NO_2 \qquad (5)$$

As seen from Table 2, a concentration NO equal to 10^{10} molec cm^{-3} would be required. While photolysis of NO_2 would return NO to the atmosphere, it should be emphasized that such levels of NO and NO_2 should be capable of detection by ultraviolet spectroscopy.

The $HO_2 + O_3$ reaction has been included since a good limit exists for O_3 concentration and both De More[19] and Simonaitis and Heicklen[20] have recently measured the rate constant at 300°K. The results in Table 2 show that the required O_3 concentration exceeds the observed upper limit by a factor of 10^2. Reaction with H should also be considered since atomic hydrogen is a known constituent of the atmosphere. Although the rate constant for the reaction

$$H + HO_2 \rightarrow 2 OH \qquad (6)$$

has not been measured, the estimated[11,21] value of $k_6 = 10^{-11} cm^3 molec^{-1} sec^{-1}$ leads to concentration of H many orders of magnitude too high.

Thus we are finally left with only two possibilities. One is the reaction[22]

$$HO_2 + O \rightarrow OH + O_2 \qquad (7)$$

which is part of the complete model proposed by McElroy and Donahue[21]. The required O atom concentration is at least not inconsistent with observations; the model does however require large eddy diffusion to transport atomic oxygen from the upper to the lower atmosphere. The final reaction is HO_2 disproportionation[23]

$$HO_2 + HO_2 \rightarrow H_2O_2 + O_2 \tag{8}$$

which is followed by photolysis of H_2O_2

$$H_2O_2 + h\nu \rightarrow 2OH. \tag{9}$$

This suggestion is due to Parkinson and Hunten[22] and the large eddy diffusion requirement of the McElroy-Donahue model is avoided. These are some minor problems with both models, but it is not inconceivable that both could be important depending on the relative concentration of O and HO_2 in the lower atmosphere. These could be variable with time. At the present writing, these two models seem to offer the most hope for eventually understanding the stability of CO_2 in the atmosphere of Mars.

Ozone Mechanism

This may be treated much more briefly. The proposed[8,18] sequence of reactions is:

$$O + O_2 + CO_2 \rightarrow O_3 + CO_2 \tag{10}$$

$$H + O_3 \rightarrow OH + O_2 \tag{11}$$

$$OH + CO \rightarrow CO_2 + H \tag{3}$$

Again, the net result is to convert the less reactive O and O_2 to OH to achieve conversion of CO and O_2 back to CO_2. However, since there are laboratory measurements for k_{11}[25] and k_1[26] and observational data for concentration of O_3, O_2 and CO_2 (see Table 1), we can readily calculate the O_3 concentration required for $H + O_3$ (11) to be competitive with $H + O_2 + CO_2$ (1).

$$k_{11}(H)(O_3) = k_1(H)(O_2)(CO_2)$$

$$O_3 = \frac{5 \times 10^{-32} \times 2 \times 10^{14} \times 1.5 \times 10^{17}}{2.6 \times 10^{-11}}$$

$$O_3 = 6 \times 10^{10} \text{ molec cm}^{-3}$$

This is well above the limit of $O_3 < 10^{10}$ molec cm^{-3} from the Mariner observations for the equitorial regions and is within a factor of 2 of the maximum values observed over the polar region in winter. In addition, for the $H + O_3$ mechanism to dominate over the HO_2 mechanism, we should assume $R_{11} = 10R_1$ which leads to O_3 concentrations of 6×10^{11} molec cm^{-3}. Clearly this mechanism is not important under the known condition of the Martian atmosphere.

Ionic Mechanism

Finally, for completeness we consider the mechanism proposed by Parks[27] which is based on the interesting observation that CO_2 shows apparent stability under radiolysis. Essentially the idea is that CO_2 is converted by ultraviolet radiation to CO and O_2 and that the photochemical products are converted back to CO_2 by means of ion-molecule reactions involving positive ions produced from CO_2. The mechanism requires positive ion concentrations of the order of 10^3 cm^{-3} and rate constants all of the order of 10^{-11} cm^{-3} $mole^{-1}$ sec^{-1}. Without going into the details of the full mechanism, we point out two serious problems. The first is that a laboratory measurement[28] of one of the key reactions of the mechanism

$$CO_4^+ + CO \rightarrow CO_3^+ + CO_2 \tag{12}$$

shows that the rate constant is much to slow ($k_{12} \leq 10^{-12}$ cm^3 $molec^{-1}$ sec^{-1}). The second problem, which applies to any ionic mechanism proposed, is that a means must be found to transport the positive ions from the upper atmosphere where they are formed to the very low atmosphere near the surface where CO and O_2 concentrations are many orders of magnitude larger.

Summary

The persistance of the CO_2 atmosphere on Mars appears to require some mechanism which converts the photolysis products CO and O_2 back to CO_2. At the present time, the mechanism most consistent with both astronomical observations and laboratory data involves the chain sequence: $OH + CO \rightarrow CO_2 + H$, $H + O_2 + CO_2$ $\rightarrow HO_2 + CO_2$ and $HO_2 + O \rightarrow OH + O_2$ and/or $HO_2 + HO_2 \rightarrow H_2O_2$ $h\nu$ $2OH$. The chain carries H and OH are supplied by photolysis of H_2O which is present to the extent of 0.1% of the CO_2.

INTERSTELLAR MOLECULES

Within the last five years nine diatomic and some eighteen polyatomic molecules have been discovered in interstellar space, mainly by radio astronomy. Prior to this, only a few diatomic species were observed, mainly by optical astronomy in the visible region. This unexpected increase in both the number and complexity of interstellar molecules is evidence of significant chemical phenomena associated with the interstellar medium. While the interaction of many scientic disciplines is required, i.e., astronomy, astrophysics, spectroscopy, photochemistry, chemical kinetics, surface chemistry, etc., it is the role of photochemistry in determining the lifetime of intersellar molecules to which I

shall address myself. A fuller account of this subject is given
elsewhere[29,30] and the reader is referred to these for additional
details.

The Interstellar Medium

The characteristic properties of the galaxy and interstellar
medium pertinent to a discussion of chemical phenomena are the
following: dimensions, time scale, mass, and temperature. Only
a very brief description can be given here; a fuller account is
to be found in any modern elementary astronomy textbook. In
general, compared to our terrestrial concepts, the dimension and
time scales are very large while densities and temperatures are
very low. The large-scale structure of the Milky Way galaxy is
thought to be that of a flattened spiral disc with the gas and
stars concentrated in the spiral arms. The flatness of the
galaxy is such that the galactic disc, which contains most of the
matter of the galaxy in the form of gas, stars, and grains, is
less than 1% of the diameter. The small scale structure of the
galaxy is an irregular distribution of interstellar matter with
gas and dust grains concentrated in clouds. An alternate to the
cloud model is the description in terms of random density fluctuations. Clouds occupy 5-10% of the volume of the galaxy. Table
3 summarizes the pertinent properties of the galaxy outside the
cloud regions and in the interstellar clouds themselves.

The component of the interstellar radiation field we will be
concerned with is the total dilute starlight coming from all stars.
Short wavelength radiation from hot stars can ionize the atomic
hydrogen in their immediate vicinity. We are not concerned here
with these relatively small regions dominated by ionized H but

Table 3:   THE GALAXY AND INTERSTELLAR MEDIUM

	Milky Way Galaxy	Interstellar Cloud
Diameter	30 kpc[a]	0.1 -10 pc[a]
Lifetime	10^9 years	10^7 years
Density	0.1 cm^{-3} [b]	10 to 10^6 cm^{-3}
Temperature[c]	100° -1000 °K[b]	20°-200 °K

a. 1 kiloparsec (kpc)= 3 x 10^{21} cm; 1 parsec (pc) = 3 x 10^{18} cm.
b. These entries refer to density and temperature outside the
   cloud regions.
c. kinetic temperature; rotational temperature = 3 °K.

with the larger remaining HI regions where hydrogen is neutral.
The short wavelength limit for interstellar radiation in HI
regions is therefore 912 Å, the ionization limit for atomic hydrogen. A more thorough discussion of the interstellar radiation
field adopted for these calculations has been given previously[29]
and will only be summarized here. For unobscured regions we
adopt the results of Habing[31] who obtains an ultraviolet energy
density U between 30 and 50×10^{-18} ergs cm^{-3} $Å^{-1}$ for wavelengths
between 1000 and 2200 Å. Lambrecht and Zimmermann's[32] results
for wavelength to 3646 Å are sufficiently consistent with this
value that we have adopted a constant $U = 40 \times 10^{-18}$ ergs cm^{-3}.
$Å^{-1}$ for $\lambda > 912$ Å. The uncertainty of the calculations and the
small effect on the lifetimes do not warrant allowing for any
wavelength dependence.

To determine the radiation field in obscuring clouds, we
combine the adopted constant flux with an interstellar extinction
curve covering the visible and ultraviolet regions. For the
visible we used Johnson's[33] curve for the Perseus region and for
the ultraviolet, Stecher's[34] curve determined from rocket observations in Perseus. Qualitatively it is observed that the extinction increases with decreasing wavelength, i.e. the ultraviolet is
more strongly attenuated. From the quantitative extinction curves,
a plot is prepared of transmissivety T as a function of wavelength
for various optical depths in the cloud.

A detailed account of the history of interstellar molecule
observations will not be attempted here. The reader is referred
to recent comprehensive review articles by Snyder and Buhl[35],
Rank et al[36], Turner[37] and Solomon[38]. Table 4 lists in approximate
chronological order all interstellar molecules observed to date.
Whenever possible, references are given to the papers in which the
observations were first reported.

The following comments may be of interest regarding the
molecules listed in Table 4. The three diatomics CH, CN, and CH^+
were detected by optical spectroscopy in the visible region, H_2
and CO by optical spectroscopy in the far ultraviolet using rockets.
CO and CN have been observed by both optical and radio spectroscopy,
although not in the same regions. All other molecules were detected by radio astronomy. The X in Table 4 represents an observed
microwave line which does not correspond to an observed or calculated transition for any known molecule. Some molecules such as
OH, CO, and H_2CO appear to have a rather general distribution and
are observed in many different regions. Other molecules such as
NH_3 and some of the recently discovered complex organic molecules
are seen only in a few sources, mainly in the direction of the
galactic center. Finally, it should be noted that the molecules
are observed in a variety of conditions: normal absorption (OH,
H_2CO, and all the optical detections), normal emission, maser

Table 4:   INTERSTELLAR MOLECULES OBSERVED AS OF AUGUST 1973

Molecule	Ref.	Molecule	Ref.	Molecule	Ref.
CH	39	HCN	50	OCS	59
CN	40,41	X	51	$CH_3C{\equiv}CH$	60
CH^+	42	$HC{\equiv}CCN$	52	HNCO	60
OH	43	CH_3OH	53	HNC	60
NH_3	44	HCOOH	54	CH_3CHO	61
H_2O	45	CS	55	H_2CS	62
H_2CO	46	NH_2HCO	56	H_2S	63
CO	47,48	SiO	57	$H_2C{=}NH$	64
H_2	49	CH_3CN	58	SO	65

emission (OH and H_2O), and the inverse maser or refrigerator absorption observed in H_2CO.

The Lifetime of Interstellar Molecules

The probability P that a molecule is decomposed by light in interstellar space free of obscuring clouds is

$$P = \frac{U\Phi}{h} \int \sigma_\lambda \lambda d_\lambda = \frac{U\Phi}{h} < \sigma\lambda > \tag{13}$$

where h = Planck's constant, U is the energy density, Φ is the primary quantum yield of dissociation, σ_λ is the absorption cross section, and the limits of integration are from 912 Å to the photodissociation threshold. The lifetime τ against photodissociation is simply the reciprocal of the dissociation probability P. From the list of interstellar molecules observed to date, we have chosen six for which sufficient data is available: NH_3, H_2O, H_2CO, CO, COS and CH_3CCH. Calculations have also been made for CH_4 since there is a suggestion[66] that CH_4 may have been detected by absorption in the infrared.

For most of the molecules considered, we have taken the quantum yield of dissociation Φ to be unity and independent of wavelength. This is consistent with the available laboratory data and is to be expected for molecules with essentially continuous absorption. Carbon monoxide is unique in that no laboratory studies of its photodecomposition have been reported for wavelengths below the decomposition threshold at 1115 Å. If long-lived excited states are involved below 1115 Å as have been shown to occur at longer wavelengths, the dissociation probability could be significantly less than unity. We will thus carry Φ as an unevaluated factor at this stage. It should also be noted that while the total Φ may be wavelength independent, quantum yields for individual primary processes are frequently wavelength dependent. While this does not affect the calculated lifetime for the molecule in question, it may have interesting consequences for the composition of the interstellar medium. Thus at increasing depths of the cloud where the short wavelengths are preferentially absorbed, there will be an increasing contribution from primary processes favored at longer wavelengths.

From equation (13) and using the laboratory data for σ, the procedure was to plot $\sigma\lambda$ vs λ and obtain $<\sigma\lambda>$ from the area under the curve. Both of these were done numerically on a computer with areas determined by the spline interpolation method[67] Table 5 gives the lifetimes in unobscured regions for the molecules in question.

The results in Table 5 suggest that in clear interstellar regions, most of the molecules under consideration have lifetimes against photodissociation of the order of 100 years or less. That these lifetimes are extremely short on the galactic time scale is evident from the following. If one assigns to the molecules velocities comparable to those of the clouds as a whole, that is velocities of the order of 10 to 100 km/sec, the total distance traveled by a molecule in its lifetime of 100 years (3×10^9 sec) is only 3×10^{15} to 3×10^{16} cm or 0.001 to 0.01 pc. Since cloud diameters are typically of the order 0.1 to 10 pc and intercloud distances even larger, it is evident that the average polyatomic molecule can not travel a significant distance in unobscured regions without being subject to destruction by interstellar photons. Thus polyatomic molecules can exist only in dense clouds which protect them from the full interstellar radiation field. This is consistent with observations. Further, these molecules can never have been exposed to the unobscured radiation from the time of formation until protected in clouds. This requirement imposes a severe restriction on possible mechanisms of formation. It implies that polyatomic molecules were formed or released in the gas phase in the cloud where they now occur.

Table 5:   LIFETIMES OF MOLECULES IN UNOBSCURED REGIONS

Molecule	τ (years)	Molecule	τ (years)
OCS	10	CH_4	40
H_2CO	30	H_2O	65
$CH_3C\equiv CH$	30	CO	$100/\Phi$
NH_3	30		

It is obvious that lifetimes will be orders of magnitude longer in clouds where there is high obscuration of interstellar radiation. In order to examine the effect of increasing depth of the cloud, we have calculated the lifetimes of the molecules discussed above as a function of increasing optical thickness. The dissociation probability for an obscured region is given by

$$P = \frac{U\Phi}{h} \int T_\lambda \sigma_\lambda \lambda d_\lambda = \frac{U\Phi}{h} < T\sigma\lambda > \qquad (14)$$

where T is the transmissivity of the cloud. The latter, as discussed above, may be determined as a function of wavelength for various optical depths using the quantitative extinction curves measured in the visible and ultraviolet. Similar to the previous calculation, the value of $< T\sigma\lambda >$ was obtained via plotting and numerical integration using a computer. The results of the calculations show that, as expected, the lifetimes of molecules in clouds with a few magnitudes extinction become the order of 10^6 years. Contrary to the lifetime of less than one hundred years estimated in the cloud-free regions, lifetimes in clouds of moderate opacity are significant compared to the time scale of the clouds themselves.

REFERENCES

1. J. R. McNesby, J. Atmos. Sci. 26, 594 (1969); M. Lenzi, J. R. McNesby, A. Mele and C. N. Xuan, J. Chem. Phys. 57, 319 (1972).
2. D. F. Strobel, J. Atmos. Sci. 26, 906 (1969); ibid. 30, 489 (1973); ibid. (in press).
3. C. A. Barth, C. W. Hord, A. I. Stewart, A. L. Lane, M. L. Dick and G. P. Anderson, Science 179, 795 (1973).
4. P. S. Julienne, D. Neumann and M. Krauss, J. Atmos. Sci. 28, 833 (1971).
5. W. B. De More and M. Patapoff, J. Geophys. Res. 77, 6291 (1972).
6. E. C. Y. Inn, J. Geophys. Res. 77, 1991 (1972).
7. T. G. Slanger, B. J. Wood and G. Black, J. Chem. Phys. 57, 233 (1972).
8. P. Harteck, R. R. Reeves, Jr., B. A. Thompson and R. W. Waldron, Tellus XVIII, 192 (1966); ibid., J. Phys. Chem. 70, 1637 (1966).
9. D. E. Anderson and C. W. Hord, J. Geophys. Res. 76, 6666 (1971); G. E. Thomas, J. Atmos. Sci. 28, 859 (1971).
10. S. W. Benson, "The Foundations of Chemical Kinetics," McGraw-Hill Book Co., Inc., New York, N.Y., 1960, p. 461.
11. A. A. Westenberg and N. de Haas, J. Phys. Chem. 76, 1586 (1972).
12. R. R. Baldwin, R. W. Walker and S. J. Webster, Combust. Flames 15, 167 (1970).
13. D. H. Volman and R. A. Gorse, J. Phys. Chem. 76, 3301 (1972).
14. D. D. Davis, W. A. Payne and L. J. Stief, Science 179, 280 (1973).

15. R. Simonaitis and J. Heicklen, J. Phys. Chem. 77, 1096 (1973).

16. D. Wyrsch, H.R. Wendt and H.E. Hunziker, preprints XI International Symposium on Free Radicals (1973).

17. W.A. Payne, L.J. Stief and D.D. Davis, J. Amer. Chem. Soc. 95, 0000 (1973).

18. M.B. McElroy and J.C. McConnell, J. Atmos. Sci. 28, 879(1971).

19. W.B. De More, Science 180, 735 (1973).

20. R. Simonaitis and J. Heicklen, J. Phys. Chem. 77, 1932 (1973).

21. M.B. McElroy and T.M. Donahue, Science 177, 986 (1972).

22. C.J. Hochanadel, J.A. Ghormley and P.J. Ogren, J. Chem. Phys. 56, 4426 (1972).

23. T.T. Pankert and H.S. Johnston, J. Chem. Phys. 56, 2824(1972).

24. T.D. Parkinson and D.M. Hunten, J. Atmos. Sci. 29, 1380(1972).

25. L.F. Phillips and H.I. Schiff, J. Chem. Phys. 37, 1233 (1962).

26. M.J. Kurylo, J. Phys. Chem. 76, 3518 (1972); D.L. Baulch, D.D. Drysdale, D.G. Horne and A.C. Lloyd, "Evaluated Kinetic Data for High Temperature Reactions," Vol. 1, Butterworth & Co., London, 1972.

27. D.A. Parkes, Nature 241, 110 (1973).

28. L.W. Sieck, R. Gorden, Jr. and P. Ausloos, Planet. Space Sci. (in press).

29. L.J. Stief, B. Donn, S. Glicker, E.P. Gentieu and J.E. Mentall, Astrophys. J. 171, 21 (1972).

30. L.J. Stief, Mol. Photochem. 4, 153 (1972); ibid., in "Molecules in the Galactic Environment" edited by M.A. Gordon and L.E. Snyder, John Wiley and Sons, New York (1973).

31. H.L. Habing, Bull. Astron. Inst. Nethr. 19, 421 (1968).

32. H. Lambrecht and H. Zimmerman, Wiss. Zeit. Friedrich-Schiller-Univ. Jena, Math. Naturwiss. Reihe, 3-4, 217 (1955).

33. H.L. Johnson, Astrophys. J., 141, 923 (1965).

34. T.P. Stecher, Astrophys. J., 157, L125 (1969).

35. L.E. Snyder and D. Buhl, Sky and Telescope, 40, 267, 345 (1970).

36. D.M. Rank, C.H. Townes, and W.J. Welch, Science, 174, 1083 (1971).

37. B.E. Turner, Sci. Amer. 228, 50 (1973).

38. P.M. Solomon, Physics Today 26, 32 (1973).

39. T. Dunham, Proc. Astron. Soc. Pacific, 53, 26 (1937); P. Swings and L. Rosenfeld, Astrophys. J., 86, 483 (1937).

40. W.S. Adams, Astrophys.J., 93, 11 (1941); A. McKellar, Publ. Dom. Astrophys. Obs., 7, 251 (1941).

41. K.B. Jefferts, A.A. Penzias, and R.W. Wilson, Astrophys. J., 161, L87 (1970).

42. W.S. Adams, Astrophys. J., 93, 11 (1941); A.E. Douglas and G. Herzberg, Astrophys. J.,94, 381 (1941).

43. A.H. Barrett, M.L. Meeks, and J.C. Henry, Nature, 200, 829 (1963).

44. A.C. Cheung, D.M. Rank, C.H. Townes, D.D. Thorton, and W.J. Welch, Phys. Rev. Letters, 21, 1701 (1968).

45. A.C. Cheung, D.M. Rank, C.H. Townes, D.D. Thorton, and W.J. Welch, Nature, <u>221</u>, 626 (1969).

46. L.E. Snyder, D. Buhl, B. Zuckerman, and P. Palmer, Phys. Rev. Letters, <u>22</u>, 679 (1969).

47. R.W. Wilson, K.B. Jefferts, and A.A. Penzias, Astrophys. J., <u>161</u>, 143 (1970).

48. A.M. Smith and T.P. Stecher, Astrophys. J., <u>164</u>, L43 (1971).

49. G.R. Carruthers, Astrophys. J., <u>161</u>, L81 (1970).

50. L.E. Snyder and D. Buhl, Astrophys. J., <u>163</u>, L47 (1971).

51. D. Buhl and L.E. Snyder, Nature, <u>228</u>, 267 (1970).

52. B.E. Turner, Astrophys. J., <u>163</u>, L35 (1971).

53. J.A. Ball, C.A. Gottlieb, A.E. Lilley, and H.E. Radford, Astrophys. J., <u>162</u>, L203 (1970).

54. B. Zuckerman, J.A. Ball, and C.A. Gottlieb, Astrophys. J., <u>163</u>, L41 (1971).

55. A.A. Penzias, P.M. Solomon, R.W. Wilson, and K.B. Jefferts, Astrophys. J., <u>168</u>, L53 (1971).

56. R.H. Rubin, G.W. Swenson, Jr., R.C. Benson, H.L. Tigelarr, and W.H. Flygare, Astrophys. J., <u>169</u>, L39 (1971).

57. R.W. Wilson, A.A. Penzias, K.B. Jefferts, M. Kutner, and P. Thaddeus, Astrophys. J., <u>167</u>, L97 (1971).

58. P.M. Solomon, K.B. Jefferts, A.A. Penzias, and R.W. Wilson, Astrophys. J., <u>168</u>, L107 (1971).

59. K.B. Jefferts, A.A. Penzias, R.W. Wilson, and P.M. Solomon, Astrophys. J., <u>168</u>, L111 (1971).

60. L.E. Snyder and D. Buhl, Bull. A.A.S., <u>3</u>, No. 3, Part I, 338 (1971).

61. J.A. Ball, C.A. Gottlieb, A.E. Lilley and H.E. Radford, IAU Circ. No. 2350, 1971; C.A. Gottlieb, in "Molecules in the Galactic Environment" edited by M.A. Gordon and L.E. Snyder, John Wiley and Sons, New York (1973).

62. M.W. Sinclair, J.C. Ribes, N. Fourikis, R.D. Brown and P.D. Godfrey, IAU Circ. No. 2362, (1971).

63. P. Thaddeus, M.L. Kuntner, A.A. Penzias, R.W. Wilson, and K.B. Jefferts, Astrophys. J., <u>176</u>, L73 (1972).

64. P.D. Godfrey, R.D. Brown, B.J. Robinson, and M.W. Sinclair, Astrophys. Lett. <u>13</u>, 119 (1973).

65. C.A. Gottlieb and J.A. Ball, Astrophys. J., <u>184</u>, L57 (1973).

66. G. Herzberg, Conf. on Lab. Astrophysics, Lunteren, The Netherlands, 1968.

67. J. Scudder, GSFC Document X-692-71-200 (1971); R.F. Thompson, GSFC Document X-692-70-261 (1970).

Ab initio calculations,correlation diagrams by, 261
    for excited states, 257–285
    for Rydberg species, 258
Absolute collision cross sections, 46
Absolute elastic cross sections, 46
Absolute intensities in electron impact spectroscopy, 43
Acetaldehyde, 239
Acetone, 51, 299, 480
Acetylene, 2, 36, 259, 321, 581
Acetyl halides, 239
ACO, 546
ACU, 546
Adiabatic ionization energies, 30, 75, 88
Alcohols, 18, 20, 301, 307
Alkali halides, 41, 95
Alkanes, 295, 496, 565
    photochemistry of, 465–480, 483–491
Alkylated olefins, 223
Alkyl halides, 4, 5, 14, 388, 398, 483
Alkyl iodides, 5, 14, 192, 388, 398, 486
    Rydberg series of, 193
All-external correlations, 340, 349
Amides, 19
Amines, 19, 203
Ammonia, 423, 531, 532, 581
Analysis of the wave function, 197–209, 407
Angular distribution of fragments, 529
Anisotropy, 528
Anthracene, 566
Ar I radiation, 27
Aromatic molecules, fluorescence from, 505
Asymmetry parameter, 553
Atmospheric molecules, 565, 571–584
Atomic calibration for Rydberg states, 320
Atomic collisions, 526
Atomic nitrogen, 85
Atomic oxygen, 82
Autoionization, 6, 8, 76, 85, 435, 437, 554
    selection rules for, 82
    and emission, 85
Average circulating current, 548

Barrier to internal rotation, 251, 256
Basis set, larger AO, 267
    double zeta quality, 269
Beam-foil spectroscopy, 363

Benzene, 2, 4, 14, 59, 68, 132, 216, 321, 324, 455, 457, 506, 528, 553, 566, 581
Benzenoid compounds, 35
Benzonitrile, 454
BEUTLER-FANO profile, 86
BH, 123, 198
Biphenyls, 35
Bond dissociation energy, 513
Bonding-σ-electrons, 149–175
BORN approximation, deviations from at small scattering angles, 62
Boron compounds, 224–231
Branching ratios, 108
    determination of, 108
Breakdown graph, 449, 452, 453
Bromobenzene, 168
Bromoethylenes, 239
Butadiene, 5, 271, 298, 321, 411
Butane, 22, 470, 472

$C_2F_5C\ell$, 160, 161

C_2F_6, 160

C_2H radical, 81

C_2H_5 radical, 78

$C_2H_6^+$ ion, 179

C_3H_7 radical, 79, 80

C_6H_5 radical, 79, 80

Cage effect, 306
Canonical orbital energy curves, 285
Carbon ring liquids, 132
Carbonyl compounds photochemistry of, 480, 528
Carbonyl group, 395
CEA, 546
Cesium halides, 11, 95
$CF_2C\ell_2$, 157, 158
CF_3CH_2OH, 22
CF_3Br, 154, 155, 156
$CF_3C\ell$, 153, 155, 156
CF_3 radical, 78
CF_4, 22, 50
Chain reactions, 416
CH_3 radical, 78
CH_3I, 4, 5, 14, 388, 398
$CH_3-C \equiv CH$, 581

Charges, 407
Charge density contours, 258
Charge wave function, 342
Chemical unstability, 77
Chemical reactivity, 407–411
Chlorobenzene, 168
Chromene derivatives, 526
Circular dichroism spectra, 211–222
   of benzene, 216, 221
     methylindan, 217
     α-pinene, 219
     butyl-benzene, 220
     ethylene, 220
     trans-cyclooctene, 221
$C\ell_2$, 125, 529
   emission of, 125
Closed shell, definition of, 371
CN, 527
CO, 425, 429, 581
CO_2, 6, 32, 52, 53, 321, 424, 564
CO_2^+, fluorescence of, 29
Collective electronic excitation, 132
Collective electron resonances, 130
Collision amplitude, 43, 45
Collision induced dipole moment, 51
Collision cross sections, 46
Complex index of refraction, 132
Concave grating spectrograph, 114, 130
Cone half-angle, 547
Configuration interaction, 197
Contact potentials, 26
Cooper minimum, 108
Core-polarization, 360
Correlation diagrams, 261, 266
Correlation energies, 268, 337
Covalent binding, 9
Cross sections, 46
Critical wavelength, 545
CS_2, 6
Cycloalkanes, 465–480
   photochemistry of, 470
Cyclohexane, 136, 253
Cyclohexadiene, 134
Cyclohexene, 135
4-cyclopentane-1,3-dione, 244
Cycloketones, 480
Cyclopropane, 488
Cylindrical analyser, 107
CZERNY-TURNER type instruments, 114

D_2, photoionization of, 101
DCI, 546
Decane, 22
Δ-catastrophe, 375, 377, 378
Δ SCF method, 238
Density matrix, 408
DESY, 546
Development of organic life, 426
Diacetylene, 5
Diazobicyclooctane, 30
Diazomethane, 489
Dibutylether, 22
Dielectric constants, 131-147
Differential cross section, 553
Differential ionization potential, 251
Differential pumping, 116
Diffusion of atoms through a rare-gas matrix, 307
Difluorodiazine, 254
Dimethylamine, 203
1,2-dimethylcyclopropane, 485, 488, 491
Dimethylether, 22
Dimethylsulfide, 242
Direct dissociation, 435, 442
    ionization, 435
Dissociation energies, 465
    processes and thermochemical quantities, 439
DORIS, 546
Duty factor, 544

Eagle mounting, 114
EBERT type instruments, 114
Elastic cross sections, 46
Electrical discharges, 77
    atoms in, 77
Electron impact spectroscopy, 43-73, 287-304, 317, 526
    absolute intensities in, 43
    classification of transitions in, 43
Electron correlation and spectroscopic properties, 339
Electron energy analysers, 26
Electron energy loss spectroscopy, 25, 43-73, 287-304
Electron spectroscopy, 251
Electron spin resonance, 467
Electronically excited photofragments, 528
Electronic relaxation, 451, 452, 453, 457, 459, 460
Emission and autoionization, 85
    from excited fragments, 467
    from Rydberg states, 503
Emission quantum yields from alkanes, 496
    from cycloalkanes, 502

Energy deposition (and photoionization), 436
    gap law, 450
    loss function, 131-147
    partitioning, 483, 525
Energy loss spectroscopy, 25, 43-73, 287-304
Ethane, 4, 7, 177-190, 427, 470, 472
Ethers, 18, 22
Ethylene, 3, 220, 259, 273, 279, 478, 507
Ethylene derivatives, 162
Ethylensulfide, 242
Excitation by metastable argon, 532
Excitation energy, 251
Excited states of ions, 28
    triplet of butadiene, 411
Exciton localizability, 209

F_2, 119, 125
    emission of, 125
FANO parameter, 86
Flash discharge apparatus, 118
Flash photolysis, 118, 527
Fluctuation potentials, 382
Fluorescein, 553
Fluorene, 36
Fluorescence from hydrocarbons, 495-512
    from n-alkane liquids, 499
    from upper electronic states, 505
    quantum yields, 496, 526
Fluorinated alkanes, 486
Fluorobenzenes, 165-167
Fluoro-chloroethylenes, 165
Fluoro-ethylenes, 162-164
Fluoroiodoethanes, 193
Fluoromethanes, 151, 154
Forbidden transitions and Franck-Condon principle, 54
    at low kinetic energies, 60
Foregrating, 116
Foreprism, 115
Formamide, 15
Formaldehyde, 2, 31, 51, 259, 270, 300
Formation of interstellar molecules, 581
Forward scattering, 48
Fragmentation threshold, 76, 438, 440
Franck-Condon factors, 101, 450
    principle, 58
    and forbidden transitions, 54
    and invariance in relative intensities, 58
Free radicals, 5, 9, 75, 77, 78, 94, 305-315, 469
    methods of production of, 77

Free radicals, photoionization of, 75
Free valence, 407
Furan, 5

Gallium monohalide, 95
Geminally substituted alkanes, 502
Generalized oscillator strength, 43, 45
Graphite surfaces, 26
Glycerol liquid, 39
Grating calibrator, 130

1-Hexene, 508
2-Hexene, 508
H_2, 101, 423
H_2O, 18, 20, 21, 301, 399, 426, 531, 532, 581
H_2O_2, photolysis of, 576
H_2S, 39, 399, 531
Halogen acids, 10, 399
HAM, 247
Harmonic generation in atomic vapors, 539
HARTREE-FOCK representation of chemical bonds, 197, 244
HCN photolysis, 532
HD, 101
HELLMAN-FEYNMAN theorem, 408
He I photoelectron spectroscopy, 7
He II photoelectron spectroscopy, 7
He I radiation, 27
He II radiation, 27
Hexafluoroethane, 159
HF, 22, 119, 125, 388
    emission of, 125
HI, 388
Higher ionization potentials, 8
Higher orders of gratings, 115
High resolution vacuum ultraviolet, 113–128
HNO_3, 88, 90
HO_2 mechanisms, 573
Holographic gratings, 116
Hot atom chemistry, 530, 531
Hot boron, 39
Hydrogen, 101, 423
Hydrogen laser, 536

Indium monohalide, 95
Inelastic electron scattering, 43
Information theory, 200, 204

Iodobenzene, 168
Ionic binding, 9
Ionic fragmentation of CO_2, 444
   of H_2O, 443
   of NO_2^+, 445

Ionicity, 98
Ions, excited states of, 28
Ionization chamber, 7, 8
Ionization energy, definition of, 26
Ionization quantum yield, 466
Ion-molecule reactions, 577
Ion scattering, 290, 293
Inner ionization potentials, 365
Inner orbitals of molecules, 9
INS-SOR, 546
Instrumental band width, 34
Interhalogen compounds, 242
Internal conversion, 436, 450
   correlations, 340
Interstates, 358
Interstellar lifetime of, 580
   medium, 578
   molecules observed, 579, 580
   radiation field, 578
Intersystem crossing, 436, 451
Intramolecular energy transfer, 526
Intravalency and Rydberg transitions, 337, 350
   states, 341, 379
Involatile systems, 94
   high temperature species, 95
   ionic molecules, 95
Isobutane, 471
Isotope effect, 450

JAHN-TELLER effect, 60, 239

Ketene, 31, 32
Ketones, 19, 480, 528
Kinetic energy distribution, 434, 435, 459, 460, 529
Kinetic shift, 441, 448, 459
Kinetic spectroscopy, 485
KL-core effects, 358
Koopmans' theorem, 35, 165, 179, 231, 238, 244, 251, 262
Krypton lamp, 420, 469

Laser, 485, 529
    induced plasmas, 118
    vacuum ultraviolet, 532-542
LCAO theory, 247
Leading event, 200
Light emission from photofragments, 527
Light sources, 27, 469
    for photoelectron spectroscopy, 27
    for vacuum ultraviolet, 116
    Lyman, 117
Lifetime of excited states, 484, 500, 530
Limiting oscillator strength, 47
Liquid water, 129, 137
Liquids in vacuum ultraviolet, 129-147
Lithium fluoride window, 8
Loge theory, 198
    and Rydberg states, 207
Lone pairs of electrons, 149-175
Lyman α, 27
    source, 117

Magnetic circular dichroism, 211-222
Maleic anhydride, 243
Many electron theory, 340
Mars, 572
    atmosphere of, 572
Mass spectra, 98
    statistical theory of, 448
    spectrometer, 433, 529
Matrix isolation technique, 305
MECKE's rule, 30
Medical diagnostics, 553
Mesitylene, 506
Metastable ions, 433, 434, 435, 446, 448, 455, 459, 460
    peak, 441
Methane, 8, 9, 21, 50, 150, 427, 470, 488, 531, 581
Methanol, 20, 301, 307
2-Methyl-2-Pentene, 509
Methyl iodide, 4, 5, 14, 388, 398
Methylamine, 203
Methylene, 488
Methylvinylether, 169, 171
Methylvinylthioether, 169, 172
Microwave discharge, 421
    powered lamps, 495
Modulated trapped electron method, 293
Molecular beams, 93
Molecular crystals, 559

Molecular ions, 305-315
   spectra of, 86
Molecular orbitals and photoelectron spectra, 35
   orbital calculations, 489
Molecular Rydberg series, assignment of, 385
   term values for, 395
Multiple-pole Wiggler magnet sections, 552

N_2, 124, 321, 423, 430, 564
Nanosecond discharge sources, 553
Naphthalene, 506
2-Naphthol, 553
NBS, 546
Ne I radiation, 27
NH, 532
NH_2, 527
NH_3, 423, 531, 532, 581
NINA, 545, 546
Nitric oxide, 123
Nitriles, 36
Nitrogen, 124, 321, 423, 430, 564
   oxydation of, 430
Nitrous oxide, 51, 321
N-N-dimethyl formamide, 15
NO, 531
NO_2, 529
$NOC\ell$, 529
Non-closed shell states, 340, 348
   many electron theory for, 340
$n \rightarrow \pi^*$ transitions, 255

O_2, 321, 423, 564
OCS, 581
OH, 527
Olefins, 19, 149, 295, 508
   by threshold electron impact, 295
   fluorescence from, 508
   UV bands of, 223
One-particle density matrix, 408
Open-dish method, 130
Operating current, 545
Optical oscillator strength, 46
Orbital polarization, 374
Order overlap with continuous source, 550
Oscillator strength, 43, 45, 46, 47, 48, 51, 85, 280, 361, 505
   and lifetime, 47
   generalized, 43, 45, 51

Oscillator strength,
  limiting, 47
  optical, 46, 48
Osmium tetroxide, 397
Oxygen, 321, 423, 564
Ozone, 77, 88

Paneth method, 528
Partial cross section, 108
Penetration effects, 17
Penning ionization and photoionization, 531
Phase shift method, 363
Phenanthrene, 409
Phenylacetylenes, 36
Phenylacetylene derivatives, 37
Phosphoryl molecules, 239
Photoabsorption, 436
Photodissociation, 513
  of electronically excited species, 513
Photoelectron spectra, 25-41, 149-180, 237-245, 251-255, 436
  from solids, 40
  of molecular beams, 93
  of radicals, 5, 9
Photofragmentation of polyatomic molecules, 433-463
Photofragments, 467, 468, 525
  high emission from, 526
  velocity spectrum of, 527
Photoionization, 7, 75, 91, 433, 463
  efficiency, 6
  of free radicals, 75
Photolysis mapping, 528
Photons, source in electron spectroscopy, 25
Photon desorption, 553
π-σ splitting, 95
Planetary atmospheres, 571
  photochemistry of, 571
Plasmas, 118
Plasmon response, 132
Point charge model, 408
Positive ion scattering, 61
  and triplets, 61
Post-luminescence phenomena, 553
Potential barrier, 440
Precursor core orbitals, 18
Predisperser, 115
Predissociation, 435, 439, 444, 446, 449, 513, 529
Preionization, 435
Pre-Rydberg transitions, 353, 357, 361, 379
Primary photochemical processes, 469, 476

Propane, 471, 472, 484, 488
Propylene, 508
Protonation of excited molecules, 409
PSL-I ring, 544, 546
PSL II, 546
Pseudo-photoelectron spectrum, 86
Pseudopotential method for Rydberg states, 318
PTB, 546
Pulsed polarized laser, 529
Pyrazine, electron impact spectrum of, 255
Pyrene, 506
Pyridine, 253, 300
Pyrolysis, 77
P-Xylene, 506, 507

QET, 451, 452, 455
Quantum defects, 120, 317, 385
Quantum yield of ionization, 465
Quasi-discrete (superexcited) states, 435, 436
Quasiequilibrium theory, 449

Radiationless transitions, 450, 451, 459, 525, 529
Radicals, 5, 9, 75, 77, 78, 94, 305-315, 469
   photoelectron spectra of, 5, 9
Radical scavengers, 472
Radius of the electron orbit, 545
Rare gas continua, 117
Rare gases, 553
   solid, 563
Reaction between O and H_2, 429
Recoiling fragment, 528
Reflection spectroscopy, 561
Reflection type polarizers, 130
Refractive index, 132, 212
Relative missing information function, 200
Retarding potential analyser, 8
Rotational predissociation, 439
ROOTHAAN's equations, 249
RRKM theory, 447, 449, 459, 460, 525
Rubidium halides, 11
Ruthenium tetroxide, 397
Rydberg states, 13, 85, 120, 122, 149-173, 207, 271, 317-332,
   337-384, 385-400, 437

Saturated hydrocarbons by threshold electron impact, 295
Scavenger techniques, 290, 292, 469
SCF calculations, 238, 250
Secondary photolysis processes, 308

Second row elements, 240
Selenium compounds, 242
Semi-cylinder method, 130
Semi-empirical theory, 247
Semi-internal correlations, 340
Semi-internal orbitals, 341, 373
Shielding, 248
Schumann ultraviolet, 417
Silicon vacuum pump oils, 139
Singlet-triplet transitions at high kinetic energy, 63
    in electron-impact spectra, 60, 69, 289
    selection rules for, 65
Singlet-triplet splittings, 350, 352
    for Rydberg states, 271
Slater-type shielding, 248
Solid rare gases, 563
Solids in vacuum ultraviolet, 129
SPEAR, 546
SPINDO, 247
Spin conservation, 441
    violation of, 445
    conversion rule, 513, 518
Spin-orbit splitting, 4, 39, 85, 95, 153, 239, 450, 452
Static indices, 407
Statistical partitioning, 490
    theory of mass spectra, 448
    weights, ratio of, 107
Stilbenes, 37
Stokes shift, 500, 510
Storage rings, 543
STRICKLER-BERG equation, 505
Styrene, 35, 36
Succinic anhydride, 254
Sulfides, 19
Sulfur 3d orbitals, 169
Sulfur compounds, 172, 242
Superconducting magnet, 216
Superexcited states, 6, 435
Supermultiplet, 362
SURF, 546
Synchrotron radiation, 5, 118, 543-555, 559-568
    polarization of, 549

Tellurium compounds, 242
Term values, for rare gases, 403
    correlation with molecular term values, 403
    for Rydberg states, 17, 21, 22, 149-173, 177-189
    for fluorinated molecules, 21
Tetrahydropyran, 169

Tetrahydrothiopyran, 169, 170
Thallium halides, 11
Thallium monohalides, 95
Thermochemical threshold, 441
Thioformaldehyde, 259, 270
Thiols, photochemistry of, 531
Thiophene, 5
Thiophosphoryl molecules, 239
Threshold electron impact spectroscopy, 287–304
Threshold measurements, 514
Through space interactions, 242
Time of flight mass spectrometer, 529
Toluene, 506
Totally symmetric broadening, 15
Transients, 525
Transient species, 39, 238
    photoelectron spectra of, 8, 16, 25–42
    and other physical parameters, 38
Transition polarizability, 63
Transmission method, 130
    (for spectra of liquids)
Trapped electron method, 290
Trimethylamine, 22
Trimethylborane, 224
Triplets, and positive ion scattering, 61
    in threshold electron impact spectroscopy, 287
    of water, 66
Two-electron excitation processes, 438
Twisting potential curves, 280

United atom, problems of, 373

Vacuum ultraviolet lasers, 535–542
Vacuum ultraviolet photolysis, 305–315
    of acetylene, 310
    of cyanides, 307
    of methane, 307
    of methanol, 307
    of NO_2, 311
Valence orbitals, 9
Valence shell MO's, 9, 13
Valence states, 119
Velocity spectrum of photofragments, 527, 529
VEPP-2', 546
VEPP-3, 546
Vibrational fine structure, 30
    in photoelectron spectra, 30, 32
Vibrational predissociation, 439

Vibrational structure of electronic transitions, 278
Vibronic coupling, 450
Vinyl iodide, 191
Violation of spin conservation, 445

Walsh diagrams, 261
Wannier exciton, 21
Water, 18, 20, 21, 301, 399, 426, 531, 532, 581
Wiggler magnets, 551
Window materials, 550

Xenon lamp, 418, 469
Xenon molecular dissociation laser, 537
X-ray microscopy, 553
X-ray photoelectron spectroscopy, 9, 553

Zeeman splitting, 213